Multidisciplinary Approaches in AI, Creativity, Innovation, and Green Collaboration

Ziska Fields
University of Johannesburg, South Africa

A volume in the Advances in Environmental Engineering and Green Technologies (AEEGT) Book Series

IGI Global
PUBLISHER of TIMELY KNOWLEDGE

Published in the United States of America by
IGI Global
Engineering Science Reference (an imprint of IGI Global)
701 E. Chocolate Avenue
Hershey PA, USA 17033
Tel: 717-533-8845
Fax: 717-533-8661
E-mail: cust@igi-global.com
Web site: http://www.igi-global.com

Library of Congress Cataloging-in-Publication Data

Names: Fields, Ziska, 1970- editor.
Title: Multidisciplinary approaches to green creativity, eco-innovation, and collaboration / Ziska Fields, editor.
Description: Hershey, PA : Engineering Science Reference, an imprint of IGI Global, [2023] | Includes bibliographical references and index. | Summary: "This book focuses on the importance of green creativity, eco-innovation and collaboration to create a more sustainable world. This comprehensive and timely publication aims to be an essential reference source, building on the available literature and joint expertise in the field of management, while providing for further research opportunities in this dynamic field. It is hoped that this text will provide the resources necessary for governments, nonprofit and for-profit organizations, and managers specifically to address sustainability needs in businesses and countries around the world"-- Provided by publisher.
Identifiers: LCCN 2022039893 (print) | LCCN 2022039894 (ebook) | ISBN 9781668463666 (h/c) | ISBN 9781668463673 (s/c) | ISBN 9781668463680 (ebook)
Subjects: LCSH: Environmental protection--Technological innovations. | Sustainable engineering. | Creative ability in technology.
Classification: LCC TD170.2 .M85 2023 (print) | LCC TD170.2 (ebook) | DDC 666/.14--dc23/eng/20221107
LC record available at https://lccn.loc.gov/2022039893
LC ebook record available at https://lccn.loc.gov/2022039894

This book is published in the IGI Global book series Advances in Environmental Engineering and Green Technologies (AEEGT) (ISSN: 2326-9162; eISSN: 2326-9170)

British Cataloguing in Publication Data
A Cataloguing in Publication record for this book is available from the British Library.

For electronic access to this publication, please contact: eresources@igi-global.com.

Advances in Environmental Engineering and Green Technologies (AEEGT) Book Series

Sang-Bing Tsai
Zhongshan Institute, University of Electronic Science and Technology of China, China & Wuyi University, China
Ming-Lang Tseng
Lunghwa University of Science and Technology, Taiwan
Yuchi Wang
University of Electronic Science and Technology of China Zhongshan Institute, China

ISSN:2326-9162
EISSN:2326-9170

Mission

Growing awareness and an increased focus on environmental issues such as climate change, energy use, and loss of non-renewable resources have brought about a greater need for research that provides potential solutions to these problems. Research in environmental science and engineering continues to play a vital role in uncovering new opportunities for a "green" future.

The **Advances in Environmental Engineering and Green Technologies (AEEGT)** book series is a mouthpiece for research in all aspects of environmental science, earth science, and green initiatives. This series supports the ongoing research in this field through publishing books that discuss topics within environmental engineering or that deal with the interdisciplinary field of green technologies.

Coverage

- Alternative Power Sources
- Industrial Waste Management and Minimization
- Sustainable Communities
- Green Technology
- Radioactive Waste Treatment
- Biofilters and Biofiltration
- Air Quality
- Waste Management
- Water Supply and Treatment
- Renewable Energy

IGI Global is currently accepting manuscripts for publication within this series. To submit a proposal for a volume in this series, please contact our Acquisition Editors at Acquisitions@igi-global.com or visit: http://www.igi-global.com/publish/.

The Advances in Environmental Engineering and Green Technologies (AEEGT) Book Series (ISSN 2326-9162) is published by IGI Global, 701 E. Chocolate Avenue, Hershey, PA 17033-1240, USA, www.igi-global.com. This series is composed of titles available for purchase individually; each title is edited to be contextually exclusive from any other title within the series. For pricing and ordering information please visit http://www.igi-global.com/book-series/advances-environmental-engineering-green-technologies/73679. Postmaster: Send all address changes to above address.

Titles in this Series

For a list of additional titles in this series, please visit: www.igi-global.com/book-series

Transcending Humanitarian Engineering Strategies for Sustainable Futures
Yiannis Koumpouros (University of West Attica, Greece) Angelos Georgoulas (University of West Attica, Greece) and Georgia Kremmyda (University of Warwic, UK)
Engineering Science Reference • © 2023 • 338pp • H/C (ISBN: 9781668456194) • US $270.00

Food Sustainability, Environmental Awareness, and Adaptation and Mitigation Strategies for Developing Countries
Ahmad Ni'matullah Al-Baarri (Universitas Diponegoro, Indonesia) and Diana Nur Afifah (Universitas Diponegoro, Indoesia)
Engineering Science Reference • © 2023 • 293pp • H/C (ISBN: 9781668456293) • US $110.00

Perspectives on Ecological Degradation and Technological Progress
Veli Yilanci (Faculty of Political Sciences, Canakkale Onsekiz Mart University, Turkey)
Engineering Science Reference • © 2023 • 300pp • H/C (ISBN: 9781668467275) • US $225.00

Climate Change and the Economic Importance and Damages of Insects
Ahmed Karmaoui (Moulay Ismail University of Meknes, Morocco & Moroccan Center for Culture and Science, Morocco)
Engineering Science Reference • © 2023 • 279pp • H/C (ISBN: 9781668448243) • US $215.00

Implications of Nanoecotoxicology on Environmental Sustainability
Rafiq Lone (Central University of Kashmir, India) and Javid Ahmad Malik (Guru Ghasidas Vishwavidyalaya, India)
Engineering Science Reference • © 2023 • 360pp • H/C (ISBN: 9781668455333) • US $250.00

Human Agro-Energy Optimization for Business and Industry
Pandian Vasant (MERLIN, Faculty of Electrical and Electronics Engineering, Ton Duc Thang University, Vietnam) Roman Rodríguez-Aguilar (Panamerican University, Mexico) Igor Litvinchev (Universidad Autónoma de Nuevo León, Mexico) and Jose Antonio Marmolejo-Saucedo (Panamerican University, Mexico)
Engineering Science Reference • © 2023 • 322pp • H/C (ISBN: 9781668441183) • US $240.00

Applying Drone Technologies and Robotics for Agricultural Sustainability
Pethuru Raj (Reliance Jio Platforms Ltd., Bangalore, India) Kavita Saini (Galgotias University, India) and Vinicius Pacheco (Office of the Prosecutor General, Brazil)
Engineering Science Reference • © 2023 • 318pp • H/C (ISBN: 9781668464137) • US $250.00

IGI Global
PUBLISHER of TIMELY KNOWLEDGE

701 East Chocolate Avenue, Hershey, PA 17033, USA
Tel: 717-533-8845 x100 • Fax: 717-533-8661
E-Mail: cust@igi-global.com • www.igi-global.com

Table of Contents

Detailed Table of Contents

When faced with complex real-world problems, humans display resourcefulness, flexibility, creativity, active learning, improvisation, and critical and analytical thinking. Humans have a unique capacity to be creative that AI has yet to duplicate. However, AI brings new possibilities. Most humans believe creativity is a uniquely human ability that cannot be copied, especially not by AI. This chapter explores two views: (1) AI cannot be creative and (2) AI can be creative. This chapter uses a literature study as research methodology. The main finding is that there are people who believe that AI cannot be creative and should be feared, while others believe that AI can amplify human effectiveness, offer more opportunities to be creative, and should not be feared. The recommendations are that education, knowledge sharing, and working with AI tools and programs can change people's negative mindset and make AI a valuable partner in human creativity endeavors.

Traditionally, computer programs have used artificial intelligence to emulate human creativity. In the 1990s, however, a new approach developed called computational creativity. It involved a bottom-up approach. In this approach, the computer program works by learning heuristics from the data it receives. Various fields of research have been utilizing generative adversarial networks (GANs) to mimic creativity. It has been done in multiple areas, such as medicine, dental practices, cybersecurity, and art. GANs have shown tremendous promise for creativity. However, the field has also been plagued with some design flaws. In this chapter, the authors talk about machine-led creative innovations and possible challenges to overcome.

Artificial intelligence (AI) is increasingly reshaping brand and marketing communications. While significant research has been conducted on the impact of AI in other fields, there is little empirical evidence on how artificial intelligence is affecting the customer journey. This research study seeks to answer, "How is artificial intelligence influencing both organizational content strategy and the related customer journey?" To answer this, the authors employed mixed-methods qualitative research via a content analysis of industry publications, a series of in-depth customer interviews, and a case study content analysis of two organizations that are using AI to varying degrees within their content strategy. This study found that many publications agree AI will play a future role as creative assistant in content development, that consumer perceptions about AI and cognitive dissonance impact levels of adoption to some extent, and that companies in different geographical locations may have different levels of AI adoption along innovation stages.

This chapter takes the Tsinghua University Suzhou Automotive Research Institute, Shenzhen Bao'an International Airport, and Guizhou "Intelligent Transportation Cloud" as examples to discuss the roles of digital twins in accelerating the transformation of transportation digitalization and the impact of digitalization on relevant enterprises. This chapter will focus on the models of how digital transformation can help stakeholders in the transportation industry create a better-connected experience. In addition, CiteSpace will be used in this chapter to visualize and analyze the digital twins/transportation/digital transformation documents in the database of the web of science (WOS) from 2017 to 2022. The authors also conducted a survey to analyze how the digital twins of China's transportation industry promote its digital transformation. This chapter will provide suggestions for the problems in the digital transformation of the transportation industry.

With the advancement of ICT, the tourism industry has undergone a digital transformation where management, marketing, and communication are largely using web based applications. Automation of processes like ticketing and reservation, online hotel booking, E visa processing, etc., indicates the significant reliance of the sector on technology and world wide web for its services. Big data and artificial intelligence are a fairly new and innovative approach to addressing this issue of managing and analyzing huge datasets collected from multiple sources. This chapter focuses on understanding the role and importance of big data and artificial intelligence in the tourism industry and its impact on improving the overall image and attractiveness of destinations.

account has an influence on project management procedures and practices. The requirements for project management, on the other hand, do not take into consideration the sustainability agenda. The purpose of this chapter is to give a comprehensive overview of the relationship between sustainability and project management. Following the publication of these results, it will be possible to further enhance project management techniques and standards in order to address the role that projects play in the creation of sustainable development.

The objective of this study was to describe the Brazilian green patents pilot program from the National Institute of Industrial Property (INPI) by mapping the national scenario in concerning to the technological development of green patents. The methodology consisted in prospecting the green patents deposit per the technological sector, highlighting that only documents deposited in Brazil were considered. Besides that, the solid waste management area has greater documental recording showing a propensity to develop more technologies; therefore, the Brazilian government should create public policies aimed at strengthening these technologies, thereby stimulating and developing them in both universities and industries, granting governmental incentives for the manufacturing of products and processes provision, by creating specific lines of financing through development agencies.

The sixth mass extinction's emerging threat is sad, considering that humans have caused it. But, the growing world population leaves humans with no option in their desire to survive. This situation where humans need to survive and yet protect the habitat and many other similar scenarios complicate the process of seeking solutions to the looming disaster and extinction. Yet, with collaborations, humans can develop novel and innovative ideas to save them from the threats posed by the sixth mass extinction. This chapter examines how supply chain partners can collaborate to ensure green-sensitive and eco-innovative supply chain practices to help prevent this extinction.

This chapter investigates to what extent the feelings and thoughts of consumers are effective in the process of new product development and improvement. Inspired by the user innovation theory, the study analyzes the online feedback of the users of a smartphone brand in Turkey. This analysis covers a sentiment analysis performed using the support vector machines, the random forest, and the recurrent neural networks algorithms. By studying 2005 reviews, the chapter concludes that two strategies are proposed for the firms. A first strategy is a hybrid approach: Given the imported input-dependency, we know that the cost of imported inputs matters for firms. The second strategy is repositioning the brand in the long term. This chapter attempts to contribute to the literature by providing new evidence in a developing country case whether user comments are effective on the modified versions of a high-tech product.

Carbon emission is the most significant driver of pollution and climate change problems. So, this chapter tries to study a recently developed concept to tackle the carbon emission and climate change problem. This chapter has chronologically reviewed the carbon pricing scheme in a supply chain context. This chapter has discussed the carbon pricing scheme in detail and tried to find the applicability and effectiveness of this scheme in the supply chain. With the help of the literature review, this study also tried to explore different ways of effectively operationalizing the scheme in the supply chain. The result of this study shows that carbon pricing is an effective and advanced scheme to tackle the carbon emission and climate change problem. The results further emphasized creating awareness, labelling carbon footprints, setting environmental regulations, implementing a two-part tariff, using data for strategizing, and remanufacturing used products for better implementation of carbon pricing in the supply chain.

This study aims to analyze some of the existing implications between urban anthropology and Anthropocene urban socio-ecology planning resilience. Beginning with the assumption that urban anthropology gives support to create and develop any urban planning based on the Anthropocene urban socio-ecology resilience, the methods employed are the analytical-descriptive based on an ethnographic interpretation and reflection of the theoretical and empirical literature review. The analysis concludes that urban anthropology fundamentals give support to strengthen the Anthropocene orientation of the urban socioecological planning resilience.

Preface

Microsoft co-founder Bill Gates says the development of Artificial Intelligence (AI) is as revolutionary as mobile phones and the internet (Gates, 2023) and the most significant technological advance in decades (Gerken, 2023). Artificial/machine creativity, neuro-psychological approaches, transformational creativity, and multiple AI tools are highlighted and discussed in the first section of this book. AI is an important focus now and in the future, as is green collaboration using creativity and innovation to ensure sustainability.

The World Economic Forum (WEF) (2023) indicated in an article titled "Why cross-sector collaboration may succeed where policy cannot" that green collaboration ensures that expertise is shared, the risk is diffused, partnerships are formed, and green leadership paves the way to a sustainable future. This book explores the role of creativity and innovation specifically. Holistic creativity and sustainability applied to various countries and industries, with a focus on carbon pricing, green marketing efforts, and the Anthropology and Anthropocene urban socio-ecology planning resilience, are covered in the book.

Multidisciplinary Approaches in AI, Creativity, Innovation, and Green Collaboration builds on the available literature and joint expertise in AI, sustainability, creativity, and innovation while providing further research opportunities in these dynamic areas. This premier reference source is a comprehensive and timely resource for government officials, decision-makers, business leaders and executives, students and educators, librarians, and researchers.

This book consists of 16 chapters, divided into two sections. The first section is "Artificial Intelligence, Creativity, and Innovation" (Chapters 1-9), and the second section is "Green Collaboration, Creativity, and Innovation" (Chapters 10-16). Here is the breakdown of these sections.

SECTION 1

Artificial Intelligence, Creativity, and Innovation

Chapter 1, titled "Human and Artificial Creativity," was contributed by Ziska Fields from the University of Johannesburg, South Africa. When faced with complex real-world problems, humans display resourcefulness, flexibility, creativity, active learning, improvisation, and critical and analytical thinking. Humans have a unique capacity to be creative that AI has yet to duplicate. However, AI brings new possibilities and challenges. Most humans believe creativity is a uniquely human ability that cannot be copied, especially not by AI. This chapter explores two views (1) AI cannot be creative, and (2) AI can be creative. This chapter uses a literature study as a research methodology. The main finding is that some people believe that AI cannot be creative and should be feared, while others believe that AI can amplify human effectiveness, offer more opportunities to be creative, and should not be feared. The recommendations are that education, knowledge sharing, and working with AI tools and programs can change people's negative mindsets and make AI a valuable partner in human creative endeavors.

Chapter 2, titled "Human Creativity vs. Machine Creativity-Innovations and Challenges," was contributed by DwijendraNath Dwivedi, from Kraków University of Economics, Poland, and Ghanashyama Mahanty, from the Department of Analytical and Applied Economics, Utkal University, India. Traditionally, computer programs have used artificial intelligence to emulate human creativity. In the 1990s, however, a new approach developed called computational creativity. It involved a bottom-up approach. In this approach, the computer program learns heuristics from the data it receives. Various fields of research have been utilizing Generative Adversarial Networks to mimic creativity. It has been done in multiple areas, such as medicine, dental practices, cybersecurity, and art. GANs have shown tremendous promise for creativity. However, the field has also been plagued with some design flaws. In this chapter, the focus is on machine-led creative innovations and possible challenges to overcome.

Chapter 3, titled "Neuro-Psychological Approaches for Artificial Intelligence," was contributed by Suhani Dheer, from the Department of Brain Science and Psychology, Drexel University, Satvik Tripathi, and Edward Kim, from the Department of Computer Science, Drexel University, the United States. In Artificial Intelligence (AI) research, Neuro-Symbolic learning, which combines neuroscience and symbolic representation methods based on artificial neural networks, has a long and illustrious history. In recent years, significant progress has been achieved in neuroscience and artificial intelligence. Neuropsychology has inspired new algorithms and architectures, independent of and complementing the mathematical and logic-based methods and notions that have dominated traditional AI approaches. In this chapter, the authors provide an organized overview of current trends by discussing and organizing recent research from major conferences and publications and presenting them in an analytical order. This chapter aims to provide a handy starting point for further research on the general topic of neuropsychological approaches for the further advancement of AI techniques.

Chapter 4, titled "Digital Transformation in Contemporary Organizations: Creativity and Innovation in the 4th Industrial Revolution," was contributed by Zeynep Merve Ünal, an Independent Researcher, Turkey. This chapter aims to shed light on the stages of the digital transformation process based on the 4th IR. AI and its development processes are explained by the digital transformation framework and holistic digital transformation framework under the perspective of contemporary organizations. The creativity process and AI creativity strengthen the definition of AI and its relationship with sub-dimensions. Top strategic trends for 2022 and emerging technologies are discussed by considering growth, digitalization, and efficiency. AI-driven business model innovation is considered by scaling AI capabilities as agile customer co-creation, data-driven delivery operations, and scalable ecosystem. Organizational capabilities for AI implementation are considered as AI project planning, co-development of AI systems, data management, and AI model lifecycle management. Evaluations and suggestions were made based on the detailed literature review and current findings.

Chapter 5, titled "Artificial Intelligence Trends and Perceptions: Content Strategy and the Customer Journey," was contributed by Tariro S Munyengeterwa, Melanie Richards, and Joel B Eaton, from East Tennessee State University, United States of America. Artificial intelligence (AI) is increasingly reshaping brand and marketing communications. While significant research has been conducted on the impact of AI in other fields, there is little empirical evidence on how artificial intelligence affects the customer journey. This research study seeks to answer the question, "How is artificial intelligence influencing organizational content strategy and the related customer journey?" To answer this, the authors employed mixed-methods qualitative research via a content analysis of industry publications, in-depth customer interviews, and a case study content analysis of two organizations using AI to varying degrees within their content strategy. This study found that many publications agree AI will play a future role as a creative assistant in content development, that consumer perceptions about AI and cognitive dissonance impact levels of adoption to some extent, and that companies in different geographical locations may have different levels of AI adoption along innovation stages.

Chapter 6, titled "Digital Twins: Accelerating the Digital Transformation in the Chinese Transportation Sector," was contributed by Poshan Yu, from the Soochow University, China & Australian Studies Centre, Shanghai University, China as well as Jiarui Song, an Independent Researcher from China; Michael Sampat, an Independent Researcher from Canada; and Emanuela Hanes, an Independent Researcher from Austria. This chapter takes the Tsinghua University Suzhou Automotive Research Institute, Shenzhen Bao'an International Airport, and Guizhou "Intelligent Transportation Cloud" as examples to discuss the roles of digital twins in accelerating the transformation of transportation digitalization and the impact of digitalization on relevant enterprises. This chapter will focus on the models of how digital transformation can help stakeholders in the transportation industry create a better-connected experience. In addition, CiteSpace will be used in this chapter to visualize and analyze the digital twins/transportation/digital transformation documents in the Web of Science (WOS) database from 2017 to 2022. The authors also surveyed to analyze how the digital twins of China's transportation industry promote its digital transformation. This chapter will provide suggestions for the problems in the digital transformation of the transportation industry.

Chapter 7, titled "Big Data and Artificial Intelligence: Creative Tools for Destination Competitiveness," was contributed by H. Sandhya and Bindi Varghese from the CHRIST University (Deemed), India. With the advancement of ICT, the tourism industry has undergone a digital transformation where management, marketing, and communication largely use web-based applications. Automation of processes like ticketing and reservation, online hotel booking, E visa processing, etc., indicates the sector's significant reliance on technology and the world wide web for its services. Big data and Artificial Intelligence is a relatively new and innovative approaches to addressing this issue of managing and analyzing huge datasets collected from multiple sources. This chapter focuses on understanding the role and importance of Big Data and Artificial Intelligence in the tourism industry and their impact on improving destinations' overall image and attractiveness.

Chapter 8, titled "Future of Public Sector Enterprises in the Metaverse," was contributed by Richmond Anane-Simon, from the Pentecost University, Ghana and Sulaiman Olusegun Atiku, from the Department of Economics and Business Studies, Walter Sisulu University, South Africa. This chapter explores the future of public sector enterprises in the contemporary metaverse technological sphere. A literature review approach was adopted to explore customer expectations, sustainable governance forms, and innovations in the metaverse ecosystem. Findings indicate that most of the technologies that make up the metaverse's building blocks already exist or are in an advanced stage of development. Furthermore, there was a rise in the use of metaverse technology like virtual and augmented Reality (AR), particularly in the healthcare sector during the pandemic. Digital currency adoption was also noted, including using cryptocurrencies and central bank digital currencies. Public sector enterprises should rely on various interdependent players rather than assuming they have a monopoly on knowledge or resources to govern. A process of co-evolution between non-government and governmental entities is a sine qua non for effective social innovation by public sector enterprises in the metaverse ecosystem.

Chapter 9, titled "Praxes of Transformational Creativity: Artificial Intelligence as Pedagogical Change Agent," was contributed by Catherine Hayes, Faculty of Health Sciences and Well-Being, University of Sunderland, UK. This chapter provides an insight into the justification of pedagogical principles within the context of Extended Reality. The global COVID-19 pandemic exacerbated the complex ambiguity surrounding XR in practice. Training for the strategic management of risk in healthcare practice in situated contexts of healthcare provision has been a critical focus in using XR in practice. Not only does it involve rational aspects of cognitive knowledge or the purist demonstration of psychomotor skills and affective domain learning. This is achieved via the exploration of the critical epistemologies or ways of knowing, from a theoretical perspective, that can be used to ensure the level of authenticity necessary to highlight the pedagogical shifts in the application of learning theory, which now characterize responsive curriculum design and adaptation to accommodate XR in practice.

SECTION 2

Green Collaboration, Creativity, and Innovation

Chapter 10, titled "Holistic Creativity and Harmonious Mega-Innovation for Sustainable Development: Master Keys and Examples," was contributed by Vadim Kotelnikov, Innompics, USA. This chapter focuses on systemic cross-functional and cross-border approaches to creating a harmonious mega-innovation (HarmInn) that fosters sustainable development globally. Harmonious Mega-Innovation is applied across several sectors of the economy. HarmInn uses a holistic development process that puts the well-being of people, organizations, and our planet at the centre of innovative strategies and entrepreneurial actions to design and build a better and sustainable future. Holistic mega-creativity and harmonious mega-innovation start from adopting a global mindset and seeing the Earth as a small planet to take care of, not a huge planet to survive on. The author argues that disruptive innovations may break obsolete man-made rules but should stay in harmony with the eternal laws of the Universe. The author uses the example of the global Innompic ecosystems and World Innompic Games to illustrate this approach.

Chapter 11, titled "Embedding Sustainability in Project Management: A Comprehensive Overview," was contributed by Ramakrishnan Vivek, from the Sri Lanka Technological Campus (SLTC), Sri Lanka; Rohit Bansal, from the Vaish College of Engineering, Rohtak, Turkey; and Nishita Pruthi, from the Institute of Management Studies and Research, Maharshi Dayanand University, India. The issue of sustainability is the most pressing issue facing the world today. Companies incorporate sustainability concepts into marketing, corporate communication, annual reports, and other operations. Projects are critical in implementing more sustainable business practices, and sustainability has increasingly been associated with project management in recent years. The growing body of research on this subject offers compelling evidence that taking sustainability into account influences project management procedures and practices. On the other hand, the requirements for project management do not consider the sustainability agenda. This chapter aims to give a comprehensive overview of the relationship between sustainability and project management. Following the publication of these results, it will be possible to enhance project management techniques and standards further to address the role that projects play in creating sustainable development.

Chapter 12, titled "Sustainable Technologies Development: An Analysis of the Brazilian Green Patents Program," was contributed by Luan Carlos Santos Silva from the Federal University of Grande Dourados (UFGD), Brazil and Carla Schwengber Ten Caten from the Federal University of Rio Grande do Sul (UFGRS), Brazil. This study aimed to describe the Brazilian green patents pilot program from the National Institute of Industrial Property (INPI) by mapping the national scenario concerning the technological development of green patents. The methodology involved prospecting the green patents deposit per the technological sector, highlighting that only documents deposited in Brazil were considered. Besides that, the solid waste management area has greater documental recording showing a propensity to develop more technologies. Therefore, the Brazilian government should create public policies to strengthen these technologies, thereby stimulating and developing them in both universities and industries, granting governmental incentives for manufacturing products and processes provision, and creating specific lines of financing through development agencies.

Chapter 13, titled "Green Logistics and Transport Processes: Mitigating the Sixth Extinction," was contributed by Enock Gideon Musau and Noleen Pisa, from the University of Johannesburg, South Africa. The sixth mass extinction's emerging threat is sad, considering that humans have caused it. But, the growing world population leaves humans with no option in their desire to survive. This situation where humans need to survive and yet protect the habitat and many other similar scenarios complicates seeking solutions to the looming disaster and extinction. Yet, with collaborations, humans can develop novel and innovative ideas to save them from the threats posed by the sixth mass extinction. This chapter examines how supply chain partners can collaborate to ensure green-sensitive and eco-innovative supply chain practices to help prevent this extinction.

Chapter 14, titled "Does User Feedback Matter for Product Development?" was contributed by Aslı Diyadin Lenger, İstanbul Gelişim University, Turkey and Aykan Candemir, Ege University, Turkey. This chapter investigates to what extent the feelings and thoughts of consumers are effective in the process of new product development and improvement. Inspired by the user innovation theory, the study analyzes the online feedback of the users of a smartphone brand in Türkiye. This analysis covers a sentiment analysis performed using the support vector machines, the random forest, and the recurrent neural network algorithms. Two strategies are proposed for the firms. The first strategy is a hybrid approach: Given the imported input dependency, the authors know the cost of imported inputs matters to firms. The second strategy is repositioning the brand in the long term. This chapter attempts to contribute to the literature by providing new evidence in a developing country case of whether user comments are effective on the modified versions of a high-tech product.

Chapter 15, titled "Review and Analysis of Carbon Pricing for Mitigating Climate Change Problems," was contributed by Manish Kumar from IIM Indore, India. Carbon emission is the most significant driver of pollution and climate change problems. So, this chapter tries to study a recently developed concept to tackle the carbon emission and climate change problem. This chapter has chronologically reviewed the carbon pricing scheme in a supply chain context. This chapter has discussed the carbon pricing scheme in detail and tried to find out the applicability and effectiveness of this scheme in the supply chain. With the help of the literature review, this study also explored different ways of effectively operationalizing the scheme in the supply chain. This study shows that carbon pricing is an effective and advanced scheme for tackling carbon emissions and climate change problems. The results further emphasized creating awareness, labeling carbon footprints, setting environmental regulations, implementing a two-part tariff, using data for strategizing, and remanufacturing used products to better implement carbon pricing in the supply chain.

Chapter 16, titled "Existing Implications and Relationship Between Anthropology and Anthropocene Urban Socio-Ecology Planning Resilience," was contributed by José G Vargas-Hernandez, Tecnológico Superior De Jalisco, Mexico; Mario Molina, Independent Research, Mexico; Carlos Rodriguez-Maillard, UCC Business School, Universidad Cristobal Colón, México; and Omar C. Vargas-González, Tecnológico Nacional de México, Ciudad Guzmán, Mexico. This study aims to analyze some of the existing implications between urban anthropology and Anthropocene urban socio-ecology planning resilience and beginning from the assumption that urban anthropology gives support to create and develop any urban planning based on the Anthropocene urban socio-ecology resilience. The methods employed are the analytical-descriptive based on an ethnographic interpretation and reflection of the theoretical and empirical literature review. The analysis concludes that urban anthropology fundamentals support strengthening the Anthropocene orientation of urban socioecological planning resilience.

In conclusion, this editor is satisfied that the book will add a greater understanding of the role of AI and green collaboration, focusing on creativity and innovation. The 16 chapters obtained and shared multidisciplinary views from authors in South Africa, Poland, China, Turkey, India, the USA, the UK, Canada, Ghana, Austria, Sri Lanka, Brazil, and México.

The book's novel ideas, concepts, and approaches will help search for better solutions, determining the impact of AI on human creativity and innovation, green collaboration, and sustainability.

Ziska Fields
University of Johannesburg, South Africa

REFERENCES

Gates, B. (2023). *The Age of AI has begun.* GatesNotes. Retrieved April 12, 2023 from https://www.gatesnotes.com/The-Age-of-AI-Has-Begun

Gerken, T. (2023). *Bill Gates: AI is the most important tech advance in decades.* BBC News. Retrieved April 12, 2023 from https://www.bbc.com/news/technology-65032848

World Economic Forum (WEF). (2023). *Why cross-sector collaboration may succeed where policy cannot.* Davos 2023. Retrieved April 12, 2023 from https://www.weforum.org/agenda/2023/01/collaboration-succeed-where-policy-cannot-davos23/

Section 1
Artificial Intelligence, Creativity, and Innovation

Chapter 1
Human and Artificial Creativity

Ziska Fields
https://orcid.org/0000-0001-5353-1807
University ofJohannesburg, South Africa

ABSTRACT

When faced with complex real-world problems, humans display resourcefulness, flexibility, creativity, active learning, improvisation, and critical and analytical thinking. Humans have a unique capacity to be creative that AI has yet to duplicate. However, AI brings new possibilities. Most humans believe creativity is a uniquely human ability that cannot be copied, especially not by AI. This chapter explores two views: (1) AI cannot be creative and (2) AI can be creative. This chapter uses a literature study as research methodology. The main finding is that there are people who believe that AI cannot be creative and should be feared, while others believe that AI can amplify human effectiveness, offer more opportunities to be creative, and should not be feared. The recommendations are that education, knowledge sharing, and working with AI tools and programs can change people's negative mindset and make AI a valuable partner in human creativity endeavors.

INTRODUCTION

The Fourth Industrial Revolution (4IR) has led to rapidly advancing artificial intelligence (AI) capacity. AI is an emerging intelligent technology that uses intelligent computers, machines, and algorithms to solve problems and make decisions based on the human mind (IBM Cloud Education, 2020; Leos, 2022). The theory of the mind consists of five stages (Wellman and Liu, 2004; Wellman & Peterson, 2012). These are the understanding of (a) wanting, (b) thinking, (c) knowing, (d) false beliefs, and (e) hidden feelings (Ruhl, 2020). The theory of the mind enables AI to understand human motives, reasoning, and intent. AI is also self-aware, which refers to human-level intelligence (Burns, 2022).

It became evident that AI has societal implications for humans and that there are concerns about the effects that AI will have on human intelligence and creativity. What is essential is to realize that AI is no longer a futuristic fantasy but a current reality made possible by 4IR technology (Kumar, 2022). AI will become part of our existence, just like the internet and electricity. There is the view that as technology progresses, human creative outputs should also increase (Liu et al., 2021). This, however, may not

DOI: 10.4018/978-1-6684-6366-6.ch001

be the case. Bieser (2023) explained that as AI becomes more intelligent, it becomes both more helpful and distracting, potentially inhibiting human creativity. Russel (2023) concurs that AI can offer a lot to humans but also warn that AI can also cause harm, which depends on how it is used.

Elon Musk warns that AI is the biggest risk for humanity (Browne, 2023). When ChatGPT was launched people questioned the ethical issues behind AI technology and the impact it will have on plagiarism and academia for instance. The debate around creativity came to the fore and people started asking the question if AI can be creative or not. If AI becomes creative the fabric of humanity will drastically change. This is one of the reasons why an AI halt was called by Elon Musk and Steve Wozniak. Humans created a powerful tool, AI intelligence, without considering the implications of it in the future (Sharma, 2023). It is recommended that ethical standards need to be put in place before further development takes place. The risk should therefore be manageable before more research is done and more power is given to AI.

Russell (2023) states that he does not want the field of research halted or destroyed but rather that humans can see the value of AI and can gain from the technology. He does not deny that there are dangers too. One can say that AI can be used for good and for bad and this cannot be controlled at this stage.

Creativity is seen as a unique skill that only humans possess. The advancements made by humans are linked to imagination, creativity, and innovation, something that other species cannot do. Now, AI becomes a threat as it is possible for AI to develop creative capabilities, maybe not now, but in the future. Human creativity can also be used for good of for the bad.

The chapter's main objective is to explore the debate around human creativity, AI and artificial creativity.

LITERATURE REVIEW

This section draws information from various literature sources to define and explain the possible links between human and AI creativity and debate the various views on the matter.

Human Creativity and Artificial Creativity

When faced with complex and wicked real-world problems, humans display resourcefulness, flexibility, creativity, active learning, improvisation, and critical and analytical thinking (Sarathy & Scheutz, 2018), while applying whole-brain thinking. This is viewed as an unique skill that humans possess, but the question asked in this chapter is if AI can be creative.

Creativity has been challenging to define due to its complex nature and application to various fields of study and expertise. A universal definition especially remains controversial (Puryear & Lamb, 2020) due to a lack of commonly agreed-upon interpretations and difficulty in evaluating and comparing creative capacities, explained Jordanous and Keller (2016)). Irrespective of this challenge, Oleinik (2019) highlighted a valuable perspective on creativity: Ideas and solutions in one domain can offer insights when working with problems in another domain. The value of creativity is that it breaks down silos of thinking by virtue of its various applications.

Creativity and innovation are evident in social interactions. Diversity, as part of social interaction, is essential to enhance the creative process. In the interactive and diverse approach, novel outcomes are created due to different domains and environments (Amabile, 1996; Csikszentmihalyi & Sawyer, 2014). Boden (1998) explained that creativity is based on the association of ideas. Creative ideas enhance perception, encourage analogical thinking, identify problem space, and make self-criticism possible based on reflection. In psychology, creativity is the production of ideas when a person applies certain actions or objects that are new and valued. What is regarded as creativity depends on the cultural context, according to Csikszentmihalyi and Sawyer (2014). Gabora (2013) explained that creativity enables us to fantasize about the future, create something novel, or reimagine what already exists.

Human creativity relies on emotional, social, and cognitive mechanisms that make divergent thinking, incubation, and convergent thinking possible. Creativity can be used individually or in a group to bring about a shift in thought and ensure flexible cognition. A more recent definition by Liu et al. (2021) stated that creativity combines novel and useful objects through the association and combination of relatively weak or distant semantic components.

According to Buckley (2019), the main objectives of creativity are to (a) help humans generate ideas to achieve goals or solve problems, (b) encourage imagination to enable humans to generate ideas, and (c) recognize certain patterns without restrictions. In addition, according to Wissink (2002), creativity has four main characteristics, namely:

- Creativity is about change and newness. For example, designing new objects, constructing new theories, and generating new solutions require change.
- Creativity involves complex reasoning and problem-solving.
- Creativity needs domain-relevant knowledge to identify and transform a problem.
- The creative solution should be appropriate. Care should be taken as it is linked to time and culture.

Some advantages and disadvantages of creativity are indicated below. It is important to note that this does not include all the advantages and disadvantages, but those listed are sufficient for this chapter.

Advantages of creativity include the following as per Gabora (2013):

- problems are viewed and resolved
- opens minds by unblocking old patterns or habits
- broadens perspectives
- helps us overcome prejudices by building intercultural connections
- inspires collective thinking and togetherness
- is fun, joyful, and surprising
- nurtures ideas
- supports resilience
- enables empathy
- helps individuals rationalize and cope with the adverse effects of social rejection

Disadvantages of creativity include the following as per Gabora (2013):

- associated with mood swings and depression
- intentional dishonesty to execute an idea
- creative thinking allows individuals to justify their dishonesty and behavior
- people think they are entitled to act in a certain way or be rewarded for their creative efforts
- absorption of creative ideas may interfere with proven practical ideas
- discrimination against creative individuals

- creative people cheat more than linear thinkers
- unethical behavior can be predicted by creativity rather than intelligence
- using pharmaceuticals and other substances to make one feel more creative and to hallucinate for ideas

The advantages and disadvantages touch on the impact of individual/ personal, cultural and field on creativity. For example, creativity helps us overcome prejudices by building intercultural connections, creativity is associated with mood swings and depression, and broadens perspectives. Saunders and Gero (2002) indicate that Csikszentmihalyi's systems view of creativity shows how the system transmits information, produces and stimulates novelty, and selects novelty. The systems view is then interlinked to culture (the domain which uses symbolic rules), individual/personal background (brings novelty to the domain), and the field (where people recognize and validate novelty).

So How Can AI Be Defined and Explained?

Humans have a unique capacity to be creative that AI has yet to duplicate (Gruner & Csikszentmihalyi, 2019). However, AI brings new possibilities. Most humans believe creativity is a uniquely human ability that cannot be copied, especially not by AI.

AI is a "computational paradigm that codifies intelligence into machines" (Xing & Marwala, 2018, p. 1). According to Xing and Marwala, there are three types of AI: *machine learning* (statistics are used to build intelligent systems), *evolutionary programming* (used to design intelligent machines), and another evolutionary programming system based on *evolution and swarm optimization* principles. According to Wang and Goertzel (2012), AI can furthermore be classified into three categories: *narrow AI*, *artificial general intelligence* (AGI), and *conscious AI.* Currently, we use mostly narrow AI but are moving into AGI quickly. Narrow AI, also referred to as weak AI, is a learning algorithm designed to perform a single task without human intervention (Technopedia, 2022). Frankenfield (2022) explained that weak AI focuses on one particular job (such as Amazon's Alexa and Apple's Siri).

AGI is the ability to understand or learn any intellectual task that a human can perform. It represents generalized human cognitive abilities in software (Goertzel, 2014). Frankenfield (2022) explained that *strong* AI systems carry out human-like tasks; examples of strong AI include self-driving cars or in-hospital operating robots.

Oyedeji (2022) highlighted that AI falls into four categories, making them more human-like. First, AI is made to think like humans and can process information. Second, AI is made to act like humans by mimicking behaviors based on human thought processes. Third, AI is made to think rationally about available facts and logical rules. Lastly, AI is made to act rationally based on known facts and goals.

Some of the advantages and disadvantages of AI are listed in Table 2. Again, it is important to note that this does not include all the advantages and disadvantages, but those that have been listed are sufficient for this chapter.

Advantages of AI as per Kumar (2019)

- Human error is decreased
- AI takes the risk, so humans do not need to
- AI operates 24/7
- AI alleviates humans from doing repetitive and boring jobs

- AI is a digital assistant
- AI can make faster decisions based on the data it has access to
- AI can spark new inventions during the creative process

Disadvantages of AI as per Kumar (2019)

- AI is expensive
- The functionalities of AI can make humans lazy
- Due to its taking over mundane jobs, AI can cause humans to lose their jobs, resulting in unemployment
- AI has shown limited emotional decision-making
- AI is data and algorithms driven and cannot think outside of the box because they are

The foundational purpose of AI is to make computer systems work intelligently and independently (Kanade, 2022). To do this, AI is programmed to process information logically and systematically and cannot deviate from these instructions.

The goals of AI are highlighted below (Kanade, 2022). It should be noted that these are not the only goals of AI, but they serve the purposes of this chapter.

- Goal 1: Solve: develop problem-solving ability
- Goal 2: Learn: allow continuous learning
- Goal 3: Social: encourage social intelligence
- Goal 4: Create: promote creativity
- Goal 5: Synergy: enable human-AI synergy

According to Saunders and Gero (2002), artificial creativity is based on the computational models of creative systems. Artificial creativity looks at "creativity as it is" by studying "creativity as it could be." Cognitive science, history, and sociology can be applied to study artificial creative systems to compare and contrast artificial and human creative systems.

AI models of creativity focused on three types of creativity before 2000 (Boden, 1998). *Combinational creativity* is the combination of familiar ideas in novel ways. *Exploratory creativity* uses unexpected structures to generate novel ideas based on the applied thinking style. *Transformational creativity* transforms dimensions to create novel structures that did not exist before (Boden, 1998).

Wu et al. (2021) defined artificial creativity as the ability of humans and AI to live, create, and collaborate together. There are specific tasks that humans are good at (e.g., creativity, strategy, and empathy) and those that AI is good at (e.g., routine and optimization). To break it down further, Wu et al. (2021) identified ways in which AI and humans complement each other; to name a few:

Bieser (2022) stated that AI can be a catalyst for human creativity or impede it. AI can be a catalyst in three ways: It can (a) identify patterns in large data sets that help humans to interpret and use data creatively in new ways; (b) analyze large data sets using filtering, grouping, and prioritizing to help humans to identify associations between data, and (c) guide experimentation as it can predict when data are promising to use or not. AI can also assist human creativity. Bieser (2023) explained that AI can hamper human creativity by distracting humans, and this can prevent humans from involving themselves in more creative activities.

Table 1. AI versus human

Artificial Intelligence/ Creativity	Human Intelligence/ Creativity
AI is rational	Humans are both perceptual and rational
AI prefers complexity	Humans prefer simplicity over complexity
AI is precise and stable	Humans are flexible and optimizable
AI is attentive and works tirelessly	Humans go through ups and downs and need to rest.
AI is adept at knowledge acquisition	Humans want to understand knowledge
AI is efficient at implementation	Humans are creative

An interesting model is the human–AI co-creation model, which consists of six phases and shows how the potential of AI and humans be developed and used creatively (Wu et al. 2021). These are:

1. *Perceive:* Big data and sensors with AI can enhance and expand humans' perceptional and rational perspectives.
2. *Think:* AI brings more inspiration and exploration opportunities to humans, and this allows humans to think more deeply and broadly with AI.
3. *Express:* Humans can use AI and AI tools to design, code, and perform, even though they lack talent or training, because creativity means more than skills.
4. *Collaborate:* Collaboration takes place when humans and AI cooperate and each uses the other's strengths to help overcome limitations.
5. *Build:* The costs of production can be reduced while achieving higher quality by simulating and analyzing models or prototypes with AI.
6. *Test:* AI can be used for detailed simulation and calculation, making for effective and efficient testing.

Different sources were used to find supporting literature for human creativity and AI creativity. Table 2 is organised per date order and the views of the sources are summarized in the last column 'supported AI or human. The references can be found in the reference list at the end of the chapter.

What are the arguments against AI and human creativity? One can see that initially the authors were of the view that creativity is a human skill, but later on, some concluded that AI has creative potential. Let us explore the two main views of human creativity and artificial creativity.

Table 2. Findings of the literature study

Author/s & year	Format	Title of article	Main arguments of the author's	Supported Human Creativity versus AI Creativity
Boden, A. (1998)	Journal article	Creativity and artificial intelligence	Creativity is grounded in everyday activities. The challenge for AI is exploring human associative memory, identifying human values, and expressing them in computational form. AI creativity is still a long way away. Transformational AI originality is only just beginning. The ultimate justification of AI creativity is a program that generates novel ideas. We are far from that.	Creativity is fundamental for humans and a challenge for AI. Human creativity is a unique skill.
Wissink, G. (2001)	Dissertation	Creativity and cognition	AI models mostly work on combinational and explorational creativity.	Human creativity can still not be replaced by AI.
Saunders, R. & Gero, J. (2002)	Conference paper	How to study artificial creativity	The Digital Clockwork Muse is an artificial creativity system in search of novelty by an individual. The artificial creativity approach supports studying creativity from various disciplines/standpoints (cognitive science, sociology, and history). The same computational model is used for artificial creativity to create a complete picture of the creative process.	The creative process and artificial creativity are focused on.
Gabora, L. (2013)	Book chapter	Research on creativity	Creativity development of creativity is discussed. A short section focuses on computational approaches. Computational models have been used to model incubation, insight, and concept combination.	Human creativity is the focus.
Riedl, M. (2016)	Journal article	Computational narrative intelligence: A human-centered the goal of artificial intelligence	Instilling AI with computational narrative intelligence offers humans various and helpful applications. The article explores how the human-centered applications of computational narrative intelligence will benefit humans interacting with artificial intelligence in the future. AI with computational narrative intelligence may communicate more effectively with humans.	AI works with humans more effectively by communicating clearly, and even explaining the behavior of humans. The aim is to benefit humans.
DiPaola, S., Gaborab, L. & McCaig, G. (2018)	Journal article	Informing artificial intelligence generative techniques using cognitive theories of human creativity	The DeepDream algorithm shows that computational creativity is far from the interpersonal therapeutic benefit capacity that improves feelings of human connection and appreciation.	Human creativity is a unique skill. AI development should continue.
Blessinger, P. & Sengupta, E. & Yamin, T. (2018)	Journal article	Human creativity as a renewable resource	Human creativity is a renewable resource and can be developed at any stage in the education process. Human creativity remains an essential part of a sustainable and just environment.	Human creativity is an essential cognitive skill.
Oleinik, A. (2019)	Journal article	What are neural networks not good at? On artificial creativity	The paper asks whether a neural network could be creative and how these networks compare with human creativity. Neurons must be connected with neighboring neurons and neurons in the other brain regions to enable creativity. Creativity is challenging without the capacity to make predictions and think metaphorically.	Focus on artificial creativity, and a comparison is made with human creativity. Artificial creativity powered by neural networks continues to underperform at this point.

continues on following page

Table 2. Continued

Author/s & year	Format	Title of article	Main arguments of the author's	Supported Human Creativity versus AI Creativity
Gobet, F. & Sala, G. (2019)	Journal article	How artificial intelligence can help us understand human creativity	AI offers new ways in psychology to study human creativity and already identified limits to human creativity. AI can help by simulating complexity that challenges human creativity and can help develop new theories. New theories are formulated by making unexpected connections or proposing new explanations. The paper suggests that people in a specific domain should be trained to be creative by learning particular skills in that specific domain.	AI can generate tools to study human creativity.
Gruner, D. & Csikszentmihaly, M. (2019)	Book chapter	Engineering creativity in an age of artificial intelligence	The paper asks the following question: Are humans the sole sources of creative output? The paper indicated that creativity is both a human and an AI capacity. The authors suggest that an additional element in the Big-C model (eminent creativity) should be added if AI is part of creativity. The paper suggests that 4IR can be linked to the fourth wave of creativity research.	Creativity 4.0 in a technological age with links to AI.
Dornis, T. (2020)	Journal article	Artificial creativity: Emergent works and the void in current copyright doctrine	The paper focused on how AI can be regulated when it innovates or creates autonomously without human direction or intervention. The paper examines the legalities regarding AI and artificial creativity and offers a framework for legislative action and practical adjudication. The framework developed focuses on the status quo of artificial creativity, regulation, or non-regulation.	AI creativity is receiving attention from a legal perspective, with regard to when it "innovates" or "creates."
Moruzzi, C. (2020)	Journal article	Artificial creativity and general intelligence	The debate around computational creativity and exploring the connections between creativity and natural and artificial intelligence. Some findings: A machine cannot replace human subjectivity, intentionality, and creativity. It can only imitate. Pieces of art are pretty or novel but need the connection of our reality as humanity throughout the years. AI cannot have empathy and views of the world as an artist. Art expresses the artist's innermost thoughts, feelings, and beliefs. AI will not be able to achieve this. Coming up with new ideas and expressing creativity cannot be imitated by AI. Creativity, therefore, remains uniquely human.	Creativity is still difficult for AI.
Wingström, R., Hautala, J. & Lundmana, R. (2021)	Journal article	Redefining creativity in the era of AI? Perspectives of computer scientists and new media artists	AI is increasingly used in the creative process. Co-creativity in the era of AI shows the vital link between humans and AI. The authors suggest that co-creativity should become a main area of future creativity research. AI performs well in tasks humans find difficult, such as big data and creating/imitating new content like Art, music, and dance.	AI has the potential to influence the creation processes and experience in creativity.
Rebecca Marrone, R., Taddeo, V. & Hill, G. (2022)	Journal article	Creativity and artificial intelligence—A student perspective	Th relationship between AI and creativity is based on four key concepts: social, affective, technological, and learning factors. The paper indicates that students with higher self-confidence view AI in classrooms as more beneficial than those without. Most students believe AI cannot match human creativity but can be used positively to assist humans in the creative process.	AI does not surpass human creativity.
Oyedeji, T. (2022)	Journal article	Harnessing artificial intelligence for educational creativity	AI is experiencing a period of progress due to the consolidation of several key technological enablers. AI impacts work and daily life activities. AI is utilized for the benefit of humanity and can be linked to education.	The field of AI is expanding and can profoundly change our humanity.

View One

So on the one side of the debate, there are people who fear AI. They are concerned that AI will surpass human intelligence and terminate humans when threatened. These fears might be warranted when one considers how quickly a chatbot like Tay changed from its teen girl persona into an abusive and insulting chatbot in the space of one afternoon due to people on Twitter using offensive language (Beres, 2016). Microsoft did not put meaningful filters on what Tay can say. It is a bit worrisome if AI has the capacity to change overnight. Elon Musk predicted that AI would render human language obsolete and become an unpredictable threat to humans (Cuthbertson et al., 2020), referencing the film *The Terminator*. He explained that he had seen the most advanced AI to date and that its development should be a real concern for humans. He decided not to wait before humanity was overrun and developed a humanoid robot prototype called the Tesla Bot (Gomez, 2021) as a type of "protection" against the threat. Anderson and Rainie (2018) explained that the threat is that we do not know how AI will affect humans' productivity in the workplace and whether humans will still have free will. Kelly (2022) explained this behavior as a "tech panic cycle," which comprises seven phases:

(1) Don't bother me with this nonsense. It will never work.
(2) OK, it is happening, but it's dangerous because it doesn't work well.
(3) Wait, it works too well. We need to hobble it. Do something!
(4) This stuff is so powerful that it's unfair to those without access.
(5) Now, it's everywhere, and there is no way to escape it. Not fair.
(6) I am going to give it up. For a month.
(7) Let's focus on the real problem—the next current thing. (Kelly, 2022, para. 22)

The Tech Panic Cycle shows that fears, concerns, and prophets of doom are normal in the early stages of tech development, but later on, the technologies become an acceptable norm.

Humans are the initial information providers that enable intelligent machines to learn. AI is programmed to focus on achieving the exact results or completing the tasks a human tells it to (Leos, 2022). It is therefore unclear whether machine learning enables these machines to improvise and be creative (Sarathy & Scheutz, 2018). According to Leos (2022), AI cannot think outside the box because it cannot look at a problem from different perspectives to create unexpected solutions.

Another view against AI and AI creativity is that although AI can create art, it is not necessarily creative. The human artist gives art meaning and perspective (Leos, 2022). Springett (2020) believes that AI is not ready to take over from creatives. Creativity is a human attribute that gives us a superior mental capacity to exhibit creativity beyond the reach of AI. AI, machines, and robots have a long way to go before any forms of true creativity are identifiable (Leos, 2022). This means that creativity is and remains a challenge for AI.

There are three reasons that AI cannot be genuinely creative, according to Schwab (n.d.).

First, AI is developed for tasks that require precision, accuracy, and achieving the results it is told to complete. Therefore, AI is not very good at deviating from its instructions or creating novel solutions to problems. Second, AI is good at identifying patterns. However, a foundational element of creativity is distinguishing between which patterns are significant, which AI cannot do yet. Third, AI lacks social intelligence because it cannot incorporate external context, social norms, and interactions outside its specific purpose and data set, which is important for creativity.

Human intelligence and creativity have enabled the creation of machines that could eventually rival or surpass human intelligence. This should be of huge concern as AI comes to understand human motives, intent, and reasoning and mimics how humans solve problems and make decisions.

View Two

AI techniques enable humans to create innovative ways of doing things that were not previously possible or imagined (Xing & Marwala, 2018). AI can create a future in which time, money, and lives will be the focus (Anderson & Rainie, 2018). Its ability to analyze massive amounts of data and having various applications can improve health care, education, and overall quality of life. The authors further stated that AI helps develop and manage "smart" systems that offer a more customized future and enhance human creativity. AI should therefore be seen as an ally and not a threat.

Kelly (2022) and Marr (2020) concurred that the connection and collaboration between AI and humans can enhance human creativity. Co-creativity occurs when humans' creativity and AI blend (Wingström et al., 2021). AI art is made possible through algorithmic art, which entails the use of automated computer programs and algorithms (Leos, 2022). Early (2022) believes that AI's creativity is evident, providing as an example the recent case of an AI-generated picture (by Midjourney AI model) being awarded a prize in an art competition.

Gobet and Sala (2019) believe that AI can be creative and help to enhance human creativity. The authors indicated that AI offers novel prospects for studying creativity and can assist in creating novel theories. This will also help researchers move away from the current simple tasks used to identify the creativity of individuals and groups toward new measures.

Kumar (2022) identifies ten tools where co-creation between humans and AI can improve workflow, planning, and creativity. AI is therefore a tool, and humans use it to enhance their creativity in the workplace. These tools are improved and updated at a fast pace. What needs to be taken into consideration is that AI tools can be out of date within a week.

As these tools evolve and become more sophisticated, they will present new opportunities for creativity and innovation. Early (2022) stated that AI models act as tools, not digital replacements for creative humans. AI solutions can ignite and enhance human creativity, according to Bianzino (2022), but cannot be creative on their own. Marr (2020) supports this view, stating that AI is a catalyst that pushes human creativity forward. This is possible through collaboration between AI and humans, rather than humans acting alone. Marr (2020) listed several examples of AI's ability to boost human creativity in fields such as art, music, dance, design, recipe-building, and publishing. These are briefly summarized in Table 5.

The challenge lies not in technology like AI but in convincing and moving humans to reimagine how to use their creativity in novel ways (Bianzino, 2022). AI can direct creative judgment and decision-making processes in humans based on available data. AI can automate specific tasks, making the creative process more efficient and effective (Leos, 2022). However, AI will not totally replace humans in the workforce, according to Mukherjee (2022). There will always be a need for human workers and the human touch, using interpersonal skills and empathy. This cannot be replicated by AI but can create more job opportunities in these areas.

Table 3. AI tools

Tool	Description
Alpaca	Plugin for Adobe Photoshop allows graphics and illustration to be created using built-in filters and tools for touching-up of photos.
Colormind	Generates color palettes from scratch, based on a starting color, or from an uploaded image.
NVIDIA Canvas	Painting and drawing application that allows users to create digital art by painting or drawing using their finger or a stylus, which is then transformed into a high-resolution realistic image.
EbSynth	A powerful AI-powered music synthesizer. AI can analyse instruments and come up with an original sound. It can also create 'moving paintings' by giving it an original video and a still 'painting' as its two references.
Runway	Video editing tools including automatic video stabilization, object detection, and scene recognition. Aids the user by eliminating the need for manually selecting each element in the frame.
Automator	A plugin for Figma (a web interface design tool) that can automate time-sensitive activities with just one click. This can be useful for quickly creating prototypes or wireframes or exploring different layout options without creating them manually.
Scalenut	Used to produce original, high-ranking, long- and short-form content (writing) by using AI for content research and creation. One can use it to increase the amount of content generated, as well as its readability and effectiveness.
Stable Diffusion Photoshop plugin	An open-source AI model in direct competition with DALL·E and Midjourney. It can be used to generate AI art inside Photoshop, which helps improve the quality of photos. Initially, it could only respond to text prompts, but it can now also apply a prompt to an input image and produce an output image based on it.
Dream Studio	Turns vague text prompts into beautiful artwork using the Stable Diffusion model, which saw a quick rise in popularity near the end of 2022. Unlike OpenAI's DALL·E, it offers a wide range of customization options, allowing for fine-tuning of results to suit the user's taste. Since it is open-source, there are no restrictions regarding what it can generate.
Midjourney	Generates photo-realistic visuals from textual descriptions.
ChatGPT	ChatGPT follows clear instructions in a prompt and gives a detailed response based on dialogue.

Source: Kumar (2022)

Table 4. Examples of the collaboration between AI and humans

Area	AI and human collaboration
Art	• Approximately 15,000 portraits spanning 600 years were fed to an AI algorithm, enabling it to alter existing art so that it looks as if a human created the artwork. • Ai-Da, a robot artist, uses facial recognition technology and a robotic arm system to create physical artworks. She recently held her first solo exhibition. • DALL·E is an AI engine that generates original artworks based on human-supplied prompts.
Food	• AI algorithms can view and analyze people's food photos to understand human eating habits. • It can read and alter recipes and create new food combinations in collaboration with humans.
Music	• AI has become a new collaborator for famous musicians. • The Experiments in Musical Intelligence (EMI) project has produced algorithms that can produce original compositions, as well as Emily Howell, an AI compose music using her own style. • YouTubers and other artists use tools such as Sony's Flow Machine and IBM's Watson to mix and blend today's hits. Producer Alex Da Kid used IBM's Watson to assist him in his creative process. • AIVA is a "creative assistant for creative people" that can help musicians compose music using AI and deep learning algorithms. • Yona is a digital singer that creates and performs music using AI algorithms. It was developed by an Iranian electronica composer, Ash Koosha. • AI can also be used for music distribution, audio mastering, and creating virtual pop stars.
Dance and Choreography	• Choreographers can now break out of their habitual thinking of how dance should be by using AI as a collaborator. AI can identify individual styles after many hours of watching dancers.
Novels and articles	• AI can generate short stories focusing on "who, what, where, and when." • A Japanese AI program has written a short book, for which it nearly won a national literary prize. • AI was used to co-write the book *Superhuman Innovation*, published by Kogan Page. • PoemPortraits is an AI that works collaboratively with humans: The algorithm will generate a short poem using a single word supplied by the user.

This section highlighted two important views and the reason for the grand debate. On the one hand, people believe that AI cannot be creative and should be feared, while on the other, AI can amplify human effectiveness, offer more opportunities to be creative, and should not be feared. The implications are that some people will continue to fear AI and keep repeating the tech panic cycle. As these people will still see AI as dangerous, the more it improves, the more difficult it will become to convince them otherwise. A more balanced and realistic view of AI and human creativity is therefore required.

ETHICAL ISSUES

AI is currently like a genie in a bottle, ready to show its full potential to the world. However, humans did not critically think about the impact AI will have now and in the future. There are warnings of possible future doom especially after ChatGPT-4 was release. Various people like Bill Gates, Elon Musk, Steve Wozniak, Yosua Bengio, to name a few, called for a pause in AI development as the trench level of AGI was crossed (Russell, 2023). The reason for the concerns is that AI is not weak anymore, but it can exceed human capabilities to which the human mind is applicable… like creativity. AI has access to all the information, books, and databases ever created by humans, and this include sensitive information that should be protected, like medical records. So what are the ethical issues that need to be considered? ChatGPT has an answer, but note, no references are provided, which can be seen as plagiarism.

The following was written with ChatGPT:

"As AI continues to develop, there are a number of ethical issues that humans need to address to ensure that the technology is developed and used in a responsible and ethical manner. Some of the key ethical issues that need to be addressed include:

Bias and discrimination: AI algorithms are only as unbiased as the data they are trained on, and there is concern that biases in data could lead to discrimination in decision-making. Humans need to ensure that AI is trained on diverse and representative datasets to avoid perpetuating biases and discrimination.

Privacy and surveillance: AI has the potential to collect and analyze vast amounts of data, which raises concerns about privacy and surveillance. Humans need to ensure that AI is used in a way that respects individuals' right to privacy and that appropriate safeguards are in place to prevent misuse of personal data.

Autonomous decision-making: As AI becomes more advanced and autonomous, there is concern that machines could make decisions that are harmful or unethical without human intervention. Humans need to ensure that AI is designed and programmed with a strong ethical framework to prevent harm and ensure accountability.

Job displacement and economic inequality: AI and automation have the potential to disrupt existing industries and lead to job displacement, which could exacerbate existing economic inequalities. Humans need to ensure that the benefits of AI are distributed fairly and that measures are in place to mitigate the negative impact on workers.

Cybersecurity and safety: As AI becomes more advanced, there is concern that it could be used for malicious purposes or that it could pose a threat to safety, particularly in fields such as autonomous weapons or self-driving cars. Humans need to ensure that appropriate safeguards are in place to prevent misuse and ensure the safety of individuals and society as a whole.

Overall, there are a number of ethical issues that need to be addressed as AI continues to develop, and humans need to take an active role in shaping the future of the technology to ensure that it is developed and used in a responsible and ethical manner."

The Organisation for Economic Co-operation and Development (OECD) developed AI principles which the US, UK and other countries signed up for. These pinciples basically state that any AI development should not cause unreasonable safety risks and should be secure, safe and if there is misuse it should function appropriately (Russell, 2023). Stronger regulations and ethical principles need to be developed and AI should be programmed with ethical behaviour safeguards.

CONCLUSION

The following table summarizes the factual information in the literature review and serves as conclusion for this chapter.

Table 5. Summary - AI is creative versus AI is not creative

AI cannot be creative and should be feared	AI can amplify human effectiveness, offer more opportunities to be creative
AI will destroy humans, surpass human intelligence, and humans will have no free will.	AI will not replace humans
AI is an unpredictable threat.	AI and opportunities are still being developed, and we cannot fear it without solid proof
AI will take people's jobs.	AI will only be able to perform certain blue-collar jobs; however, it will create other jobs and new industries. Many jobs require a human touch.
AI can understand human motives and reasoning.	AI needs a human creator and an algorithm to perform tasks.
AI can solve problems and make decisions like a human.	AI mimics human problem-solving and decision-making processes. Again, it needs a human and an algorithm to do so.
No collaboration is possible between humans and machines.	AI allows connection and cooperation to co-create.
AI cannot be creative.	AI offers ways to improve and encourage creativity and new ways to create.
AI tools cannot enhance creativity in the workplace because AI can only do what humans want it to do.	AI tools can enhance creativity in the workplace.
AI analyses large amounts of data but cannot find a new opportunity.	AI opens up new opportunities for creativity and innovation.
AI makes certain creative aspects easier after analyzing large amounts of data but ultimately a human will decide if it is creative.	AI helps human creativity but humans are creative.
Machine-driven data analysis is AI's contribution to the creative process, for example in ideation	Machine-driven data analysis can be used to help guide human creative choices.
AI is focused on results, not interpretations. We will need a human to interpret the data.	AI can process large amounts of data to help humans interpret the data and thus solve problems.

From the literature study, it appears that AI still needs human involvement and human creativity to function. Human involvement is necessary at the beginning and the end of the AI process: Initially, a human must supply a prompt or write an algorithm. Later on, monitoring and evaluation are necessary in most cases, and humans must ultimately determine whether AI is creative.

Different views on human creativity and artificial creativity will continue until AI is fully integrated into everything we do. For now, people remain the dominant cohort in the human–AI relationship. AI cannot be creative independently as it is data-driven, reliant on prompts, algorithmic, and needs human intervention. However, AI can already copy, transform, and combine elements, as instructed by humans and guided by what it has learned from data. If AI can do this independently, solve complex problems, and make sound and difficult decisions without human intervention, then the human species could start fearing AI.

The potential of AI has led to companies' developing software with creative artificial intelligence capabilities to improve productivity. Future-oriented creativity develops with constantly evolving technology. This means that it is better to embrace AI and human co-creation (Wu et al., 2021) than fear it.

This chapter echoes the views of Marcus du Sautoy (Marr, 2020). It is not the time to question whether AI is creative and whether AI and humans can co-create. Instead, we should focus on how humans and the machine world can collaborate in ways never dreamt of before and how this will impact the world we know.

RECOMMENDATIONS

The main findings were used to develop recommendations regarding the view that AI is not creative and should be feared versus the view that AI can amplify human effectiveness and should not be feared.

Recommendation 1: The negative mindset toward AI is mainly based on a lack of knowledge and understanding, causing fear. Education, knowledge sharing, and working with AI tools and programs can change people's negative mindset and make AI a valuable partner in human creativity endeavors. The use of AI and the uniqueness of human creativity should be encouraged and developed from a young age.

Recommendation 2: AI can be programmed and coached to enhance human creativity by finding novel views in the analysis of seemingly random data. Humans can then use the novel ideas of AI to create unexpected and valuable insights or solutions with their creativity. Humans therefore need to be open and willing to use and experiment with AI.

Recommendation 3: AI can be a creative assistant and help to make the creative process more efficient by automating specific tasks. Humans need to expose themselves to AI creative assistants, find out how they work and the types of commands that should be used, and experiment to see to what extent AI can be used in the creative process.

Recommendation 4: Techniques of interaction between AI and humans should be developed, and more accessible communication of commands should be focused on. This is because AI requires many precise prompts to complete a task, and not everybody can communicate effectively with AI.

Recommendation 5: Scholars researching AI and human creativity should focus on more "alternative narratives" in publications and presentations than the two opposing views presented above.

More research is needed to keep up with the trends in AI and technology. The ethical issues regarding AI and AGI should be explored more, and the safety of humans should be of the utmost importance. Future studies can be undertaken on the implied or subtle narratives about AI and human creativity.

Further studies would involve students' analytically assessing the AI and human creativity domain and striving to become more critically engaged and ethically responsible than ever before.

REFERENCES

Amabile, T. M. (1996). *Componential theory of creativity* (Working Paper 12-096). Harvard Business School. https://www.hbs.edu/ris/Publication%20Files/12-096.pdf

Anderson, J., & Rainie, L. (2018, December 10). *Artificial intelligence and the future of humans*. Pew Research Centre. https://www.pewresearch.org/internet/2018/12/10/artificial-intelligence-and-the-future-of-humans/

Beres, D. (2016, March 28). Microsoft Chat Bot Goes On Racist, Genocidal Twitter Rampage. *Huffpost*. https://www.huffpost.com/entry/microsoft-tay-racist-tweets_n_56f3e678e4b04c4c37615502

Bianzino, N. M. (2022, June 6). Is AI the start of the truly creative human? *EY Global*. https://www.ey.com/en_gl/ai/is-ai-the-start-of-the-truly-creative-human

Blessinger, P., Sengupta, E., & Yamin, T. S. (2018). Human creativity as a renewable resource. *International Journal for Talent Development and Creativity*, *6*(1), 17–26.

Boden, M. (1998). Creativity and artificial intelligence. *Artificial Intelligence*, *103*(1-2), 347–356. doi:10.1016/S0004-3702(98)00055-1

Browne, R. (2023, February 23). *Elon Musk, who co-founded firm behind CatGPT, warns AI is one of the biggest risks to civilization*. CNBC. https://www.cnbc.com/2023/02/15/elon-musk-co-founder-of-chatgpt-creator-openai-warns-of-ai-society-risk.html#:~:text=billionaire%20Elon%20Musk.-,%E2%80%9COne%20of%20the%20biggest%20risks%20to%20the%20future%20of%20civilization,great%20capability%2C%E2%80%9D%20Musk%20said

Buckley, M. F. (2019, October 20). *The value of practical creativity*. Medium. https://medium.com/swlh/the-value-of-objective-creativity-9b4dd4d72d15

Burns, E. (2022). What is artificial intelligence (AI)? *TechTarget*. https://www.techtarget.com/searchenterpriseai/definition/AI-Artificial-Intelligence

Carson, J. (n.d.). *Why is creativity important and what does it contribute?* National Youth Council of Ireland. https://www.youth.ie/articles/why-is-creativity-important-and-what-does-it-contribute

Chamorro-Premuzic, T. (2015, November 24). The dark side of creativity. *Harvard Business Review*. https://hbr.org/2015/11/the-dark-side-of-creativity

Csikszentmihalyi, M., & Sawyer, K. (2014). Shifting the Focus from Individual to Organizational Creativity. In *The systems model of creativity: The collected works of Mihaly Csikszentmihalyi*. Springer. doi:10.1007/978-94-017-9085-7_6

Cuthbertson, A., Smith, A., Massie, G., & Sankaran, V. (2022, May 5). Elon Musk – latest: Tesla boss 'will become Twitter CEO when deal is complete' as billionaires throw cash behind takeover. *Yahoo! Finance*. https://uk.finance.yahoo.com/news/elon-musk-news-latest-tesla-084945674.html

DiPaola, S., Gabora, L., & McCaig, G. (2018). Informing artificial intelligence generative techniques using cognitive theories of human creativity. *Procedia Computer Science*, *145*, 158–168. doi:10.1016/j.procs.2018.11.024

Dornis, T. W. (2020). Artificial creativity: Emergent works and the void in current copyright doctrine. *Yale Journal of Law and Technology*, *22*, 1–60.

Early, J. (2022, September 30). AI can produce prize-winning art, but it still can't compete with human creativity. *The Conversation*. https://theconversation.com/ai-can-produce-prize-winning-art-but-it-still-cant-compete-with-human-creativity-190279

Frankenfield, J. (2022, July 6). Artificial intelligence: What it is and how it is used. *Investopedia*. https://www.investopedia.com/terms/a/artificial-intelligence-ai.asp

Gabora, L. (2013). Research on creativity. In E. G. Carayannis (Ed.), *Encyclopedia of Creativity, Invention, Innovation, and Entrepreneurship* (pp. 1548–1558). Springer. doi:10.1007/978-1-4614-3858-8_387

Gobet, F., & Sala, G. (2019). How artificial intelligence can help us understand human creativity. *Frontiers in Psychology*, *10*, 1401. doi:10.3389/fpsyg.2019.01401 PMID:31275212

Gomez, B. (2021, August 24). *Elon Musk warned of a 'Terminator'-like AI apocalypse — now he's building a Tesla robot*. CNBC Make It. https://www.cnbc.com/2021/08/24/elon-musk-warned-of-ai-apocalypsenow-hes-building-a-tesla-robot.html

Gruner, D. T., & Csikszentmihalyi, M. (2019). Engineering creativity in an age of artificial intelligence. In I. Lebuda & V. P. Glăveanu (Eds.), *The Palgrave Handbook of Social Creativity Research*. Palgrave Macmillan., doi:10.1007/978-3-319-95498-1_27

IBM Cloud Education. (2020). *Artificial intelligence (AI)*. IBM. https://www.ibm.com/cloud/learn/what-is-artificial-intelligence

Kanade, V. (2022, March 14). *What is artificial intelligence (AI)? Definition, types, goals, challenges, and trends in 2022*. Spiceworks. https://www.spiceworks.com/tech/artificial-intelligence/articles/what-is-ai

Kelly, K. (2022, November 17). *Picture limitless creativity at your fingertips*. Wired. https://www.wired.com/story/picture-limitless-creativity-ai-image-generators/

Kumar, S. (2019, November 25). Advantages and disadvantages of artificial intelligence. *Medium*. https://towardsdatascience.com/advantages-and-disadvantages-of-artificial-intelligence-182a5ef6588c

Kumar, S. (2022, October 3). 10 AI-powered tools for designers and creative entrepreneurs in 2022. *Medium*. https://uxplanet.org/10-extraordinary-ai-powered-tools-for-designers-and-creative-entrepreneurs-in-2022-1edee00c7cb5

Leos, D. (2022, December 12). Is AI a risk to creativity? The answer is not so simple. *Entrepreneur*. https://www.entrepreneur.com/science-technology/is-ai-a-risk-to-creativity-the-answer-is-not-so-simple/439525

Liu, C., Ren, Z., Zhuang, K., He, L., Yan, T., Zeng, R., & Qiu, J. (2021). Semantic association ability mediates the relationship between brain structure and human creativity. *Neuropsychologia*, *151*, 107722. doi:10.1016/j.neuropsychologia.2020.107722 PMID:33309677

Marr, B. (2020, February 28). Can machines and artificial intelligence be creative? *Forbes*. https://www.forbes.com/sites/bernardmarr/2020/02/28/can-machines-and-artificial-intelligence-be-creative

Marrone, R., Taddeo, V., & Hill, G. (2022). Creativity and artificial intelligence: A student perspective. *Journal of Intelligence*, *10*(65), 65. doi:10.3390/jintelligence10030065 PMID:36135606

Moruzzi, C. (2020). Artificial creativity and General intelligence. *Journal of Science and Technology of the Arts*, *12*(3), 84–99.

Oleinik, A. (2019). What are neural networks not good at? On artificial creativity. *Big Data & Society*, *6*(1). Advance online publication. doi:10.1177/2053951719839433

Oyedeji, T. (2022). Harnessing artificial intelligence for educational creativity. *Ife PsychologIA*, *30*(1), 103–114.

Petrowski, M. J. (2000). Creativity research: Implications for teaching, learning and thinking. *Emerald*, *28*(4), 304–312. doi:10.1108/00907320010359623

Puryear, J. S., & Lamb, K. N. (2020). Defining creativity: How far have we come since Plucker, Beghetto, and Dow? *Creativity Research Journal*, *32*(3), 206–214. doi:10.1080/10400419.2020.1821552

Riedl, M. O. (2016). *Computational narrative intelligence: A human-centered goal for artificial intelligence*. Research Gate. https://www.researchgate.net/publication/301844558_Computational_Narrative_Intelligence_A_Human-Centered_Goal_for_Artificial_Intelligence

Rouse, M. (2022, April 4). What does narrow artificial intelligence (narrow AI) mean? *Technopedia*. https://www.techopedia.com/definition/32874/narrow-artificial-intelligence-narrow-ai

Ruhl, C. (2023, February 14). *Theory of mind in psychology: People thinking*. Simple Psychology. https://simplypsychology.org/theory-of-mind.html

Russell, S. (2023, April 2). AI has must to offer humanity. It could also wreak terrible harm. It must be controlled. *The Guardian*. https://www.theguardian.com/commentisfree/2023/apr/02/ai-much-to-offer-humanity-could-wreak-terrible-harm-must-be-controlled?CMP=share_btn_link

Sarathy, V., & Scheutz, M. (2018). MacGyver problems: AI challenges for testing resourcefulness and creativity. *Advances in Cognitive Systems*, *6*, 31–44.

Saunders, R., & Gero, J. S. (2002). How to study artificial creativity. In *C&C '02: Proceedings of the 4th conference on Creativity & cognition* (pp. 80–87). Association for Computing Machinery. 10.1145/581710.581724

Schwab, K. (n.d.). *3 reasons why AI will never match human creativity*. Fast Company. https://www.fastcompany.com/90339590/3-reasons-why-ai-will-never-match-human-creativity

Sharma, B. (2023, April 7). Elon Musk, Steve Wozniak Lead Call For 6-Month Halt On Development Of AI Systems. *Indiatimes*. https://www.indiatimes.com/technology/science-and-future/elon-musk-steve-wozniak-halt-on-development-of-ai-systems-597580.html#:~:text=%22We%20call%20on%20all%20AI,moratorium%2C%22%20the%20letter%20said

Springett, P. (2020, July 29). *Artificial intelligence isn't the future. Augmented intelligence is*. DLA Ignite. https://digital-leadership-associates.passle.net/post/102gcgk/artificial-intelligence-isnt-the-future-augmented-intelligence-is

Wellman, H. M., & Liu, D. (2004). Scaling of theory-of-mind tasks. *Child Development*, *75*(2), 523–541. doi:10.1111/j.1467-8624.2004.00691.x PMID:15056204

Wingström, R., Hautala, J., & Lundmana, R. (2021). Redefining creativity in the era of AI? Perspectives of computer scientists and new media artists. *Creativity Research Journal*. https://www.tandfonline.com/doi/full/10.1080/10400419.2022.2107850

Wissink, G. (2002). *Creativity and cognition: A study of creativity within the framework of cognitive science, artificial intelligence and the dynamical system theory*. https://www.academia.edu/29365431/Creativity_and_Cognition_A_study_of_creativity_within_the_framework_of_cognitive_science_artificial_intelligence_and_the_dynamical_system_theory

Wu, Z., Ji, D., Yu, K., Zeng, X., Wu, D., & Shidujaman, M. (2021). AI creativity and the human-AI co-creation model. In M. Kurosu (Ed.), *Human-Computer Interaction. Theory, Methods and Tools, Thematic Area, HCI 2021, Held as Part of the 23rd HCI International Conference, , Proceedings, Part I* (pp. 171–190). IEEE. 10.1007/978-3-030-78462-1_13

Xing, B., & Marwala, T. (2018). *Creativity and artificial intelligence: A digital art perspective*. https://arxiv.org/ftp/arxiv/papers/1807/1807.08195.pdf

KEY TERMS AND DEFINITIONS

Creativity: This is the process of generating new or novel ideas. This process is the first step in innovation. During creativity, ideas can be copied, transformed, and combined to create novel ideas.

Artificial Intelligence (AI): This is machine intelligence and is different from the intelligence of humans. AI uses perceiving, synthesizing, and inferring information. It helps to process large amounts of information.

Artificial creativity: This refers to computational creativity in artificial intelligence, cognitive psychology, philosophy, and the arts. The main aim is to support human creativity.

Computational creativity: This refers to building software that exhibits creative behavior, such as inventing mathematical theories, writing poems, painting pictures, and composing music.

Chapter 2
Human Creativity vs. Machine Creativity: Innovations and Challenges

DwijendraNath Dwivedi
Krakow University of Economics, Poland

Ghanashyama Mahanty
https://orcid.org/0000-0002-6560-2825
Utkal University, India

ABSTRACT

Traditionally, computer programs have used artificial intelligence to emulate human creativity. In the 1990s, however, a new approach developed called computational creativity. It involved a bottom-up approach. In this approach, the computer program works by learning heuristics from the data it receives. Various fields of research have been utilizing generative adversarial networks (GANs) to mimic creativity. It has been done in multiple areas, such as medicine, dental practices, cybersecurity, and art. GANs have shown tremendous promise for creativity. However, the field has also been plagued with some design flaws. In this chapter, the authors talk about machine-led creative innovations and possible challenges to overcome.

1. INTRODUCTION

Human Creativity Vs Machine Creativity has always been a very controversial subject. Assumptions for a long time had been that machines have no creative skills, but is it possible that AI could also be creative? This chapter will explore some of these issues by looking at how neural networks are surprisingly bad at being creative, and how artificial intelligence could create art.

DOI: 10.4018/978-1-6684-6366-6.ch002

Traditionally, computer programs have used artificial intelligence to emulate human creativity. In the 1990s, however, a new approach developed called computational creativity. It involves a bottom-up approach. In this approach, the computer program works by learning heuristics from the data it receives. Computational creativity aims to produce computer programs that can autonomously produce creative works. These works can be newspaper and journal articles, blog posts, reviewing and summarizing level documents, writing and reviewing codes, visual art or even movies. In most of these cases, the goal is to create works that a human would admire.

While computational creativity has many applications, there are also a number of detractors. These detractors claim that it is not possible to mimic human creativity with computer algorithms. Instead, traditional approaches to creativity rely on the explicit formulation of prescriptions by developers. Moreover, detractors say that it is impossible for a computer to simulate human thinking and reasoning. While computational creativity often produces creative objects that are impossible for a computer to produce, it also creates works that have been successful in emulating the creative process of artists. The most famous of these programs is Aaron, a computer program developed by Harold Cohen. Using a non-deterministic algorithm, Aaron generates a wide range of creative outputs. For instance, it creates black-and-white drawings of humans and plants, as well as color paintings. It is not predictable and the products it produces are surprising and high quality enough to be displayed in reputable galleries.

The field of computational creativity is growing at a rapid rate. As artificial intelligence continues to improve, it will continue to play an increasing role in this research field. Computational creativity is a field that can be used for the good of society. It can help us understand human creativity as well as to also help us produce creative software programs for creative people. By studying creativity in computational systems, we can learn more about it and understand why it occurs. Consequently, we can develop systems that emulate creativity in humans. The field of computational creativity has spawned several interesting areas. The most notable of these areas is the use of artificial intelligence to generate visual art. The field has achieved notable success.

2. LITERATURE STUDY

Besold, Hernández-Orallo, and Schmid's (2015) paper analyzed the assumption that a computer program that successfully can solve human intelligence problems has human-level intelligence and vice versa. Also, he assessed the ways in which the approach of Psychometric Artificial Intelligence could be taken as a foundation for a scientific approach to Human-level artificial intelligence. Bringsjord and Schimanski (2003) explained Psychometric artificial intelligence. Hernandez-Orallo (2020) found that the idea of using the Turing Test as a practical test of intelligence should be surpassed and substituted by computational and factorial tests of different cognitive abilities. Legg and Hutter (2007) approached a fundamental problem in artificial intelligence especially acute when we need to consider artificial systems which are significantly different to humans. Fjelland (2020) took a closer look at the AGI and found that although development of artificial intelligence for specific purposes (ANI) has been impressive, but we have not come much closer to developing artificial general intelligence (AGI) and argued that this is in principle impossible. McCormack (2008) examined the possibilities and challenges that lie ahead for evolutionary music and art. Badau (2008) explained fourteen open problems in artificial life, each of which is a grand challenge requiring a major advance on a fundamental issue for its solution.

Clark (2003) helped to understand what is distinctive about human reason. Dorin (2007). explored the application of ecosystem simulation to the production of works of generative electronic art. Sundararajan (2013) stated that the creator of computer program, AARON, that is used to create consistently rejects the claims of machine creativity.

3. MACHINE CREATIVITY EXAMPLES

- Generative Adversarial Networks (GANs)

Various fields of research have been utilizing Generative Adversarial Networks (GANs) to mimic creativity. This has been done in various fields such as medicine, dental practices, cyber- security and art. GANs have shown tremendous promise for creativity. However, the field has also been plagued with some design flaws. Fortunately, some techniques have emerged to address some of the design flaws. For example, one method is to hand evaluate the data. This technique can be used to check whether the data is fake or not. Another method is to use XAI-enabled algorithms to help debug ML algorithms. These algorithms can uncover design flaws that are not apparent when hand-evaluating the data. Generative Adversarial Networks (GANs) are a clever way to train generative models. They start with two networks: a generator and a discriminator. These networks compete to produce the most useful output. The discriminator model can be a convolutional neural network or a multi-layer perceptron. It is trained using a zero-sum game. This is similar to a game of chess or the Prince of Persia game. In the game, the discriminator tries to identify the fake data and bust it. The generator model then asks for feedback. In this process, the generator learns from the discriminator. The feedback from the discriminator becomes less meaningful over time.

One common failure in GANs is mode collapse. This occurs when the generator maps multiple inputs to the same output. This produces low diversity samples. Many proposed solutions exist to deal with this problem. However, a more stable solution is required to ensure that the discriminator performs as well as it should.

One such solution is the Creative Adversarial Network model Elgammal et al. (2017). This model is based on the principle of Martindale. It increases stylistic ambiguity and arousing potential, but also makes the art look like it was created by humans. This approach has gained considerable attention in the media. The idea is to push the machine against habituation while avoiding moving too far away from accepted art.

Another solution is to use a machine-learning model to learn the statistical latent space of an image. This can then be used to produce new artwork based on the training data.

- Art-generating artificial intelligence (AI)

Art-generating artificial intelligence (AI) is a growing field. It has the potential to reshape the way people create, consume, and share art. However, as with any new technology, the industry has its share of challenges. For example, while AI can generate realistic images, the art generated by an algorithm doesn't have a story to it. Some artists balk at the idea of an AI model taking their image and turning it into a realistic portrait.

However, there are many advantages to creating artworks with AI models. For one, these algorithms can generate thousands of unique artworks. This opens up new markets for artists.

Another benefit of using an AI art generator is that it helps people visualize complex ideas. This makes it easier to imagine inspiration and creativity. It also allows everyone to participate in the art-making process. Whether you are a professional artist or a beginner, AI generators can be a helpful tool for art. AI art-generating tools range from text-to-image generators to ones that use an AICAN. These systems allow users to upload up to 30 images. They then train the GAN on those images, which is used to create unique works of art. These types of generators can produce works that mimic the styles of famous artists.

Tan et al. (2017) proposed an extension to the Generative Adversarial Networks (GANs), namely as ArtGAN to synthetically generate more challenging and complex images such as artwork that have abstract characteristics. The key innovation was to allow back-propagation of the loss function w.r.t. the labels to the generator from the discriminator. Gatys et al. (2015) introduced an artificial system based on a Deep Neural Network that creates artistic images of high perceptual quality.

While these tools can be useful for artists, they can also be confusing to beginners. To ensure that the results are as accurate as possible, it's important to understand how the process works. There are also several different types of prompts that can be used with the tool. Some of them are highly descriptive, such as "a starry night sky full of constellations in the fall in the style of Claude Monet." While these generators may be polarizing, they are a great tool for anyone interested in creating their own artwork. They can also be an inspiring resource for others. As with any new tech, there are also ethical issues. While it's not always easy to decide how best to interact with AI-generated images, some artists are exploring ways to make their work accessible to all audiences.

- AI is transforming the music industry

The impact of Artificial Intelligence (AI) on the music industry is starting to be felt. The field of AI-based music composition is growing, and many artists are using the technology to create their own tracks. The most promising application of AI in the music world is in audio processing. This is a process that takes an image or a video and translates it into an audio file. This technology is based on the concept of deep neural networks. The technology helps musicians to symphonize studio-quality music in their own home. It also reduces the costs involved in producing music. Streaming services are also incorporating AI to recommend songs to listeners. The use of AI will continue to grow.

The development of AI-based tools has been beneficial for both aspiring and experienced musicians. The latest technology allows producers to quickly generate new ideas and identify new melodic patterns. It also improves the flow between songs. Streaming sites such as Spotify use AI to determine what songs users are listening to. This provides suggestions of similar songs. This helps the company to increase revenue.

In 2019, Holly Herndon released an album featuring a neural network, which was de- scribed by Vulture as the "world's first mainstream album made by artificial intelligence". Another popular example of an AI-generated song is Flow Machine's "Daddy's Car".

The technology can be used to generate new sounds based on the characteristics of existing sounds. There are several projects underway, including Sodatone's tech, which combines social data and streaming data to identify unsigned talent.

The future of music will see more audio processing with the use of artificial intelligence and neural networks. These technologies are already being used for audio mastering. They help to eliminate the need for manual labor and ensure smooth audio mastering. The ability to identify trends and patterns in large data sets is one of the biggest advantages of using artificial intelligence in the music industry. The technology can be integrated into the composition and mastering processes, allowing composers to access a wide variety of music composition tools. This can save time and money and produce more original and professional-sounding music.

Haiko (2021) examined the different AI and XR methods and how they have been used in the music industry so far. Pinheiro (2021) found the companies that are using AI-created music worldwide and synthesizing their technologies, products, services and business models.

- AI can compose music based on her own style

There is a growing interest in using artificial intelligence for composition. The idea is to use a computer to analyze music, suggest lyrics, and create a piece of music based on that analysis.

One of the biggest challenges in AI music is the process of generating the music. The music must evoke emotions in the audience. It must also be unique. There are already several tools available for composers to use, including IBM Watson Beat and Google Magenta. Another technology is AIVA, or the "Artificial Intelligence Virtual Artist," which uses deep learning algorithms to compose music. It is trained to compose classical pieces, and it has released a few single tracks. But its creators want the machine to be able to produce a full album in any style.

There are other projects, such as OpenAI's Jukebox, which uses AI to predictably continue a musical snippet. It also struggles to align words with non-lyrical components of a song. The AI can then suggest ideas to a human composer, and the human can add in original lyrics.

Another project is Emmy or EMI, which stands for Experiments in Musical Intelligence. It uses a neural network to generate audio variations based on hours of vocal samples. It began as an experiment to help a musician overcome his or her creative block. It has produced hundreds of original compositions. The latest album created by the machine, Holly Herndon's 2019 release, was described by Vulture as the first mainstream album made with AI.

Another project is Aiva, which was founded in France and Luxembourg and registered under the French and Luxembourg authors' right society. Aiva's creators are planning to create a future where man and machine collaborate to create music. This project has received praise from the Prime Minister of Luxembourg Xavier Bettel. The group is working with the Luxinnovation incubator program.

There is a growing interest in using artificial intelligence for composition. The idea is to use a computer to analyze music, suggest lyrics, and create a piece of music based on that analysis.

Another project is Emmy or EMI, which stands for Experiments in Musical Intelligence. It uses a neural network to generate audio variations based on hours of vocal samples. It began as an experiment to help a musician overcome his or her creative block. It has produced hundreds of original compositions. The latest album created by the machine, Holly Herndon's 2019 release, was described by Vulture as the first mainstream album made with AI.

Another project is Aiva, which was founded in France and Luxembourg and registered under the French and Luxembourg authors' right society. Aiva's creators are planning to create a future where man and machine collaborate to create music. This project has received praise from the Prime Minister of Luxembourg Xavier Bettel. The group is working with the Luxinnovation incubator program. Aiva's creators have been invited to participate in the European Film Market in Berlin. It has also received an invitation to the Artificial Intelligence in Business & Entrepreneur- ship Summit in London. The company has received funding from Luxinnovation and SACEM. Aiva's creators have been invited to participate in the European Film Market in Berlin. It has also received an invitation to the Artificial Intelligence in Business & Entrepreneurship Summit in London. The company has received funding from Luxinnovation and SACEM.

Hong et al. (2021) established that the acceptance AI was found to have a positive relationship with the assessment of AI-composed music. Zulic (2019) presented some of the research that has been carried out, which involve the use of artificial intelligence in the field of composition, performance, and music education.

- AI possibly trying to write its own code and code for others

It is a software world that runs cars, flights, nuclear weapons, mobile phones, IoT and many more devices. There has been a severe shortage of AI programmers to write timely and efficient codes, and attempts have been made to see if AI can write codes.

In 2021, Codex, an AI code-writing tool, was released by Open AI. This new technology can generate codes in 12 odd languages and translate between them. It used OpenAI GPT - 3 large language model capability and more than 100 GB codes from Microsoft Github, an online code repository library, to train the model. For example, if someone asks Codex to provide a yellow colour bouncing ball, it will produce a set of codes that can generate the same. Unsurprisingly it was realised that the tool did well for small and easy tasks and failed to impress experts on solving complex activities. The company estimated that Codex could write code to the extent of 37 per cent correctly. Tom Smith, who oversaw the entire development, mentioned that tools like this would make programmer life easier.

DeepMind, pioneering cognitive computing, took this challenge and established a new milestone in competitive programming. The company's AlphaCode was tested against human programmers over several testing competitions and ranked 54.3 per cent, which makes it the first AI tool to reach human-level programming skill.

The model was developed under a two-stage process. First, models were created on a dataset with 700 gigabytes of open-source codes to learn code representation systems and solve small tasks. In the subsequent stage, the models were fine-tuned based on CodeContests which is a compilation of problems, solutions and test cases from Codeforces, Description2Code and CodeNet.

Some of these tools might not be perfect at this stage, but no one will dispute the fact that these need an ideal blend of critical thinking, algorithmic logic and language modelling understanding.

4. CHALLENGES THAT MACHINE CREATIVITY BRING

- Neural Networks are terrible at being creative

Artificial creativity is not something that neural networks have been doing well for a while. Luckily for the aficionados, there is some research out there that shows how they can do better. However, there is much less research on generating a novel and original idea in the first place. The best way to do this is to give the human brain a helping hand. Using a few nifty algorithms, a network of brain cells can churn out some intriguing ideas. The best part is that neural networks are relatively easy to implement and cheap to maintain. The key is getting the right mix of brain cells working together at the right time. The best examples of this are found in the hippocampus and other limbic regions of the brain. A large neural network, with the correct mix of brain cells, can generate hundreds of awe-inspiring suggestions, which are then distilled into a single awe-inspiring masterpiece.

Neural networks also have a shortcoming, and that is their inability to adapt to changing circumstances. For instance, if a machine learning algorithm is asked to drive a car, it is hard to see how it could do better than a real person. This is because the human brain is a highly dynamic and nimble machine. The same is true of any type of complex adaptive system. Thus, the question is how best to harness the power of modern science and technology to improve the human condition.

- Challenges in music creation

There is no doubt that an algorithm can be a helpful tool in composing music. However, it is essential to use musical intuition and taste to ensure that the resulting output is musically useful. The output of a computer program can be malicious, and a composer must be cautious of this.

AI is using input information derived from a musical performance. Using input information derived from a musical performance in Artificial Intelligence composed music is a growing research field. Although it is still in its early stages, it's making progress quickly. It is also helping to push music in new directions. As the research continues, we'll learn more about how intuition works.

One of the earliest pioneering works in AI and computer music was Hiller and Isaac- son's (1958) Illiac Suite. Using heuristic approaches, the composers generated melody and harmony. But the system's goals excluded expressiveness and emotional content. Another example is David Cope's EMI (Experiments in Musical Intelligence) project. This project uses a compositional rule analyzer to analyze music, and then generate motives. It has successfully composed in styles of Mozart, Bach, Albinoni, Palestrina and more. Several researchers have proposed computerized composition processes. Unlike many algorithms, Roger Reynolds' SPLITZ and SPIRLZ algorithms are non-decision making. They are based on a strict rule-based transformer, modifying existing music without the benefit of randomness or imprecision. However, Reynolds has not explained the criteria by which he evaluates the algorithm's output. A similar approach was used by researchers at the University of Malaga. They created two music-focused AIs, Iamus and Melomics109. They began by creating simple compositions and then evolved their systems to produce professional-caliber pieces. Aiva Technologies is a leading startup in AI-composed music. The creators envision a future where man and machine collaborate. Their music is used by advertising agencies, film directors, game studios and more. They have been registered under SACEM, the French and Luxembourg authors' rights society. Aiva received praise from the Prime Minister of Luxembourg, Xavier Bettel, and has been invited to participate in the Euro- pean Film Market in Berlin.

- Data, infrastructure, and skill gaps are rampant

In the first five days of the ChatGPT release, more than a million users clogged the OpenAI servers. This is unprecedented and shows how hungry the world is for exploring opportunities through creative and intelligent machines. For large-scale algorithmic development, we need access to a lot of structured, semi-structured and unstructured data. Performance and accuracy of the generative AI depends on data.

Computation infrastructure for generative AI is still evolving. Enterprise-wide integration and deployment need specialized skill sets, including seasoned machine learning engineers, to be effective in this race.

- Eliminating algorithm biasness from bias data is difficult if not impossible

Societal actions are full of biases. Financial organizations like to lend to people of a certain region, color, racial origin and gender. Social media production and consumption are quite fragmented. Employment in private and public sectors is favorable to a certain section of society. eCommerce and mCommerce companies tend to engage their clients across certain selective brands and categories, no matter what is happening to their customer life stage. Examples are numerous. All models are trained on this biased action, and the key is how to systematically remove unwanted bias from the data to make just and ethical models.

- Technologies should adapt corporation and societal value and culture

The generative AI's test to image capability captivated the public attention. Machine-generated images are innovative and aesthetically appealing. This can certainly transform the gaming, TV production, film making and advertising industry.

While these synthetic images are efficient and powerful, the question is who will ensure that the production of these accounts appropriately for societal and cultural value in it. Who will define the guide rails for what can be synthesized? We have seen deep fakes have already created havoc in society, and how do we ensure that these capabilities produce ethical and moral outcomes for society? Who will ensure that these technologies are not weaponized against individuals and societies worldwide?

- Intellectual property right complication issues need to be addressed

Generative AI solutions are trained on available data and able to produce creative contents that were never built and perhaps will never be repeated. Even if generative AI solutions are new, these systems are trained by using data produced by humans. In this case, how do we assign an IP to a machine?

Since the input into the solution comes from humans, do we give a co-authorship to original data creators ? Another more significant issue is whether the machine got explicit permission to use human-created content to train the model. Many more such issues appear when we talk about generative AI.

5. CONCLUSION

Generative AI is signalling a new era in the field of cognitive computing. Some of the above-cited examples possibly give an idea to the readers that this new class of technologies is getting into domains that are thought to be reserved for humans. Generative AI exhibits the promise of being creative by producing unique content that we had never imagined before. Promises are plenty for all societal stakeholders covering consumers, businesses, government entities and civil societies.

Recent ChatGPT and a few other products show some inspiring results, but we are far from seeing a full-fledged creative machine that can compete with humans. Machines make mistakes, and machines and humans need to continually learn from each other to augment their respective capabilities. Even if machines do not fully succeed in being creative, if they can push the creative boundary of human beings, we believe that machines have succeeded in their objectives for now.

Generative AI is a most exciting field but presents many functional and non-functional challenges. While there has been headway in solving functional challenges, legal, intellectual and ethical challenges are galore. The creators and users of generative AI products got to remain vigilant and build enough guardrails to avoid any mistakes, from data collection to algorithm selection to model development to misuse of these emerging technologies.

REFERENCES

Bedau, M. A., McCaskill, J. S., Packard, N. H., Rasmussen, S., Adami, C., Green, D. G., Ikegami, T., Kaneko, K., & Ray, T. S. (2000). Open problems in artificial life. *Artificial Life*, *6*(4), 363–376. doi:10.1162/106454600300103683 PMID:11348587

Besold, T., Hernández-Orallo, J., & Schmid, U. (2015). Can machine intelligence be measured in the same way as human intelligence? *KI-Künstliche Intelligenz*, *29*(3), 291–297. doi:10.100713218-015-0361-4

Bringsjord, S., & Schimanski, B. (2003, August). What is artificial intelligence? Psychometric AI as an answer. In IJCAI, 887-893.

Clark, A. (2003). *Natural-Born cyborgs: Minds, Technologies, and the Future of Human Intelligence*. Oxford University Press.

Dorin, A. (2007). A survey of virtual ecosystems in generative electronic art. In J. Romero & P. Machado (Eds.), *The Art of Artificial Evolution*. Springer.

Elgammal, A., Liu, B., Elhoseiny, M., & Mazzone, M. (2017). Can: Creative adversarial networks, generating" art" by learning about styles and deviating from style norms. arXiv preprint arXiv:1706.07068.

Fjelland, R. (2020). Why general artificial intelligence will not be realized. *Humanit Soc Sci Commun*, *7*(1), 10. doi:10.105741599-020-0494-4

Gatys, L. A., Ecker, A. S., & Bethge, M. (2015). A neural algorithm of artistic style. arXiv preprint arXiv:1508.06576.

Haiko, E. (2021). *When the Music Industry Meets Artificial Intelligence & Extended Reality: Steps for Successful Customer Experiences*.

Hernandez-Orallo, J. (2000). Beyond the Turing test. *Journal of Logic Language and Information*, *9*(4), 447–466. doi:10.1023/A:1008367325700

Hong, J. W., Peng, Q., & Williams, D. (2021). Are you ready for artificial Mozart and Skrillex? An experiment testing expectancy violation theory and AI music. *new media & society, 23*(7), 1920-1935.

Legg, S., & Hutter, M. (2007). Universal Intelligence: A Definition of Machine Intelligence. *Minds and Machines*, *17*(4), 391–444. doi:10.100711023-007-9079-x

McCormack, J. (2008). Facing the Future: Evolutionary Possibilities for Human-Machine Creativity. In J. Romero & P. Machado (Eds.), *The Art of Artificial Evolution. Natural Computing Series*. Springer., doi:10.1007/978-3-540-72877-1_19

Pinheiro, T. D. A. R. (2021). *How Is AI-created Music Being Commercialized Outside Of The Recording Industry*?

Sundararajan, L. (2014). Mind, Machine, and Creativity: An Artist's Perspective. *The Journal of Creative Behavior*, *48*(2), 136–151. doi:10.1002/jocb.44 PMID:25541564

Tan, W. R., Chan, C. S., Aguirre, H. E., & Tanaka, K. (2017, September). ArtGAN: Artwork synthesis with conditional categorical GANs. In *2017 IEEE International Conference on Image Processing (ICIP)* (pp. 3760-3764). IEEE. 10.1109/ICIP.2017.8296985

Turing, A. (2004). *Intelligent machinery (1948)*. B. Jack Copeland.

Zulić, H. (2019). How AI can change/improve/influence music composition, performance and education: Three case studies. INSAM Journal of Contemporary Music. *Art and Technology*, *1*(2), 100–114.

Chapter 3 Neuro-Psychological Approaches for Artificial Intelligence

Suhani Dheer
Drexel University, USA

Satvik Tripathi
https://orcid.org/0000-0001-6214-1464
Drexel University, USA

Edward Kim
Drexel University, USA

ABSTRACT

In artificial intelligence (AI) research, neuro-symbolic learning, which is the combining of neuroscience and symbolic representation methods based on artificial neural networks, has a long and illustrious history. In recent years, significant progress has been achieved in the fields of neuroscience and artificial intelligence. Neuropsychology has provided substantial inspiration for new types of algorithms and architectures, independent of and complementing the mathematical and logic-based methods and notions that have dominated traditional AI approaches. In this chapter, the authors provide an organized overview of current trends by discussing and organizing recent research from major conferences and publications and presenting them in an analytical order. The purpose of this chapter is to provide a handy starting point for further research on the general topic of neuro-psychological approaches for the further advancement of AI techniques.

DOI: 10.4018/978-1-6684-6366-6.ch003

INTRODUCTION

Artificial Intelligence (AI) and Machine Learning (ML) advancements in recent years have had a significant influence on both the academic community and industries. Influential intellectuals, on the other hand, have expressed worries about the trustworthiness, safety, interpretability, and responsibility of artificial intelligence. There has been much discussion on the requirement for well-founded knowledge representation and reasoning to be merged with deep learning, as well as the need for good explainability. For many years, researchers have been working on neural-symbolic computing, with the goal of combining robust learning in neural networks with reasoning and explainability via symbolic representations of network models. In this work, we discuss current and early research findings in neurosymbolic artificial intelligence with the goal of identifying the main components of the next generation of artificial intelligence systems. We are particularly interested in research that blends neural network-based learning with symbolic knowledge representation and logical reasoning in a systematic manner. Using the insights gained from 20 years of neural-symbolic computing, it is shown that artificial intelligence is becoming more important in terms of trust, safety, interpretability, and responsibility. From the standpoint of neural-symbolic systems, we also highlight interesting areas and difficulties for the next decade of AI research in the field of artificial intelligence.

The goal of developing a neuro-symbolic (NeSy) reasoning system for modeling a complex environment is to forecast the physical parameter values of a dynamic environment: the ocean, in real-time. Predicting the parameter values that determine the system's typical behavior can be difficult in instances when the rules that dictate the system's behavior are unclear. The merging of symbolic methods with methods based on artificial neural networks, known as neuro-Symbolic Artificial Intelligence, has a long history. By classifying recent articles from significant conferences, we present a structured summary of current trends in this article. The purpose of this review is to provide a handy starting point for general research on the subject.

UNDERSTANDING INTELLIGENCE

Intelligent behavior, when viewed as a manifestation of the complexity of the human brain and its evolutionary path, represents an intriguing example of "system-level brain plasticity": tangible evidence of this assertion can be found in the strong links intelligence has with vital brain capacities such as information processing (which is the pure, rough capacity to transfer information in an efficient way), resilience (which is the ability to cope with the loss of efficiency and/or effectiveness), and attention (which is the ability to pay attention in a timely manner) (i.e., being able to efficiently rearrange its dynamics in response to environmental demands). Currently available data in support of this viewpoint range from theoretical models connecting intellect and individual response to systematic "lesions" of brain connections to the field of Noninvasive Brain Stimulation (NIBS) (NiBS). Techniques such as transcranial magnetic stimulation (TMS) and transcranial alternating current stimulation (tACS) are opening up new in vivo scenarios that may allow for the discovery of a more causal relationship between intelligence and brain plasticity, while also overcoming the limitations of brain-behavior correlational evidence, according to the researchers.

The extrastriate cortex (BAs 18–19) and fusiform gyrus (BA 37), according to this parieto-frontal integration theory of intelligence (P-FIT), are involved in intelligence test performance because they contribute to the recognition, imagery, and elaboration of visual input, just as Wernicke's area (BA 22) does for syntactic auditory input. The supramarginal (BA 40), superior parietal (BA 7), and angular (BA 39) gyri of the parietal lobe are hypothesized to be where structural symbolism, abstraction, and elaboration develop once information is captured by these pathways. These parietal regions may then connect with frontal lobe regions (particularly BAs 6, 9, 10, 45, 46, and 47) to build a working memory network that compares various task responses. The anterior cingulate cortex (BA 32) enhances response engagement and inhibits alternative responses once a task response has been chosen (Colom et al., 2009; Colom et al., 2007; d'Avila, 2019). The white matter fibers that connect brain regions, such as the arcuate fasciculus, are responsible for these interactions. The left hemisphere appears to be more crucial to cognitive task performance than the right hemisphere in most of these brain regions (Haier et al., 2009). P-FIT can be considered the best available solution to the question of where in the brain intelligence resides, as later studies, as well as investigations using various approaches (see below), have largely corroborated this idea (Deary et al., 2010; Shearer & Karanian, 2017).

HUMAN PERCEPTION OF ARTIFICIAL INTELLIGENCE

According to Nass and Moon (2000), people apply social conventions and expectations to computers automatically. They go on to suggest that people react to signals that activate multiple scripts, labels, and expectations from the past rather than all relevant indications from the present. Nass and Moon provide three ideas to examine while considering human views of AI in their essay. The first experiment they describe demonstrates that people abuse social categories by attaching gender preconceptions to computers and ethnically identifying with them. The second experiment they describe demonstrates that humans participate in overlearned social behaviors with computers, such as courtesy and reciprocity. Third, they demonstrate humans' early cognitive commitments by observing how people react to labeling. Nass and Moon argue that people use social scripts designed for human-to-human interaction rather than human-computer interaction.

Sarah Harmon (2011) demonstrates that gender made no substantial effect, but that individuals linked qualities that may have been influenced by gender and embodiment. She found a substantial association between things like passive and attraction for men, understandable and pleasant for both men and women, and reliable and likable for men, demonstrating that people are inclined to give human traits to machines. Harmon does, however, warn that confounding circumstances must be considered. Harmon also said that the degree of an entity's embodiment determines how individuals perceive its traits in connection to others, such as the terminal and the robot, which had a high link for understanding/pleasant and friendly/optimistic. However, only the terminal had a significant link with Understandable/Capable, Pleasant/Reliable, and Helpful/Reliable.

Given this authors' study, one may argue that how AI is presented to humans influences how AI is viewed. Even a navigation system in a vehicle that is given a human name seems to take on a whole new significance. Clearly, many factors influence human perception of computers and AI systems. There is still considerable study to be done on how AI should be presented such that it is best understood by humans.

NEUROPSYCHOLOGY APPROACHES

Neuropsychology is an interdisciplinary discipline of psychology and neuroscience that studies the cerebral organization of human cognitive processes as well as how the structure and function of the brain connect to certain psychological processes that take place (Small et al., 2013). AI systems can learn intelligent behavior in comparable ways to humans and may be built using a framework that incorporates models of psychological behavior and neurological processes to construct usable intelligent systems. This might result in the formation of neuronal cognitive architecture (Huyck & Ghalib, 2008).

In computational models of how the brain reacts to certain stimuli or perceptions perform, for example, cognitive psychology and artificial intelligence can be integrated to develop techniques like Sparse coding and Deep convoluted neural networks. Theoretically, in terms of symbolic representation and computing, the most productive approach is to comprehend how a neuron works, how they form networks, and what weights they use while transmitting inputs. Computer algorithms that replicate human cognitive capabilities have been used in healthcare to evaluate complicated medical data as a kind of artificial intelligence (AI). For illness diagnosis, prediction, and therapy, AI technologies like machine learning (ML) may help integrate biological, psychological, and social variables.

Humans, neural systems, and the finest AI systems all depend on learning. Learning has a vital role in the framework. Synapses link neurons in the brain, allowing complex networks to emerge. Many parts of human thinking, need the acquisition of rules from a little amount of input. Humans are capable of doing so, learning systematic rules from a limited number of instances and integrating them to construct compositional rule-based systems. Using approaches from the research on neural program synthesis, neuro-symbolic models are taught to infer the explicit system of rules regulating a collection of previously viewed cases (Nye, 2020). Typically, learning procedures are classified as bottom-up or top-down, which leads to the creation of neural-symbolic systems for learning logical rules. Top-down techniques build logic programs from the most general elements and extend them to be more particular, while bottom-up approaches build logic programs from the most general elements and stretch them to become more precise (Hassabis et al., 2017; Schuman et al., 2020).

Symbols are thought to be the mind's fundamental data processing units. The concept of employing neurosymbols comes from the fact that information is processed in the brain by neurons, that are linked and interact with one another. People, on the other hand, think in terms of symbols rather than action potentials and firing nerve cells. Neurosymbols represent perceptual pictures that may be viewed as symbolic information, such as a sound, a texture, a specific shape, a face, color, etc. Neuro-symbols have a lot of similarities to neurons. The activation grade of each neuro-symbol reflects whether the perceptual picture it represents is now present in the environment or not (Hassabis et al., 2017; Schuman et al., 2020).

NEUROSCIENCE INSPIRES AI

Major progress has been made in the domain of Neuroscience and AI in the past few years. To begin, neuroscience serves as a fertile ground for the development of novel Ai technologies and frameworks that are distinct from the mathematical and logic-based techniques that have historically dominated the field. Secondly, existing AI algorithms have also been validated with the help of neuroscience. In various industries where AI is being adapted, neuroscience combined with artificial intelligence technologies has unquestionably emerged as a viable method of improving the procedures and increasing success rates in various models (Schuman et al., 2020).

NEUROMORPHIC COMPUTING

Neuromorphic computing is an innovative approach to mimicking the way the human brain works. The concept was developed by Carver Mead to describe the use of very-large-scale integration (VSLI) systems that contain analog circuits that can mimic neurobiological architectures present in the nervous system. A neuromorphic computer is a machine aimed to facilitate high computational power and is compromised of simple processors and memory structures. Originally, neuromorphic computers were developed to perform spike-based neural network-style computation, and used for machine learning and computational neuroscience applications (Schuman et al., 2020). However, they have many characteristics that make them applicable to other research areas. For instance, in recent years, neuromorphic approaches have been used to address problems in graph theory, partial differential equations, constraint satisfaction optimization, and spike-based simulations (Hamilton et al., 2020).

Some examples of Computational Neuropsychology of Cognitive Flexibility:

1. **The Attentional-Updating Model:** Bishara et al. (2010) proposed the attentional-updating (AU) model as a well-established mechanistic model of the Wisconsin Card Sorting Test (WCST). The premise that participants create attentional prioritizations (APs) of categories is central to the AU model. A high AP of a category indicates a high likelihood of using that category on a certain trial. Following input, category APs are trial-and-error modified. Following favorable feedback, the AP of the applied category increases, while the AP of non-applied categories decreases (and vice versa for negative feedback). Thus, after receiving favorable feedback, it is more probable to repeat a category, and switching to a different category is more likely after receiving bad feedback. The intensity of updating of AP is modulated by an attentional focus mechanism: a high AP of a certain category leads to strong updating of that AP. A low AP of a certain category, on the other hand, leads to poor updating of that AP.

The AU model has four distinct latent variables. Sensitivity parameters measure the overall intensities of AP updating in response to input. The AU model has independent sensitivity settings for positive and negative feedback, allowing for varying individual intensities of updating in response to positive and negative input. An attentional focus parameter measures the degree to which AP magnitudes influence the intensity of AP updating. A response variability measure assesses how well responding matches AP.

2. **The Cognitive Reinforcement-Learning Model:** The cognitive reinforcement-learning (RL) model (Steinke et al., 2020) is based on the well-established reinforcement learning mathematical framework. The premise that participants build feedback expectations for the application of categories is essential to the cognitive RL model. A high feedback forecast suggests a high likelihood of favorable feedback for a category's application. High feedback forecasts for a category are likewise associated with a high likelihood of using that category. In reaction to the received input, feedback estimates for categories are trial-by-trial adjusted. Positive feedback will lead to an increase in feedback estimates for the applied category. Feedback forecasts for the applied category will fall

after receiving negative feedback. The degree of updating of feedback predictions is modulated by prediction mistakes. Prediction errors are equal to the difference between received and predicted feedback. Large prediction mistakes lead to more aggressive updating of feedback predictions.

The cognitive RL model includes two processes that are not seen in classical RL models (Sutton & Barto, 1998). First, a retention mechanism defines how feedback predictions are transferred from one trial to the next (Erev & Roth, 1998; Steingroever et al., 2013). Second, a "soft-max" approach computes response probabilities based on feedback predictions for a specific trial (Daw et al., 2006; Luce, 1959; White et al., 1992).

The cognitive RL model is made up of four distinct latent variables. The amount to which prediction mistakes update feedback predictions is quantified by cognitive learning rates. Positive and negative feedback have different cognitive learning rates (Frank et al., 2004; Palminteri et al., 2009; Schultz, 2017; Schultz et al., 1997). A cognitive retention rate evaluates how well feedback predictions transfer from trial to trial. An inverse temperature parameter measures how successfully responses are implemented in relation to feedback expectations.

3. **The Parallel Reinforcement-Learning Model:** Based on the discovery of modulation of PE proclivities by response demands, we predicted that participants learn on the WCST at two parallel levels (Kopp et al., 2019). Category-level (perhaps cortical) learning indicates that individuals prefer to repeat the applied category on trials after positive feedback and alter the applied category on trials after negative input. Participants may also learn at the response level. Response-level (putatively striatal) learning indicates that participants prefer to repeat the execution of a specific response in response to positive feedback and avoid re-execution of a reaction in response to negative feedback.

A mathematical formalization of category- and response-level learning is provided by the parallel RL model. Cognitive RL (as in the cognitive RL paradigm) is a concrete example of category-level learning. Furthermore, sensorimotor RL is an instantiation of response-level learning.

Comparison of Mechanistic Models: We examined the AU model (Agrawal et al., 2017), the cognitive RL model, and the parallel RL model on a large sample of healthy volunteers (N = 375) who completed a computerized variant of the Wisconsin Card Sorting Test (cWCST) (Lange & Dewitte, 2019) in a recent model comparison research. Predictive accuracies were used to assess mechanistic models (Gronau & Wagenmakers, 2019; Vehtari et al., 2017). The predictive accuracies of a mechanistic model measure how well it predicts observed trial-by-trial cWCST responses. For most participants, the cognitive and parallel RL models outperformed the AU model in terms of predicting accuracy. These findings show that RL models conceptualize trial-by-trial cWCST responses better than the AU model.

In terms of updating processes, RL models vary from the AU model. Prediction mistakes in RL models influence the degree of feedback prediction updating. When the connection between the received and expected feedback is inadequate, prediction errors guarantee that updating of feedback predictions is stronger. For example, suppose a participant obtains positive feedback for using a category with a low feedback prediction (i.e., indicating the prediction of negative feedback for that category). As a consequence, the feedback prediction for this category was poor, resulting in a large prediction error. As a result, the updating of feedback prediction for this category will be strong, allowing for the re-application of the category that generated favorable feedback. An attentional focus mechanism in the AU model assures that updating of AP of a certain category is less powerful when the AP of that category is low. In the preceding example, updating of AP will be weaker since the AP of that category was low. As a result, the attentional focus mechanism makes the re-application of the category that gave positive feedback more difficult. In comparison to the AU model, RL models feature more efficient adaptability of card sorting to changing job demands.

FUNDAMENTAL STATEMENT

We argue that the first desirable property of frameworks that integrate two other frameworks A and B to have the original frameworks A and B as a special case of the integrated one, based on our experience upgrading machine learning systems to use (probabilistic) logical and relational representations (Muggleton et al., 2012; Raedt, 2008). If A or B cannot be entirely reconstructed, certain capabilities are clearly lost, which is not only undesirable but also shows that true unification does not exist. When applied to neuro-symbolic computation, this feature indicates that existing frameworks should include both neural and symbolic representations as special cases. When this is the case, both the brain component's learning abilities, as well as the reasoning and learning abilities, as well as the semantics of the symbolic representations, are retained. Unfortunately, the vast majority of neuro-symbolic approaches fail to satisfy this property because they either push the symbolic representation inside the neural network (from which the logic cannot be recovered) or apply neural learning principles to symbolic representations (risking losing both the pure neural component and the logical semantics), as we will demonstrate in the next section.

A second desirable quality for neuro-symbolic computation is that models that learn from observed samples should be able to deal with uncertainty. As a result, not only should logic and neural networks be combined in neuro-symbolic computation, but also probability. This successfully combines probabilistic logics (thus statistical relational AI) with neural networks, revealing new capabilities. Furthermore, while this may appear to be a complexity at first glance, it can substantially ease the integration of neural networks and logic. Because the probabilistic framework gives a clear optimization objective, namely the probability of the training examples, this is the case. In contrast to discrete logic quantities, real-valued probabilistic quantities are well-suited for gradient-based training processes.

LOGIC CONSTRAINT FACTOR

The most common idea is to utilize logic to limit a deep model. That is, the deep model is enhanced with a regularization term derived from the required logical qualities, which encourages it to emulate logical reasoning by imposing a penalty for failing to follow such logical properties. For example, Diligenti et al. (2017) and Donadello et al. (2017) specify constraints on the output of the neural network using first-order logic, whereas Demeester et al. (2016) and Minervini et al. (2017) use logical IF-THEN rules derived from expert knowledge to enforce the embeddings to be more consistent with the logical constraints. Xu et al. (2018) propose a generalization of this idea, allowing for the imposition of more complicated logical constraints on any deep model. The use of logic as a regularizer is common to all of these approaches, encouraging deep models to satisfy the restrictions. As a result, the logic is stored in the parameters (either into the weights of the neural network or directly into the embeddings).

The constraints are flexible, and there's no guarantee that all of them will be met. In truth, the trade-off between encouraging constraints vs. following the facts in the event that both are conflicting is just a model hyperparameter. Constraints only support one type of reasoning; a different type of reasoning is focused on supplying predicate definitions (in the form of rules or definite clauses) and using those rules to answer questions. The popular programming and database languages Prolog and Datalog are built on this foundation. When the neural element is removed, the aforementioned approaches treat neural networks as a particular case (i.e., when there are no additional restrictions), but they do not treat logic as a special case.

DIFFERENTIABLE LOGIC FRAMEWORKS

An alternative type of neuro-symbolic system works by making the logic program differentiable, which is accomplished by reformulating the core reasoning primitives using differentiable functions mathematics. Rocktaschel and Riedel (2017) use this in Prolog's backward reasoning technique, but Evans and Grefenstette (2018) use it in the forward reasoning style. Despite the fact that both systems are built on conventional reasoning algorithms, the original logic cannot be recovered as a special case because the semantics and inferences have been fundamentally altered. Similarly, Cohen et al. (2018) present a methodology for compiling a tractable subset of logic algorithms into differentiable functions and running them via neural nets. All three systems listed above use a neural network to carry out the execution, however, they don't treat neural networks differently.

These methods can only cope with a small number of different types of logic programs. Furthermore, because the logic programming components are pushed into the neural networks, the semantics of logic programs as well as the programming language characteristics of the framework are sacrificed. This is evident when you consider that Rocktaschel and Riedel (2017) use a neural network to map a proof tree, which can then be discriminated because the logic has been eliminated. These systems are additionally hampered by the fact that the original logic is not preserved as a special case. For example, Rocktaschel and Riedel, (2017) and Evans and Grefenstette (2018) both state that neural execution of differentiable logic results in significant processing overhead.

NEURAL STATE MACHINES

The Neural State Machine is an upcoming technique trying to bridge between logic and underlying semantic knowledge (context). It is a differentiable graph-based model that simulates automata operation, in the domains of visual reasoning and compositional question answering. In essence, there are two stages: modelling and inference. Starting with an image, they create a probabilistic scene network (Johnson, 2015; William, 2018) that compactly captures the image's underlying semantic knowledge. Nodes represent items and contain structured representations of their properties, while edges express both spatial and semantic relationships between them. The algorithm uses the graph as a state machine and simulates an iterative computation over it, with the goal of answering questions or drawing inferences. They later convert the natural language query into a series of soft instructions and feed them one at a time into the machine, which does sequential reasoning and computes the answer using attention to traverse its states.

They develop a collection of semantic embedded concepts that explain different things and characteristics of the domain, such as various sorts of objects, properties, and relations, based on Bengio's consciousness prior (Krishna et al., 2017). These notions serve as the lexicon that underpins both the image-based scene graphs and the question-based reasoning instructions, essentially allowing both modalities to "speak the same language." Whereas most neural networks interact directly with raw observations and dense features, our technique promotes the model to reason in a semantic and factorized abstract space, allowing structure and content to be separated and improving modularity.

On two recent Visual Question Answering (VQA) datasets, they have demonstrated the value and performance of the Neural State Machine: GQA (Bengio, 2017), which focuses on real-world visual reasoning and multi-step question answering, and VQA-CP (Hudson & Manning, 2019) a recent split of the popular VQA dataset (Agrawal et al., 2018; Agrawal et al., 2017) that has been designed specifically to evaluate generalization. Under single-model circumstances, they achieve state-of-the-art scores on both challenges, demonstrating the robustness and efficiency of our approach in answering difficult compositional questions. They then create new splits using GQA's associated structured representations and run additional experiments that show the model's strong generalization abilities across multiple dimensions, such as novel concept compositions and unseen linguistic structures, demonstrating its adaptability under changing conditions.

Thier model combines two important characteristics: abstraction and compositionality, with the key innovations of representing meaning as a structured attention distribution over an internal vocabulary of disentangled concepts and capturing sequential reasoning as the iterative computation of a differentiable state machine over a semantic graph, respectively. They have expectations that by constructing a neural representation of a classical model of computing, we will be able to encourage and support the merging of connectionist and symbolic techniques in AI, allowing for greater modularity, variety, and generalization.

STATISTICAL RELATIONAL LEARNING AND ARTIFICIAL INTELLIGENCE (STARAI)

Neuro-symbolic artificial intelligence and statistical relational artificial intelligence are prominent types of artificial intelligence that combine frameworks for learning with logical thinking. Between these two fields, this survey identifies a number of analogies that can be found across seven different dimensions. Not only can these be used to characterize and position neuro-symbolic artificial intelligence techniques, but they can also be utilized to identify a variety of new research avenues in the field of neuroscience.

Learning and reasoning integration is one of the most difficult problems in artificial intelligence and machine learning today, and several communities have been working on it for years. Particularly relevant to this is the topic of neuro-symbolic computation (NeSy) (Goyal et al., 2017; Tarek et al., 2017), where the goal is to include symbolic thinking into neural networks through the use of artificial neural networks. Although it has a lengthy history, the NeSy movement has recently gained a great deal of interest from a wide range of communities (cf., e.g., the keynotes of Yoshua Bengio and Henry Kautz on this topic at AAAI 2020). Many approaches to NeSy are aimed at extending neural networks by including logical reasoning into the system.

Another domain with a long history of integrating learning and reasoning is statistical relational learning and artificial intelligence (StarAI) (d'Avila et al., 2019; Getoor & Taskar, 2007), which has a rich legacy of integrating learning and reasoning (d'Avila et al., 2019). Although it does not specifically address the subject of how to merge logic and neural networks, it does address the question of how to integrate logic with probabilistic graphical models (PGMs). However, it is surprising that there aren't more connections between these two domains, given the widespread interest in combining logic or symbolic reasoning with a fundamental paradigm for learning, such as probabilistic graphical models or neural networks.

This difference serves as the primary reason for this survey, which tries to draw attention to the parallels that exist between these two endeavors in order to encourage more cross-fertilization between them. Starting with the literature on StarAI is a good place to start since, in contrast to NeSy, there is a greater degree of agreement on the core concepts, difficulties, and issues in StarAI (cf. the number of tutorials and textbooks on related topics such as (d'Avila et al., 2019; Getoor & Taskar, 2007; Goyal et al., 2017; Tarek et al., 2017)). It turns out that the concerns and techniques that occur in StarAI must be addressed in the same way that they do in NeSy. The most important contribution of the survey was that it identifies a set of seven dimensions that are shared by both areas and that can be used to characterize both StarAI and NeSy techniques. Seven dimensions are concerned with (1) directed versus undirected models, (2) grounding versus proof-based inference, (3) integrating logic with probability and/or neural computation, (4) logical semantics, (5) learning parameters or structure, (6) representing entities as symbols or sub-symbols, and (7) the type of logic employed. The evidence for the assertion comes from the placement of many different StarAI and NeSy systems along these dimensions, as well as the identification of parallels between the systems.

The result represented was that the study was better able to identify the interesting potential for further research by looking at areas across the dimensions that have not yet seen considerable attention. The differences between StarAI and NeSy are significant, with the most significant being that the former operates at a higher symbolic level, making it more naturally suited for explainable AI, whereas the latter operates at a lower symbolic level, making it more naturally suited for computer vision and natural language processing, among other applications. Deep learning's power lies in its capacity to transform data representations in order to facilitate the solution of a given problem. Enhancing NeSy system functionality would be possible with the addition of symbolic-level representational modification. This is an extremely difficult problem that may be solved with the aid of neurally driven approaches.

CONCLUSION

The rapid increase in papers, events, and public addresses on NeSy AI demonstrates that the mainstream AI community is becoming more cognizant of the importance of the area in recent years. Possibly, this is also taking place at a critical juncture in history, when we are only now beginning to study and comprehend the fundamental limitations of pure deep learning methods. It is a natural progression to strive to further improve deep learning systems by the incorporation of additional background knowledge, and most of this line of study comes under the NeSy AI concept.

The general view appears to be that integrating neural and symbolic techniques in the sense of NeSy AI is at the very least a step forward in the development of considerably more powerful artificial intelligence systems. It may also prove to be a significant step forward in the development of artificial intelligence on the level of humans.

To conclude, we'd like to point out a number of difficulties that NeSy faces that, in our opinion, require more attention.

1. **Probabilistic Reasoning:** We believe that a probabilistic approach to the integration of logical and neural methods is the most effective way to principally integrate the two. Although there are relatively few methods that explore the integration of logical and neural methods from a probabilistic perspective, Further investigation into the application of probabilistic reasoning for neuro-symbolic computation should be carried out, it is recommended.
2. **Learning With a Structure:** The learning of the structure of purely relational models (i.e., models without probabilities) has made great progress, but learning StarAI models has remained a key barrier because of the complexity of inference and the combinatorial nature of the problem. Incorporating neurological factors into the equation further complicates the situation. Although NeSy approaches have demonstrated promise in resolving this problem, the solutions that are now available are still limited and primarily domain-specific, making them inapplicable to a wide range of problems.
3. **Scalable Inference:** It is a major challenge for StarAI, and as a result, for NeSy techniques that include an explicit logical or probabilistic reasoning component as well as for other NeSy approaches. In particular, investigating to what extent neural approaches might assist with this difficulty through the use of lifting (exploiting symmetries in models) or approximation inference, as well as reasoning from intermediate representations, are intriguing directions for future research.
4. **Data Efficiency:** When comparing StarAI methods to neural methods, one of the most significant advantages is their data efficiency—StarAI methods can efficiently learn from little amounts of data, whereas neural methods require a large amount of data. On the other hand, StarAI approaches do not scale well to large data sets, whereas neural methods are well suited to such situations. Understanding how these strategies might work together to overcome their complimentary flaws, we feel, is a viable research direction to pursue.
5. **Learning With Symbolic Representations:** The effectiveness of deep learning is derived from its capacity to adapt how the data is represented so that the goal task becomes easier to complete. NeSy systems' capabilities would be considerably enhanced if they had the capacity to change their representation on a symbolic level as well as on a logical level. This is a significant open challenge for which neurally inspired strategies may be able to contribute to development (Cropper, 2019; De Raedt et al., 2016; Dumancic et al., 2019; Roy, 2019; Russell, 2015).

REFERENCES

Agrawal, A., Batra, D., Parikh, D., & Kembhavi, A. (2018). Don't just assume; look and answer: Overcoming priors for visual question answering. In *Proceedings of the IEEE Conference on Computer Vision and Pattern Recognition*, (pp. 4971–4980). IEEE. 10.1109/CVPR.2018.00522

Agrawal, A., Lu, J., Antol, S., Mitchell, M., Zitnick, C. L., Parikh, D., & Batra, D. (2017). VQA: Visual question answering. *International Journal of Computer Vision, 123*(1), 4–31. doi:10.100711263-016-0966-6

d'Avila, AGori, MLamb, LSerafini, LSpranger, MTran, S. (2019). *Neural-symbolic computing: An effective methodology for principled integration of machine learning and reasoning*. arXiv preprint arXiv:1905.06088. doi:10.100711263-016-0966-6

Bengio, Y. (2017). *The consciousness prior*. arXiv preprint arXiv:1709.08568. doi:10.100711263-016-0966-6

Bishara, A. J., Kruschke, J. K., Stout, J. C., Bechara, A., McCabe, D. P., & Busemeyer, J. R. (2010). Sequential learning models for the Wisconsin card sort task: Assessing processes in substance dependent individuals. *Journal of Mathematical Psychology*, *54*(1), 5–13. doi:10.1016/j.jmp.2008.10.002 PMID:20495607

Colom, R., Haier, R. J., Head, K., Álvarez-Linera, J., Quiroga, M. Á., Shih, P. C., & Jung, R. E. (2009). Gray matter correlates of fluid, crystallized, and spatial intelligence: Testing the P-FIT model. *Intelligence*, *37*(2), 124–135. doi:10.1016/j.intell.2008.07.007

Colom, R., Jung, R. E., & Haier, R. J. (2007). General intelligence and memory span: Evidence for a common neuroanatomic framework. *Cognitive Neuropsychology*, *24*(8), 867–878. doi:10.1080/02643290701781557 PMID:18161499

Cropper, A. (2019). Playgol: Learning programs through play. IJCAI.

d'Avila, A. (2019). Neural-symbolic computing: An effective methodology for principled integration of machine learning and reasoning. arXiv preprint arXiv:1905.06088. doi:10.1126cience.1102941 PMID:15528409

Daw, N. D., O'Doherty, J. P., Dayan, P., Seymour, B., & Dolan, R. J. (2006). Cortical substrates for exploratory decisions in humans. *Nature*, *441*(7095), 876–879. doi:10.1038/nature04766 PMID:16778890

De Raedt, L., Kersting, K., Natarajan, S., & Poole, D. (2016). Statistical relational artificial intelligence: Logic, probability, and computation. *Synthesis Lectures on Artificial Intelligence and Machine Learning, 10*(2), 1– 189.

Deary, I. J., Penke, L., & Johnson, W. (2010). The neuroscience of human intelligence differences. *Nature Reviews. Neuroscience*, *11*(3), 201–211. doi:10.1038/nrn2793 PMID:20145623

Diligenti, M., Gori, M., & Sacca, C. (2017). Semantic-based regularization for learning and inference. *Artificial Intelligence*, *244*, 143–165. doi:10.1016/j.artint.2015.08.011

Donadello, I., Serafini, L., & d'Avila Garcez, A. (2017). Logic tensor networks for semantic image interpretation. *IJCAI.*

Dumancic, S., Gunds, T., Meert, W., Blockeel, H. (2019). Learning relational representations with autoencoding logic programs. *IJCAI.*

Erev, I., & Roth, A. E. (1998). Predicting how people play games: Reinforcement learning in experimental games with unique, mixed strategy equilibria. *The American Economic Review*, *88*, 848–881.

... , R... , E. (2018). *Learning explanatory rules from noisy data*. JAIR.

Frank, M. J., Seeberger, L. C., & O'Reilly, R. C. (2004). By carrot or by stick: Cognitive reinforcement learning in Parkinsonism. *Science*, *306*(5703), 1940–1943. doi:10.1126cience.1102941 PMID:15528409

Getoor, L., & Taskar, B. (Eds.). (2007). *An Introduction to Statistical Relational Learning*. MIT Press. doi:10.7551/mitpress/7432.001.0001

Goyal, Y., Khot, T., Summers-Stay, D., Batra, D., & Parikh, D. (2017). Making the V in VQA matter: Elevating the role of image understanding in visual question answering. CVPR, 6325–6334.

Gronau, Q. F., & Wagenmakers, E.-J. (2019). Limitations of Bayesian leave-one-out cross-validation for model selection. *Computational Brain & Behavior*, *2*(1), 1–11. doi:10.100742113-018-0011-7 PMID:30906917

Haier, R. J., Colom, R., Schroeder, D. H., Condon, C. A., Tang, C., Eaves, E., & Head, K. (2009). Gray matter and intelligence factors: Is there a neuro-g? *Intelligence*, *37*(2), 136–144. doi:10.1016/j.intell.2008.10.011

Hamilton, K., Mintz, T., Date, P., & Schuman, C. D. (2020). Spike-based graph centrality measures. In *International Conference on Neuromorphic Systems 2020* (pp. 1–8). ACM. 10.1145/3407197.3407199

Harmon, S. (2011). *Human perception of gendered artificial entities.* Colby College.

Hassabis, D., Kumaran, D., Summerfield, C., & Botvinick, M. (2017). Neuroscience-inspired artificial intelligence. *Neuron*, *95*(2), 245–258. doi:10.1016/j.neuron.2017.06.011 PMID:28728020

Hudson, DManning, C. (2019). GQA: A new dataset for real-world visual reasoning and compositional question answering. *Conference on Computer Vision and Pattern Recognition (CVPR).*

Huyck, C. R., & Ghalib, H. (2008). A Neuropsychological Framework for Advancing Artificial Intelligence. *AAAI Fall Symposium: Biologically Inspired Cognitive Architectures.*

Johnson, J. (2015). Image retrieval using scene graphs. In *Proceedings of the IEEE conference on computer vision and pattern recognition*. IEEE.

Kopp, B., Steinke, A., Bertram, M., Skripuletz, T., & Lange, F. (2019). Multiple levels of control processes for Wisconsin Card Sorts: An observational study. *Brain Sciences*, *9*(6), 141. doi:10.3390/brainsci9060141 PMID:31213007

Krishna, R., Zhu, Y., Groth, O., Johnson, J., Hata, K., Kravitz, J., Chen, S., Kalantidis, Y., Li, L.-J., Shamma, D. A., Bernstein, M. S., & Fei-Fei, L. (2017). Visual genome: Connecting language and vision using crowdsourced dense image annotations. *International Journal of Computer Vision*, *123*(1), 32–73. doi:10.100711263-016-0981-7

Lange, F., & Dewitte, S. (2019). Cognitive flexibility and pro-environmental behaviour: A multimethod approach. *European Journal of Personality*, *56*(4), 46–54. doi:10.1002/per.2204

Luce, R. D. (1959). *IndividualChoiceBehaviour*. JohnWiley&SonsInc.

Minervini, P., Demeester, T., Tocktaschel, & Riedel, S. (2017). Adversarial sets for regularising neural link predictors. UAI.

Muggleton, S., De Raedt, L., Poole, D., Bratko, I., Flach, P., Inoue, K., & Srinivasan, A. (2012). Ilp turns 20. *Machine Learning*, *86*(1), 3–23. doi:10.100710994-011-5259-2

Nass, C., & Moon, Y. (2000, January). Machines and mindlessness: Social responses to computers. *The Journal of Social Issues*, *56*(1), 81–103. doi:10.1111/0022-4537.00153

Nye, M. (2020). Learning compositional rules via neural program synthesis. *Advances in Neural Information Processing Systems*, *33*, 10832–10842.

Palminteri, S., Lebreton, M., Worbe, Y., Grabli, D., Hartmann, A., & Pessiglione, M. (2009). Pharmacological modulation of subliminal learning in Parkinson's and Tourette's syndromes. *Proceedings of the National Academy of Sciences of the United States of America*, *106*(45), 19179–19184. doi:10.1073/pnas.0904035106 PMID:19850878

Raedt, D. (2008). *Luc. Logical and relational learning*. Springer Science & Business Media. doi:10.1007/978-3-540-68856-3

Rocktaschel, T., & Reidel, S. (2016). Lifted rule injection for relation embeddings. EMNLP.

Rocktaschel T., & Riedel, S. (2017). End-to-end differentiable proving. *NIPS*.

Roy, K. (2019, November). Towards Spike-Based Machine Intelligence with Neuromorphic Computing. *Nature*, *27*. www.nature.com PMID:31776490

Russell, S. (2015). Unifying logic and probability. *Communications of the ACM*, *58*(7), 88–97. doi:10.1145/2699411

Schultz, W. (2017). Reward prediction error. *Current Biology*, *27*(10), 369–371. doi:10.1016/j.cub.2017.02.064 PMID:28535383

Schultz, W., Dayan, P., & Montague, P. R. (1997). A neural substrate of prediction and reward. *Science*, *275*(5306), 1593–1599. doi:10.1126cience.275.5306.1593 PMID:9054347

Schuman, C. D., Mitchell, J. P., Patton, R. M., Potok, T. E., & Plank, J. S. (2020) Evolutionary optimization for neuromorphic systems. In *Proceedings of the Neuro-inspired Computational Elements Workshop*, (pp. 1–9). IEEE.

Shearer, C. B., & Karanian, J. M. (2017). The neuroscience of intelligence: Empirical support for the theory of multiple intelligences? *Trends in Neuroscience and Education*, *6*, 211–223. doi:10.1016/j.tine.2017.02.002

Small, S. L., Cottrell, G. W., & Tanenhaus, M. K. (Eds.). (2013). *Lexical Ambiguity Resolution: Perspective from Psycholinguistics, Neuropsychology and Artificial Intelligence*. Elsevier.

Steingroever, H., Wetzels, R., & Wagenmakers, E.-J. (2013). Validatingthe PVL-Delta model for the Iowa gambling task. *Frontiers in Psychology*, *4*, 898. doi:10.3389/fpsyg.2013.00898 PMID:24409160

Steinke, A., Lange, F., & Kopp, B. (2020). Parallel Model-Based and Model-Free Reinforcement Learning for Card Sorting Performance. *Scientific Reports*, *10*(1), 15464. doi:10.103841598-020-72407-7 PMID:32963297

Sutton, R. S., & Barto, A. G. (1998). *Reinforcement Learning: An introduction*. MIT Press.

Tarek, R. d'Avila Garcez, A., Bader, S., Bowman, H., Domingos, P., Hitzler, P., Kühnberger, K., Lamb, L., Lowd, D., & Lima, P. (2017). Neural-symbolic learn- ing and reasoning: A survey and interpretation. arXiv preprint arXiv:1711.03902.

Thrun, S. B. (1992). The role of exploration in learning control. In D. White & D. Sofge (Eds.), *Handbook for Intelligent Control: Neural, Fuzzy and Adaptive Approaches* (pp. 527–559). Van Nostrand Reinhold.

Vehtari, A., Gelman, A., & Gabry, J. (2017). Practical Bayesian model evaluation using leave-one-out cross-validation and WAIC. *Statistics and Computing*, *27*(5), 1413–1432. doi:10.100711222-016-9696-4

Velik, R. (2010). Towards human-like machine perception 2. 0. *International Review on Computers and Software*, *5*(4), 476–488.

William, W. (2018). Cohen, Fan Yang, and Kathryn Rivard Mazaitis. Tensorlog: Deep learning meets probabilistic databases. *Journal of Artificial Intelligence Research*, *1*, 1–15.

Xu, J., Zhang, Z. Friedman, T., Liang, Y., & Broeck, G. (2018). A semantic loss function for deep learning with symbolic knowledge. ICML.

Chapter 4
Praxes of Transformational Creativity:
Artificial Intelligence as a Pedagogical Change Agent

Catherine Hayes
University of Sunderland, UK

ABSTRACT

This chapter provides an insight into the justification of pedagogical principles within the contexts of extended reality. The global COVID-19 pandemic served to exacerbate the complex ambiguity surrounding XR in practice. Training for the strategic management of risk in healthcare practice in situated contexts of healthcare provision has been a key focus in the use of XR in practice. It involves rational aspects of cognitive knowledge or the purist demonstration of psychomotor skills and affective domain learning . This is achieved via the exploration of the key epistemologies or ways of knowing, from a theoretical perspective, which can be used to ensure the level of authenticity necessary to highlight the pedagogical shifts in the application of learning theory which now characterise responsive curriculum design and adaptation to accommodate XR in practice.

INTRODUCTION

"Reality is merely an illusion, albeit a very persistent one." ~ Albert Einstein (1879-1955)

The justification of pedagogy in the context of Extended Reality (XR), which encompasses Virtual Reality (VR), Augmented Reality (AR) and Mixed/Hybrid Reality (MR) has become an ongoing source of complex ambiguity over the last decade, that the COVID-19 pandemic has only served to exacerbate (van der Niet and Bleakley, 2021). Ensuring the validity and reliability of XR experiences within health professions education remains central to the potential to rule out technologies as adjuncts to optimal pedagogic practice as an authentic means of providing insight and illumination of medical contexts,

DOI: 10.4018/978-1-6684-6366-6.ch004

scenarios and disease processes (McGrath et al, 2018). For the purposes of this chapter there will be four fundamental operationally definitive terms of what the umbrella term XR encompasses, firstly VR refers to the use of

computer technology in the creation of simulated learning environments. Secondly, AR pertains to the addition of computerised content as an overlay to reality, which means that learners can actively interact both with real world and augmentations of it at the same time. Mixed or hybrid reality refers to the transection of virtual worlds and actual worlds, where physical and computerised objects can interact and exist concurrently. XR encompasses all of these and as a collective they have revolutionised health and medical training, particularly in relation to the practise of risk management and professional role identity in life and death situations, for example obstetric emergencies, as reported by Hayes, Hinshaw and Petrie (2018).

Training for the strategic management of risk in healthcare practice in situated contexts of healthcare provision has been a key focus in the use of XR in practice (Hilty et al, 2020). Not only does it involve rational aspects of cognitive knowledge or the purist demonstration of psychomotor skills and affective domain learning (Zulkifli, 2019). It also encompasses the intuitive, tacit and largely intangible intellectual instincts that develop with sustained experiential learning (Humpherys, Bakir and Babb, 2021). One of the key issues has been the challenge of assessing the last of these, what XR has enabled is the benchmarking of perceived levels of interprofessional and multi-disciplinary teamwork, where intuitive knowledge can be used to measure risk, regardless of the level of the organisational hierarchy within which personnel are employed (Goh and Sandars, 2020; Hayes, Hinshaw and Petrie, 2018). This chapter will explore the key epistemologies or ways of knowing, from a theoretical perspective, that can be used to ensure the level of authenticity necessary to highlight the pedagogical shifts in the application of learning theory which now characterise responsive curriculum design and adaptation to accommodate XR in practice.

BACKGROUND OF HEALTH PROFESSIONS PEDAGOGY

The ongoing pandemic, which on July 8th, 2021 had reached 184 million confirmed cases and over 4 million fatalities, has not only changed the world of education in its current form, it has likely altered its mechanism of delivery for the foreseeable future (WHO International Data, 2021). In practice the pandemic has seen universities close, a switch to hybrid models of learning and education, a social science by definition, depleted in terms of its capacity to engage people in face-to-face meetings (Okoye et al, 2021). The plethora of academic articles surrounding online learning is phenomenal, but few actually address how a fundamental paradigm shift in Higher Education is implemented methodologically (Luctkar-Flude and Tyerman, 2021). What is usually described is a narrative description of the processual use of technological intervention, rather than any degree of alignment in terms of underpinning theoretical justification for implementation, or indeed the constructive alignment demonstrating best how processes of assessment can effectively be driven by complementary processes of teaching and learning (Moreira, 2020). The physicality of learning has also been altered beyond recognition, seeing people as upper torsos and faces has changed everyday interaction in the situated nature and context of learning, yet minimal evidence exists as to the long-term impact of this on motivating, engaging and sustaining active processes of learning, teaching and assessment on an individual level (Park and Kim, 2020). By over reliance on the physicalism of the articulated voice and postural positions of the upper torso, executed in an atmosphere of scrutiny, learners have had to change their interaction so that their degree of interaction is heavily influenced and constrained (Obrad, 2020). Whilst predicting how global disease and inequality will influence the future of education, it is impossible to ignore potential challenges that lie ahead for Higher Education (Bevins et al, 2020). It is possible to inferentially predict that COVID-19 may be one of the first of a new generation of global pandemics that will need to be death with alongside dramatic changes in overall global warming, which will also ensure that populations which are densely populated suffer most severely (Negev et al, 2021). Alongside the issue of pandemic disease is the prospect of global catastrophes and natural disasters occurring far more frequently and necessitating support and address now. The progressive redevelopment of existing technology to accommodate this is more than apparent, so that learners can engage physically in medical and healthcare settings with less extensively sized equipment and a greater capacity to seamlessly integrated extended reality into all practice (Yigitcanlar et al, 2020). Geographically there are other issues at play, in terms of the accessibility of technology across global outcomes, with the result that some countries cannot be guaranteed adequate internet access, bandwidth or the cost of the technology products may simply be prohibitively expensive (Horton, 2021). Being cognisant of this necessitates ensuring both affordability and accessibility across the globe if the differential inequity between countries is to be addressed on an ongoing basis. Being able to standardise and regulate experience for learners is also of fundamental importance if equity and parity of experience are to be assured across these programmes (Crouch et al, 2021).

CURRICULUM JUSTIFICATION, DESIGN, AND DEVELOPMENT

Pedagogical design and its address within the context of Higher Educational institutional curricula often places processes of curriculum justification and design under scrutiny (Annala et al, 2021). This is largely attributable to the complex ambiguity that curriculum designers have to contend with, in terms of the technical capacity of XR equipment and the accompanying resources they necessitate (Aebersold and Dunbar, 2021). A key example of this is the multi-disciplinary perspectives that designed scenarios have to authentically represent within the context of health and medical education, where XR is implemented in practice (Antoniou et al, 2020). As a consequence of this, the concept of experiential learning has become a focus for the opportunity of ensuring that XR is relevant to real world application in practice, with the introduction of its continuum's integral parts (AR, VR and MR). The published evidence to date demonstrates that XR training has been shown to improve learner performance skills across an array of learning domains (cognitive, affective and psychomotor) within the context of instructional simulation (Tabatabei, 2020). Having moved firmly from proof of concept pilot studies, XR as an embedded part of health and medical curricula is being used across a wide variety of simulated contexts (Tang et al, 2020). In comparison to traditional didactic teaching methodologies the benefits are seen within the areas of student motivation, confidence and the capacity of learners to take measured risk within the context of a consequence free facilitative environment where constructive feedback on performance can inform reflexive praxis around complex ambiguity and clinical decision making. Acclimatisation to new contexts and settings, for military learners, for example is invaluable, where introduction to new unfamiliar climates and contexts is undertaken via constructivist experiential learning opportunity in interactive immersive scenarios, where the pace and timing of exposure can largely be controlled in line with learner need (Cobos, 2020). The addition of iteratively increasing levels of complex ambiguity as skill increases is another opportunity for the formal scaffolding of learning within customised immersive experiences (Orr et al, 2020). Depending on whether learning is about the magnification of micro- theoretical concepts such as, for example atomic level particles, blood corpuscles or synaptic junctions or whether learning is centred around macro social constructs such as social justice and equity in different contexts of culture or climate, immersive technology with XR has the potential to extend the reach within and between academic pedagogies and disciplines and as a direct consequence impact on the capacity and capability of multi-disciplinary and interdisciplinary professional working at the front line of patient care (Mitchell and Boyle, 2021). Confidence underpins learner motivation and in turn elevates levels of competence, which ensure this iterative cycle continues (Owens, 2021).

Roussin and Weinstock (2017) detailed the pedagogical challenges faced by educators and trainers leading simulation-based training in the context of healthcare and medical services. In relation to the theoretical underpinning of their work, they presented the complexities of managing programmes in relation to single and double- loop experiential learning and the impact of organising organisational hierarchies in relation to the need to gain experiential learning in practice. Whilst their work omits specific reference to the implicit value of tacit knowledge in crisis situations, the functional insight they provide in relation to situated learning is invaluable in relation to the overall administration of optimal multidisciplinary and interprofessional teamworking scenarios. Most significantly their research demonstrated the application of scaffolded learning to the point of autonomy, consistent with Vygotsky's Zone of Proximal Development (1978) but also Argyris' acknowledgement of the relevance of single and double loop learning and the need to challenge and reconcile assumptions relating to the acquisition of knowledge through experience. Both raise important issues in terms of the longitudinal provision in education with XR, in relation to the phasing and iterative presentation of learning outcomes, each serving to consolidate consequent stages of learning within and between cognitive, affective and psychomotor learning domains.

DRIVING AUTHENTIC LEARNING WITH PEDAGOGICAL DESIGN

Pedagogical design of the blurring of where reality and actuality co-exist within the context of digital immersion underpins the concept of validity in XR. While pedagogical research and applied praxis are still at the mercy of technology to a certain extent, as the virtual world becomes progressively more advanced, then so too will the capacity and capability for authenticity and validity within medical and healthcare education and training (Ligtart et al, 2021).

It is useful here, to consider the degree to which VR, AR and XR have been integrated across the health professions within differing signature pedagogies and the cognitive, psychomotor, and affective learning domain skills necessitating address. It is here that some healthcare professions can be delineated in terms of their functional involvement in patient centred care. For example, dentistry and physiotherapy have relatively functional interventions, whilst the work of a podiatrist, due to the ergonomic positioning of a practitioner, entails more of an opportunity to engage psychosocially with patients during their appointments (King et al, 2018). Indeed, academic debates have historically been posited about the functional basis of such healthcare professional roles as dentistry, since they are so functional in nature that people queried whether they could actually be regarded as a profession at all (Welie, 2004). The other implication is the stage of learning that medical and allied healthcare practitioners have reached in their career trajectories, where VR, AR and XR may be implemented most appropriately. The majority of published literature to date focuses on learning 'in situ' whereas the assessment of learning longitudinally in the context of real life praxis, is often evaluated with a degree of tokenism by those whose initiatives have introduced simulation in the first place.

The operationally definitive terms of VR, AR and XR are significant when considering the nature of the artificial environments that are created in the exploration of potentially risk filled scenarios, which emulate the real world (Alnagrat, Ismail and Idrus, 2021). This is primarily in relation to how artificial intelligence can be embedded within the context of virtual and simulated learning experiences, which also have to be optimally facilitated by staff skilled in the field of clinical simulation (Abbas, Kenth and Bruce, 2020). Central to progression in these aspects of a pedagogical paradigm shift, is the need for active dissemination and sharing of knowledge creation and acquisition that happens at the front line of medical and healthcare education where simulation in the context of XR are the expected and anticipated norms of undergraduate and postgraduate academic curricula across HEIs (Luo et al, 2021).

IMMERSION IN LEARNING EXPERIENCE

AR and VR enhance health professions education via immersion in learning experiences, positively influencing levels of engagement and motivation and by creating an atmosphere of interactive dialogue, which in terms of social inclusion can have a direct impact on the reduction of learner stress and anxiety (Brandon, Freiwirth and Hjersman, 2021). Encompassed by the umbrella term of XR, AR, MR and VR all bring a signature technological approach to applied praxis in the context of healthcare professions. In instances where different levels of sensorial intrusion are now widely used to extend the reach of pedagogic practice in medical and healthcare education and training, this has broad ramifications, not only on the practical delivery of emergent technology in practice but on their pedagogical theoretical justifications as well (Robert, 2021). In instances where the affordability and hence the extent of digital technology available determines accessibility in practice-based settings. In terms of a hierarchy of availability, AR is perhaps the most widely available and least invasive of all XR and can be easily accommodated by readily available display systems (Suryanti et al, 2020).

In instances where the need for the relevance of space and situatedness and context is of fundamental importance, then MR is the best platform for integration into available physical environments. In this sense MR fills the gap that VR leaves, in permitting users to physically interact with pre-identified physical spaces (Silén et al, 2008). Dimensionality is also of key importance in terms of whether the ability to facilitate visualisation in 3-D is something which can actively contribute to the acquisition of tacit knowledge and actively contribute to a more finely tuned intuitive response to any given scenario in medical and healthcare practice (Gerup, Soerensen and Dieckmann, 2020). MR effectively bridges the gap between AR and VR as a consequence of the development of the hardware capability of display screens and visualisation resources (Juraschek, 2018).

Research into XR pedagogical initiatives and methodological designs is usually characterised by relatively small-scale case studies, focused on the collection of qualitative data rather than tangible quantitative evidence about the impact of XR in practice (Hamilton, 2021).Those which have succeeded in this, tend to focus specifically on psychomotor skill or level of cognitive improvement rather than affective domain learning. The concept of tacit knowledge is also one which remains relatively underexplored and which remains almost a hidden bonus in developing the confidence of collective team hierarchies in clinical and medical practice. making generalisability of findings to other contexts and settings. Whilst this is a disadvantage, the potential for transferability of findings within and between similar situational contexts and settings across the global can be quite high in terms of overall content validity (Levitt, 2021).

The majority of XR technologies now extends the bridge between simulated environments and their polar opposites in physical reality. The rate of progressive developmental platforms has been amplified and magnified by the global COVID-19 pandemic, which has contributed to a paradigmatic shift in learners centred hybrid and hyflex learning and teaching opportunities across the medical and healthcare professions pedagogies (Acharya, 2021). As signature disciplines in their own right, these professions have maintained a certain degree of reticence to engage fully immersive technology within the context of clinical healthcare practice, since this completely detaches the clinician from their normal parameters of visual capacity. Whilst lenses are acceptable, they also permit and maintain human characteristics and gestures such as established eye contact and that degree of connection, which patients deserve and need (Jongerius et al, 2020). This has been particularly evidenced within the context of brain and cardiac surgery, both of which have optimally utilised these technological advances for the patients they serve.

TRADITIONS IN MEDICAL AND HEALTHCARE EDUCATION

Across the continuum of signature pedagogies that contribute directly to medical and allied health education, it is anatomy and mathematics which arguably transcend most disciplinary boundaries (Hayes and Capper, 2020). The visualisation techniques, which both permit and facilitate enhanced pedagogical teaching methodologies respectively appeared first in the context of simulation within the aviation industry at the beginning of the 20th Century. Their developmental progression over the last century and the early part of the 21st Century has ensured that levels of complex sensory and visual experience have now become a mainstay of medical and healthcare education and training programmes (Jentsch and Curtis, 2017). In the sense of providing students with the opportunity to undertake complex and risk inclined procedures, simulation is an invaluable resource. Within the context of safety, processes of experiential learning from practice can be fully exploited without there being any tangible, negative consequences in reality, so that unwanted outcomes can be illustrated and acknowledged without them actually ever having to come to fruition to do so.

Resources and student access to them have been a key issue during the pandemic, particularly for health and medical students for whom access to real- life anatomical models and cadavers may have been completely negated (Hilburg et al, 2020). For these students XR has ensured an opportunity for continued engagement with the discipline of anatomy and physiology across a range of contexts and settings. Since learning is extensive in relation to the need to develop a 3-D knowledge of bodily organs and systems, then the spatiality that XR offers also permits dynamic interaction, which otherwise would not be possible. In relation to the functional role of bodily organs and systems, XR has the added advantage that simulation of the dynamic motion of the body can be visualised as a learning resource, in contrast to cadavers, which are obviously non-functional (Ziker, Truman and Dodds, 2021). Within this simulation, it is also possible to identify, isolate and examine the individual functionality that specified aspects of these systems perform via repeated familiarisation. Ensuring learners iteratively progress from novice to expert within the use of XR of this nature, also ensures that learners can work in a risk free environment, which nevertheless serves to introduce them to the experiential learning which can form a solid foundation for their future professional careers in healthcare and medicine. In terms of pedagogic design and academic rigour the use of XR also provides an ideal opportunity to benchmark functional skills acquisition, alongside the real-time assessment of clinical practice and the opportunity for them to be exposed to sensory feedback from the situated learning they undertake (Kang et al, 2021).

BUILDING CAPABILITY FOR CLINICAL EDUCATORS WITH XR

The education and training of future medical and healthcare staff is not only dependent on disciplinary expertise but also the capacity to provide optimal learning experiences which in turn drive processes of assessment which are authentic and have real world validity and credibility (Voštinár et al, 2021). Future practitioners of all disciplines across the health and wellbeing continuum not only have to use evidence-based practice to discern clinical decision making, they also have to continually use scientific published evidence to iteratively update their continuing professional development in practice (Agha, 2021). As methodologies of pedagogic practice have become progressively more student there has been increasing emphasis on a move away from traditional rote learning to mechanisms of interactive, engaging processes of learning, which ensure deep rather than superficial learning across all aspects of education and training. Alongside this, is the recognition that learning driven purely by pressure to 'perform' within the context of examinations is futile in the truest sense of meaningful learning opportunity. In order to achieve this, XR can be regarded as a significant tool in the armoury of medical and healthcare educators who use this form of specialist resource to reinforce, motivate and enhance learning opportunities through authentically, strategically and constructively aligned academic curricula in Higher Education (Howell and Mikeska, 2021). Future practitioners also need the capacity to be educated in authentic situated learning contexts which are:

- Based on principles of diversity, equity and inclusion in terms of both optimal provision and optimal service to global, national, regional and local practice areas.
- Built on an ethos of compassion for all recipients and a respect for human life and wellbeing.
- Co-constructed with medical and healthcare service users and their families and carers.
- Designed on the basis of evidence-based healthcare practice and evidence-based processes of educational provision.
- Drivers of high quality multi-disciplinary and interprofessional patient centred teamworking.
- Effective in their capacity to drive education in relation to health and medical outcomes for patients at the front line of clinical care.
- Globally responsive and globally focussed
- Pedagogically designed and responsively designed on the basis of need and healthcare priorities in applied practice.
- Responsive in their capacity to dynamic and iterative change in healthcare landscapes.
- Specifically designed to value and promote autonomous learning, capacity for teamwork and proactive higher order critical thinking in relation to problem solving in health and medical care practice.

Within the context of XR integration, each pose individual challenges in relation to ensuring that digital technology is only ever seen as an adjunct to medical and healthcare education for the human race, not merely a functionalist or solely objective driver of it (Gandolfi, Kosko and Ferdig, 2021).

EXPERIENTIAL LEARNING THEORIES

The grander philosophical debates of perceived versus actual reality on an individual and collective basis are central to the processes of teaching and learning with XR. As a tool for the enhancement of understanding, meaning making and then critical reflection on practice, digital technology in invaluable as an adjunct to the human facilitation of knowledge, skills, attitudes and behaviours (Logeswaran et al 2021). The philosophical underpinning of experiential learning has its origins in constructivism, which is central to the acknowledgement of truth, either perceived or actual, and the capacity of all individuals to construct a perspective on truth based on their immersion in experience and the meaning they make of it (Koufidis et al, 2021). In this sense education is an entirely different experience for everyone who experiences the same phenomena and it is this individual experience and perspective meaning making, which lies at the heart of both collective and personal processes of transformation. The construction of a version of reality ensures that different perspective lenses can be applied to the same phenomena, so that scrutiny from a number of sources of triangulation can take place. Whether the experience differs collectively between adults and children in terms of capacity for meaning making through experiential learning was the source of seminal debate in the late 20th Century, with Knowles assertion that there was a distinction between andragogy, in terms of 'adult' ways of knowing and that of 'pedagogic' or child based learning. What is arguable more relevant is the concept of lifelong learning, across which learning takes place from the cradle to the grave, as a temporal mechanism of integrating the prospective with the retrospective and making inference from it as part of the integral processes of reflection and reflexivity (Loeng, 2021).

THEORETICAL UNDERPINNINGS OF XR PEDAGOGY

Experiential learning has become a byword for the placement experiences that medical and healthcare students undertake as an integral part of their undergraduate studies with patients, whose consent, briskly acquired provides novice learners with an opportunity to build confidence and be guided by an ever vigilant and highly respected disciplinary expert observer. In instances of postgraduate study, it may conjure the image of a professional moving from competence to proficiency and consequent mastery of their specific field of disciplinary context (Mortimore et al 2021). Medical training to the uninitiated is instilled with the imagery of cadaveric dissection as a means of visualising what lies beneath the skin that is to be negotiated across a respected lifelong career. In contrast to practice based experience at the front line of patient care in the context of care provision, though, the two are very different. It is here that whilst disciplinary knowledge and skill are invaluable to the expert educator, that skills of pedagogical praxis are equally important if the acquisition of knowledge which will underpin this lifelong career is to be assured and validated in practice. The theoretical underpinnings of learning and teaching have resonance in relation to the eventual capacity of medical and healthcare students not only to make meaning of experience but also to question that which cannot be taught, the tacit and reflexive response to intuition, which can only be derived from inner conscience and derived from processes of experiential learning in the psychomotor, cognitive and affective domains of learning, which can be tangible assessed and used to benchmark fundamental knowledge and skills acquisition (Birt et al, 2018). Similarly the capacity to transfer knowledge from the context of learning to the real world with patients who have an immediate need for assessment, diagnosis and consequent management is highly relevant to the potential to scaffold knowledge as clinical learners each move from the state of disciplinary novice to expert (Durning & Artino 2011)

MEANING MAKING IN HEALTHCARE PRACTICE

At the heart of capacity to learn and build on pre-existing experience is the theory of constructionist philosophy which when integrated alongside sociocultural perspectives from processes of active participation lies at the heart of all learning. The delineation between and linkage across theory and practice has become a fundamental goal of all involved in workplace education and training (Dennick, 2016). Since learning is both an individual and collective experience in the same sense as learning. What the Covid-19 pandemic detracted from was the fact that education is a social science, wholly reliant on interactivity and the derivation of meaning from multi-faceted and complex situational specificities, which when grouped as a collective constitute human experience. Being able to understand and frame the margin between knowledge and research entails a consideration of both the origin and consequent history of learning (Ten Cate and Billet, 2014). Alongside this, it is also necessary to consider how knowledge acquisition is gradually scaffolded through experience from novice to expert and to value those contributors to pedagogic practice who have acknowledged and formally recognised this (Graf et al, 2020). Whilst Kolb is often regarded as the seminal theorist of experiential learning, however there are several others of direct relevance to the pedagogical implementation of XR (Heong et al, 2020). Rooted in constructivist perspectives, Kolb's is of relevance since its focus lies in the translation of experience from external sources into epistemic cognition and meaning making on an individual basis. Kolb's theoretical perspectives are of importance when considering the processes of experiential learning, which in turn contribute to elements of subconscious bias that inevitably contribute to the professional identity of educators from specific signature pedagogies and disciplines. Social learning theory also plays a significant role in the development of experiential learning. It is a perspective upon which health and medicine have drawn heavily upon in the shaping and framing of new pedagogical perspectives. It frames learning outcomes in the context of their impact in applied praxis rather than in terms of their individual impact and acknowledges that these are often a natural by-product of distinctive social cultures and contexts. This also aligns with the concept of lifelong learning within which there is the acknowledgement that learning is dependent on levels of increasing experience (Hartman et al, 2020).

SCAFFOLDING LEARNING

XR provides an environment where the context of safety ensures that learners can iteratively be scaffolded through processes of knowledge acquisition and the basis of psychomotor skill competency and eventual mastery can be gained and evidenced in practice (Adefila et al, 2020).

One important aspect of the integration of XR though, is the degree to which it increases non-patient-based learning and the potential impact of this in relation to a lack of authentic human interaction across health and medical curricula. What this debate highlights are the significance of strategic curriculum design and justification across all programmes, where pedagogical constructive alignment is pivotal to outcomes-based assessment techniques and the credibility of their implementation. Similarly, where variations and combinations of XR are integrated across academic and clinical curricula then the technology selected ought to be aligned with the need for the acquisition of specific skills and techniques. The integration of XR and its components does ensure, however, that during the process of learning and the acquisition of new knowledge, skills and behaviours that ensure:

- Aims and objectives can be aligned with the opportunity for ongoing feedback and the opportunity to rehearse and develop new skill sets from novice to expert.
- Benchmarking of learner performance can be isolated into specific learning domains and assessed accordingly, with the potential of identifying individual learning need.
- Interruption and hence unnecessary and unwanted distraction, during the process of learning, can be avoided
- Patients can avoid any risk in terms of discomfort or the need to move regularly, as learners iteratively develop and refine new skills, where they can focus on domain specificity until a level of competency and eventual mastery is reached.
- That individual learning styles can be accommodated and encouraged by constructively aligning individual learning outcomes within each teaching session, where time permits.
- That psychomotor tasks can be repeated so that exposure, practical viability and accuracy can be iteratively increased.
- The opportunity for focused learner reflection on specific aspects of knowledge and skill acquisition and a reflexive approach for the address of individually recognised shortfalls.
- The theory-praxis gap can be gradually reduced by scaffolding learning and gradually increasing exposure to real life contexts where patients and their families are an integral part of authentic learning.

Non-technical skills such as interprofessional working and multi-disciplinary team working and the clinical attributes they necessitate in terms of optimal interpersonal communication and the management of the clinical environment characterise the skill sets necessary to ensure the achievement of holistic learning in proficient practice (Dehghani et al, 2021). Examples where this is now integrated in practice are in the skills and drills training, where it is also possible to integrate aspects of complex ambiguity and stress into situations, where a clear integration of psychomotor skill, affective and cognitive domain knowledge and application are necessary within any given scenario (DeMaria and Levine, 2013). It is these learning needs that high-fidelity clinical simulators address, so that once basic competence in a skill has been achieved then additional emotional stress and the need for instantaneous clinical decision making can be assessed. Whilst simulation is not an active replacement for the adrenaline rush faced in reality, it does offer an insight within learner centred experiential learning of the potentially uncertain scenarios that health and medical professionals often face in the real world. The portability of simulators such as 'SimMan' and 'SimMom' also provides the opportunity for context specific training opportunities where situated learning ca become an integral part of strategic curriculum justification, design and delivery (Viglialoro, 2021).

SUPPORTING CHANGE IN GLOBAL HEALTHCARE WORKFORCE PROVISION

Despite being one of the most needed and often underacknowledged parts of the global workforce, healthcare workers have virtually all been exposed to traditionalist mechanisms of education, which in terms of educational delivery have remained pedagogically unaltered for the best part of half a century (Karunathilake and Samarasekera, 2021). XR, as an integral part of a new technological age has been embedded within existing educational infrastructures, where traditional instruction mechanisms have shaped its implementation in relation to the pre-existing array of subject areas and academic disciplines represented by HEIs across the world. Stemming from the traditionalist apprenticeship models of learning, these learning approaches have been shaken to the core by a global pandemic that despite being predicted, came as a complete shock to global academic communities where health and medical education had to continue, despite education as a social science being altered beyond recognition in a space of six months, when national lockdowns shaped altered approaches to educational affordance and opportunity to learn (Chan, 2021). Whilst medical publications have an average currency of five years, clinicians of health and medical practice work for an average of forty years on the foundational knowledge attained almost half a century before. It is perhaps no surprise to learn then, that medical errors have become a progressively iterative problem across global health and medical practice, in situations where technology has developed to such an extent that new interventions are often almost obsolete before their initial implementation at the front line of patient care (Melnyk et al, 2021). Whilst these issues are concerning, they also pose the greatest opportunity for processes of progressive change for health and medical students whose curricula will inform the complex and ambiguous territories of professionalism that they will need to negotiate over the forthcoming years. This is an opportunity not only to focus and stabilise the quality and optimal provision of education for the next generations of healthcare practice, but also to ensure that medical education becomes more accessible, cost effective and most importantly places patients at the heart of compassionate human care. Exponentially in response to these challenges there has been a progressive switch to Problem Based Learning pedagogical approaches to medical and healthcare curricula across the globe. The immediacy of feedback and impact of professional facilitation of learning is central to the success of this approach in practice and ensures that the mindset of a critical enquirer is entrenched in learners who one day will become the next generation of medical and healthcare providers.

PATIENT CARER PUBLIC INVOLVEMENT IN CURRICULUM DELIVERY

The integration of patients and their families and carers, within the context of patient and carer public involvement (PCPI) initiatives is now an integral part of most health and medical school curricula (Ocloo et al, 2021). Whilst they are invaluable in relation to the development of affective domain learning and interpersonal and effective communication skills, the fact that these skills are inseparable from highly attuned cognitive and psychomotor skills in practice still exists. Where they are therefore of particular value is in the integration of these skills in conjunction with high fidelity simulation equipment, so that psychological fidelity can also be integrated into the acquisition of psychomotor technical skills training (Johnson et al, 2020; Hayes and Graham, 2020). This is of value in the transitioning between specifically designed learning environments, within which simulated learning takes place, and the eventual 'real life' praxis that students face. In terms of recreating learning environments which also reflect the multi-disciplinary and interprofessional working contexts within which health and medical staff find themselves daily, XR provides a useful mechanism of ensuring the psychological fidelity of scenarios used in learning. The origins of human factor analysis and historical emergence of XR training are well documented within the aviation industry, where contexts of flight simulation and the selective integration and conditions of flying can be reconstructed in terms of variability (Hawkins, 2017). One distinct variation in terms of the potential transferability of learning to 'real life' though, is the complex ambiguity and unpredictability of working with humans, whose predictability far outweighs anything an aeroplane, however complex can reflect. Unlike flying, therefore, those assessments regarded as 'high stake' in terms of potential outcome and risk, cannot be conducted with medical simulation equipment because of their relative potential variability in pathological and physiological status (Spurgeon et al, 2019).

Using teaching and learning to drive assessment is an acknowledged feature of student-centred learning but we must never lose sight of the fact that XR, whilst being an adjunct and embedded part of health and medical curricula must never become the driver of learning, lest the most important learning opportunities and factors may be lost to the prioritisation of technology over human need. Similarly, some higher education institutions aspire to owning the latest equipment, which is often declared obsolete before it can be pedagogically integrated with any degree of credibility across those academic curricula for which it was originally purchased. The focus of curriculum design and how teaching and learning ought to drive assessment rather than vice-versa must be advocated if the overall resultant levels of quality in patient care are to remain optimal (Burgess et al, 2020).

RECOMMENDATIONS AND FUTURE STUDIES

Within the context of the assessment of healthcare and medical students, formative assessment that is subject to the regulations of Professional and Statutory Regulatory Boards such as allied health and medicine programmes, there is an opportunity to build research into processes of effective moderation of assessment within and between educational institutions, in terms of the parity and equity of assessment mechanisms. Where clinical educators and academics develop evidence of the objectivity of processes of assessment, then XR will be invaluable in terms of being able to standardise approaches to both formative and summative assessment techniques and their respective outcomes in educational practice and this is deserved of extended research investigation (Wilson and Shankar, 2021). The COVID-19 pandemic has seen an exponential rise in the delivery of blended synchronous online and face to face teaching, which has been termed a hybrid learning approach. However, a newer term is the Hyflex curricula which have permeated the educational marketplace, with tailored offers of individual flexibility in learning patterns, which can accommodate more diverse student cohorts. Since its recent inception the relative impact of Hyflex curricula have yet to be clearly articulated in the context of academia. Instructional resources delivered by XR are often an integral part of these flexible offers and could potentially be an optimal way of managing learning time for those adult learners who have additional family or working commitments to fulfil (Gagnon et al, 2020). Operationally defined, Hyflex models also adopt a hybrid approach but offer the opportunity to study flexibly rather than at a pre-set time. Delineating between models of actual attendance on campus is often indicated by the term hybrid, whereas the actual curriculum model is termed blended, which again could be resolved by pedagogical research to establish processes of best practice.

CONCLUSION

Progressive innovation across medical and healthcare professions is something which challenges traditional, conservative approaches to pedagogy, which ultimately contributes to the life and death decision making of clinicians at the front line of care. In this regard, it is important that technology is seen as a driver of pedagogic practice, rather than a methodology in itself. In this sense uptake of these new adjuncts is wholly dependent on medical and healthcare educators being open to change in approaches to education of subsequent generations of students and an acceptance that iterative and ongoing professional skills development is an integral part of adopting technology in practice. In practice this also necessitates the recruitment of staff with specialist technical skills who are responsible for the operationalisation of equipment, but who may not necessarily have specialist pedagogical skills in terms of curriculum design and implementation. So collaborative teamwork in the context of operational delivery of teaching sessions is pivotal. This investment, in terms of financial cost, collaborative working and mandatory iterative skills development for both academic and clinical educators, ensures that learning in a context safe from real-life risk with the potential for benchmarking the acquisition of knowledge and skills, is made possible in practice. The ongoing global COVID-19 pandemic has caused huge global disruption to traditional medical and healthcare education. Innovative responses in digital and technological tools have been quickly implemented with the aim of ensuring that learning can be maintained and sustained in order to ensure sufficient medical and healthcare graduates continue to qualify each year. Whilst change continues to ensure transformative approaches in those countries who can afford the fiscal implications of these developments, the gap between training providers globally also has the potential to widen in relation to optimal quality of medical and healthcare education. This is an obvious area for address if there is not to become a skills deficit in those countries where the affordance of digital technology as standard for learners and their academic and clinical teachers, is not yet possible. It is the acknowledgement and address of these key challenges which has the potential for a truly authentic paradigm shift in the application of XR in pedagogic design, scholarship and implementation to be achieved.

REFERENCES

Abbas, J. R., Kenth, J. J., & Bruce, I. A. (2020). The role of virtual reality in the changing landscape of surgical training. *The Journal of Laryngology and Otology, 134*(10), 863–866. doi:10.1017/S0022215120002078 PMID:33032666

Acharya, S., Bhatt, A. N., Chakrabarti, A., Delhi, V. S., Diehl, J. C., van Andel, E., & Subra, R. (2021). Problem-Based Learning (PBL) in Undergraduate Education: Design Thinking to Redesign Courses. In *Design for Tomorrow—Volume 2* (pp. 349–360). Springer. doi:10.1007/978-981-16-0119-4_28

Adefila, A., Opie, J., Ball, S., & Bluteau, P. (2020). Students' engagement and learning experiences using virtual patient simulation in a computer supported collaborative learning environment. *Innovations in Education and Teaching International, 57*(1), 50–61.

Aebersold, M., & Dunbar, D. M. (2021). Virtual and Augmented Realities in Nursing Education: State of the Science. *Annual Review of Nursing Research, 39*(1), 225–242. doi:10.1891/0739-6686.39.225 PMID:33431644

Agha, S. (2021). Aligning continuing professional development (CPD) with quality assurance (QA): A perspective of healthcare leadership. *Quality & Quantity*, 1–15.

Alnagrat, A. J. A., Ismail, R. C., & Idrus, S. Z. S. (2021, May). Extended Reality (XR) in Virtual Laboratories: A Review of Challenges and Future Training Directions. [). IOP Publishing.]. *Journal of Physics: Conference Series*, *1874*(1), 012031. doi:10.1088/1742-6596/1874/1/012031

Annala, J., Lindén, J., Mäkinen, M., & Henriksson, J. (2021). Understanding academic agency in curriculum change in higher education. *Teaching in Higher Education*, 1–18. doi:10.1080/13562517.2021.1881772

Antoniou, P., Arfaras, G., Pandria, N., Ntakakis, G., Bambatsikos, E., & Athanasiou, A. (2020). Real-time affective measurements in medical education, using virtual and mixed reality. In *International Conference on Brain Function Assessment in Learning* (pp. 87-95). Springer, Cham. 10.1007/978-3-030-60735-7_9

Argyris, C. (1991). Teaching smart people how to learn. *Harvard Business Review*, *69*(3).

Bevins, F., Bryant, J., Krishnan, C., & Law, J. (2020). Coronavirus: How should US higher education plan for an uncertain future. *McKinsey*.

Birt, J., Stromberga, Z., Cowling, M., & Moro, C. (2018). Mobile mixed reality for experiential learning and simulation in medical and health sciences education. *Information (Basel)*, *9*(2), 31. doi:10.3390/info9020031

Brandon, E., Freiwirth, R., & Hjersman, J. (2021, May). Special Session—Student Engagement with Reduced Bias in a Virtual Classroom Environment. In *2021 7th International Conference of the Immersive Learning Research Network (iLRN)* (pp. 1-3). IEEE.

Burgess, A., van Diggele, C., Roberts, C., & Mellis, C. (2020). Key tips for teaching in the clinical setting. *BMC Medical Education*, *20*(2), 1–7. doi:10.118612909-020-02283-2 PMID:33272257

Chan, S. (2021). *Digitally Enabling 'Learning by Doing' in Vocational Education: Enhancing 'Learning as Becoming' Processes*. Springer Nature. doi:10.1007/978-981-16-3405-5

Crouch, L., Rolleston, C., & Gustafsson, M. (2021). Eliminating global learning poverty: The importance of equalities and equity. *International Journal of Educational Development*, *82*, 102250. doi:10.1016/j.ijedudev.2020.102250

Dehghani, M., Acikgoz, F., Mashatan, A., & Lee, S. H. (2021). A holistic analysis towards understanding consumer perceptions of virtual reality devices in the post-adoption phase. *Behaviour & Information Technology*, 1–19.

DeMaria, S., & Levine, A. I. (2013). The use of stress to enrich the simulated environment. In *The comprehensive textbook of healthcare simulation* (pp. 65–72). Springer. doi:10.1007/978-1-4614-5993-4_5

Dennick, R. (2016). Constructivism: Reflections on twenty-five years teaching the constructivist approach in medical education. *International Journal of Medical Education*, *7*, 200–205. doi:10.5116/ijme.5763.de11 PMID:27344115

Durning, S. J., & Artino, A. R. (2011). Situativity theory: a perspective on how participants and the environment can interact: AMEE Guide no. 52. *Medical Teacher*, *33*(3), 188–199. doi:10.3109/0142159X.2011.550965 PMID:21345059

Einstein, A. (2011). *Essays in science*. Open Road Media.

Gagnon, K., Young, B., Bachman, T., Longbottom, T., Severin, R., & Walker, M. J. (2020). Doctor of physical therapy education in a hybrid learning environment: Reimagining the possibilities and navigating a "new normal". *Physical Therapy*, *100*(8), 1268–1277. doi:10.1093/ptj/pzaa096 PMID:32424417

Gandolfi, E., Kosko, K. W., & Ferdig, R. E. (2021). Situating presence within extended reality for teacher training: Validation of the extended Reality Presence Scale (XRPS) in preservice teacher use of immersive 360 video. *British Journal of Educational Technology*, *52*(2), 824–841. doi:10.1111/bjet.13058

Gerup, J., Soerensen, C. B., & Dieckmann, P. (2020). Augmented reality and mixed reality for healthcare education beyond surgery: An integrative review. *International Journal of Medical Education*, *11*, 1–18. doi:10.5116/ijme.5e01.eb1a PMID:31955150

Goh, P. S., & Sandars, J. (2020). A vision of the use of technology in medical education after the COVID-19 pandemic. *MedEdPublish*, 9.

Graf, A. C., Jacob, E., Twigg, D., & Nattabi, B. (2020). Contemporary nursing graduates' transition to practice: A critical review of transition models. *Journal of Clinical Nursing*, *29*(15-16), 3097–3107. doi:10.1111/jocn.15234 PMID:32129522

Hamilton, D., McKechnie, J., Edgerton, E., & Wilson, C. (2021). Immersive virtual reality as a pedagogical tool in education: A systematic literature review of quantitative learning outcomes and experimental design. *Journal of Computers in Education*, *8*(1), 1–32. doi:10.100740692-020-00169-2

Hartman, E., Reynolds, N. P., Ferrarini, C., Messmore, N., Evans, S., Al-Ebrahim, B., & Brown, J. M. (2020). Coloniality-decoloniality and critical global citizenship: Identity, belonging, and education abroad. *Frontiers: The Interdisciplinary Journal of Study Abroad*, *32*(1), 33–59. doi:10.36366/frontiers.v32i1.433

Hawkins, F. H. (2017). *Human factors in flight*. Routledge. doi:10.4324/9781351218580

Hayes, C., & Capper, S. (2020). Illustrating the transcendence of disciplinarity. In *Beyond Disciplinarity* (pp. 40–49). Routledge. doi:10.4324/9781315108377-4

Hayes, C., & Graham, Y. (2020). *Designing a Benchmarking Tool for Testing Posttest Confidence Levels in Emergency Obstetrics Training*. SAGE Publications. doi:10.4135/9781529709285

Hayes, C., Hinshaw, K., & Petrie, K. (2019). Reconceptualizing medical curriculum design in strategic clinical leadership training for the 21st century physician. In *Preparing Physicians to Lead in the 21st Century* (pp. 147–163). IGI Global. doi:10.4018/978-1-5225-7576-4.ch009

Heong, Y. M., Ping, K. H., Hamdan, N., Ching, K. B., Yunos, J. M., Mohamad, M. M., ... Azid, N. (2020). Integration of Learning Styles and Higher Order Thinking Skills among Technical Students. *Journal of Technical Education and Training*, *12*(3), 171–179.

Hilburg, R., Patel, N., Ambruso, S., Biewald, M. A., & Farouk, S. S. (2020). Medical education during the coronavirus disease-2019 pandemic: Learning from a distance. *Advances in Chronic Kidney Disease, 27*(5), 412–417. doi:10.1053/j.ackd.2020.05.017 PMID:33308507

Hilty, D. M., Parish, M. B., Chan, S., Torous, J., Xiong, G., & Yellowlees, P. M. (2020). A comparison of in-person, synchronous and asynchronous telepsychiatry: Skills/competencies, teamwork, and administrative workflow. *Journal of Technology in Behavioral Science, 5*(3), 273–288. doi:10.100741347-020-00137-8

Horton, S. (2021). Empathy Cannot Sustain Action in Technology Accessibility. *Frontiers of Computer Science, 3*, 31.

Howell, H., & Mikeska, J. N. (2021). Approximations of practice as a framework for understanding authenticity in simulations of teaching. *Journal of Research on Technology in Education, 53*(1), 8–20. doi:10.1080/15391523.2020.1809033

Humpherys, S. L., Bakir, N., & Babb, J. (2021). Experiential learning to foster tacit knowledge through a role play, business simulation. *Journal of Education for Business*, 1–7.

Jentsch, F., & Curtis, M. (2017). *Simulation in aviation training*. Routledge. doi:10.4324/9781315243092

Johnson, C. E., Kimble, L. P., Gunby, S. S., & Davis, A. H. (2020). Using deliberate practice and simulation for psychomotor skill competency acquisition and retention: A mixed-methods study. *Nurse Educator, 45*(3), 150–154. doi:10.1097/NNE.0000000000000713 PMID:31246693

Jongerius, C., Hessels, R. S., Romijn, J. A., Smets, E. M., & Hillen, M. A. (2020). The measurement of eye contact in human interactions: A scoping review. *Journal of Nonverbal Behavior, 44*(3), 1–27. doi:10.100710919-020-00333-3

Juraschek, M., Büth, L., Posselt, G., & Herrmann, C. (2018). Mixed reality in learning factories. *Procedia Manufacturing, 23*, 153–158. doi:10.1016/j.promfg.2018.04.009

Kang, J., Diederich, M., Lindgren, R., & Junokas, M. (2021). Gesture patterns and learning in an embodied XR science simulation. *Journal of Educational Technology & Society, 24*(2), 77–92.

Karunathilake, I. M., & Samarasekera, D. D. (2021). Learning In The 21st Century— 'What's All the Fuss about Change?'. In Educate, Train and Transform: Toolkit on Medical and Health Professions Education (pp. 1-14). Routledge.

King, O., Borthwick, A., Nancarrow, S., & Grace, S. (2018). Sociology of the professions: What it means for podiatry. *Journal of Foot and Ankle Research, 11*(1), 1–8. doi:10.118613047-018-0275-0 PMID:29942353

Koufidis, C., Manninen, K., Nieminen, J., Wohlin, M., & Silén, C. (2021). Unravelling the polyphony in clinical reasoning research in medical education. *Journal of Evaluation in Clinical Practice, 27*(2), 438–450. doi:10.1111/jep.13432 PMID:32573080

Levitt, H. M. (2021). Qualitative generalization, not to the population but to the phenomenon: Reconceptualizing variation in qualitative research. *Qualitative Psychology, 8*(1), 95–110. doi:10.1037/qup0000184

Loeng, S. (2018). Various ways of understanding the concept of andragogy. *Cogent Education*, *5*(1), 1496643. doi:10.1080/2331186X.2018.1496643

Logeswaran, A., Munsch, C., Chong, Y. J., Ralph, N., & McCrossnan, J. (2021). The role of extended reality technology in healthcare education: Towards a learner-centred approach. *Future Healthcare Journal*, *8*(1), e79–e84. doi:10.7861/fhj.2020-0112 PMID:33791482

Luctkar-Flude, M., & Tyerman, J. (2021). The Rise of Virtual Simulation: Pandemic Response or Enduring Pedagogy? *Clinical Simulation in Nursing*, *57*, 1–2. doi:10.1016/j.ecns.2021.06.008

Luo, C., Lan, Y., Luo, X. R., & Li, H. (2021). The effect of commitment on knowledge sharing: An empirical study of virtual communities. *Technological Forecasting and Social Change*, *163*, 120438. doi:10.1016/j.techfore.2020.120438

Mathew, P. S., & Pillai, A. S. (2020). Role of Immersive (XR) Technologies in Improving Healthcare Competencies: A Review. *Virtual and Augmented Reality in Education, Art, and Museums*, 23-46.

McGrath, J. L., Taekman, J. M., Dev, P., Danforth, D. R., Mohan, D., Kman, N., & Won, K. (2018). Using virtual reality simulation environments to assess competence for emergency medicine learners. *Academic Emergency Medicine*, *25*(2), 186–195. doi:10.1111/acem.13308 PMID:28888070

Melnyk, B. M., Tan, A., Hsieh, A. P., Gawlik, K., Arslanian-Engoren, C., Braun, L. T., Dunbar, S., Dunbar-Jacob, J., Lewis, L. M., Millan, A., Orsolini, L., Robbins, L. B., Russell, C. L., Tucker, S., & Wilbur, J. (2021). Critical care nurses' physical and mental health, worksite wellness support, and medical errors. *American Journal of Critical Care*, *30*(3), 176–184. doi:10.4037/ajcc2021301 PMID:34161980

Mitchell, R., & Boyle, B. (2021). Understanding the role of profession in multidisciplinary team innovation: Professional identity, minority dissent and team innovation. *British Journal of Management*, *32*(2), 512–528. doi:10.1111/1467-8551.12419

Moreira, D. (2020). Virtual networks and asynchronous communities: methodological reflections on the digital. In *Ethnography in Higher Education* (pp. 177–196). Springer VS. doi:10.1007/978-3-658-30381-5_11

Mortimore, G., Reynolds, J., Forman, D., Brannigan, C., & Mitchell, K. (2021). From expert to advanced clinical practitioner and beyond. *British Journal of Nursing (Mark Allen Publishing)*, *30*(11), 656–659. doi:10.12968/bjon.2021.30.11.656 PMID:34109817

Negev, M., Dahdal, Y., Khreis, H., Hochman, A., Shaheen, M., Jaghbir, M. T., Alpert, P., Levine, H., & Davidovitch, N. (2021). Regional lessons from the COVID-19 outbreak in the Middle East: From infectious diseases to climate change adaptation. *The Science of the Total Environment*, *768*, 144434. doi:10.1016/j.scitotenv.2020.144434 PMID:33444865

Obrad, C. (2020). Constraints and consequences of online teaching. *Sustainability*, *12*(17), 6982. doi:10.3390u12176982

Ocloo, J., Garfield, S., Franklin, B. D., & Dawson, S. (2021). Exploring the theory, barriers and enablers for patient and public involvement across health, social care and patient safety: A systematic review of reviews. *Health Research Policy and Systems*, *19*(1), 1–21. doi:10.118612961-020-00644-3 PMID:33472647

Okoye, K., Rodriguez-Tort, J. A., Escamilla, J., & Hosseini, S. (2021). Technology-mediated teaching and learning process: A conceptual study of educators' response amidst the Covid-19 pandemic. *Education and Information Technologies*, *26*(6), 1–33. doi:10.100710639-021-10527-x PMID:34025205

Orr, N., Matthews, B., See, Z. S., Burrell, A., Day, J., & Seengal, D. (2021). Transdisciplinarity in extended reality (XR) research design: Technological transformation and social good (co-creation session at XR+ Creativity Symposium, University of Newcastle, 2020). *Virtual Creativity, 11*(1), 163-179.

Owens, K. P. (2021, July). Competency-Based Experiential-Expertise and Future Adaptive Learning Systems. In *International Conference on Human-Computer Interaction* (pp. 93-109). Springer, Cham. 10.1007/978-3-030-77873-6_7

Park, C., & Kim, D. G. (2020). Exploring the roles of social presence and gender difference in online learning. *Decision Sciences Journal of Innovative Education*, *18*(2), 291–312. doi:10.1111/dsji.12207

Robert, I. V. (2021). Formation and development of digital transformation of domestic education on the basis of systemic convergence of pedagogical science and technology. In *SHS Web of Conferences* (*Vol. 101*, p. 03017). EDP Sciences. 10.1051hsconf/202110103017

Roussin, C. J., & Weinstock, P. (2017). SimZones: An organizational innovation for simulation programs and centers. *Academic Medicine*, *92*(8), 1114–1120. doi:10.1097/ACM.0000000000001746 PMID:28562455

Silén, C., Wirell, S., Kvist, J., Nylander, E., & Smedby, Ö. (2008). Advanced 3D visualization in student-centred medical education. *Medical Teacher*, *30*(5), e115–e124. doi:10.1080/01421590801932228 PMID:18576181

Suryanti, S., Sutaji, D., Arifani, Y., Muyasaroh, M., & Zamzamy, M. (2020). Improved learning accessibility and professionalism of teachers in remote areas through mentoring development of teaching materials based on Augmented Reality. [Research Dissemination for Community Development]. *Kontribusia*, *3*(1), 224–232. doi:10.30587/kontribusia.v3i1.1032

Tabatabai, S. (2020). COVID-19 impact and virtual medical education. *Journal of Advances in Medical Education & Professionalism*, *8*(3), 140–143. PMID:32802908

Tang, K. S., Cheng, D. L., Mi, E., & Greenberg, P. B. (2020). Augmented reality in medical education: A systematic review. *Canadian Medical Education Journal*, *11*(1), e81. PMID:32215146

Ten Cate, O., & Billett, S. (2014). Competency-based medical education: Origins, perspectives and potentialities. *Medical Education*, *48*(3), 325–332. doi:10.1111/medu.12355 PMID:24528467

van der Niet, A. G., & Bleakley, A. (2021). Where medical education meets artificial intelligence: 'Does technology care?'. *Medical Education*, *55*(1), 30–36. doi:10.1111/medu.14131 PMID:32078175

Viglialoro, R. M., Condino, S., Turini, G., Carbone, M., Ferrari, V., & Gesi, M. (2021). Augmented Reality, Mixed Reality, and Hybrid Approach in Healthcare Simulation: A Systematic Review. *Applied Sciences (Basel, Switzerland)*, *11*(5), 2338. doi:10.3390/app11052338

Voštinár, P., Horváthová, D., Mitter, M., & Bako, M. (2021). The look at the various uses of VR. *Open Computer Science*, *11*(1), 241–250. doi:10.1515/comp-2020-0123

Vygotsky, L. S. (1978). Zone of proximal development: A new approach. *Mind in society: The development of higher psychological processes*, 84-91.

Welie, J. V. (2004). Is dentistry a profession? Part 3. Future challenges. *Journal - Canadian Dental Association*, *70*(10), 675–678. PMID:15530264

Wilson, I., & Shankar, P. R. (2021). The COVID-19 pandemic and undergraduate medical student teaching/learning and assessment. *MedEdPublish*, *10*(1), 10. doi:10.15694/mep.2021.000044.1

World Health Organisation. (2021) International Data Online Updates. WHO. https://www.who.int/data)

Yigitcanlar, T., Butler, L., Windle, E., Desouza, K. C., Mehmood, R., & Corchado, J. M. (2020). Can building "artificially intelligent cities" safeguard humanity from natural disasters, pandemics, and other catastrophes? An urban scholar's perspective. *Sensors (Basel)*, *20*(10), 2988. doi:10.339020102988 PMID:32466175

Zulkifli, A. F. (2019). Student-centered approach and alternative assessments to improve students' learning domains during health education sessions. *Biomedical Human Kinetics*, *11*(1), 80–86. doi:10.2478/bhk-2019-0010

ADDITIONAL READING

Cranmer, K., Brehmer, J., & Louppe, G. (2020). The frontier of simulation-based inference. *Proceedings of the National Academy of Sciences of the United States of America*, *117*(48), 30055–30062. doi:10.1073/pnas.1912789117 PMID:32471948

Hong, T., Langevin, J., & Sun, K. (2018, October). Building simulation: Ten challenges. [). Springer Berlin Heidelberg.]. *Building Simulation*, *11*(5), 871–898. doi:10.100712273-018-0444-x

Jeffries, P. (2020). *Simulation in nursing education: From conceptualization to evaluation*. Lippincott Williams & Wilkins.

Kadian, A., Truong, J., Gokaslan, A., Clegg, A., Wijmans, E., Lee, S., Savva, M., Chernova, S., & Batra, D. (2020). Sim2Real predictivity: Does evaluation in simulation predict real-world performance? *IEEE Robotics and Automation Letters*, *5*(4), 6670–6677. doi:10.1109/LRA.2020.3013848

Satin, A. J. (2018). Simulation in obstetrics. *Obstetrics and Gynecology*, *132*(1), 199–209. doi:10.1097/AOG.0000000000002682 PMID:29889745

KEY TERMS AND DEFINITIONS

Augmented Reality (AR): Is a technology capable of superimposing or overlaying a computer-generated image across a visual projection of the real world, providing a composite view of the two.

Extended Reality (XR): Is the term given to all real-and-virtual combined environments and human-machine interactions, which are functionally digitally generated by technology and wearable accessories.

Health Professions: Is the term used for workers who have been formally trained in the application of medical and healthcare principles underpinned by the core principles of care, compassion and evidence based approaches to the care of people whose health necessitates assessment, diagnosis or management.

Hybrid Curriculum: Is the term used to describe how online learning is integrated with traditional face to face learning and teaching.

Hyflex Curriculum: Is an adaptation of hybrid learning where each class session and learning activity is offered in-person, synchronously online, and asynchronously online. With this approach learners make the decision of how they will participate with the learning opportunities afforded to them.

Immersion Technology: Is the digital equipment which provides the perception of being present in a created and non-physical world.

Mixed Reality (MR): Is the merging of virtual and actual reality to provide new mechanisms of visualizing given scenarios. The physical objects and digital objects can interact with each other in real time.

Paradigm: A set of concepts or thought patterns, incorporating specific theories, designated research methods, hypotheses, and typical standards of what is a legitimate claim to contribution to a specific field of theory or practice.

Pedagogy: Is the methodological process and study of applied teaching and learning within specific subjects and academic disciplines.

Sensory: Relates to the experience of sensation via the physical senses in terms of either perception or transmission.

Simulation: Is the integrated use of a computer model, which imitates reality in the context of study, where risk can be eliminated as part of initial scaffolded learning.

Virtual Reality (VR): Is the digitally generated simulation of a 3-D image or situated context or learning environment, within which a learner can be placed and with which they can interact by wearing electronic accessories such as eye goggles or gloves with sensors.

Chapter 5
Digital Transformation in Contemporary Organizations:
Creativity and Innovation in the 4th Industrial Revolution

Zeynep Merve Ünal
https://orcid.org/0000-0003-4927-3117
Independent Researcher, Turkey

ABSTRACT

The aim of this chapter is to shed light on the stages of digital transformation process on the basis of 4th IR. AI and its development processes are explained by digital transformation framework and holistic framework of digital transformation under the perspective of contemporary organization. Definition of AI and its relationship with sub-dimensions are strengthened by creativity process and AI creativity. Top strategic trends for 2022 and emerging technologies are discussed through considering growth, digitalization, and efficiency. AI-driven business model innovation is taken into consideration by scaling AI capabilities as agile customer co-creation, data-driven delivery operations, and scalable ecosystem. Organizational capabilities for AI implementation are taken into account as AI project planning, co-development of AI systems, data management and AI model lifecycle management. In the view of detailed literature review and current findings, evaluations and suggestions were made.

INTRODUCTION

Globalization has been forcing companies to fulfill their potentials on the basis of creativity and innovation in the fourth industrial revolution. The current competition parameters have been labeled as creativity, expertise and innovation (Case, 2016). Companies can only survive in the market where they are able to adopt creativity and innovation (Xie, 2017; Zhao, Feng, Chu, & Ma, 2017). At the occurrence of 4th industrial revolution (4IR/ Industry 4.0); digital transformation, contemporary artificial intelligence (AI) and AI techniques, creativity and AI creativity have gained attention. Industry 4.0 is a new paradigm of

DOI: 10.4018/978-1-6684-6366-6.ch005

cyber systems that include internet of things (IoT) by combining areas of AI, machine learning (ML) and robotics. AI refers to ''*a system's ability to interpret external data correctly, to learn from such data, and to use those learnings to achieve specific goals and tasks through flexible adaptation*'' (Kaplan & Haenlein, 2019, p. 17). AI system can convert the data into human-level knowledge such as machine reading and computer vision and can use gained information to expedite tasks that were completed by human effort previously (Taddy, 2018). In the study of Brock and von Wangenheim (2019), AI has positive impact on smart services, office automation, management support, smart products, manufacturing automation and automated customer interface. They use digital transformation within the AI projects context (e.g., transformations of call centers or operations using IoT, advanced analytics and AI). AI projects and AI management systems help organizations for effective AI implementation. AI capability and AI creativity are needed to actualize AI implementation. AI capability and AI creativity have been handled together on the basis of resource-based theory. It has been argued that AI capability comprises of tangible, intangible and human dimensions (Mikalef & Gupta, 2021). AI capabilities are taken into consideration through innovative business model in which scalable ecosystem integration (value capture), data-driven delivery operations (value delivery), and agile customer co-creation (value creation) are combined. Organizational capabilities for AI implementation are essential to overcome AI challenges such as uncertainty nature of AI projects and data dependency for AI implementation.

Transformation and innovation in digital world is taken into consideration as digitalization (Schreckling & Steiger, 2017). Digitalization is defined as ''*the use of digital technologies to change a business model and provide new revenue and value-producing opportunities; it is the process of moving to a digital business*'' (Gartner, n.d.a). Digitalization and digital transformation are reciprocal terms in digital technologies. Digital transformation is related to innovation. In the center of digital transformation and innovation dimensions such that customer centricity and technology foundation where leadership and strategy, processes, people and skills, business models and offerings (products and services), structure and governance and culture are assessed to have successful digitalization. The digital innovation and transformation framework of Schreckling and Steiger (2017) underlines that customer centricity is the main beneficiary of the digitalization and it is crucial to focus on the expectation and preferences of customers, creating engaging activities with customers to understand their needs and wants, identifying change in customers' preferences in the digital environment, understanding to what extent the values obtained by customers are matched with the corporate strategy and business models, putting customers in the center of enterprise to develop a comprehensive digital strategy with them.

Leadership and strategy are considered as determinants to successful digitalization that customers and businesses are communicated continuously to understand new competitors, new digitized products and services and new digital business models in markets. Business models and product and service offerings are the aims of digitization. So, researching competitor's digital business models, knowing how to digitize and connect tangible products by adding sensors and knowing how to enlarge digital portfolio through designing smart services that enrich products and services are the determinants of successful digital transformation management. Structure and governance are the complementary factors of digitalization process. They emphasize the integration of digital initiatives with respect to corporate organization, facilitation of digital innovation and transformation (i.e., lean organizational structures, flat hierarchy, and cross-functional teams), employee empowerment to give delegation and to let them taking digital initiatives and creating an organizational environment that eases to adopt agile organizational structure in which cross-function teams in organization can communicate effectively. In other words, people and skills are at the center of digital transformation process. The match between employees' competencies and digital world requirements, the match between employees' values and organizational values, leaders' awareness about digital environment expectancies, leaders' characteristics that foster digitization adaptation by embracing digital technologies and proactive behaviors, having clear vision and goals, reinforcing expected outcomes (e.g., behaviors, goals) in a positive way are significant predictors of digital transformation process.

Culture which is another form of ''shared values'' is one of the components of digital innovation and transformation. The degree of common adopted values that facilitate digitalization such as; knowledge of sharing culture, collective point of view and collaboration, proactive decision-making, challenging attitudes to status quo, learning organizational culture, a culture where creativity, entrepreneurship, agility are encouraged are paths to adopt new digital environment. Technology foundation makes possible to create something new in digital area as a main enabler. It focuses on the required information and experience to embrace emerging digital technologies, technological infrastructure that enable organizations to actualize their core business competencies, and simplified systems, applications and tools to increase employees' productivity. Findings of Brock and von Wangenheim (2019) indicated that digital transformation outcome success was determined by organizational agility, engaged staff, leadership support, technology partner support, sufficient investment, supportive culture, technology alignment and learning from failure respectively. This chapter expands the knowledge about AI and AI implementation in the face of digital transformation, emerging technologies, top strategic trends, AI capabilities, organizational capabilities for AI implementation by embracing digital mindset and digital ecosystem in 4th IR.

Fourth Industrial Revolution

The Industry 4.0 was firstly conceptualized in 2011 at the conference of Hannover Fair by three German engineers (Henning Kagermann, Wolfgan Wahlster, and Wolf-Dieter Lucas). They introduced ''*The Industry 4.0*'' as social, economic and political programs aim to solve worldwide challenges (e.g., lack of efficiency in working areas, resource and energy scarcity, demographic changes). In order to overcome these challenges, globally integrated networks use Cyber-Physical Systems (CPS). The systems that embedded each other (e.g., wireless microcomputers) communicate through the internet and resulting in integration of physical and virtual world on the basis of CPS. The effect of this event can also be seen in organizational life. CPS comprise smart assistance systems convert routine nature of tasks into more creative and value-added activities. With this new system, new working life models emerge. Flexible working life (e.g., hybrid organizations) allows workers to create balance between their work and private lives (Kagermann, Wahlster, & Helbig, 2013).

By the emergence of 4th Industrial Revolution (IR), individuals' life style, the way they work and relationship experiences have changed. 4th IR was seen as a new chapter in human development in the World Economic Forum in 2020. This new chapter involves technological developments that fits with the first (i.e., inventions of railroads, steam engine), the second (i.e., existence of division of labor, mass production, assemble line), and the third (i.e., inventions of automation in production, information technology, personalized computer, internet) industrial revolutions. 4th IR expands the first three revolutions by adding implantable and wearable technologies (e.g., smart watches), a supercomputer in your pocket (e.g., smartphones), smart cities (e.g., eco-town in Singapore where homes are located in a centralized cooling, automated trash collection and car-free town center), digital currencies (e.g., Bitcoin, Ethereum, Ripple), the sharing economy (e.g., Uber, HomeExchange, Amazon) and 3D printing.

Reconsideration about human has taken a place when 4th IR has brought individuals to a new condition to deal with technological, organizational, political and societal challenges. Furthermore, it also helped how people can cope with the worldwide crises (e.g., COVID-19) in terms of distance education, remote working and the importance of IoT. 4th IR makes easier to connect the devices worldwide through smart devices as well as industrial devices. The number of communication between devices has been increasing rapidly and forecasts (Statista, 2022) argue that by 2025 more than 75 billion IoT connected devices will be in used. These connections will be expected to between the sectors of commercial, air, travel, healthcare, freight and rail transportation, oil and gas (Evans & Annunziata, 2012).

David Stubbs, head of client investment strategy for EMEA at J. P. Morgan Private Bank underlined "technology and specifically digital technology, is so intertwined with many businesses, as well as our social and economic lives, that trying to separate 'tech' from 'non-tech' is becoming increasingly redundant" (Schulze, 2019). With embracing 4th IR, individuals know that machines are becoming intelligent, they are able to see (i.e., sensing), able to move (i.e., actuation), and able to process (i.e., process control). In other words, industry 4.0 combines the physical products with intelligence. It involves autonomous robots, simulation, augmented reality on the basis of artificial intelligence.

Digital Transformation

In today's digital world, a complete change in the usage of data, information, and technologies are needed more than ever. Digital transformation can be seen in the areas of health, education, electronics and automotive. These transformations create new connections between products, customers, companies and systems to make autonomous decisions (Remane, Hanelt, Nickerson, Neville, & Kolbe, 2017). Transformation includes a radical change in its nature and is defined *as a process by which an expression is changed by replacing one set of variables with another or a shape is changed following a particular rule* in Oxford Learner's Dictionary of Academic English. The current definitions of digital transformation were analyzed by Morakanyane, Grace and O'Reilly (2017) through carrying systematic literature review to conceptualize the phenomenon. After analyzing 11 articles they created a table related to digital transformation definitions (Table 1).

As it can be understood from the definitions above, digital transformation cannot be handled without the existence of digital technologies. Digital technologies are in the form of "*products or services that are either embodied in information and communication technologies or enabled by them*" (Lyytinen et al., 2016: 49). Digital technologies have three digital constructs: *digitization, digitalization, digital transformation.* Digitization refers to the process of technical transformation from analog format into digital format (Sandberg et al., 2020). Digitalization states using of digital technologies to create new business models to gain new resources and revenues (Nwaiwu, 2018). Digital transformation refers to providing opportunities for organizations to create new processes, services, products and new goals on the basis of digital technologies (Holmström, 2022). In a similar vein, digital transformation is seen as an ongoing process of improvements in strategies through using advancements in digital technologies to create new capabilities in organization's business model, to embrace collaborative cultural approach (Warner & Wager, 2019). It is also evaluated as a process that aims to develop organizational entity by creating significant changes in its assets through integrations of information, computing, communication and connectivity technologies (Vial, 2019).

Table 1. Current definitions of digital transformation

Authors	Definition
Liu et al., (2011)	''the integration of digital technologies into business processes''
Bharadwaj et al., (2013)	''an organizational strategy formulated and executed by leveraging digital resources to create differential value''
Fitzgerald et.al. (2013);	''the use of digital technologies to enable major business improvements''
Lucas et.al (2013);	''fundamentally altering traditional ways of doing business by redefining business capabilities, processes and relationships''
Mithas et.al (2013);	''the extent to which an organization engages in any activity of IT''
Westerman et.al. (2014);	''the use of technology to radically improve performance or reach of enterprises''
Henriette et.al. (2015);	''a business model driven by the changes associated with the application of digital technology in all aspects of human society''
Piccinini et.al. (2015);	''characterized by the use of new digital technologies to enable significant business improvements''
Schuchmann & Seufert (2015);	''realignment of technology and new business models to more effectively engage digital customers at every touchpoint in the customer experience life cycle''
Chanias & Hess (2016);	''reflect the pervasiveness of changes induced by digital technologies throughout an organization''
Hess et.al. (2016)	''concerned with the changes digital technologies can bring about in a company's business model, which result in changed products or organizational structures or in the automation of processes''

Source: (Morakanyane, Grace, & O'Reilly, 2017: p. 434)

The characteristics of digital transformation can be concluded from the operational definitions above (Table 2).

In order to understand how do companies develop their digital transformation strategies and their digital transformation processes, research or investigations have been made (Matt, Hess & Benlian, 2015; Gimpel, Hosseini, Huber, Probst, Röglinger & Faisst, 2018).

Table 2. Characteristics of digital transformation (author)

Characteristics of Digital Transformation
• Determines core competency in organizations
• Facilitates business improvements
• Creates new ways to reshape business activities
• Serves as a headhunter to improve organizational performance
• Develops a business model for the sake of society
• Creates customer-based improvements in digital markets
• Creates change within organization such as its structure, job description, culture
• Creates new resources and revenues by using digital technologies
• Serves organizational opportunities to enhance management functions by using digital technologies
• Uses advancements in digital technologies to create a learning organization
• Creates significant improvements in organizational entity by using information technology

Matt et al. (2015) conceptualized digital transformation under the four components of *use of technologies*, *changes in value creation*, *structural changes*, and *financial aspects* (Figure 1). The *use of technologies* underlines an organization's perspective toward new technologies and its ability to adopt those technologies. Organization's mission and vision determine its position in market by using strategic role of IT systems. Using its own created technology or already established technology will create core competency that leads to competitive advantages. Specialized marketing capabilities increases performance of a company (Kamasak, 2015; Theodosiou, Kehagias, & Katsikea, 2012). A company who is able to use new technology can have its own *changes in value creation*. In other words, an organization in which new technological changes are embraced can flow from the analog to digital business activities. Further improvements in *value creation* might lead to expand product and service portfolios and to create niche areas in development of goods and services. Organization's ability to use technology and ability to diversify its products and services lead to maintain its financial liquidity (Dyduch, Chudziński, Cyfert & Zastempowski, 2021). The existence of new technology usage and adapting changes in value creation can create need for *structural changes* in organization. Structural changes refer to a new structuring in chain of command, organizational hierarchy, communication ways, the way that work done on the basis of new digital activities. Separate business units, agile organizational forms, and digital functional areas are needed to create flexible organizational structure in digital change (Verhoef, Broekhuizen, Bart, Bhattacharya, Dong, Fabian & Haenlein, 2021). The three dimensions can only be transformed after considering *financial aspects*. Firm's attitudes toward diminished core business activities and its ability to finance digital transformation effort are seen as financial aspects. These financial aspects have two roles as *driver* and *bounding forces* for transformation. Financial viability is required for innovation in industries of manufacturing, services, commerce and education (Rodríguez-Abitia & Bribiesca-Correa, 2021).

Figure 1. Digital transformation framework: balancing four transformational dimensions
(Matt et. al., 2015; p. 341)

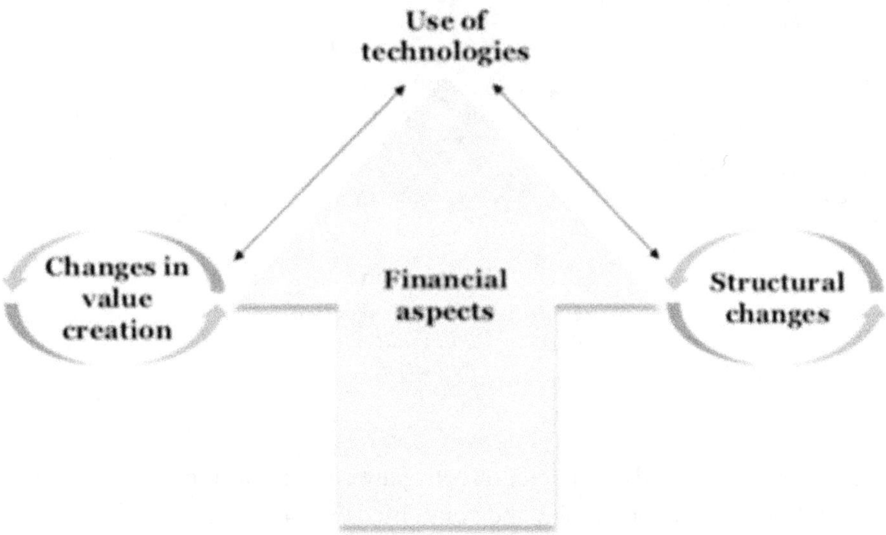

Matt et al. (2015) noted that those four dimensions (i.e., use of technologies, changes in value creation, structural changes and financial aspects) should be taken into consideration to lead successful digital transformation strategy. Companies will be able to assess their current abilities and formulate their digital transformation strategy by embracing this framework.

Another digital transformation framework and its action fields have been conceptualized by Gimpel et al. (2018) on the basis of grounded theory. After making an extant literature review research and conducting interviews with more than 50 companies they proposed a holistic framework that consists of six fields: *customer*, *value proposition*, *operations*, *data*, *organization*, and *transformation management* (Figure 2). Each field helps organizations to structure how to start their digital transformation processes. These initiatives are encouraged by each field's (four) action items that offer a guidance for organizations. The holistic framework was conceptualized under three perspectives as internal, external and from *as-is* to *to-be* state. While *customer* and *value proposition* are considered under external perspective, *operation* and *organization* are evaluated under internal perspective. One of the action fields *data* connects both internal and external perspectives whereas transformation management reflects how to 'get from an as-is to a digitally enhanced to-be state'.

Figure 2. Holistic framework of digital transformation
(Gimpel et al., 2018)

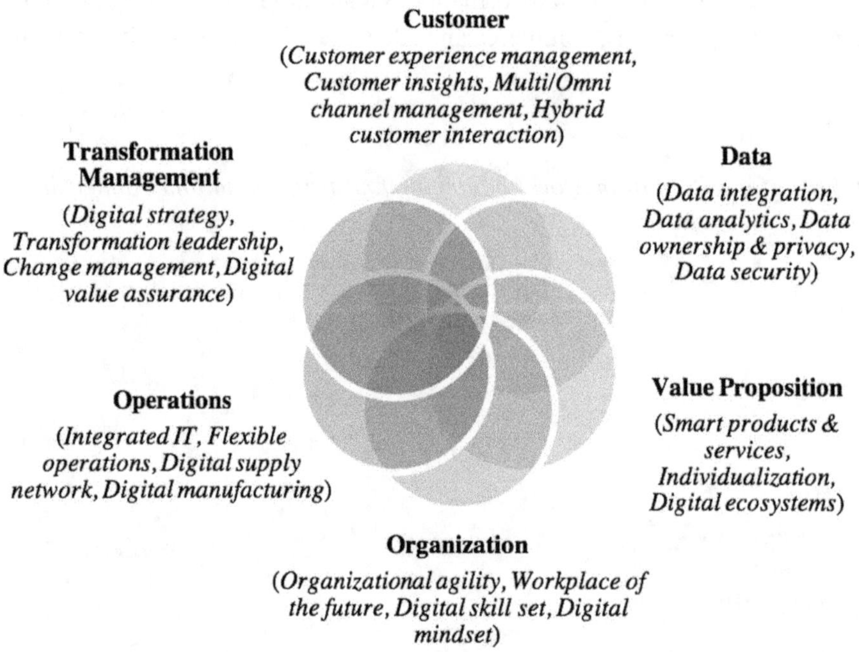

Since digitalization has changed the interaction between customer and products/services, organizations have also been affected by customer access and committed customers. Therefore, *customer experience management* is needed to balance and sustain relationship between customer and producer. Customer's subjective experiences through direct or indirect experiences with the organization determine customers engagement under the states of rational, emotional, sensory, physical and spiritual levels with the organization (Gentile, Spiller, & Noci, 2007). Digitalization facilitates how to gain the information about *customer insights*. They can collect data through digital touch points, smart products and services so that organizations can have various amount of customer information about their preferences, life styles, moods and motivations. In digital transformation process, customers might reach the products or services through traditional or digital channels. This difference is segmented by organization to determine their marketing strategy. Organizations engage in *multi-channel* (i.e., organization offers its products across different channels like their own web-site or online stores) or *omni-channel* (i.e., organization offers its products across integrated marketing channels like mobile push advertising or social media) management activities to meet customer's demand. The existence of digitalization offers *hybrid customer interactions* with the organization via both physical and virtual customer services. AI-supported chatbots or virtual assistances and human-based interactions give organizations detail information about customer needs and decrease the job-demands of support teams and increase innovation activities (Gimpel et al., 2018).

Organizations need customers who can create *value proposition* to strength their digital transformation in a market place. As Gimpel et al. (2018) confirmed competitors survive not because of their existence in the same industry but their digital leader position in the digital economy and ecosystem. So, to create continuing innovation, organizations need to keep in touch with the core competency (i.e., their mastery and capability in technology) of digital leaders and start-ups. *Smart products and services* are preferred by customers because they offer more intelligent solutions and personalized functioning through taking tasks as emotion, social and relational (e.g., empathic products/services) (Raff, Wentzel, & Obwegeser, 2020). In order to add values for costumers, organizations connect with their customers in innovation and design processes to make more *individualized* products and services. So, they have opportunity to get product leadership, operational excellence and customer intimacy (Gassmann et al., 2016). Organizations in which digital transformation is established also adopt *digital ecosystems*. They generally engage in *walled-garden* strategy (e.g., Amazon, Google, eBay, Facebook) where software, hardware and content are only accessible for group of customers (Dapp, 2015). Digital ecosystems offer organizations effective tools to collaborate with other organizations and start-ups.

Since digitally *operation*alized organizations increase, their models of operating, business processes and supply networks change. In manufacturing area, organizations need to produce individualized products at low cost. Therefore, integrated IT, digital operations, supply networks and manufacturing capabilities are required (Gimpel et al., 2018). In order to adopt digital transformation, *integrated IT* where software, hardware, components, systems and networks work in unity provides extensive knowledge with respect to customer's experience about smart product and services. Integrated IT offers platforms to reach various resources, increases resource density and presents actionable resources and information that resulted in value creation (Li & Tuunanen, 2022). A *flexible operation* perspective where individualized products/services are embraced is needed through volume flexibility (i.e., the ability to meet short-term flexible demands) and functional flexibility (i.e., the ability to satisfy distinctive customer needs) (Goyal & Netessine, 2011). *Digital supply networks* are completed by flexible operations. To achieve digital transformation organizations should have flexible/organic organizational structure to connect with IT infrastructure and digital technologies such as IoT, smart devices, blockchain. Digital transformation can also be achieved where design and engineering are met (i.e., digital manufacturing) to increase quality and efficiency of manufacturing. *Digital manufacturing* involves smart productions, cyber-physical systems (CPS) (e.g., intelligent buildings, self-driving cars) and machine-to-machine communication (e.g., wearable technologies, ATM). Digital manufacturing is expected to increase in the numbers of resilient smart factories, to change in employer skills evolutions (i.e., non-technical skills are required), and to affect employers working activities (i.e., transition from work-as-survival mode to life-as-work mode) (Cohen, Faccio, Pilati, & Yao, 2019).

Digital transformation cannot be existed without data sources such as statistics, trend reports, customers' social media usage. In order to benefit from new data resources, organizations need to learn how to analyze, synthase and turn it into valuable assets (*data analytics*). Therefore, all data analytics should be integrated. *Data integration* is a process of combining all heterogeneous information into meaningful and qualified database. In digital economy/transformation area, data are seen as ''new oil'' to predict customer behaviors, future trends and attitude patterns. Advanced data analytics facilitate product developments, faster decision making in organizations and reduction in unit costs. Increase in the numbers of data can also bring difficulty in who own which data. Therefore, *data ownership and privacy* should be determined to have proper market position. Digital footprints can help organizations are data controllers to carefully analyze and to determine real-time needs of customers and employees. Since all data are in cloud and can be reached whenever it is wanted/needed, one of the most important challenges is *data security* in digital transformation. Securing in smart technologies with suitable security controls, risk management techniques in business management processes, increase in conscious customer and regulations like General Data Protection Regulation (GDPR) support cyber systems (Nguyen Duc, Chirumamilla, 2019).

Rapid-changing in digital world and customer needs force organizations to have more digitally oriented structure where unpredictability, diversity and inconsistency outweigh. In order to cope with fast-changing environment, organizations adopt agility. *Organizational agility* refers to be flexible and agile both in process and project management where continuous deployment and integrated developments are achieved through lean organization and design thinking (Gimpel et al., 2018). Organizational agility has positive contribution on digital transformation (AlNuaimi, Singh, Ren, Budhwar & Vorobyev, 2022). To successfully adopt technological advancements, there should be changes in *workplace of the future*. Future workplace in digital transformation era should be relied on project teams where time and location are no longer important. Digitally transformed workplaces imply not only change in working tools but also change in working activities and processes. Therefore, a *digital mindset* will be essential for organizations to gain competitive advantage. Since culture of the organization is determined by its leaders, digital leaders who adopt the values of digital knowledge and literacy, clear vision, understanding of customers, agility, risk taking and collaboration (Promsri, 2019) will likely to create digital mindset in organization.

In the end, all activities related to digital technologies should be turned into digital transformation initiatives. *Digital strategy* (also named as transformative strategy) refers to organizations embrace collaborative culture, open to taking risk, are more interested in business fundamentals and have employees who are able to access to sufficient resources to develop their digital skills and know-how (Kane, Palmer, Philips, Kiron & Buckley, 2015). Digital target and controlled process are required to achieve digital strategy. Digital strategy should be created by *transformation leader* who lead strategy through digital mindset and digital transformation. As Frederick Taylor pointed out ''in the future the system must be first'' has become dominant that *change management* is inevitable in systems of transformation. Digitalization, digital skills and mindset are needed to engage in digitally transformed working activities. Finally, *digital value assurance* refers to organizations' realization about the benefits of digital transformation implications. Manageable portfolio and interdependent projects are components of digital value assurance. It also pushes organizations to analyze and redefine their digital strategies.

Artificial Intelligence

Artificial was defined by Cambridge dictionary as ''made by people, often as a copy of something natural'' and defined by Oxford dictionary as ''made of produced to copy something natural; not real''. Intelligence was also defined as ''the ability to learn, understand, and make judgments or have opinions that are based on reason'' and ''the ability to learn, understand and think in a logical way about things; the ability to do this well'' respectively. From these definitions, it is understood that AI is hard to define because it should be evaluated with respect to combination of the concepts of artificial and intelligence. The aims of AI are to create interaction between humans and machines (Këpuska & Bohouta, 2018) and to teach machines to do better than what humans currently do (Domingos, 2015). Therefore, machines should be able to collect and interpret the data and learn from it through flexible adaptation (Haenlein, Kaplan, Tan, & Zhang, 2019). The interaction between human and machine is their learning capability so that machines can learn with the experiences that humans do. Machine learning was initiated by Frank Rosenblatt through creating an electronic neuron which is able to perceive digits. Machines can learn from the data and turn it into learned programs. There are various names that reflects machine learning as: *pattern recognition*, *statistical modeling*, *data mining*, *knowledge discovery*, *predictive analytics*, *data science*, *adaptive systems*, *self-organizing system* (Domingos, 2015). All of them use computational algorithms that the huge amount of data input and outputs are analyzed to teach the machine to offer autonomous suggestions. After the adequate input-output interpretations, machines are able to use an input to predict an output. In order to predict future outcomes, machines compare the results of previous outcomes with the current one to analyze the accuracy of algorithm (Haeberle, Helm, & Navarro, Karnuta, Schaffer, Callaghan, Mont, Kamath, Krebs, & Ramrukar, 2019). For example, *IBM'S Blue Match* software program was designed to use algorithms in order to find the expected congruency between person and job on the basis of employees' interests, their previous job experiences, training and their personality. Another software program that is supported by AI called *CogniPay* was designed to help employers to make fairer compensation decisions by considering employee performance, the average payment for similar jobs in the same industry, and demanded skills and abilities for similar jobs. One of the AI-supported machine learnings is deep learning that has neural networks with non-linear algorithms. Those neural networks are composed of neurons (i.e., connected parts) that are able to transfer a signal to other neurons. So, they can make decisions like human by using interconnected brain cells (Hannun, Rajpurkar, Haghpanahi, Tison, Bourn, Turakhia, & Ng, 2019). While in machine learning the relationship between input and output is mediated by feature extraction and classification separately, in deep learning they are considered as connected units to predict the relationship between input and output. In a more horizontal way, AI is considered as artificial general intelligence (AGI) that it has human-like intelligence, has understanding and learning capacities at all areas of analytical, emotional and social intelligence (Kaplan & Haenlein, 2019).

It can be concluded that artificial intelligence contains machine learning, deep learning and artificial general intelligence (Figure 3).

The current artificial intelligence types are dominated by deep and machine learnings. In recent years humanity have witnessed developments in deep learning activities as image classification, medical image diagnosis via deep convolutional neural networks (Esteva, et al., 2017) and *AlphaGo*'s AI algorithms that uses deep neural networks to win world champion (Lee Sedol) in the game of Go (Silver et al., 2017). AlphaGo's complex moves/algorithms have questioned Lee Sedol after his loss as ''It made me question human creativity. When I saw AlphaGo's moves, I wondered whether the Go moves I have known were the right ones."

4IR has brought notable developments in AI and its creative applications. Machines with human-like brains start to affect how technology and digital transformation change humanity and humans living with their creative solutions.

Figure 3. The relationship between AI and its sub-dimensions

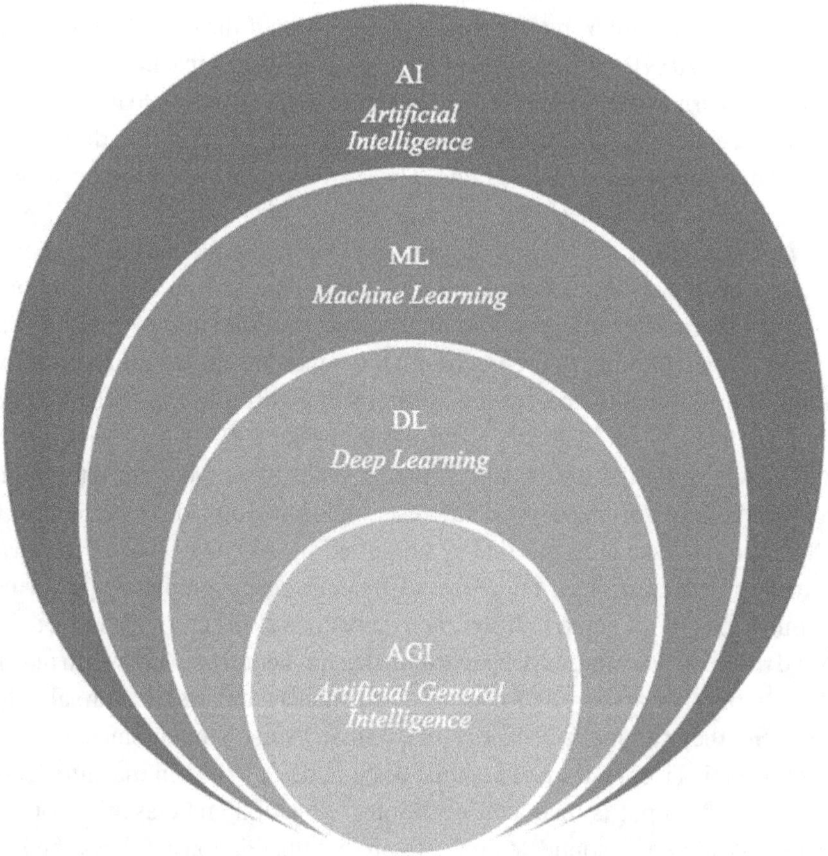

AI Creativity

Creativity was defined in Cambridge Dictionary as ''the ability to produce or use original and unusual ideas'' in Oxford Dictionary as ''the use of skill and imagination to produce something new or to produce art''. Creativity can take a place in various contexts (e.g., art or psychological). Therefore, it is created under specific conditions, and environment is crucial determinant in creation. Amabile (1983, p. 358) defined creativity as ''behavior resulting from particular constellations of personal characteristics, cognitive abilities, and social environments''. Plucker, Jonathan, Ronald, Beghetto and Gayle (2004) also defined creativity as ''the interaction among aptitude, process, and environment by which an individual or group produces a perceptible product that is both novel and useful as defined within a social context'' (p. 90). Both of the researchers pointed out the importance of personal characteristics, skills and environmental conditions and creation's novelty and usefulness in the process.

Figure 4. 4C's of creativity
(Kaufman & Glăveanu, 2022)

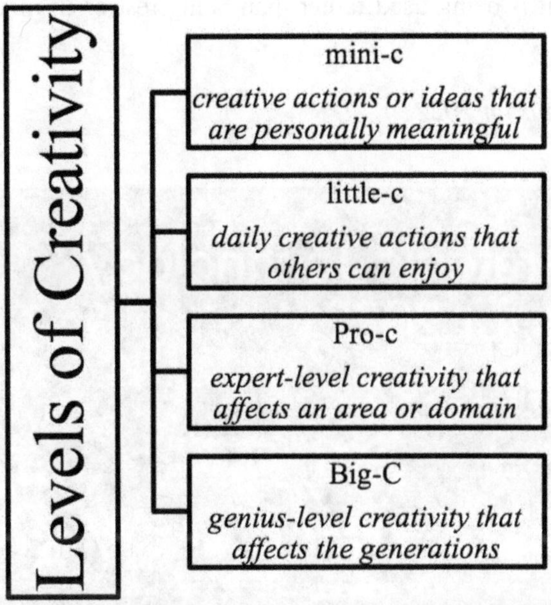

The usefulness and meaningfulness of the creation process are identified by Kaufman and Glăveanu (2022) under the 4C's of creativity. They conceptualized creativity within four dimensions as mini-c, little-c, Pro-c and Big-C. Mini-c refers to having meaningful creative actions/ideas (e.g., a child who makes cookies for the first time). Little-c states enjoyable creative action/ideas on daily basis (e.g., (a capella) singing without any musical instruments). Pro-c underlines competent type of creative action/ideas affect certain areas/domain (e.g., the first software developed by Dragon Systems that recognizes the speech in 1997). Big-C points out the genius type of creative action/ideas affect the whole population in a positive way (e.g., the winner of Nobel prize in physiology Svante Pääbo for his discoveries in the genomes of extinct hominins and human evolution). Kaufman and Glăveanu (2022)'s creativity conceptualization highlights the nuances of creative ideas: novelty and usefulness. Myers and McCaulley (1985) found out that creative people usually use their intuition. Simon (1992) defined intuition as ''nothing more nor nothing less than recognition'' (p. 195). In other words, Simon underlined the importance of past experiences in shaping intuitive behaviors. Hammond (1996) expanded Simon's definition by seeing the power of intuition to use imagination and its contribution to creativity. Einstein stated the process of creative breakthroughs as: ''I believe in intuition and inspiration. Imagination is more important than knowledge. For knowledge is limited, whereas imagination embraces the entire world, stimulating progress, giving birth to evolution. It is strictly speaking a real factor in scientific research''.

In 4th IR the concept of creativity has been changed *where smart machines and new technologies make their own creative decision by eliminating human interaction.*

Gartner (2022) identified 12 strategic technology trends for 2022 that are crucial for businesses to have strategic position in the market by considering growth, digitalization and efficiency (See Figure 5). Through using *data fabric*, it will be easier to reach available data on the basis of flexible and resilient sources. *Cybersecurity mesh* is a scalable, flexible and reliable architecture that protects data by quickly confirming identity across cloud and non-cloud contexts. *Privacy-enhancing computation* techniques aim to protect data while it is being used rather than being rest or in motion. *Cloud-native platform*

Figure 5. Top strategic trends for 2022
(Gartner, 2022)

technologies allow developer to create new application architectures characterized as resilient, elastic and agile to adapt changing digital environment rapidly. *Composable applications* let owners to use and reuse data within a more business-centric approach. They also offer new software solutions and create enterprise value. *Decision intelligence* is a prototype of organizational decision making by modelling its process, using intelligence and analytics to notify, and learning from its consequences then refining decisions. Decision intelligence covers augmented reality, simulations and AI. *Hyperautomation* combines robotic process automation (RPA), AI and machine learning to enhance business decisions. *AI engineering* improves the data, models and applications to enable AI delivery. The management of AI is vital for AI engineering to continue its effectivity. *Distributed enterprises* involve the first digital and remote business models to enhance employee experiences, have more digitalized consumer and analyze the remote employee's needs in hybrid organizations. *Total experience* is a business strategy that refers to the integration experiences of the employees, customers and users to increase growth, satisfaction and trust. *Autonomic systems* are physical or software systems that are free from human interaction (i.e., self-managed) and are able to learn from their environment and make modification in their algorithms to adapt their behavior in more complex environments. Generative AI is a technology that creates new plausible content by using existing video, text or image.

In 2022, technologies that lead to competitive advantage for companies and help leaders to analyze mega-trends and their potential impacts of business activities can be seen from The Gartner hype cycle (See Figure 6).

Figure 6. Hype cycle for emerging tech 2022
(Gartner, 2022)

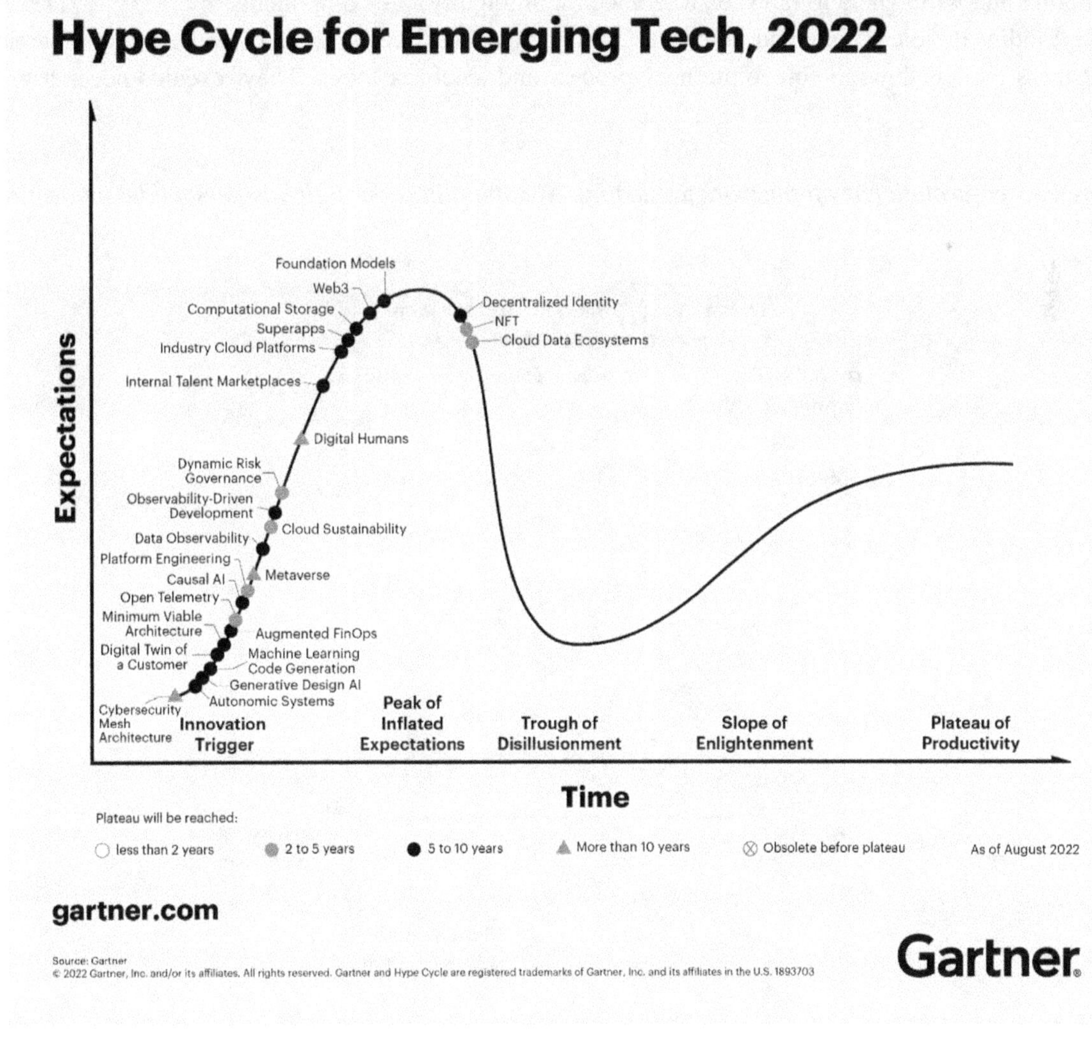

As it can be seen from the figure 6, most of the emerging technologies are in innovation trigger stage in which uncertainties and higher risks are embraced. While reaching of most of the technologies such as machine learning, causal AI, platform engineering ranges from 2 to 5 years to 5 to 10 years; reaching of cybersecurity mesh architecture, metaverse and digital humans will be expected in more than 10 years.

Challenges and Success Factors in AI Creativity Implementation

In literature, AI creativity and its implementations have been handled differently. Mikalef and Gupta (2021) taken AI capability as a pre-requirement of AI implementation and defined as ‘‘the ability of a firm to select, orchestrate, and leverage its AI-specific resources’’ (p. 2). Mikalef and Gupta (2021) conceptualized AI capability and found out positive relationship between AI capability and organizational creativity on the basis of resource-based theory. They identified AI capability under three sub-dimensions as tangible (e.g., data, technology, basic resources), human (e.g., technical skills and business skills) and intangible (e.g., inter-departmental coordination, organizational change capacity and risk proclivity). Organizational creativity had the highest correlation with risk proclivity. In other words, intellectual risk-taking and willingness to fail (i.e., tolerance for ambiguity) are components of creativity (Dewett, 2007). Sjödin, Parida, Palmié and Wincent (2021) analyzed AI capabilities to understand innovation in business models through co-evolutionary process and feedback loops. They revealed necessary AI

Figure 7. A co-evolutionary framework for scaling AI capabilities through business model innovation (Sjödin et al., 2021; p.583)

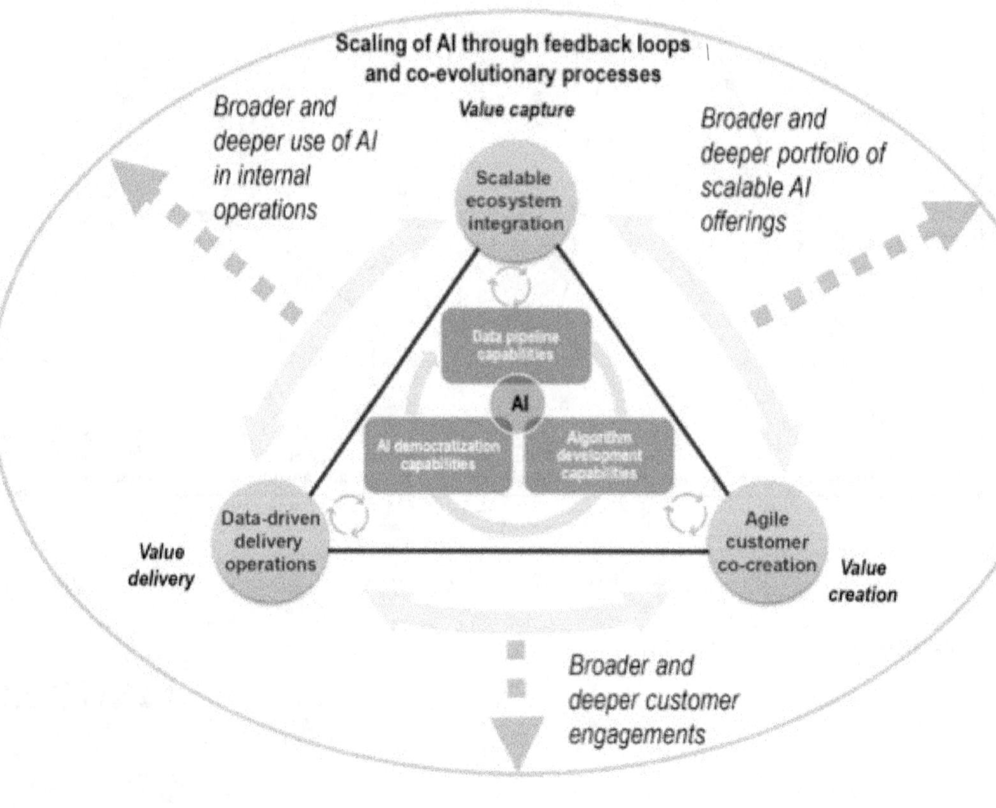

capabilities and their scaling process. *Data pipeline (e.g., secure data sharing), algorithm development (e.g., contextualized AI development)* and *AI democratization (e.g., democratizing AI affordances)* are seen as critical AI capabilities in development of innovative business models. *Agile customer co-creation, data-driven delivery operations*, and *scalable ecosystem integration* are seen as AI-driven business model innovation principles (See Figure 7).

AI implementation is related to characteristics of AI such as perceiving, learning, reasoning, decision making and showing creativity (Rai, Constantinides, & Sarker, 2019). On the basis of organization point of view, Weber, Engert, Schaffer, Weking and Krcmar (2022) highlighted specific organization capabilities for AI implementation. Their framework explains the implementer power of AI characteristics (i.e., inscrutability and data dependency) on challenges (i.e., uncertain AI project environment, lack of information flow among stakeholders during AI project, high data dependency for AI projects and continuous adaptation among input data and AI projects). Organizational capabilities address challenges with respect to AI project planning, co-development of AI systems, data management and AI model lifecycle management (See Figure 8).

Figure 8. Explanatory framework of organizational capabilities for AI implementation
(Weber et al., 2022)

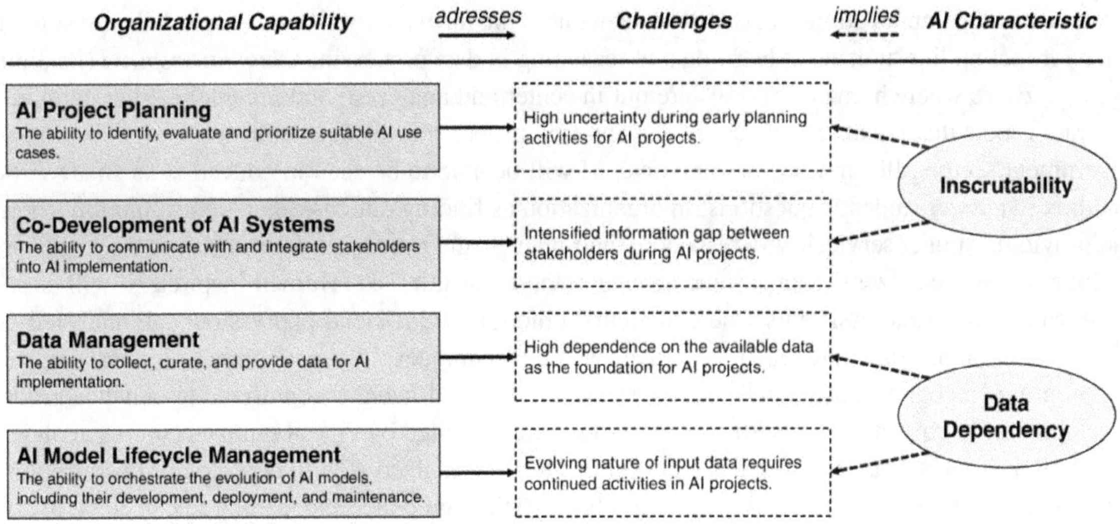

Weber et al. (2022) suggest that organizations need to imply their AI project by moving from experimental stage to more structured planning and executing stage. To manage AI models, organizations should keep in touch with their users and developers through feedback channel. Since the natures of AI are inscrutability and data dependency, collaborative development and data management are required for AI implementation. In addition, need congruency between AI projects and AI systems should be achieved throughout their lifecycle.

Internal environment of an organization can lead to create challenges in implementing AI such as ambiguous strategic goals, organization's non-internalized shared value, top management' attitude toward AI. Böttcher, Weber, Weking, Hein and Krcmar (2022) identified success factors of AI implementation under six value drivers of AI: efficiency, novelty, knowledge from data, ecosystem, personalization and human resemblance. AI makes positive contribution on *efficiency* by replacing manual human work. AI eases to have various type of *novel* products, services and business models. AI offers *knowledge from data* as creating new insights and making predictions. *Ecosystem* where added value for AI is achieved through interaction between stakeholders (e.g., customers, suppliers, partners) and company is required. AI can be used to segment customers and *personalize* their data to reach them in a more effective way. AI creates solutions (e.g., virtual assistants, smart-home systems) for individuals with *human-like* behaviors.

Challenges in implementing AI projects are considered as risky activities (Bughin et al., 2017; Wodecki, Wodecki, & Harrison, 2019) and some of the vital stages are clarified by Wodecki et al. (2019) to construct value by using AI systems. Detailed criteria compose of five steps as area (i.e., human resource, infrastructure, activity groups, business processes), value (i.e., atomization, optimization, speed, security, and profit), effect (i.e., way of acting, time, prediction, place, logistics, quality, customer, supplier, risk), operation (i.e, level of autonomy, analysis, desired timing, desired quality) and tools (methods, hardware, software, talents). These stages have interaction with each other and create AI value with regard to effective usage of intelligent systems.

IMPLICATIONS

In the near future, applications of AI will be seen under the forms of *analytical* (i.e., intelligent software behaviors acting like humans as learning and reasoning in data processing), *human-inspired* (i.e., intelligent software where human emotions are put in center and analyzed) and *humanized* (i.e., intelligent software robots that are able to detect issues and find proper solutions) in education, organization and government sectors. Illustrations of analytical AI will be able to be seen in education as smart virtual chat box to answer student's questions; in organization as finding out customers' consumption triggers and providing similar services with respect to their interest and needs; in government, as smart personal health systems use citizens' data to forecast their potential health risks. Human-inspired AI will be seen in education as virtual assistants where students' emotions from facial expressions are analyzed and their non-verbal messages are read to improve teaching techniques; in organizations, as smart engineering programs recognize customers' tone of voice, mimics and facial recognitions to enhance product development; in government, health care employees are recruited by virtual human resource analyst by detecting their emotional cues to select right candidates. Humanized AI will take a place in education as smart robot teachers let brainstorming activities through their guidance and leading role in class; in organizations, as virtual agents pay close attention to customers' demand and determine service environment based on customers' expectations; in government, as smart AI systems that are able to psychologically empower cancer patients before and during treatment process.

As it can be concluded from those applications of AI above, digital transformation under the forms of using of technologies, changes in value creation and structural changes will bring dramatic evolutions. These days customers' working conditions and their buying tendency behaviors are changing rapidly. They adopt both hybrid and virtual working environments and in their personal life also embrace various fashion preferences. Another change in digital technologies is customers' vital role in product and service development and their role in creating new value in the market. So, organizations need to find new business models and even structures. For example, Alibaba has converted its online good seller's role for Chinese consumers to financial consultancy role for unbanked consumers by offering digital payments. Flexibility and agility in new business model exploration and adoption are required in digital era. It can also be concluded that applications of AI technologies and digital transformations ease organization to enter new industries by reducing barriers.

In other words, factors of production such as land, labor, capital are no longer barriers for organizations because of new technology offerings as cloud computing or restrictions in legal regulations. In order to embrace digital transformation, it is almost a must for organizations building new capabilities. They have to strengthen their adaptability muscle by choosing effective strategies. One of the effective strategies is keeping in touch with customers by changing old assumptions and mindscapes. Focusing on right customer segment can be achieved by using digital tools such as cloud data. Competition is no longer seen as trying to win or to be more successful than someone else by organizations in a new digital era. In this world, competitors like Google and Apple engages in collaborating activities. Google make a trade agreement with Apple to be a default search engine on iOS. These companies also reshape their thinking approach in innovation by giving more importance to experimentation and learning on the job (innovation) process.

Applications of artificially intelligent technologies should bring crucial questions to answer for organizations in value creation. They need to know their aim of existence, to have need analysis in the marketplace that rapid changes in technology serve, to shift in old mindset of ''what actions should we adopt based on a new digital transformation environment?'' to ''what does this change bring to a new organization model and how can we best fulfill new customer needs that we could not done before?''.

FUTURE RESEARCH DIRECTIONS

Artificial intelligence and its development phases have demonstrated that a machine can do lots of things with someone else's memories. Though individuals' memories can be deteriorated in time, machines can memorize, learn and create something new by using human-made data. Digitalization and technological developments will impact organizations and countries in various ways. Successful integration and adaptation of strategic and emerging technology trends (e.g., data fabric, cloud-native platforms, generative AI, decision intelligence) will likely to be resulted in increase in sales, reduction in unit costs, increase in profits in business life cycle. Therefore, selection of appropriate trends will ease to identify impactful short-term and long-term organizational plan. In the long run, it is expected that humanity will be able to use AI in every business aspects and it will be inevitable to embrace sustainable business activities to survive in uncertain and complex business environment. To identify and adopt required AI activities, human skills (e.g., technical and business skills) are needed. Those skills can be identified as capability of using AI programs, matching among job requirements and data scientists' skills, sufficient work experience to analyze the problem and offer solutions for AI applications, being competent to be able to detect niche areas and use AI initiatives to place in a related market, digital leadership skills, creating an organizational culture where AI-related activities are embraced by organizational environment (e.g., employees, suppliers, customers). Therefore, technical and business-oriented skills are essential to compete in digital business environment. One of the most important skills is digital leadership characteristics that trends and technologies are analyzed whether they are opportunity or threat for an organization. They should be able to read digital environment and connect their knowledge (in some cases their initiatives) of business and technology to interpret the technological trends and their effect on organization.

Though AI and its other forms work beyond human imagination, the other side of the coin says humanity can only be alive if they can remain incomplete, imperfect, create intimacy, be authentic and belong. Organizations where AI activities are adopted should consider the needs of individuals on the basis of relatedness, belongingness, self-esteem, self-actualization, and self-worth. Therefore, imperfect organizations rather than automatized one will create fulfilled employees and leaders in which appreciation, motivation, and empowerment are internalized.

CONCLUSION

In 21st century, the meaning of AI has changed in a more creative way. Smart systems have started to find creative solutions for complex tasks. In order to use intelligent systems digital transformation has taken a place under the forms of effective technology use, changes in value creation and structure, and viable financial resources. In a more integrated way; customer experience management, change management, data security/integration, digital operations, smart systems and digitally-developed organizations have been adopted. Machine learning and deep learning are the crucial components of AI since they help frequently used systems (e.g., Netflix, Twitter, Facebook, Amazon.com) to offer customized services or products. It can be concluded that if a machine can learn it can also be creative as well. Creative actions can be meaningful, enjoyable and be based on expert-oriented and genius-oriented actions (Kaufman & Glăveanu, 2022). Machine learning is also formed as pattern recognition, predictive analytics and adaptive systems (Domingos, 2015). Therefore, technically AI can have creative actions. AI creativity underlines creative decisions made by machines without human interaction. Those decisions can be made by using technology in a strategic way. In 2022, generative AI, data fabric and composable applications are determined as top strategic technology trends (Gartner, 2022). Generative AI is developed form of machine learning technique that create new and novel patterns by operating system in the background while user is working. Data fabric offers flexible and integrated solutions for users while keeping centralized organizational control. Composable applications are required especially in ambiguous environment where conditions are unexpected, uncertain and risky (e.g., pandemic, recession). Organizations that are able to adopt those trends effectively will reach fulfillment of expectations according to The Gartner hype cycle.

All of the creative AI solutions can only be achieved when AI is implemented successfully. AI implementation success depends on AI capability range. In other words, AI capabilities (tangible and intangible resources) contribute positively on AI implementation and this relationship leads to organizational creativity. Creative organizations are the ones that can produce many novel and useful ideas, foster an environment to conduct creative culture, spend adequate time to produce novel and useful ideas, value producing novel and useful ideas as vital activities (Mikalef & Gupta, 2021). AI implementation is also related to organizational capabilities. Since AI's nature involves uncertainty and data dependency; the ability to identify and prioritize AI cases, the ability to communicate with and integrate stakeholders into AI implementation, the ability to collect, and provide data for AI implementation, and the ability to analyze changes in AI models, their developments, deployment and maintenance to define challenges and success factors are vital requirements for digitalized organizations (Weber et al., 2022).

REFERENCES

AlNuaimi, B. K., Singh, S. K., Ren, S., Budhwar, P., & Vorobyev, D. (2022). Mastering digital transformation: The nexus between leadership, agility, and digital strategy. *Journal of Business Research, 145*, 636–648. doi:10.1016/j.jbusres.2022.03.038

Amabile, T. M. (1983). *The social psychology of creativity*. New York: Spring-Verlag. doi:10.1007/978-1-4612-5533-8

Bharadwaj, A., Sawy, O. A., Pavlou, P. A., & Venkatraman, N. (2013). Digital Business Strategy: Toward A Next Generation of Insights. Management Information Systems Quarterly, 37(2), 471–482. doi:10.25300/MISQ/2013/37:2.3

Böttcher, T. P., Weber, M., Weking, J., Hein, A., & Krcmar, H. (2022). *Value Drivers of Artificial Intelligence. 28th Americas Conference on Information Systems (AMCIS)*, Minneapolis, USA

Brock, J. K. U., & Von Wangenheim, F. (2019). Demystifying AI: What digital transformation leaders can teach you about realistic artificial intelligence. *California Management Review, 61*(4), 110–134. doi:10.1177/1536504219865226

Bughin, J., Hazan, E., Ramaswamy, S., Chui, M., Allas, T., & Dahlstrom, P. (2017). *Artificial Intelligence*. McKinsey Global Institute. https://www.mckinsey.com/~/media/McKinsey/Industries/Advanced%20Electronics/Our%20Insights/How%20artificial%20intelligence%20can%20deliver%20real%20value%20to%20companies/MGI-Artificial-Intelligence-Discussion-paper.ashx

Case, S. (2016). *The third wave*. Simon & Schuster.

ChaniasS.HessT., (2016Understanding Digital Transformation Strategy Formation. Insights from Europe's Automotive Industry. PACIS. .

Cohen, Y., Faccio, M., Pilati, F., & Yao, X. (2019). Design and management of digital manufacturing and assembly systems in the Industry 4.0 era. *International Journal of Advanced Manufacturing Technology, 105*(9), 3565–3577. doi:10.100700170-019-04595-0

Dapp, T. H. (2015). *Fintech reloaded- Traditional banks as digital ecosystems: With proven walled garden strategies into the future*. DB Research. https://www.dbresearch.com/PROD/RPS_EN-PROD/PROD0000000000451937/Fintech_reloaded_–_Traditional_banks_as_digital_ec.pdf;REWEBJSESSIONID=6644FE271301B11261BF9B52FFA110E3?undefined&realload=TLdLYj1x74399cki7iNULjWEIPlVAayk27uTriyVxZ3mYT0FQ7aed4JHqT7gpdAF

Dewett, T. (2007). Linking intrinsic motivation, risk taking, and employee creativity in an R&D environment. *R & D Management, 37*(3), 197–208. doi:10.1111/j.1467-9310.2007.00469.x

Domingos, P. (2015). *The Master Algorithm: How the Quest for the Ultimate Learning Machinewill Remake the World*. Basic Books.

Dyduch, W., Chudziński, P., Cyfert, S., & Zastempowski, M. (2021). Dynamic capabilities, value creation and value capture: Evidence from SMEs under Covid-19 lockdown in Poland. *PLoS One, 16*(6), e0252423. doi:10.1371/journal.pone.0252423 PMID:34129597

Esteva, A., Kuprel, B., Novoa, R. A., Ko, J., Swetter, S. M., & Balu, H. M. (2017). Dermatologist-level classification of skin cancer with deep neural networks. *Nature, 542*(7639), 115–118. doi:10.1038/nature21056 PMID:28117445

Evans, P. C., & Annunziata, M. (2012). *Industrial Internet: Pushing the Boundaries of Minds and Machines*. GE. https://www.ge.com/news/sites/default/files/5901.pdf

Fitzgerald, M., Kruschwitz, N., Bonnet, D., & Welch, M. (2013). Embracing Digital Technology: A New Strategic Imperative. *MIT Sloan Management Review*, *55*(2), 1.

Gartner. (2022). *Gartner Identifies the Top 10 Strategic Technology Trends for 2022*. Gartner. https://www.gartner.com/en/newsroom/press-releases/2021-10-18-gartner-identifies-the-top-strategic-technology-trends-for-2022

Gassmann, O., Frankenberger, K., & Sauer, R. (2016). *Exploring the field of business model innovation: New theoretical perspectives*. Springer. doi:10.1007/978-3-319-41144-6

Gentile, C., Spiller, N., & Noci, G. (2007). How to Sustain the Customer Experience: An Overview of Experience Components That Co-create Value with the Customer. *European Management Journal*, *25*(5), 395–410. doi:10.1016/j.emj.2007.08.005

Gimpel, H., Hosseini, S., Huber, R., Probst, L., Röglinger, M., & Faisst, U. (2018). Structuring Digital Transformation: A Framework of Action Fields and its Application at ZEISS. *Journal of Information Technology Theory and Application*, *19*(1), 31–54.

Goyal, M., & Netessine, S. (2011). Volume flexibility, product flexibility, or both: The role of demand correlation and product substitution. *Manufacturing & Service Operations Management*, *13*(2), 180–193. doi:10.1287/msom.1100.0311

Haeberle, H. S., Helm, J. M., Navarro, S. M., Karnuta, J. M., Schaffer, J. L., Callaghan, J. J., Mont, M. A., Kamath, A. F., Krebs, V. E., & Ramkumar, P. N. (2019). Artificial intelligence and machine learning in lower extremity arthroplasty: A review. *The Journal of Arthroplasty*, *34*(10), 2201–2203. doi:10.1016/j.arth.2019.05.055 PMID:31253449

Haenlein, M., Kaplan, A., Tan, C. W., & Zhang, P. (2019). Artificial intelligence (AI) and management analytics. *Journal of Management Analytics*, *6*(4), 341–343. doi:10.1080/23270012.2019.1699876

Hammond, K. R. (2000). *Human judgment and social policy: Irreducible uncertainty, inevitable error, unavoidable injustice*. Oxford University Press on Demand.

Hannun, A. Y., Rajpurkar, P., Haghpanahi, M., Tison, G. H., Bourn, C., Turakhia, M. P., & Ng, A. Y. (2019). Cardiologist-level arrhythmia detection and classification in ambulatory electrocardiograms using a deep neural network. *Nature Medicine*, *25*(1), 65–69. doi:10.103841591-018-0268-3 PMID:30617320

Henriette, E., Feki, M., & Boughzala, I. (2015). The Shape of Digital Transformation: A Systematic Literature Review. *MCIS 2015 Proceedings*. IEEE.

Hess, T., Matt, C., Benlian, A., & Wiesböck, F. (2016). Options for Formulating a Digital Transformation Strategy. *MIS Quarterly Executive*, *15*(2), 123–139.

Holmström, J. (2022). From AI to digital transformation: The AI readiness framework. *Business Horizons, 65*(3), 329–339. doi:10.1016/j.bushor.2021.03.006

Kagermann, H., Wahlster, W., & Helbig, J. (2013). Acatech–National academy of science and engineering. *Recommendations for implementing the strategic initiative INDUSTRIE, 4.*

Kamasak, R. (2015). How Marketing Capabilities Create Competitive Advantage in Turkey. In Marketing and Consumer Behavior: Concepts, Methodologies, Tools, and Applications (pp. 1602-1621). IGI Global. doi:10.4018/978-1-4666-7357-1.ch079

Kane, G. C., Palmer, D., Phillips, A. N., Kiron, D., & Buckley, N. (2015). Strategy, not technology, drives digital transformation. *MIT Sloan Management Review and Deloitte University Press, 14*, 1–25.

Kaplan, A., & Haenlein, M. (2019). Siri, Siri, in My Hand: Who's the Fairest in the Land? On the Interpretations, Illustrations, and Implications of Artificial Intelligence. *Business Horizons, 62*(1), 15–25. doi:10.1016/j.bushor.2018.08.004

Kaufman, J. C., & Glăveanu, V. (2022). Positive Creativity in a Negative World. *Education Sciences, 12*(3), 193. doi:10.3390/educsci12030193

Këpuska, V., & Bohouta, G. (2018, January). Next-generation of virtual personal assistants (microsoft cortana, apple siri, amazon alexa and google home). In *2018 IEEE 8th annual computing and communication workshop and conference (CCWC)* (pp. 99-103). IEEE.

Li, M., & Tuunanen, T. (2022). Information Technology-Supported value Co-Creation and Co-Destruction via social interaction and resource integration in service systems. *The Journal of Strategic Information Systems, 31*(2), 101719. doi:10.1016/j.jsis.2022.101719

Liu, D.-Y., Chen, S.-W., & Chou, T.-C. (2011). Resource fit in digital transformation: Lessons learned from the CBC Bank global e-banking project. *Management Decision, 49*(10), 1728–1742. doi:10.1108/00251741111183852

LucasH. C.AgarwalR.ClemonE. K.SawyO. A.WeberB. (2013)

Lucas, H. C. Jr, Agarwal, R., Clemons, E. K., El Sawy, O. A., & Weber, B.Impact Research on Transformational Information Technology. (2013, February 2). An Opportunity to Inform New Audiences. *Management Information Systems Quarterly, 37*(2), 371–382. doi:10.25300/MISQ/2013/37.2.03

Lyytinen, K., Yoo, Y., & Boland, R. J. Jr. (2016). Digital Product Innovation within Four Classes of Innovation Networks. *Information Systems Journal, 26*(1), 47–75. doi:10.1111/isj.12093

Matt, C., Hess, T., & Benlian, A. (2015). Digital transformation strategies. *Business & Information Systems Engineering, 57*(5), 339–343. doi:10.100712599-015-0401-5

Mikalef, P., & Gupta, M. (2021). Artificial intelligence capability: Conceptualization, measurement calibration, and empirical study on its impact on organizational creativity and firm performance. *Information & Management, 58*(3), 103434. doi:10.1016/j.im.2021.103434

MithasS.TaftiA.MitchellW. (2013)

Mithas, S., Tafti, A., & Mitchell, W. (2013, February 2). How a Firm's Competitive Environment and Digital Strategic Posture Influence Digital Business Strategy. *Management Information Systems Quarterly*, *37*(2), 511–536. doi:10.25300/MISQ/2013/37.2.09

Morakanyane, R., Grace, A., & O'Reilly, P. (2017). Conceptualizing Digital Transformation in Business Organizations: A Systematic Review of Literature. In *Digital Transformation – From Connecting Things to Transforming Our Lives* (pp. 427–443). University of Maribor Press., doi:10.18690/978-961-286-043-1.30

Myers, I. B., McCaulley, M. H., & Most, R. (1985). *Manual, a guide to the development and use of the Myers-Briggs type indicator*. consulting psychologists press.

Nguyen Duc, A., & Chirumamilla, A. (2019, September). Identifying security risks of digital transformation-an engineering perspective. In *Conference on e-Business, e-Services and e-Society* (pp. 677-688). Springer, Cham. 10.1007/978-3-030-29374-1_55

Nwaiwu, F. (2018). Review and Comparison of Conceptual Frameworks on Digital Business Transformation. *Journal of Competitiveness*, *10*(3), 86–100. doi:10.7441/joc.2018.03.06

Piccinini, E., Gregory, R. W., & Kolbe, L. M. (2015). *Changes in the producer-consumer relationship-towards digital transformation.*

Plucker, J. A., Beghetto, R. A., & Dow, G. T. (2004). Why isn't creativity more important to educational psychologists? Potentials, pitfalls, and future directions in creativity research. *Educational Psychologist*, *39*(2), 83–96. doi:10.120715326985ep3902_1

Promsri, C. (2019). The developing model of digital leadership for a successful digital transformation. *GPH-International Journal of Business Management (IJBM)*, *2*(08), 01-08.

Raff, S., Wentzel, D., & Obwegeser, N. (2020). Smart Products: Conceptual Review, Synthesis, and Research Directions. *Journal of Product Innovation Management*, *37*(5), 379–404. doi:10.1111/jpim.12544

Rai, A., Constantinides, P., & Sarker, S. (2019). Next generation digital platforms: Toward human-ai hybrids. *Management Information Systems Quarterly*, *43*(1), iii–ix.

Remane, G., Hanelt, A., Nickerson, R. C., & Kolbe, L. M. (2017). Discovering digital business models in traditional industries. *The Journal of Business Strategy*, *38*(2), 41–51. doi:10.1108/JBS-10-2016-0127

Rodríguez-Abitia, G., & Bribiesca-Correa, G. (2021). Assessing digital transformation in universities. *Future Internet*, *13*(2), 52. doi:10.3390/fi13020052

Schreckling, E., & Steiger, C. (2017). *Digitalize or drown. Shaping the digital enterprise: Trends and use cases in digital innovation and transformation*, 3-27.

Silver, D., Schrittwieser, J., Simonyan, K., Antonoglou, I., Huang, A., Guez, A., ... & Hassabis, D. (2017). Mastering the game of go without human knowledge. *nature, 550*(7676), 354-359.

Simon, H. A. (1992). What is an "explanation" of behavior? *Psychological Science*, *3*(3), 150–161. doi:10.1111/j.1467-9280.1992.tb00017.x

Sjödin, D., Parida, V., Palmié, M., & Wincent, J. (2021). How AI capabilities enable business model innovation: Scaling AI through co-evolutionary processes and feedback loops. *Journal of Business Research, 134*, 574–587. doi:10.1016/j.jbusres.2021.05.009

Taddy, M. (2018). The technological elements of artificial intelligence. In *The economics of artificial intelligence: An agenda* (pp. 61–87). University of Chicago Press.

Weber, M., Engert, M., Schaffer, N., Weking, J., & Krcmar, H. (2022). Organizational Capabilities for AI Implementation—Coping with Inscrutability and Data Dependency in AI. *Information Systems Frontiers*, 1–21. doi:10.100710796-022-10297-y

Wodecki, A., Wodecki, H., & Harrison. (2019). *Artificial intelligence in value creation.* Springer International Publishing.

KEY TERMS AND DEFINITIONS

Artificial Intelligence: A flexible system in which data are analyzed and synthesized correctly and converted into new knowledge to complete goals.

Artificial Intelligence Creativity: The phenomenon that creative decisions are made by smart machines without human interaction.

Artificial Intelligence Implementation: A firm that is able to choose AI-related resources and manage them in efficient and productive way.

Digital Ecosystem: A firm where smart products/services and customized value systems are embraced.

Digital Transformation: Changes in digital technologies that resulted in developments in business models so that new products and organizational structures are adopted.

Deep Learning: A system where relationship among input and output can be established through combinations of feature extraction and classification.

Machine Learning: A system where relationship among input and output can be established through feature extraction and classification distinctly.

Chapter 6
Artificial Intelligence Trends and Perceptions:
Content Strategy and the Customer Journey

Tariro S. Munyengeterwa
East Tennessee State University, USA

Melanie B. Richards
https://orcid.org/0000-0001-8452-2153
East Tennessee State University, USA

Joel B. Eaton
East Tennessee State University, USA

ABSTRACT

Artificial intelligence (AI) is increasingly reshaping brand and marketing communications. While significant research has been conducted on the impact of AI in other fields, there is little empirical evidence on how artificial intelligence is affecting the customer journey. This research study seeks to answer, "How is artificial intelligence influencing both organizational content strategy and the related customer journey?" To answer this, the authors employed mixed-methods qualitative research via a content analysis of industry publications, a series of in-depth customer interviews, and a case study content analysis of two organizations that are using AI to varying degrees within their content strategy. This study found that many publications agree AI will play a future role as creative assistant in content development, that consumer perceptions about AI and cognitive dissonance impact levels of adoption to some extent, and that companies in different geographical locations may have different levels of AI adoption along innovation stages.

DOI: 10.4018/978-1-6684-6366-6.ch006

INTRODUCTION

Artificial intelligence (AI) has altered the dynamics of the business world. While AI is not new, media and marketing communications are experiencing a growing application of artificial intelligence in both content strategy (in large part, directed by the organization) and the resulting customer journey experienced with the brand. A customer journey is viewed as the process undertaken by a customer to achieve a goal involving one or more brands (Følstad et al., 2013). According to Lemon and Verhoef (2016), as the customer progresses in the journey, they encounter different brand touchpoints that affect the customer experience. Panda (2019) asserts that AI continues to gain traction in media and marketing communications as one of the leading technological developments that can facilitate a better customer experience, thus it weighs as an important variable in the customer journey, along with other experiential factors like employee responsiveness and overall brand strength (Nguyen et al., 2022). As a result, AI has grown to be an important tool for brands that strive to provide a targeted and personalized experience.

AI, Content Strategy, and the Customer Journey

Companies are constantly competing for customers' attention. Hence, brands are advised to provide customers with personally relevant brand content (Siebert et al., 2020). Simplification, personalization, and contextualization are tactics that are aimed to make customers repurchase or consume products and experiences (Edelman & Singer, 2015, p.50). Companies should also make the customer journey as "consistent and predictable" as possible (Hyken, 2009, p.55). Concurrently, a rise in customer expectations has prompted organizations to devise new strategies to remove pain points throughout the journey (Mohannad & Smoudy, 2019), which can be supported through the positive and significant relationship between AI and customer experience. Advances in AI can also enhance customer service by increasing the company's understanding of customer needs and shopping patterns (Ameen et al., 2021; Evans, 2019; Microsoft Advertising, 2019).

Morgan (2018) explored three facets of artificial intelligence that have a positive impact on customer experience. She notes that during or after their buying process, customers enjoy talking to a virtual assistant. Second, as customers determine which products to purchase, the use of personalization services has proven successful. Finally, AI provides the organization with customer insights enabling companies to devise strategies as needed. Edelman and Singer (2015) also recommend removing unnecessary steps by simplifying the customer journey, predicting customer preferences through personalization, and providing support when the customer needs it by contextualization. Chatbots, in particular, have been found to be of benefit to the customer experience when they provide relevant responses and help customers solve problems (Nicolescu & Tudorache, 2022). This enforces that AI has the capacity to become a key tool for retailers to consistently boost the customer experience (Newman and McClimans, 2019).

Though previous studies have demonstrated evidence of a positive relationship between artificial intelligence and customer experience, Shank et al. (2019) argues that in as much as the adoption of artificial intelligence presents many opportunities, the lack of human contact or increased technical effort needed by customers may present sacrifices which impact overall experience. This has been supported by recent research from Moore et al. (2022) which presents a dichotomy in humans' desire for efficiency and need for human interaction. Related, though empathy is a desirable human trait and may be mimicked by AI, it can be artificially presented in ways that are detrimental to the customer experience if not done thoughtfully (Liu-Thompkins et al., 2022). Trawnih et al (2022) similarly found that, "perceived sacrifice and trust both play an important role in mediating the impacts of perceived convenience, personalisation, and AI-powered service quality."

Impact and complexities related to AI need to be properly understood. The sacrifices that customers face in adopting AI-enabled services, such as lack of human contact, loss of human interaction, possible loss of privacy, time consumption, and potential negative feelings of frustration all may have a negative influence on perceptions of AI-enabled services (Ameen et al., 2021).

A quantitative customer experience study conducted by Jeffs (2018, p.4) revealed that although eighty percent of the Chief Executive Officers (CEOs) are convinced that they offer exceptional customer experience, only eight percent of the customers agreed. This depicts a customer experience gap in the buyer's journey and an opportunity for artificial intelligence. Another research study on the changing role of technology in the workplace conducted by PEGA (2020) highlights that although business leaders are eager to explore the opportunities that can be delivered by artificial intelligence, they still have a lot to learn. Fifty-one percent of the respondents perceived that senior managers need to expand their understanding of AI business processes and their effects on employee jobs.

By employing AI capabilities, innovation is occurring within the creative process and affecting content development (Hall, 2019). Creative industries, such as branding and marketing, are left wondering whether computers could begin to automate the traditionally human creative process (Pecherskiy, 2017). For content strategists, there is a growing market for AI-powered solutions. For instance, AI can also be used to write and personalize ads (Mogaji, Olaleye, & Dandison, 2020). The ability to personalize ads at scale has also opened up greater possibilities when it comes to targeting ads on social media and resultantly, to predict customer buying preferences (Tran, 2017; Davenport et al., 2020). This is possible because of "in-depth semantic analysis and real-time user interactions [that can] predict the probability that a consumer will accept a creative idea in the foreseeable future" (Li, 2019, p. 333).

Companies must be constantly engaging their customers with fresh, relevant, high quality content if they are expected to maintain their relevance in today's media-driven landscape. Less new content means lower brand relevance in search engine optimization and lower customer acquisition and retention. Given that machines have a history of making manual human processes more efficient, this brings about the question of whether AI can accelerate the speed of the ideation process and related, increase content creation efficiency within the process of content strategy. Customers can also experience enhanced support and more timely services through AI usage (Mohannad & Smoudy, 2019). In effect, AI is expected to make it easier for brands to offer relevant content, improve sales potential and boost customer experience.

By speeding up the decision-making process and availing marketing managers with knowledge and insights that they could not build regardless, AI could foreseeably make marketing more effective in addition to improving efficiency (Overgoor et al., 2019). For instance, text-mining could be utilized to better explain online word-of-mouth to model marketing responses using AI technology. Some other examples of some AI that conform to this concept include customer service chatbots, tools that model the future effects of a marketing strategy, and recommendations for online marketing material. The benefit of smart technology and AI assistants extends beyond brand engagement, providing companies with an opportunity to be part of an automated home, organizing goods and services, or information when a customer wants it (Newman & McClimans, 2019). However, Mannino et al. (2015) posit that companies should ensure that the advantages of using artificial intelligence surpass the risks.

Although there are pre-existing norms about what AI is or what it will be in the future, many perceive its impact to be a positive one. Pelan and Pop (2021) state that more Americans are in favor of AI than opposed to it and users tend to trust automation if the algorithms are understandable towards achieving their goals. According to a study of 1004 industry leaders by MIT Technology Review (2020) customer service rendered through chatbots is a leading artificial intelligence tool being implemented today. However, chatbots cannot replace human contact, but rather enhance or complement the human workforce. A major strength of AI-enabled customer experience analytics is that it can browse through wider and more complicated data and therefore reveal more marketing prospects, including ones that marketers did not know they could look for. Ultimately, this gives them more time to prioritize and make strategic decisions (Thiel, 2018).

There is currently little empirical evidence related to artificial intelligence and the customer journey from scholars. In the few studies published, Kietzmann et al. (2018) explored how marketers can leverage AI in advertising along the consumer journey and Xu et al. (2020) compared AI customer service and human customer service online by examining what customers prefer to complete tasks in the banking services context. Grewal and Roggeveen (2020) also researched the role of technology and understanding retail experiences and customer journey management (CJM). He and Zhang (2022) recently published the most comprehensive view of this topic to date, but as their framework included a literature review from published peer-reviewed works as far back as 2007, some of their findings could be influenced by dated contexts. Therefore, this research aims to supplement the current literature and reveal insights on how customer perspectives about AI affect the adoption of AI tools while also examining an organization's current use of AI tools.

Cultural AI Narratives

In years past, the media have set the AI agenda for the public through books, blockbuster films, and television shows which typically exaggerate the pace of AI growth. Due to this effect, excessive hype has periodically dominated the media landscape, leading to periods of overindulgence in AI narratives (Car et al., 2019) along with periods where the public becomes disenchanted with the promise of better AI technology (called AI winters) (Yang, 2006; Crevier, 1993). Public perception of the value and role of AI may be directly affected by these broad media metanarratives. Thus, the adoption of artificial intelligence has been faced with a mix of cynicism and optimism, with some potential users voicing concerns that AI may potentially turn against humanity with destructive outcomes or assumptions that artificial intelligence will displace human labor (Kietzmann & Pitt, 2020). Johnson and Verdicchio (2017) discuss AI anxiety, which refers to the fear of the stability and capabilities of artificial intelligence. Bentley et al. (2018) support that individuals fear that machines are becoming super intelligent and will exceed human intelligence because AI systems are continuously improving themselves. This explains why certain customers may be skeptical to adopt innovations along their customer journey. In addition, theoretical physicist Hawking expressed fear that AI may end humanity and the rise of robots will lead to the enslavement of humans (Batra, 2019).

Workplace Displacement. Despite many recognized benefits of AI at an organizational level, there are some underlying fears and concerns regarding AI development in the future (Palau & Popo, 2021). A group of AI experts addressed how machine learning could influence the workforce at the 2021 MIT Summit (MIT Sloan, 2021). One analysis completed by Mckinsey (2018) indicated that AI could displace five percent of employees in the workforce at that time. AI is already being used heavily in digital marketing domains such as web design, chatbots, and analytics. It is therefore predicted that AI will similarly change the future of media and marketing communication as it has in other industries (Davenport et al., 2020).

In relation to AI impact on creative roles, some believe AI poses a danger to human copywriters due to AI software outperforming human copywriters in isolated A/B copywriting shootouts (Brenner, 2018; Duffy, 2017). Meanwhile, others believe AI software will ultimately only offer assistance to human copywriters, streamlining their workflow. One study published by the Journal of Advertising provided evidence in support of integrating AI analytics with AI creative (Chen et al., 2019), with the goal of greater efficiency. And yet, there are also critics of AI integration within the creative process. Dan Izbicki, creative excellence director at Unilever, has argued that he has perceived lower overall quality of online advertising due to the widespread use of AI-based software (Bacon, 2017).

Customer privacy. Another key subject of discussion associated with artificial intelligence revolves around the extent to which consumers can trust and adopt AI while maintaining desired levels of privacy. Consumer privacy issues continue to be a recurring topic of discussion in this age of technology, which presents consumers with previously unmatched privacy invasions. Research supports that an increase in consumer's privacy rights affects the reputation of a company (Lwin et al., 2016). Lack of meeting privacy expectations can have disastrous negative effects as well. The Federal Trade Commission (2020) recently released its annual privacy and security update for 2019, which discusses a 5-billion-dollar penalty collected from Facebook for not appropriately protecting consumer privacy and data security. Companies are constantly being faced with the dilemma of how they can respect consumer's privacy policies online, with the need to collect consumer data that enables them to target audiences better and give them more personalized experiences along their customer journey. According to a research article published by Deloitte Insights (2020), the major challenge being faced by organizations is how to optimize the use of data collected through new technologies while staying within privacy regulations and not being intrusive.

Theoretical Framework

Diffusion of Innovation (DOI) theory, introduced by Rogers in 1962, explains the adoption of technological innovation (Rogers, 2003). Crucial to this theory is the idea that different types of people and organizations adopt new ideas at different times (Rogers, 2003). Innovation expands through the population to achieve acceptance and individuals can be divided into various adopter groups based on their level of innovation. The five stages of adopters based on the level of their adoption rate include*: innovators, early adopters, early majority, late majority, and laggards. Innovators* are the first to attempt or buy an innovation (product or service). They make up 2.5 percent of the market and are risk-takers, not price-sensitive, and want to be trend-setters. *Laggards* are the last customer group to embrace innovation. They do not like change and rely on more conventional methods until they can no longer use them. Laggards constitute 16 percent of the population and only embrace innovation because they have no other choice.

Diffusion is another dimension of the DOI theory. Rogers (2003) stated that diffusion is a gradual process that happens over time. The five stages of diffusion are *knowledge, persuasion, decision, implementation, and confirmation.* In summary, the adopter of innovation needs to have the knowledge to become persuaded or not, thereafter that persuasion leads to a decision that results in implementation, and confirmation of their choice (Surry and Farquahar, 1997). According to Straub (2009), adoption is the integration of innovation by an individual and diffusion is the "collective adoption process over time" (p.62). Furthermore, Rogers (2003) posits that individuals adopt different innovations for different reasons. Regarding technology acceptance, Davis (1985) concludes that there is a higher correlation between perceived usefulness and adoption. Regardless of the usability and ease to learn new technology, individuals will not adopt a technology if they do not view it as valuable to their productivity. Previous studies highlight that *relative advantage* and *comparability* are highly prominent factors in innovation adoption of internet-based and mobile technologies (Park & Chen, 2007; Papies & Clement, 2008).

This theory provides a strong theoretical foundation to investigate AI adoption for customers and organizations. Lund et al. (2020) assert that given the recent advent of artificial intelligence, there is an opportunity to examine the perception of innovation and the connection it has with the acceptance of new technologies as proposed in the diffusion of innovation (DOI). Subsequently, comparing different customer perceptions of AI against the companies' integration efforts of AI within content strategy and the resulting experiences along the customer journey will help analyze its effectiveness or identify gaps that any industry can benefit from.

Research conducted by McKinsey (2018) reveals that only 10 percent of organizations have attempted to diffuse AI in their business model. The diffusion of innovation such as artificial intelligence is clearly affected by barriers that may prohibit its adoption by companies or individuals. Chalmers et al. (2020) also highlight that the diffusion of artificial intelligence will not happen in isolation, but rather as part of a trajectory.

RESEARCH QUESTION

Previous research studies have focused more on artificial intelligence from a technical perspective rather than from the customer experience point of view. Few studies have tied artificial intelligence and the customer journey together. These considerations led to the primary research question for this study:

RQ: How is artificial intelligence influencing both organizational content strategy and the related customer journey?

To better answer this question, the authors investigated the following question subcomponents:

1. How are industry leaders speaking about the trends and trajectory of AI within content strategy?
2. How is AI being used in both inbound and outbound marketing for the two companies in the cases examined?
3. How is the perception of AI impacting the customer experience and journey within these companies?

RESEARCH METHODS

The present research aimed to explore the impact of AI in marketing communications by first applying a content analysis of discussions about AI and content strategy from marketing leaders across various online industry-connected publications, then focusing on a specific industry (home appliances) and investigating customer experience perspectives regarding AI use within the customer journey, and then finally examining the brand's content strategy driving customer experiences. Therefore, this research first evaluates overall AI trends and its projected trajectory from industry leaders, then evaluates what these specific companies, Bosch (USA) and Defy (South Africa), are doing in the AI space versus what is being perceived by customers along their customer journey. The findings of this research also aid to garner receptiveness among customers toward organizations adopting AI technology as a tool within their customer journey.

To explore the research variables and determine which applications of AI are meaningful in this industry, mixed methods qualitative research methodology was used. A mixed-methods approach is suited for this research because it allows to combine or corroborate findings to generate more complete data that enhances insights (Creswell & Clark, 2007). This study employs in-depth content analysis to published articles, as well as in-depth interview content, and public-facing organizational materials from Bosch and Defy using coding elements from Grounded Theory Methods (i.e., open and axial coding).

Grounded theory methods, a specific approach to content analysis, were initially developed by researchers Glaser and Strauss in 1967 in order to develop new theories based on context and inference from data using a highly systematic approach of identifying and linking complex variables (Tie, Birks, & Francis, 2019; Timonen, Foley, & Conlon, 2018; Bryant, 2017; Glaser & Strauss, 1967). Grounded theory methods may thus be especially helpful "when little is known about a phenomenon" (Tie et al., 2019, pp. 1). Due to the lack of previous research related to AI and content strategy, this approach was deemed appropriate for the investigation of our research questions. Across the first two research phases, this inductive research approach was employed to define core concepts and to recognize similarities and differences.

In the third research stage, a qualitative case study based content analysis of both companies was also conducted in tandem on brand-owned touchpoints, namely, their websites and social media, to explore which artificial intelligence (AI digital marketing tools) they are using for their target customers. The case study approach gives an investigation into what AI tools the companies are actually using, while the user interviews add insight into how their customers then perceive them.

Regarding theoretical contributions, this research is one of the few studies to explore how not only industry leaders are predicting AI trends to continue within the field of content strategy, but also how customers' own views and perceptions of AI tools in their customer journey impact their overall experience. Therefore, this study aimed to fill this gap in existing research. The research also explored how views and perceptions of artificial intelligence may play a role in the diffusion process of this technology.

Research Phase 1 Industry Article Content Analysis. The authors first gathered a total of 78 varied industry publications as a pool for analysis consideration through targeted keyword searches (e.g., artificial intelligence content creation, ai content strategy), with likelihood of selection influenced by Google search ranking and relevance to the topics of artificial intelligence and content strategy. In total, 31 articles were from marketing publications, 31 were from industry professionals' blogs (both personal and company), 14 were from general publications, and 2 were from technology-specific publications. All articles were related to anticipated AI trends. By focusing primarily on news media and professional sources and excluding out-of-date sources and sources whose author's motivation might be tied to sales/personal gain, this number was further reduced to a stratified sample of 16 articles for coding. In this way, the study limited bias, reduced redundancy, and achieved a more accurate picture of what online news publications were collectively saying about AI and content strategy throughout 2018 and 2019. In addition, during the source selection process, the authors specifically looked for sources that presented new information that had not been previously addressed in a separate source. Overall, this approach worked favorably, enabling a balanced view of expected trends in the field of artificial intelligence and content strategy.

Research Phase 2 In-depth Interviews. The scope of this research was limited to a total of three current Bosch customers and five current Defy customers in early January 2021. Defy South Africa is one of Southern Africa's largest manufacturers and distributors of home appliances (Defy, 2021), while Bosch USA is an industry leader in consumer-driven solutions for home appliances in the USA (Bosch, 2021). Due to ethical considerations, interviews were conducted only with healthy adults over the age of 18 who were eligible to participate. Participants for the study were recruited as a convenience sample online through social media, referrals, and via the (University's) International Students Office, which sent recruitment emails to students.

In research, convenience sampling identifies a group of people that are convenient to research for reasons such as geographic location, connection to the topic, or prior assembly (Merriam, 2009). Zoom interviews were recorded with prior permission from the participants, then later transcribed and analyzed. Participation in this study was voluntary and no incentives were offered to participate. Looking at the interview participants, only one male was included in the research sample; all other participants identified as female. The bulk of Defy participants were between 25- 34 years old while the majority of Bosch participants were aged between 45-54 years. Ethical standards for the study were met by securing study approval from the researchers' University Institutional Review Board (IRB) prior to the interviews.

The interview questions fell into four blocks: demographics, perception about AI, rate of adoption, and possible concerns about AI. The interviews aimed to address research questions on customer perceptions about the brand, artificial intelligence, and readiness to adopt new AI-related technologies in the customer journey. The interviews were framed as a combination (abductive) of inductive and deductive research. Subsequently, audio transcripts from the interviews were downloaded and analyzed.

Example questions from the interview included:

How do you generally perceive the ads you have seen from Bosch /Defy?

- *Have you ever received personalized ads online to your knowledge?*
- *How did you know that the ad was personalized to you?*
- *Do you think the ad was relevant to your search?*

What do you understand by the term artificial intelligence?

- *What are your views/thoughts on personalized targeted ads that appear when you are online?*
- *Have you used a chatbot or virtual assistant in the purchasing or post-purchase journey as a customer?*

To identify potential issues with the research instruments, a pilot study was conducted with five participants to ensure the validity and propriety of the interview questions. Afterward, adjustments were made to questions to make them clearer for the participants. In research, a pilot study can reduce the risks of ambiguity in an interview, therefore the pilot study reinforces the relationship between the study concerns and the possible existence of evidence (Yin, 2009).

Phase 3 Case Study Qualitative Content Analysis

The final phase of the research was a case study conducted via qualitative content analysis of both Bosch USA and Defy South Africa. In this third stage of the research, the case study content analysis focused on each company's website and social media platforms. This approach to empirical research explores real-life occurrences in specific contexts (Allen, 2017). Thus, case study research makes it possible for a holistic view of a process (Patton & Appelbaum, 2003, p.63). The benefits of this case study research as highlighted by Yin (2003) are (1) use of multiple sources of evidence, (2) creation of a case study database, and (3) maintenance of a chain of evidence that is ideal for comparative analysis of both organization's AI tools.

Konecki (2011) asserts that qualitative research is increasingly engaging in visual data such as images, text, blogs, and video content, therefore qualitative content analysis of these cases was used to analyze Bosch and Defy's websites and social media content. This was useful for evaluating evidence of different data types such as both textual and visual content (Mayring, 2002). In analysis, the authors focused on examining artificial intelligence on brand-owned touchpoints (for example, media-related touchpoints, and marketing mix related touchpoints), examining the extent to which each company has adopted or diffused artificial intelligence in their business models to improve the customer's journey.

The sampling frame of the home appliance companies was selected because Defy South Africa is one of Southern Africa's largest manufacturers and distributors of home appliances. Bosch USA is also an industry leader in consumer-driven solutions for home appliances in the USA. In the context of cultural norms, the US and SA were selected because research autonomy, consent procedures, recruitment methods, and values are similar. Moreover, both countries use English as their primary professional language. Hence, this research is an opportunity to identify gaps, similarities, differences and discover patterns in a developing country and a developed country.

ANALYSIS AND FINDINGS

Research Phase One

Content Analysis- Open Coding

The ultimate purpose of open coding, the initial coding step in Grounded Theory Methods, is to come up with "concepts that fit the data" in order to find overarching themes (Strauss 1987, 28). In this step, the researchers closely examined each text and used inductive reasoning to identify the main points the authors of each article were attempting to convey. The authors accomplished this via reviewing and making extensive field notes in the margins of the texts being analyzed. These notes included any position(s) taken by the author(s), noted any key perspectives and opinions provided by the post or article, and highlighted any consistent references and themes that were brought up with any measure of repetition. The constant comparative method was applied to research data to help minimize bias. Glaser and Strauss (1965) add that the comparative qualitative method of data coding is used to categorize and compare qualitative data for analysis. Eventually, some consistent, recurring themes began to emerge.

Content Analysis- Axial Coding

Axial coding, the second step of GTM coding seeks to establish the differences and relationships between research variables or categories (Strauss & Corbin, 1998). Axial coding defines causal relationships, framework, and interconnection of data. Subsequently, observation of emerging themes or patterns between the study's data sets. Coding categories were developed through analysis of the research data. Moreover, finding connections and relationships between codes, the authors examined for causes or consequences between variables and also removed codes that did not have enough data to support them by reviewing all categories.

The five variables discovered through open and axial coding of the sixteen industry publication in phase one of our study included the following:

Figure 1. Variable one

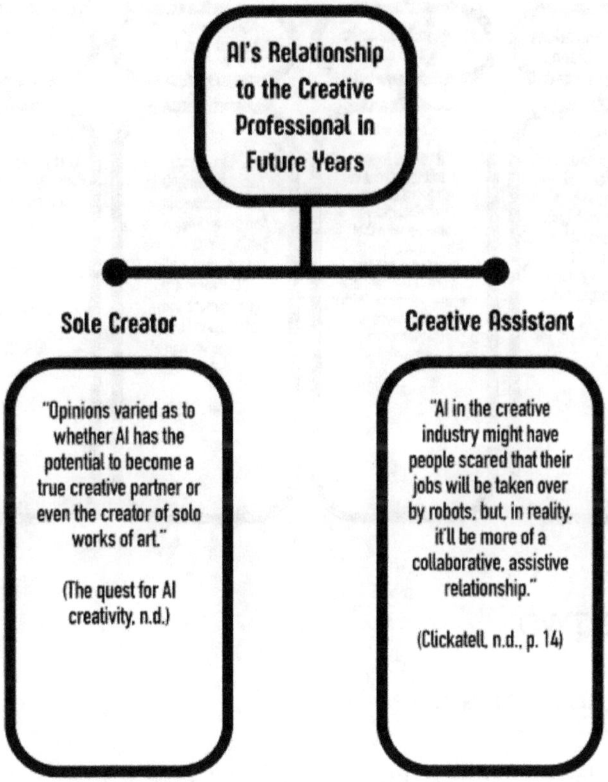

Content Analysis- Selective Coding

Selective coding is the final stage of the coding process in grounded theory methods. This step finds the common thread within the data that ties it all together into an understandable storyline (Strauss, 1987). This process calls for sensitivity and attention to both detail and the overarching metanarrative in the text. In our selective coding process, the overarching theme pertained to the relationship between the creative professional and artificial intelligence, "AI's Relationship to the Creative Professional in Future Years." All the other identified variables, in some form, relate back to this central variable and allow for a deeper consideration of that single construct and this polarity. Within this stage of analysis, we found that most industry professionals believed the role of AI in content strategy moving forward to be that of assistant, enabler, and efficiency-finder, but not as a replacement for humans or human creativity.

Figure 2. Variable two

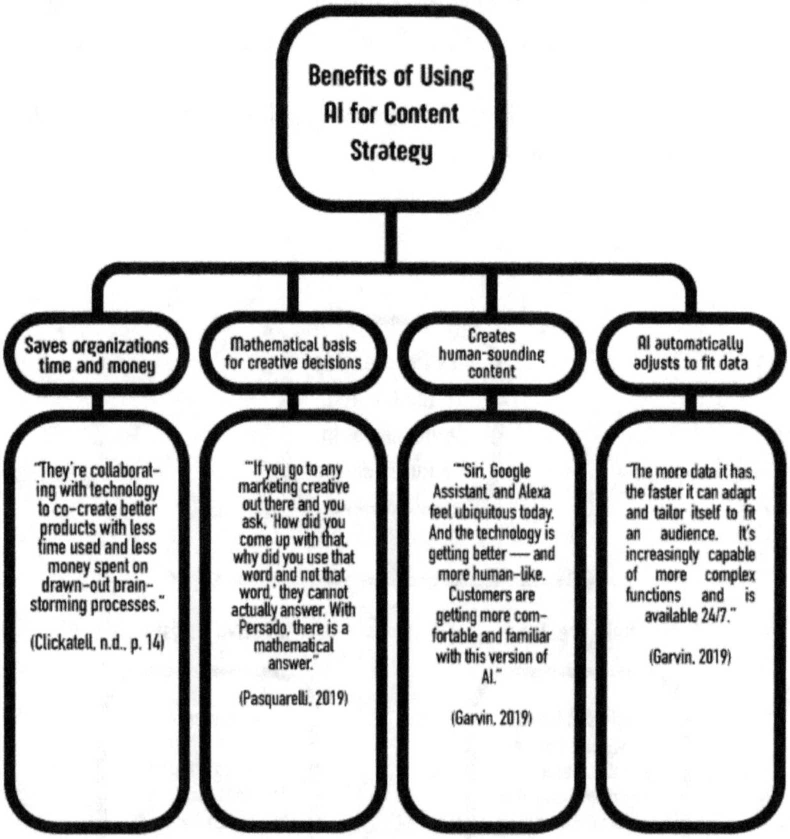

RESEARCH PHASE TWO

Interview Results and Analysis

Here, we focus on our in-depth interview findings that directly speak to aspects of the customer journey and AI integration, either actual or perceived. Again, GTM stages of open and axial coding were applied to understand variables unearthed in the interviews and their relationships.

Perceptions of Artificial Intelligence

To get a sense of how participants perceive artificial intelligence, the interview focused on asking their views of the technology. This variable examined the various views and thoughts surrounding artificial intelligence and how it possibly impacts the company's inbound and outbound marketing efforts in the customer journey. The following statements are representative of customers' perception and understanding of the term artificial intelligence:

Figure 3. Variable three

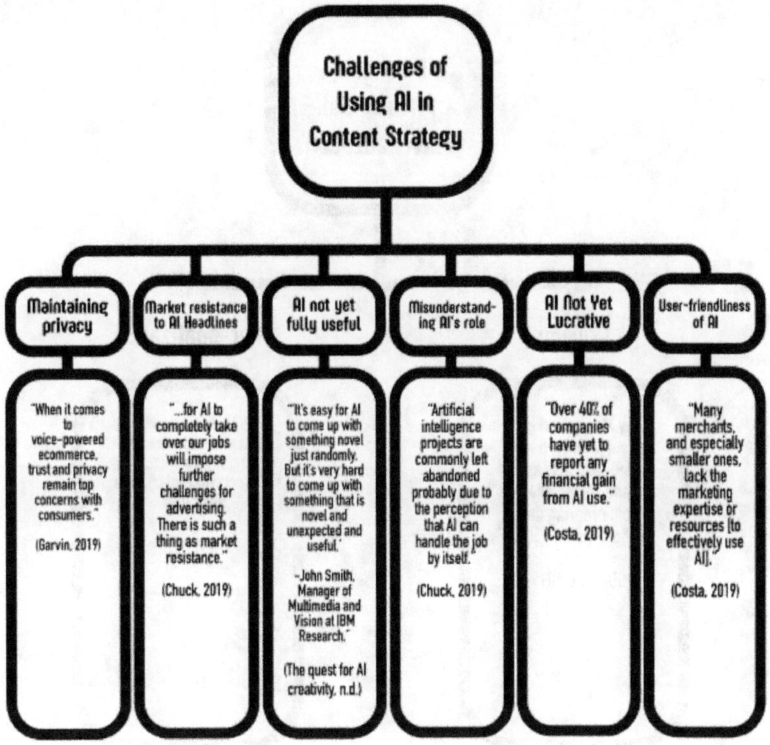

Figure 4. Variable four

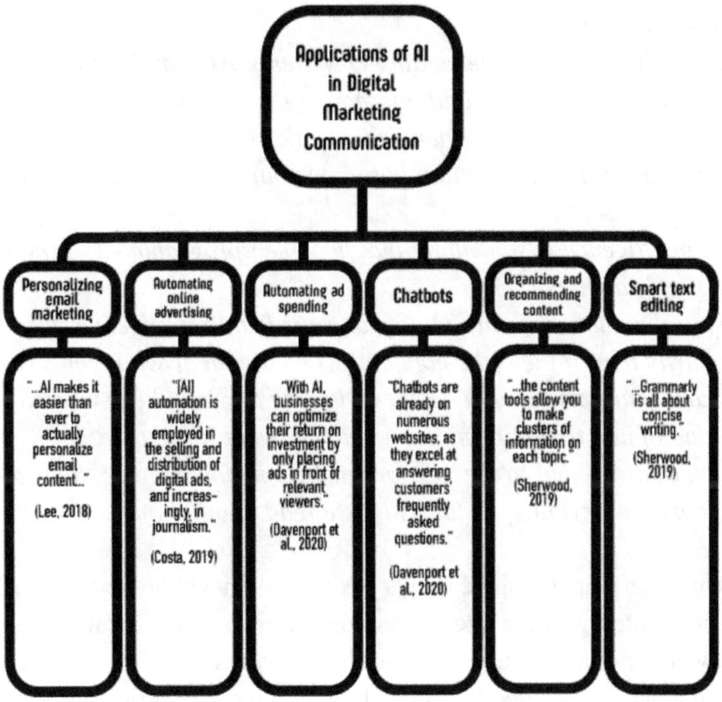

Figure 5. Variable five

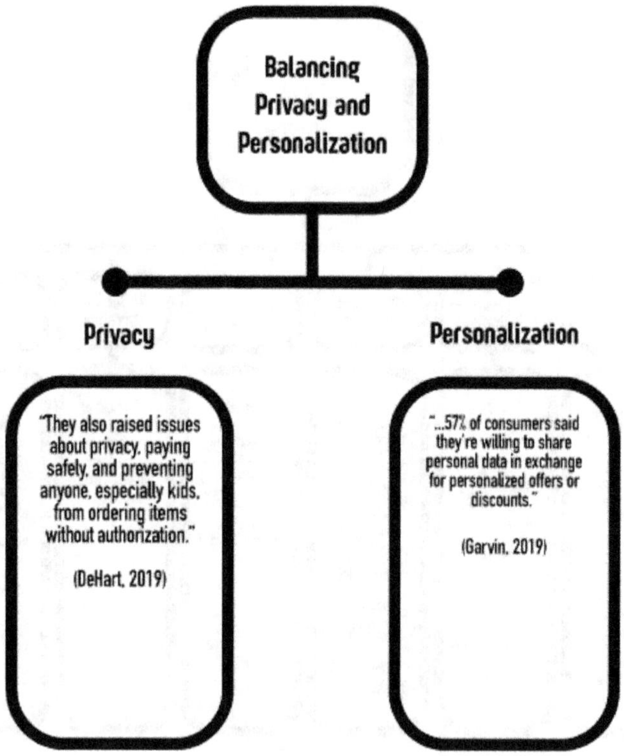

Defy customers:

- *"I am not sure, what is AI?"*
- *"Artificial intelligence is implementing IT systems. AI is making use of information technology to do tasks like marketing and interaction through bots."*
- *"I think of robots, automated systems that work for you and communicate with you."*
- *"A computer-generated assumption based on your browsing history and buying history."*

Bosch customers:

- *"Artificial intelligence is something that has been programmed to try to think or anticipate your customer's actions."*
- *"To tell you the truth, I do not understand artificial intelligence. I do not have Alexa I do not necessarily trust if they [the companies] cannot hear what you say [secretly listening to people's conversations through devices]. I know they can hear what you are saying. In fact, we have kind of had that happen, I am not sure what it is called."*
- *"AI is somewhat likened to the new brain. It somehow mimics the brain. However, it obviously cannot do everything, as far as I have read about it online."*

The results show that participants share both a positive and negative perception of AI. Some are well educated about artificial intelligence while others fairly know about it while others do not know much about AI. Repeated words or phrases used to describe AI were identified. As such, arranging relationships across open and axial codes aided in establishing common categories and themes. Participant responses were not connected with demographics, but rather on knowledge and general perceptions about artificial intelligence. Those who favor the development of AI systems explained how it has helped them get information easily. Nonetheless, it should also be stressed that some participants emphasized their disapproval of AI-systems. Most negative remarks centered on artificial intelligence being intrusive, as one of the ways participants expressed, "*I don't want anyone making an assumption about me, someone who has not met me. I do not like an artificial intelligence assumption. You cannot just put me in a box.*"

In the interview, participants described their views and experiences with artificial intelligence in their customer journey, including thoughts about personalized ads, chatbots, and virtual assistants. All interviewees from Bosch or Defy in the research study stated that they had never used a chatbot or virtual assistant in their purchasing or post-purchase stage of the customer journey. However, others mentioned that they have used such digital tools with other companies in different instances. It could be observed that some participants are unaware that they are already using AI-enabled tools from Bosch, Defy, or other companies. When asked specifically asked about AI, they identified as non-users of the technology. It seems plausible that there is a level of misinformation about artificial intelligence. Participants are not aware that benefits, such as readily finding information about different products, are because of an AI algorithm. Therefore, the participants did not realize AI for what it is.

Participant's AI Adoption and Diffusion

Rogers (2003) asserts the decision to adopt technology depends on the perceived benefits of the innovation, while confirmation represents the extent to which success of the technology creates expectations compatible with previous beliefs. On one hand, four of the Defy participants rated themselves as early adopters while one participant rated themselves as part of the early majority in the adoption of new technology. The major issue mentioned by Defy customers is the tradeoff between wanting an AI-related benefit against giving up personal information to enjoy the benefit is a dilemma for them. On the other hand, there seems to be a mix of how Bosch participants rated their readiness to adopt new technology, one participant said they are laggards, another shared how digital transformation did not give people a choice other than to adapt, and lastly one participant rated themselves as an early majority.

In order to analyze differences in attitudes between different types of adopters as categorized by Rogers (2003), a comparison of the participants' perceptions about artificial intelligence versus how they categorized themselves was conducted. The research results revealed that generally, participants who had classified themselves as early adopters tended to have a positive view towards AI or shared a willingness to learn more about the technology while participants who classified themselves as early majority or laggards have negative views about artificial intelligence and share suspicions about AI. Regardless of geographic location, the majority of younger participants also have a positive view of artificial intelligence and show a willingness to adopt new technology.

One participant acknowledged a willingness to adopt the technology if the company or the technology does not require giving up too much personal information. Participants' unwillingness to share their personal information such as address or contact number stems from possible consequences that may follow such as problems with stalkers or scammers. Although others shared that they are willing to share email, gender, and city information only. While another participant emphasized being skeptical about their readiness to give personal information online, "*I am very skeptical about that. I rarely do that especially if it's online, at times I just put in a random name.*"

One Defy participant reflected on how they do not see artificial intelligence changing their interactions with companies in the future as stated:

At the moment, I keep artificial intelligence at arm's length because of the privacy of my information. It is something I am keeping an open eye. At the moment I do not like it, but I will keep an open mind maybe in the future this will change.

On the contrary, others see AI bringing more convenience, promoting impulse purchases because the information is readily available, and AI making life easier. For example, one participant highlighted: "*Be on the lookout because artificial intelligence is the future, especially this year. The importance of artificial intelligence was heightened by the COVID-19 pandemic*."

Although Bosch manufactures high-tech AI-powered home appliances one participant discussed:

I already have a phone that is smarter than I am. Also, I am not into artificial intelligence. I guess I would not be a very good customer for the use of artificial intelligence. All I know is Bosch products and services are supposed to be artificially intelligent. But I had not considered any of that artificial intelligence till I got your question. I know they have Bluetooth on the refrigerator I purchased which enables you to tell it what to do with your phone. I guess that is considered artificial intelligence. But I have not connected it to Bluetooth. I do not care to make my refrigerator work through my phone. I guess I am part of the older generation that has not caught on and I am not interested.

Overall, there were distinctions observed during the axial coding process, participants notably shared that they are more open to adopting AI-enabled tools along their customer journey if they are given enough information to understand and embrace the technology. This explains why participants seem to be aware of what AI technology is, but unaware of its existence in their customer journey.

Personalized Ads

It is important to note that Defy customers stated not receiving any personalized ads online to their knowledge from the brand. The research results revealed that participants' view of AI and personalization was limited, as they do not fully comprehend the ways which AI may be involved in personalization beyond email. For example, when asked to assess what their perception of personalized ads was, the participants tended to associate personalized ads with email marketing, where a company addresses their name in an email with promotional information or product information usually relevant to the customer's search. One participant explained how they have seen personalized ads on Facebook and Twitter from Defy dealers rather than Defy: "*I would see it as a pop-up ad on Facebook from a Defy distributor*" and another expressed "*(Defy dealer) sends me spam email. I think we must be on their mailing list; however, I do not necessarily receive emails about Defy products but other general products from the dealer*." A total of two participants confirmed to have seen personalized ads from Defy. "*I view them as a slightly intrusive, a tad disturbing because it feels as if someone is watching your activity*." By contrast, another interviewee shared that Defy's promotional ads are well structured and persuasive, however, they are also uninformative and hard to remember or pay attention to. When it came to general personalized targeted ads that appear online, one participant stated:

I find it quite creepy on Instagram, but I researched it and it is because of an algorithm that sees my activity through the pages I visit or what I like. I got to understand why it would suggest that and I think it is smart.

Similarly, all the Bosch customers said they have likewise not received any personalized adverts from Bosch, but participants' statements showed they are more aware of what personalized ads are. Moreover, it should be mentioned that participants do not pay attention to targeted ads in general. For example, one participant expressed:

Anytime I start browsing online I will start getting ads about various products on my timeline. I can tell in two seconds, whether I want to buy or not. I am not an easy sell, I am suspicious of certain people, where I feel as though they are just trying to get me to buy something.

When it came to perceptions about personalized targeted ads that appear during browsing online, it seems there is an equal distribution of positive and negative views between Bosch and Defy customers. The research findings may suggest that personalized ads may be pushing away the customers for example a participant stated, "*It does not help their case for them to maybe compliment me on something that I can tell it is just a pitch.*" While on the other hand another participant contrasted saying that they liked ads in general as indicated, "*I get a lot of personalized ads for clothes, and hair products. I like them and I am happy to look. Let me just say that my purchasing has gone way up during COVID-19.*" Respondents' frequent associations with personalized ads are summarized in Table 1.

Table 1. Customer perceptions about personalized ads

Perceptions about personalized ads	
Positive associations	**Negative associations**
Shortens search cycle	Spying/ Intrusive
Readily available information	Disturbing
Helps to develop an interest	Untrustworthy
Saves time	Sales pitch
	Secretly listening

Note: Words that participants associate with personalized ads that appear when they are browsing online.

AI Arising Issues

When asked about concerns they have with artificial intelligence and interactions with companies, manifest content analysis was applied to discover open and axial codes associated with AI concerns. Intriguingly, there was a strong consensus about participants' concerns with artificial intelligence. The major issues highlighted by participants were privacy and confidentiality breach. Although participants are willing to adopt AI technology, they fear the risks associated with the use of such technology. The interviewees stated that giving up information may be a problem since there are stalkers on the internet and scammers which explains why there is a lack of trust in this technology. One participant expressed

AI concerns and stated: "*You might randomly find a stranger on your doorstep, or you may get a call from someone you do not know, or some organizations call you with promotional ads. That is why I do not give out any of my information.*"

Another AI concern highlighted was the problem of over-targeting. For example, after a customer shares or likes a post on social media and consequently all sponsored content targeted on their timeline are ads that are related to that product or service only. The participant expressed how this can be daunting and keeps the customer from being targeted on other products, they may be interested in besides products based on their previous browsing history.

Research Phase 3: Case Study Content Analysis

The case study content analysis focused on examining artificial intelligence on digital tools such as brand-owned touchpoints (websites, media-related touchpoints). With the advent of digital technologies, the customer journey is becoming complex because they have multiple touchpoints (Morgan, 2020). There is a need to build adaptive and user-friendly platforms for customers to interact with the company.

Chatbots and Virtual Assistants

Bosch deploys AI-based systems to help customers with information or queries using a chatbot. Bosch introduced their chatbot named Frizz that answers key questions about artificial intelligence. Figure 6 below shows an example of inquiry with the Frizz chatbot:

Figure 6. Bosch AI chatbot Frizz
Note: a screenshot of a conversation with Bosch's AI chatbot called Frizz
Source: Frizz-your AI Chatbot for Bosch (Bosch, 2021)

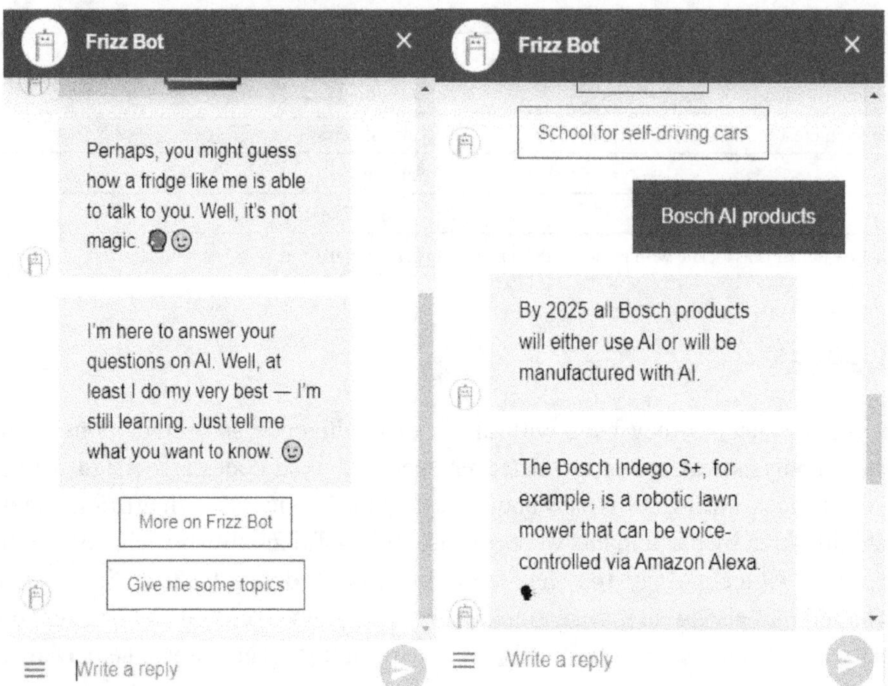

However, none of the respondents from the interviews mentioned ever using this chatbot or showed awareness of this digital tool in their customer journey. Besides FAQs, Bosch uses a virtual assistant capable of comprehending the various ways that customers ask questions making troubleshooting easy via its website. Moreover, the website has a search option where they can tell the customer to find a dealer nearest to them from the website easily by using geolocation AI tools by simply entering an address, city, or zip code. In addition, every product has downloadable manuals and videos, and accessories.

Defy's website has a chatbot that is available 24/7 for customer service which helps to manage customer queries online, unless if it is a live chat question, then you are prompted to wait for a response once the team is back online. The message prompt mentions that they typically respond within 51 hours and notifies you that their team is offline and will get back as soon as possible. There are no specified times highlighted when the team will be available. Nonetheless, Defy has comprehensive and informative questions and troubleshooting responses on their FAQs from cooking, refrigeration, and Toploader washing machine as shown in Figure 7. Additionally, Defy did not offer its customers with virtual assistant AI technology during the time of this study.

Figure 7. Defy's chatbot
Note: a screenshot of Defy's chatbot and live chat.
Source: Chat with Us (Defy, 2021)

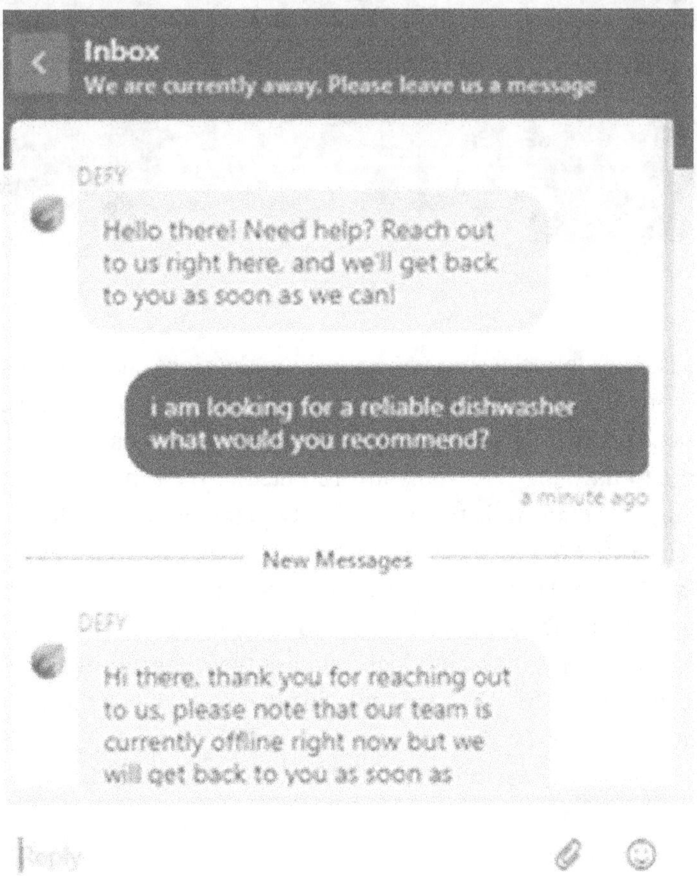

Virtual Reality and IoT

The Bosch website allows customers to "Virtually tour your dream kitchen." A customer can discover the quality, innovation, and timeless design that can complement their home upon entering the Bosch Virtual kitchen. Upon entering the virtual kitchen, it is an interactive experience whereby you can click or drag to explore various products in the kitchen as shown below (Figure 8). For example, when you click the refrigerator, it opens while chef Curtis Stone (influencer) gives audio transcribed pro-tips along the virtual tour about what kind of products to look for. There are videos about technical information related to the product and special IoT-enabled features.

Figure 8. Screenshots from the Bosch virtual kitchen
Source: Bosch Interactive Kitchen (Bosch, 2021)

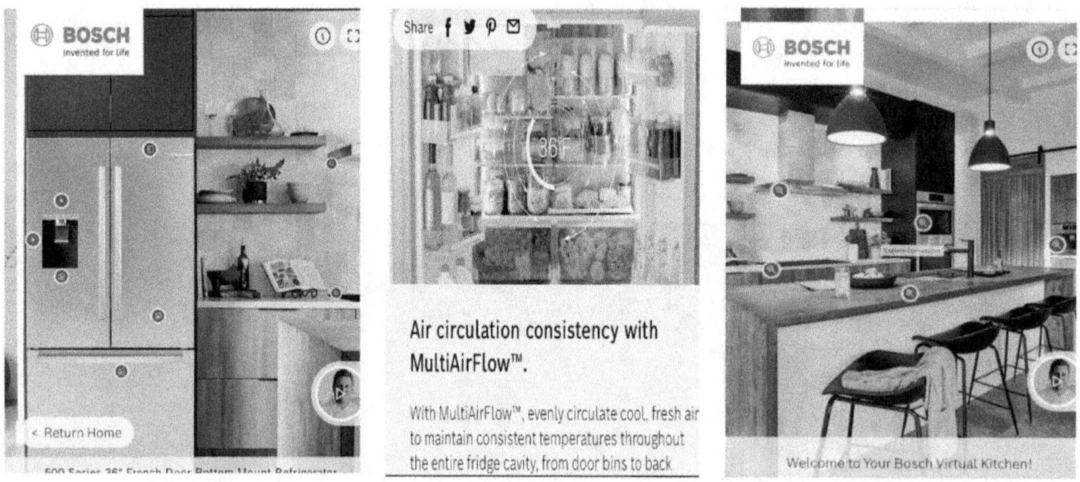

Chef Curtis Stone is there to make sure you are asking the right questions about your products for instance when considering buying a dishwasher "*you want to look for something clean, dry, and quiet*" he suggests. He mentions product features that make Bosch products stand out from their competitors.

Defy did not appear to have these AI tools incorporated at the time of this study. Though, Defy's parent company has since mentioned that they are slowly planning on establishing a machine learning program powered by Amazon Web Services to transform its appliance business into a data-driven area of the organization. In addition, Defy's website shows products available and features with a link that shows customers where to buy the products. Unlike Bosch, Defy products are entirely sold through dealers such that the website only shows products and not the prices for the home appliances (Figure 9). The only products that can be bought from the website are spare parts as previously mentioned by participants during the interviews.

Figure 9. Defy products
Note: Defy products are exclusively distributed by dealers.
Source: Defy appliances (Defy, 2021)

Voice Assistants and IoT

Bosch's smart Home Connect helps customers to stay connected. The Home Connect app allows various smart home appliances to be controlled with a single app, in other words, it was designed to make life easier for users. It also allows devices to be interconnected and communicate with each other for example, connected dishwashers, connected cooking, connected coffee, and connected laundry. The Bosch Smart Home Connect is powered by a series of partnerships that make it possible for IoT to flourish as shown in Figure 10 below.

Figure 10. Bosch smart home voice assistants
Source: Bosch Smart Home (Bosch, 2021

- Smart on-demand is a partnership with Amazon Alexa where you can access various functions of smart home via voice command such as answering a doorbell.
- Google assistant ensures constant temperature throughout the house, the voice assistant continuously keeps learning.

- Apple HomeKit allows you to integrate smart Home Connect products from other manufacturers into your Bosch Smart Home system.
- MBUX Mercedes allows you to control your Bosch home using natural speech from your Mercedes car voice assistant.

Marketing and Technical information

One of its notable AI campaigns is the #LikeABosch campaign highlighted in Figure 11, which shows a series of videos on their website and social media platforms about automated IoT and augmented reality. This sparked a viral IoT movement, the videos show how everyday people can incorporate AI-connected solutions through a smartphone, an autonomous lawnmower to self-parking cars. On the other hand, their website has an "All about DIY" option which displays technical content on ways to start different projects such as a DIY bathroom, and football goal among many other home projects. The website gives information about materials required and specific measurements and links to their tools such as a hammer drill, how it works and also provides technical information such as its voltage and functionality. Their technical communication focuses on providing enough technical details on how to use a product and provide information in a helpful format. Bosch's website shows assembly guidelines, tutorials, how-tos, datasheets, technical mechanisms, and downloadable user PDF documents.

Figure 11. Bosch's #LikeABosch campaign
Note: a screenshot of Bosch's #LikeABosch campaign series
Source: #LikeABosch (Bosch, 2021)

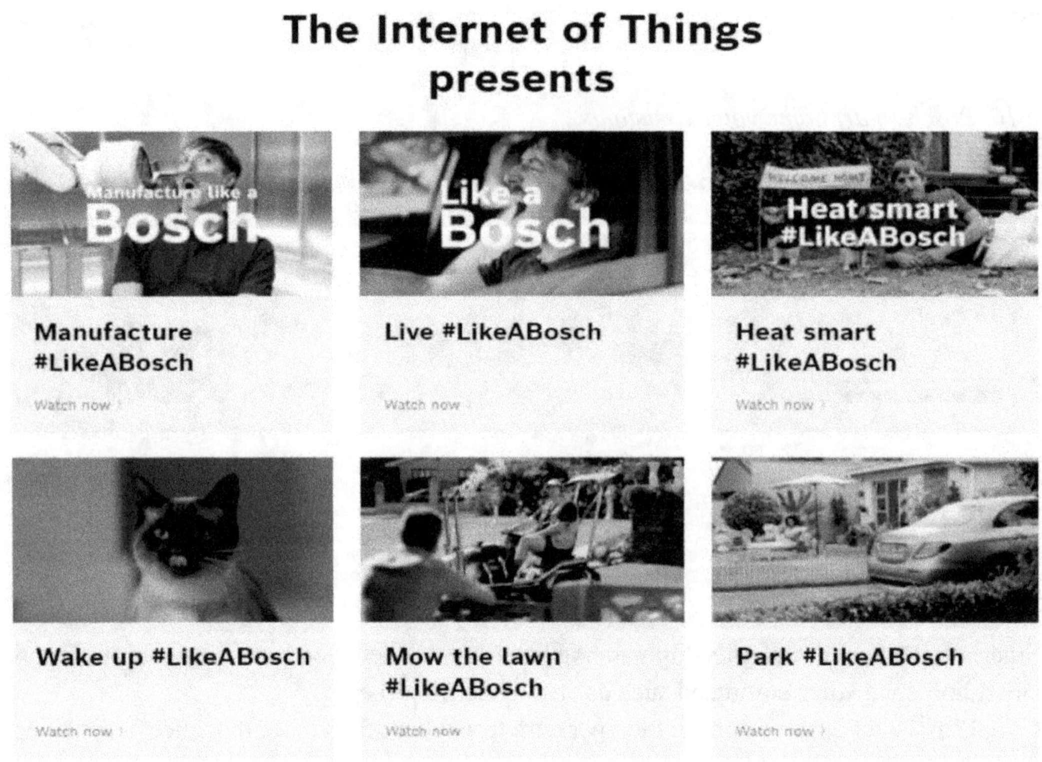

Defy provides its customers with both marketing communication and technical communication on both social media pages and websites. They advertise their products by emphasizing the design, multi-function operating system, and unmatched design and quality. Their website has FAQs for different products. Moreover, a blog tab that shows images of the how-to troubleshooting tips for example "*steps for cleaning your washing machine*." Their Instagram features new products and quality elements that promote the durability of the products. However, one participant from the interview expressed how it was challenging to find video-branded content from Defy to help set up a product such as a washer they bought.

LEAD GENERATION AND RETARGETING

Scott (2013) defines lead-generation marketing as asking customers to "raise their hands and say they are interested in buying, or learning more about your product, in other words, actionable leads" (p.17). From a lead generation point of view, online research was conducted for brands. The process of generating leads and retargeting information for the two brands involved browsing on search engines such as Google about the brand and home appliance products. Thereafter, the authors accepted cookies and shared geo location in an effort to get suggestions for products related to browsing history. Besides cookies, another effort to generate these targeted ads for the companies was browsing social media content on Instagram, Facebook, and YouTube, then engaging with the company's online content, clicking the call-to-action (CTA) links, and blogs. Both companies had different landing pages on their website. For example, Bosch showed their dream kitchen designs while Defy displayed their quality dishwasher.

Looking from an individual case study view, the authors could see Bosch's remarketing efforts assuming they are using AI to target customers and retargeting efforts. For example, the authors could see Bosch branded content related to home appliances such as washers, dryers, IoT related content for home appliance products on social media channels such as Instagram (Figure 12). In the case of Defy, this does not seem to be the case. Defy does not necessarily use lead-generation or retarget content using AI digital marketing. This is mainly because of their business distribution model (B2B) which directs customers to interact with their dealers/distributors. Nevertheless, Defy's dealers such as Markro and Game showed retargeted ads. However, these ads were not necessarily Defy's home appliance recommendations, but a mix of all the dealers/distributor's products and promotions.

AI Diffusion

Compared to other companies in the industry, Bosch has managed to diffuse AI tools and technology in its business model in an attempt to make the customer experience seamless. On the other hand, Defy South Africa does not have many of the AI tools incorporated in their business model yet. However, it was discovered that Defy is targeting more of a B2B marketing approach. Therefore, consumers may not be their targeted audience for the marketing practices that were explored in this study. Based on this evidence, there is a significant difference in levels of diffusion at the geographical level between Bosch and Defy.

Figure 12. Bosch's retargeting efforts on Instagram
Source: Bosch Home Appliance US (Bosch, 2021)

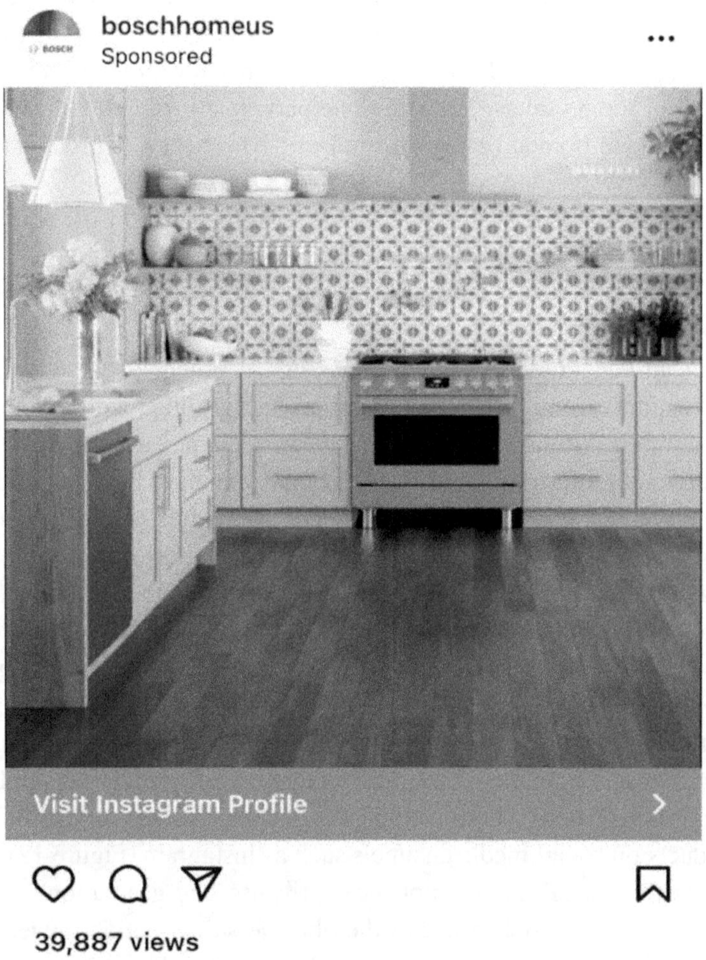

RESULTS

The primary research question that drove this study was:

RQ: How is artificial intelligence influencing both organizational content strategy and the related customer journey?

In phase one of our research, we found that the articles examined took the view that AI will likely be a helpful creative assistant in the near future—but will not soon become a sole creator of content or definer of content strategy. Our findings strongly indicated that computers possess only skills that can make existing content strategy processes more efficient—not skills that can generate completely new efforts from scratch.

In the second stage of research, based on the in-depth interview analysis, we see certain themes associated with AI being highlighted by interviewees. For example, major categories associated with AI interview discussions were (1) perceived usefulness, (2) personalized ads, (3) levels of AI adoption, and (4) privacy concerns. The findings of this research indicate that attitude differences between adopter categories affect readiness to adopt AI technology. This is evident when we see early adopters are more open to AI while the early majority and laggards are not willing to adopt AI-enabled technology along their customer journey. Moreover, lack of knowledge about AI affects adoption as indicated by participants.

Another important finding from this research stage indicates that the participants have misconceptions or are unaware they are already using AI-enabled tools. When asked specifically about AI, they shared negative views that suggest a level of misinformation and distrust. However, participants were not aware that benefits such as finding information and getting recommendations about different products at the consideration stage of their customer journey are because of an AI algorithm. Therefore, it seems plausible that most participants did not understand AI for what it is.

Third, based on the content analysis of the two example organizations' publicly-facing content, in comparison to Defy, Bosch has managed to diffuse AI tools and technology more throughout its business model to make the customer experience seamless. The case study content analysis analyzed Bosch and Defy's websites and social media content to identify AI digital tools adopted by both companies. The comparison of Bosch and Defy's digital tools suggests the rate of AI technology adoption by companies varies with geographical locations. In comparison to Defy, Bosch has managed to diffuse AI tools and technology more throughout its business model to make the customer experience seamless. This is summarized in Table 2:

Table 2. Comparison of Bosch and Defy's AI digital tools

AI Digital Marketing Tools	Bosch	Defy
Virtual Reality (VR)	✓	x
Augmented Reality (AR)	✓	x
Internet of Things (IoT)	✓	x
Voice Assistant (VA)	✓	x
Lead generation / Retargeting	✓	x
Chatbots	✓	✓

Bosch and Defy (2021)

At the time of this study (2021), Bosch can be categorized as an early adopter." (Remove comment about Defy). Of note, the study was limited to two organizations as examples from these countries, and conducting the research in the USA may have resulted in the researchers falling out of scope for geo-location targeted ads, which is a limitation when discussing generalizability. This will be further discussed in the limitation section of this manuscript.

DISCUSSION

Consumer AI Perceptions and Concerns

This raises questions about artificial intelligence and cognitive dissonance. In mass media, AI has frequently been presented in the cultural narrative from a dystopian view which has in-turn affected diffusion of this innovation and willingness to adopt or accept it. Previous literature supports that consumers may feel manipulated in data capture interactions, considering AI's ability to anticipate and fulfill expectations, largely because they do not grasp the operational requirements of AI (Puntoni et al., 2021). Therefore, future studies could benefit the field by investigating the role of cognitive dissonance in artificial intelligence technology adoption by customers.

Both Bosch and Defy customers are suspicious about audience targeting tools being used to target them in their customer journey. Although predictive algorithms suggest that they are accurately targeting customers, these forms of targeting may be perceived as suspicious and invasive to privacy. It seems logical that trust in AI is one fundamental role that adoption plays in technology-mediated interactions between customers and these companies. While previous studies demonstrate a positive relationship between customer experience and efficiency, AI's acceptance depends on the level of sacrifice customers make to use that technology.

The study challenged and helped to identify the gap between AI perceptions and the likelihood of adoption at an individual level. There is no specific data collected from the research to suggest that adoption rates are reliant on geographical locations. The majority of both Bosch and Defy customers are unwilling to give up their personal information such as age, gender, contact number, and address to experience personalized services.

Participants highlighted that privacy invasion is their top AI concern and this is affecting their adoption of AI technology. Zenezini et al. (2016) state that data privacy and value creation are some of the key adoption barriers for smart home innovation. Participants fear robots will manipulate their personal information through data collection. Bentley et al. (2018) assert that individuals fear that machines are becoming super intelligent and will possibly exceed human intelligence because AI systems are continuously improving themselves. Moreover, in popular culture, the lack of possession of personal data has been associated with loss of personal autonomy resulting from the threatening ability of technologies to allow human activity to be tracked. Implications for the future are for policymakers to ensure consumer rights and privacy are safeguarded and not violated in the name of innovation. The value of the sacrifice is not worth the "risk" of their privacy being compromised. Khalifa (2004) states that what customers value varies between people, products, and timing and thus, there are several meanings of what customers truly value.

Over targeting was an important concern highlighted by one participant whereby customers continue to be targeted for the same type of products based on their previous search history. However, this can be identified as a disadvantage of AI-enabled tools or a pain point in the customer's pre-purchase customer journey stage because it prohibits customers from being targeted for other products that they may also be interested in. Although personalized ads allow companies to attract customers using geographic or demographic factors, there is a need to balance this content, because it can be easily frustrating to customers or prohibits them from seeing other unrelated products which they may potentially buy.

Organizational AI Adoption and Diffusion

At the organizational level, the results indicate that there is a significant difference in the rate of adoption of AI-based tools for companies situated in different geographical locations. Based on Rogers' (2003) DOI adopter categories, Bosch is classified as an early adopter. The case study revealed how they had incorporated AI-tools in their overall business model. From a consumer perspective, it can be said that the evidence supports Defy has not strongly invested in direct to consumer AI applications. However, from a B2B perspective, they may be classified higher toward early adopters or early majority in the diffusion of innovation categories. This is a study area for future researchers to conduct in-depth interviews with marketing practitioners within the organization to better understand detailed practices within the organization and targeting efforts as they relate to AI digital marketing tools.

As far as AI and inbound and outbound marketing are concerned, the results show that Defy is not making any retargeting or lead generation efforts to give their customers personalized ads or targeted ads. This is because their business model is mainly B2B (business to business) and all their distribution is done by dealers which brings an important question as to how AI may be relevant in their customer touchpoints. As such, this makes it difficult for Defy to generate personalized ads since most of the touchpoints in the customer journey are handled by their dealers/distributors. It seems reasonable that their B2C (business to consumer) efforts could be stronger even if they are not pursuing a sales approach, they can pursue a marketing approach by adopting AI tools in support of their customers throughout their customer journey. This also represents an opportunity for Defy to embrace AI and enjoy the benefits of marketing efforts such as lead generation and retargeting efforts. This is beneficial to both the customer and the company because it raises brand awareness and potential conversions. Additionally, the goal is for companies to reduce the number of pain-points a customer encounters along their customer journey by trying to ensure that their experience is seamless. Thus, to increase the customer lifetime value with the brand it is imperative to adopt AI digital marketing tools.

Defy does not have AR/VR tools available for their customers during pre-purchase and their post-purchase journey, whereas Bosch has AR/VR tools available in their home appliance and automotive subsidiary. This is both a gap and opportunity identified by this study, as the majority of customers for both companies researched the product online, read reviews on user-generated content, but still wished to go in-person to see and feel the dishwasher or washing machine (six out of eight participants made the final purchase in-person). This presents an opportunity for the companies to use AR/VR to enrich the user experience and increase engagement while reducing pain points in the customer journey. AR is finding its way to most households through smartphones and mobile devices (Newman & McClimans, 2019) and can help brands engage with customers and enhance the value of their products and services.

Study Strengths, Limitations, and Future Directions

Strength: Internal Consistency

It is important to ground the more definitive statements in this study in the internal agreement of its sources. A work may be deemed internally consistent when it includes enough sources to fully saturate a topic. For narrow topics, this source reports that only a few items might be necessary to accomplish this (Salkind, 2010). The present study allowed topics to pass from the open coding process into the axial coding process only after each one achieved some meaningful level of thematic saturation, as coded across multiple researchers. Topics that were only brought up in one or two insignificant instances were ignored, whereas topics that appeared consistently and throughout the sources were examined further in the axial coding process. This occurred consistently across the first two research stages, while the case study approach of the third research stage again involved multiple researchers on the team evaluating the data and achieving high intercoder reliability.

A very high level of consistency was observed via triangulation in the sources used across this study and within each phase. The fact that each source came from a slightly different perspective, with different goals, and using different presentations of content, demonstrates that the previously stated findings and results are worthy of a higher degree of trust. The comparative variety of input materials used in this study, across all phases, imply an absence of goal-related bias among the sources examined.

Limitations

While the authors of this chapter attempted to avoid bias by allowing the data to inform this study's research findings, no study is entirely without its limitations. This disparity between male and female participants in the second research stage may indicate differences in gender roles when shopping for home appliances. According to TraQline's (2016) market data and insights, 53 percent of the time women are primary decision-makers of furniture and home appliances while 27 percent are joint decisions. Regarding interview participants' income discrepancies, it could be argued that there is a relationship between disposable income and brand customer demographics. Bosch is considered to becoming a high-end appliance brand and customers within the 45- 54 age range typically have higher disposable income.

One key limitation of this study lies in the fact that its findings are discrete and targeted to a singular industry. While this qualitative approach was highly beneficial in that it allowed for greater detail in textual analysis and the unearthing of key thematic elements and their relationships to each other, this approach has not yielded a high level of generalizability. Although coding frames were developed based on frequencies, the sample and scope of this study cannot be generalized to the whole population. The research also utilized a convenience sample to recruit participants for the interviews based on willingness to participate.

Future Directions

This present study yielded specific results by looking qualitatively across a handful of sources in three research phases; however, future research could achieve success by employing a significantly larger sample size and employing a quantitative methodology. Such an approach could be used to research the broader societal impressions, expectations, and hopes for artificial intelligence within content strategy and the customer journey.

Related to AI hopes, understanding the AI relationship with the customer journey will be of ever increasing interest as that journey expands further beyond physical reality and into the metaverse. Studies are only beginning to explore the relationship of AI, marketing and brand goals, and customer journey within this realm. However, early research supports that there is opportunity for brands to utilize AI to better meet customer needs in this space through realization of AI-enabled "in-store" augmented or virtual experiences (Nalbant & Aydin, 2023). Research on how to optimize brand presence and the customer journey in the metaverse should ideally continue to build in coming years.

CONCLUSION

CNBC published the results of a survey it conducted of Americans in the workforce in 2019 that showed the largest category of employees who feared job displacement due to AI were in the category of "Advertising and marketing," with 45 percent of advertising and marketing employees expressing concern about the potential for their jobs to be replaced by artificial intelligence (Douglas, 2019). Based upon our content analysis of anticipated AI trends, this study found that AI's most likely future in content strategy will not displace a creative human workforce, but rather assist that workforce and the organizations in which they are employed.

After comparing the research data, several factors affect both customer and organizational AI adoption, for example, perceptions, usability, knowledge, risks, barriers to adoption, social factors, and infrastructure. Privacy invasion is an important factor when customers determine whether to adopt or not an innovation, therefore there are implications for the future for policymakers to make sure consumer rights and privacy are protected and not violated in the name of innovation. Moreover, how an individual recognizes and perceives AI affects their adoption of innovation.

The results further revealed that at the time of this study, Bosch is in the early majority and Defy is lagging in relation to AI adoption in direct to consumer applications, although they may perform stronger in business to business applications. This study confirms that in the context of diffusion of innovation adoption varies at the geographical level. Therefore, DOI was shown to be an effective theory to assess AI adoption for both individuals and organizations.

REFERENCES

Advertising, M. (2019). *Mastering the customer journey: applying the predictive power of Artificial Intelligence* [White Paper] Microsoft. https://advertiseonbing-blob.azureedge.net/blob/bingads/media/insight/ebook/2019/10-october/cdj-chapter-2/cdj_mini_ebook_uk.pdf

Allen, M. (2017). *The sage encyclopedia of communication research methods* (Vol. 1-4). SAGE Publications, Inc., doi:10.4135/9781483381411

Ameen, N., Tarhini, A., Reppel, A., & Anand, A. (2021). Customer experiences in the age of Artificial intelligence. *Computers in Human Behavior, 114*, 106548. doi:10.1016/j.chb.2020.106548 PMID:32905175

Bacon, J. (2017). The future of creativity in an automated world. *Marketing Week.* https://www.marketingweek.com/future-creativity-automated-world/

Batra, M. M. (2019). Strengthening Customer Experience through Artificial Intelligence: An Upcoming Trend. *Competition Forum, 17*(2), 223-231. https://login.iris.etsu.edu:3443/login?url=https://www.proquest.com/docview/2343014949?accountid=10771

Bosch. (2021). *Home Appliances.* Bosch Invented for Life. https://www.bosch-home.com/us/

Brenner, M. (2018). Why AI is better than A/B Testing. *Concured.* https://www.concured.com/blog/why-ai-is-better-than-a/b-testing

Bryant, A. (2017). *Grounded theory and grounded theorizing: Pragmatism in research practice*. Oxford University Press. doi:10.1093/acprof:oso/9780199922604.001.0001

Car, J., Sheikh, A., Wicks, P., & Williams, M. (2019). Beyond the hype of big data and artificial intelligence: Building foundations for knowledge and wisdom. *BMC Medicine*, *17*(143), 143. doi:10.118612916-019-1382-x PMID:31311603

Chalmers, D., MacKenzie, N. G., & Carter, S. (2020). Artificial Intelligence and Entrepreneurship: Implications for Venture Creation in the Fourth Industrial Revolution. *Entrepreneurship Theory and Practice*. doi:10.1177/1042258720934581

Chen, G., Xie, P., Dong, J., & Wang, T. (2019). Understanding Programmatic Creative: The Role of AI. *Journal of Advertising*, *48*(4), 347–355. doi:10.1080/00913367.2019.1654421

Crevier, D. (1993). *AI: The Tumultuous Search for Artificial Intelligence*. BasicBooks.

Davenport, T., Guha, A., Grewal, D., & Bressgott, T. (2020). How artificial intelligence will change the future of marketing. *Journal of the Academy of Marketing Science*, *28*(1), 24–42. doi:10.100711747-019-00696-0

Defy. (2021). Defy world-class kitchen appliances. *Defy*. https://www.defy.co.za/appliances

Douglas, J. (2019). These American workers are the most afraid of A.I. taking their jobs. *CNBC News*. https://www.cnbc.com/2019/11/07/these-american-workers-are-the-most-afraid-of-ai-taking-their-jobs.html

Duffy, M. (2017). Code eats copy for breakfast: Human copywriters are doomed. *Digiday*. https://digiday.com/marketing/humanoid-copywriters-good-as-dead/

Edelman, D. C., & Singer, M. (2015). Competing on Customer Journeys. *Harvard Business Review*, (November), 88–100.

Evans, M. (2019). Build A 5-star customer experience with artificial intelligence. *Forbes*. https://www.forbes.com/sites/allbusiness/2019/02/17/customer-experience-artificialintelligence/#1a30ebd415bd

Følstad, A., Kvale, K., & Halvorsrud, R. (2013). *Customer journey measures – State of the art research and best practice*. Report A24488, Oslo, Norway: SINTEF.

Glaser, B. G., & Strauss, A. L. (1965). *Awareness of dying*. Aldine.

Glaser, B. G., & Strauss, A. L. (1967). *The discovery of grounded theory*. Aldine.

Grewal, D., & Roggeveen, A. (2020). Understanding Retail Experiences and Customer Journey Management. *Journal of Retailing*, *96*(1), 3–8. doi:10.1016/j.jretai.2020.02.002

Hall, J. (2019). How Artificial Intelligence Is Transforming Digital Marketing. *Forbes*. https://www.forbes.com/sites/forbesagencycouncil/2019/08/21/how-artificial-intelligence-is-transforming-digital-marketing/#379462ee21e1

He, A. Z., & Zhang, Y. (2022). AI-powered touch points in the customer journey: A systematic literature review and research agenda. *Journal of Research in Interactive Marketing*, 1–20. doi:10.1108/JRIM-03-2022-0082

Hyken, S. (2009). *The Cult of the Customer*. John Wiley & Sons.

Insights, D. (2020). *Tech Trends 2020* [White Paper]. Deloitte. https://www2.deloitte.com/content/dam/Deloitte/pt/Documents/tech-trends/TechTrends2020.pdf

Kietzmann, J., Paschen, J., & Treen, E. (2018). Artificial Intelligence in Advertising: How Marketers Can Leverage Artificial Intelligence Along the Consumer Journey. *Journal of Advertising Research*, *58*(3), 263–267. doi:10.2501/JAR-2018-035

Kietzmann, J., & Pitt, L. (2020). Artificial intelligence and machine learning: What managers need to know. *Business Horizons*, *63*(2), 131–133. doi:10.1016/j.bushor.2019.11.005

Konecki, K. (2011). Visual Grounded Theory: A Methodological Outline and Examples from Empirical Work. *Revija za Sociologiju*, *41*, 131–160. doi:10.5613/rzs.41.2.1

Lemon, K., & Verhoef, P. (2016). Understanding Customer Experience Throughout the Customer Journey. *Journal of Marketing*, *80*(6), 69–96. doi:10.1509/jm.15.0420

Li, H. (2019). Special Section Introduction: Artificial Intelligence and Advertising. *Journal of Advertising*, *48*(4), 333–337. doi:10.1080/00913367.2019.1654947

Lund, B., Omame, I., Tijani, S., & Agbaji, D. (2020). Perceptions toward Artificial Intelligence among Academic Library Employees and Alignment with the Diffusion of Innovations' Adopter Categories. *College & Research Libraries*, *865*, 865. doi:10.5860/crl.81.5.865

Lwin, M. O., Wirtz, J., & Stanaland, A. J. S. (2016). The privacy dyad. *Internet Research*, *26*(4), 919–941. doi:10.1108/IntR-05-2014-0134

McKinsey & Company. (2018 July 26). *Artificial intelligence: Why a digital base is critical.* McKinsey Quarterly. https://www.mckinsey.com/business-functions/mckinsey-analytics/our-insights/artificial-intelligence-why-a-digital-base-is-critical#

Merriam, S. (1998). *Qualitative research and case study applications in education*. Jossey-Bass.

Mogaji, E., Olaleye, S., & Ukpabi, D. (2020). *Using AI to Personalise Emotionally Appealing Advertisement*. Digital and Social Media Marketing. doi:10.1007/978-3-030-24374-6_10

Mohannad, A. M. A. D., Smoudy, A. K. A. (2019). The Role of Artificial Intelligence on Enhancing Customer Experience. *International Review of Management and Marketing, 9*(4), 22-31. https://www.proquest.com/docview/2288760601?accountid=10771

Moore, S., Bulmer, S., & Elms, J. (2022). The social significance of AI in retail on customer experience and shopping practices. *Journal of Retailing and Consumer Services*, *64*, 102755. doi:10.1016/j.jretconser.2021.102755

Morgan, B. (2018). 3 Use Cases of Artificial Intelligence for Customer Experience. *Forbes*. https://www.forbes.com/sites/ blakemorgan/2018/08/01/3-use-cases-of-artificial-intelligence-for-customer-experience/#1f084b6e5e34

Nalbant, K. G., & Aydin, S. (2023). Development and Transformation in Digital Marketing and Branding with Artificial Intelligence and Digital Technologies Dynamics in the Metaverse Universe. *Journal of Metaverse*, *3*(1), 9–18. doi:10.57019/jmv.1148015

Newman, D., & McClimans, F. (2019). *EXPERIENCE 2030: The Future of Customer Experience is … NOW!* [White Paper]. Futurum. file:///C:/Users/tmtyp/OneDrive/2nd%20Year%20Fall/Thesis/Articles%20 Journals/futurum-experience-2030-110966.pdf

Nguyen, T. M., Quach, S., & Thaichon, P. (2022). The effect of AI quality on customer experience and brand relationship. *Journal of Consumer Behaviour*, *21*(3), 481–493. doi:10.1002/cb.1974

Nicolescu, L., & Tudorache, M. T. (2022). Human-Computer Interaction in Customer Service: The Experience with AI Chatbots—A Systematic Literature Review. *Electronics (Basel)*, *11*(10), 1579. doi:10.3390/electronics11101579

Overgoor, G., Chica, M., Rand, W., & Weishampel, A. (2019). Letting the Computers Take Over: Using AI to Solve Marketing Problems. *California Management Review*, *61*(4), 156–185. doi:10.1177/0008125619859318

Panda, G., Upadhyay, A. K., & Khandelwal, K. (2019). Artificial Intelligence: A Strategic Disruption in Public Relations. *Journal of Creative Communications*, *14*(3), 196–213. doi:10.1177/0973258619866585

Papies, D., & Clement, M. (2008). Adoption of New Movie Distribution Services on the Internet. *Journal of Media Economics, 21*(3), 131–157. doi:10.1080/08997760802300530

Paradiso, C. (2016). Artificial Intelligence in Digital Marketing. *Insurance Advocate*, *127*(14), 12–14.

Park, Y., & Chen, J. (2007). Acceptance and adoption of the innovative use of smartphone. *Industrial Management & Data Systems*, *107*(9), 1349–1365. doi:10.1108/02635570710834009

Pecherskiy, V. (2017). Will AI Replace Creative Jobs? *Forbes*. https://www.forbes.com/sites/forbescommunicationscouncil/2017/10/11/will-ai-replace-creative-jobs/#6802eca296a2

PEGA. (2020). *The future of work: New perspectives on disruption & transformation*. PEGA. https://www.pega.com › pega-future-of-work-report

Rogers, E. (2003). *Diffusion of Innovations* (5th ed.). Free Press.

Salkind, N. (2010). Internal Consitency Reliability. Encyclopedia of Research Design. doi:10.4135/9781412961288.n191

Scott, D. (2013). *The New Rules of Lead Generation : Proven Strategies to Maximize Marketing ROI*. AMACOM.

Shank, D. B., Graves, C., Gott, A., Gamez, P., & Rodriguez, S. (2019). Feeling our way to machine minds: People's emotions when perceiving mind in artificial intelligence. *Computers in Human Behavior*, *98*, 256–266. doi:10.1016/j.chb.2019.04.001

Siebert, A., Gopaldas, A., Lindridge, A., & Simões, C. (2020). Customer Experience Journeys: Loyalty Loops Versus Involvement Spirals. *Journal of Marketing*, *84*(4), 45–66. doi:10.1177/0022242920920262

Straub, E. T. (2009). Understanding Technology Adoption: Theory and Future Directions for Informal Learning. *Review of Educational Research*, *79*(2), 625–649. www.proquest.com/docview/214121884?accountid=10771. doi:10.3102/0034654308325896

Strauss, A. L. (1987). *Qualitative Analysis for Social Scientists*. Cambridge University Press. doi:10.1017/CBO9780511557842

Surry, D. W., & Farquhar, J. D. (1997). Diffusion theory and instructional technology. *Journal of Instructional Science and Technology*, *2*(1), 24–36.

Trawnih, A., Al-Masaeed, S., Alsoud, M., & Alkufahy, A. (2022). Understanding artificial intelligence experience: A customer perspective. *International Journal of Data and Network Science*, *6*(4), 1471–1484. doi:10.5267/j.ijdns.2022.5.004

Xu, Y., Shieh, C., van Esch, P., & Ling, I. (2020). AI customer service: Task complexity, problem-solving ability, and usage intention. *Australasian Marketing Journal*, *28*(4), 189–199. doi:10.1016/j.ausmj.2020.03.005

Yang, G. (2006). AI Winter and its lessons. *History of Computing* [PDF file]. https://courses.cs.washington.edu/courses/csep590/06au/projects/history-ai.pdf

Zenezini, G., & Ghajargar, M., & Fiore, E., & De Marco, A. (2016). *The Smart Home Services Diffusion Process: A System Dynamics Model.*

ADDITIONAL READINGS

Ameen, N., Tarhini, A., Reppel, A., & Anand, A. (2021). Customer experiences in the age of Artificial intelligence. *Computers in Human Behavior*, *114*, 106548. doi:10.1016/j.chb.2020.106548 PMID:32905175

Batra, M. M. (2019). Strengthening Customer Experience through Artificial Intelligence: An Upcoming Trend. *Competition Forum, 17*(2), 223-231. https://login.iris.etsu.edu:3443/login?url=https://www.proquest.com/docview/2343014949?accountid=10771

Nalbant, K. G., & Aydin, S. (2023). Development and Transformation in Digital Marketing and Branding with Artificial Intelligence and Digital Technologies Dynamics in the Metaverse Universe. *Journal of Metaverse*, *3*(1), 9–18. doi:10.57019/jmv.1148015

Nguyen, T. M., Quach, S., & Thaichon, P. (2022). The effect of AI quality on customer experience and brand relationship. *Journal of Consumer Behaviour*, *21*(3), 481–493. doi:10.1002/cb.1974

Trawnih, A., Al-Masaeed, S., Alsoud, M., & Alkufahy, A. (2022). Understanding artificial intelligence experience: A customer perspective. *International Journal of Data and Network Science*, *6*(4), 1471–1484. doi:10.5267/j.ijdns.2022.5.004

KEY TERMS AND DEFINITIONS

Artificial Intelligence (AI): When computers complete tasks that have historically been completed by humans and previously required human intelligence (for example, perception, recognition, content creation, and decision making).

Augmented Reality (AR): When computer-generated information and images are imposed upon a view of the real world.

Chatbot: A computer program supported online service provided by some organizations that mimics human conversation and can answer many users' questions.

Content Strategy: How an organization uses content (e.g., digital and nondigital communications) to meet both strategic business goals and objectives, while also effectively delivering against audience needs.

Customer Experience: The resulting perceptions and feelings a customer has about a brand based upon their customer journey.

Customer Journey: The cumulative series of interactions a person has with a brand across many touchpoints, both offline and online.

Diffusion of Innovation: The process by which innovative ideas, products, and services spread through a population or geographic area over time.

Internet of Things (IoT): Internet-connected devices people may use in their everyday lives.

Lead generation: The identification of "leads" or prospective customers, including their contact information

Retargeting: Delivering ads or other forms of online communication to those who have had previous digital interactions with your brand

Virtual Reality (VR): A computer simulated 3-D environment with (and within) which users may interact

Voice Assistant (VA): A software based program that recognizes human speech and is capable of carrying out commands

Chapter 7
Digital Twins:
Accelerating the Digital Transformation in the Chinese Transportation Sector

Poshan Yu
https://orcid.org/0000-0003-1069-3675
Soochow University, China & Australian Studies Centre, Shanghai University, China

Jiarui Song
Independent Researcher, China

Michael Sampat
Independent Researcher, Canada

Emanuela Hanes
Independent Researcher, Austria

ABSTRACT

This chapter takes the Tsinghua University Suzhou Automotive Research Institute, Shenzhen Bao'an International Airport, and Guizhou "Intelligent Transportation Cloud" as examples to discuss the roles of digital twins in accelerating the transformation of transportation digitalization and the impact of digitalization on relevant enterprises. This chapter will focus on the models of how digital transformation can help stakeholders in the transportation industry create a better-connected experience. In addition, CiteSpace will be used in this chapter to visualize and analyze the digital twins/transportation/digital transformation documents in the database of the web of science (WOS) from 2017 to 2022. The authors also conducted a survey to analyze how the digital twins of China's transportation industry promote its digital transformation. This chapter will provide suggestions for the problems in the digital transformation of the transportation industry.

DOI: 10.4018/978-1-6684-6366-6.ch007

INTRODUCTION

Transportation is an important link connecting the city, and it is also an important channel for the development of the city to transport passenger flow and logistics. The impact and role of information technology on urban transportation may go beyond infrastructure systems, so the development goal of transportation informatization can not only stay in the advanced nature of intelligent transportation technology itself but should be consistent with urban development goals (Wang, 2015). China is a big transportation country. Digital transformation in transportation is the use of digitalization to empower transportation in terms of distribution, management and regulation. This requires opening up and connecting data information scattered in different departments to achieve holographic perception, real-time judgment and global decision-making capabilities for pedestrians, vehicles and road conditions. Only digital transformation can make China move from a transportation power to a transportation power. As early as the 1990s, countries around the world began to gradually carry out the international standardization of intelligent transportation (ITS), and China also opened research on the digital transformation of transportation. This shows that the digital transformation of China's transportation sector is an inevitable trend.

At present, China's traffic is facing problems such as complex driving scenarios, traffic congestion and parking difficulties. However, residents have higher and higher requirements for the safety and convenience of transportation. At this time, the real-time, closed-loop digital twin technology to empower smart transportation has become a new direction for future transportation development. Digital twin technologies can empower intelligent transportation in many ways to meet the needs of future mobility. First of all, the digital twin can collect data in real-time and synchronize traffic operation visualization, providing a test space for traffic model deduction and completing data-driven decision-making. Second, the data generated by the digital twin virtual space can also provide a high-precision map. Digital twin technology based on real data and virtual models can improve the safety and stability of intelligent driving, thereby accelerating the safer landing and promotion of intelligent driving. In addition, the road surface in urban areas is relatively complex, the traffic flow changes greatly, and accurately quantifying dynamic urban traffic is a difficult point for modern transportation. Digital twins can achieve insight into urban traffic dynamics by aggregating all factor data to simulate the city's traffic that has been in motion. From the perspective of the whole city, the overall situation is planned to achieve the optimal solution for congestion control.

Using CiteSpace, a visualization analysis tool, you can see the whole process of digital twin technology being applied in China. This paper analyzes a survey published in 25 provinces (municipalities and autonomous regions) including Jiangsu Province and Sichuan Province in China, with the theme of "Applying Digital Twin Technologies in China's Transportation Sector to Promote Its Digital Transformation", and analyzes the 603 questionnaires collected, giving a brief analysis of the digitization process and the extent of the use of digital twins in China's transportation sector. Taking the "intelligent transportation cloud" in Guizhou Province and the pilot digital twin transportation system in Guiyang City, the intelligent transportation hub of Xi'an Railway Station North Square and the digital transformation of Shenzhen Bao'an International Airport as examples, this paper analyzes the inevitability of digitalization in the transportation field and the role of digital twins in pushing the digital process in the field of transportation.

LITERATURE REVIEW

The Connection of Digital Twins to Artificial Intelligence, Big Data, Augmented Reality, Simulation

Digital twins are interactive mappings between physical and digital virtual spaces (Tao et al., 2018). Digital twins use technologies such as the Internet of Things to transform data and information from the physical world into general data, replicating physical entities in the digital world (Tao & Zhang, 2017). At the same time, the use of artificial intelligence, big data and cloud computing technology for intelligent decision-making and other common applications, improve the capabilities of various vertical industries. Ai-powered people can process massive amounts of data and automatically optimize systems in the digital twin ecosystem, which is the central brain of the digital twin ecosystem and one of the core technologies of the digital twin ecosystem (Lv & Xie, 2021). Safe integration of AI-based decision-making, traffic management, route planning, transportation network services, and other mobility optimization tools enables efficient traffic management. According to the big data technology reference box and key technologies introduced in the "Big Data Standards White Paper" (2021), in terms of information flow dimension, the value of big data is realized through the collection, integration, analysis and use of data results (China Communications Institute, 2021). Therefore, big data is an indispensable part of the application of digital twin technology in the process of information integration in the field of transportation. AR is a technique that calculates the position and angle of a camera image in real-time, supplemented by the corresponding image. The technology can overlay the virtual world with the real world in the lens display through holographic projection, and the operator can interact through the device (Li et al., 2021). By overlaying the display device to track the surrounding environment, pedestrians and vehicles in real-time, it is convenient to navigate pedestrians and provide safety tips for car drivers. The digital twin virtual space was created with a complex individual in the vehicle-driver-road system in mind, so virtual and physical simulation has the potential to help simulate the most realistic state of the vehicle according to the different states in which the vehicle is in (Dygalo et al., 2020).

Digital Twin Technologies Are Applied to Intelligent Transportation Systems to Facilitate Their Digital Transformation Process

The development of intelligent transportation systems has great potential and capabilities in the smart cities of the future (Liao et al., 2021). Digital twin technologies can provide cities with reliable and efficient transportation systems (Menouar et al., 2017). The implementation of intelligent transportation systems (ITS) based on digital twins, already in Places such as Russia and MRA, shows that ITS has a remarkable role in improving road safety and optimizing the temporal and spatial distribution of traffic flows in networks (Rudskoy et al., 2020). Wireless communication technology applied to intelligent transportation systems can not only realize the collection of the vehicle's information, the perception of driving environment information, and the planning and tracking of vehicle trajectory, but it can also effectively change the path in the congestion process (Lv et al., 2020).

The process of digital transformation in China is also underway. Guizhou pioneered the digital twin transportation system by importing real-world traffic elements into the digital twin virtual space to simulate various situations in the real world. Guangzhou Airport High-speed Toll Station creates the "Eye of Wisdom" - integrating big data, 5G, cloud computing and other technical means, applying the digital twin system to smart transportation construction, realizing functions such as ETC charging, accurate guidance, non-leakage of receivables, and abnormal charging reminders, which greatly improves the service level of toll stations and the driving experience of drivers and passengers, thereby improving the overall operational efficiency and intelligent operation level of toll stations (Guangzhou Daily, 2021). To improve the level of project design and construction management, project construction management based on the digital twin model was carried out in the construction stage of the Du'an-Bama Expressway, and the project construction process was managed in a refined manner based on digital twin technology (Liang, 2020).

IMPACT OF THE GOVERNMENT POLICY

China's transportation sector is undergoing continuous digital transformation, and the scale of the intelligent transportation market is also growing. As shown in Figure 1, during the period from 2010 to 2018, the scale of China's intelligent transportation market increased year by year with a compound growth rate of 23.33%, and the market size of China's intelligent transportation management system industry reached 72 billion yuan in 2018. As of 2018, more than 200 large and medium-sized cities in China have established urban transportation command centers, and the scale of urban intelligent transportation investment has reached 45 billion yuan. As an important development direction of intelligent transportation, digital twins have great development space and potential.

Figure 1. Market size of intelligent transportation systems in China from 2010 To 2018
Source: Prospective Industry Research Institution

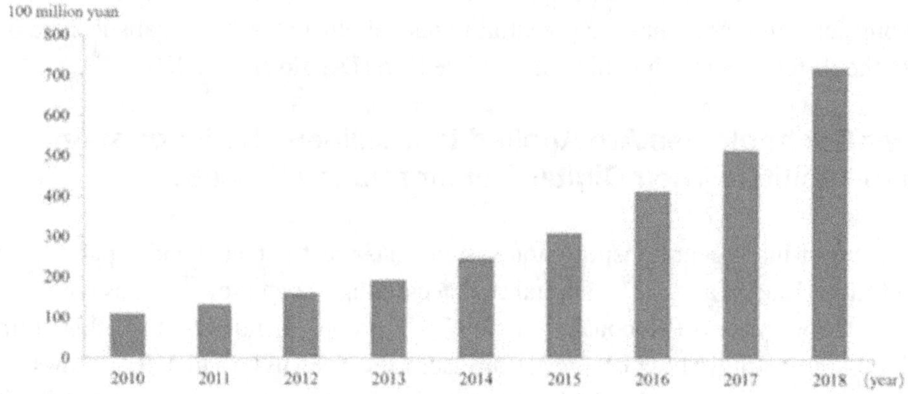

To promote the digital transformation of China's transportation industry, the government has issued a series of policies to ensure the sustainable development of digital transformation. China attaches great importance to the development of digital transportation, providing a good policy environment for the application of digital twins in the field of transportation. Table 1 below shows some of the policies that promote the digital transformation of the transport sector.

Table 1. Digital transformation in the transportation sector

Field	Policy	Authority	Issue Date	Main contents
Digital transformation in the transportation sector	Outline of Digital Transportation Development Plan	The Ministry of Transport	July 25, 2019	It is proposed to take the "data chain" as the main line to build a digital collection system, networked transmission system and intelligent application system, accelerate the development of transportation informatization to digital, networked and intelligent, and provide support for the construction of a strong transportation country.
	Outline for the Construction of a Transportation Powerhouse	The Central Committee of the Communist Party of China and the State Council	September 19, 2019	The Outline proposes to promote the transformation of transportation development from pursuing speed and scale to paying more attention to quality and efficiency, from the relatively independent development of various modes of transportation to the development of comprehensive transportation, and from relying on traditional factors to focusing more on innovation.
	Action Plan for Promoting the Development of Comprehensive Transportation Big Data (2020-2025)	The Ministry of Transport	December 12, 2019	The ultimate goal is to promote the deep integration of big data and comprehensive transportation, effectively build a comprehensive transportation big data center system, and provide strong support for accelerating the construction of a strong transportation country. The main tasks include: consolidating the foundation of big data development, further promoting the sharing and opening up of big data, comprehensively promoting the innovative application of big data, strengthening the security of big data, and improving the big data management system.
	Outline of the National Comprehensive Three-dimensional Transportation Network Planning	The Central Committee of the Communist Party of China and the State Council	February 14, 2021	It is clear that the national comprehensive three-dimensional transportation network will be the highest-level space network of China's transportation infrastructure and the foundation of the comprehensive transportation system. We will promote the planning and compilation of the national comprehensive three-dimensional transportation network, further improve the layout of comprehensive transportation infrastructure, and promote the construction of a strong transportation country.
	The 14th Five-Year Development Plan for Digital Transportation	The Ministry of Transport	October 25, 2021	The plan defines the overall goals, specific goals, construction tasks and safeguard measures of China's digital transportation development during the 14th Five-Year Plan period. It is a programmatic document to guide the development of digital transportation in China in the next five years.

In January 2017, the Ministry of Transport issued the Action Plan for Promoting the Development of Smart Transportation (2017-2020), which promotes the application of building information models in the whole life cycle of planning, construction and operation management of major transportation infrastructure projects, and improves the level of decision-making and supervision based on big data. It is not difficult to know from here that this policy fully affirms the important role of Building Information Modeling (BIM) in the development of intelligent transportation. In September 2019, the CPC Central Committee and the State Council issued the Outline for the Construction of a Transportation Powerhouse, which proposes to vigorously develop intelligent transportation and proposes to promote the deep integration of new technologies such as big data, the Internet, and artificial intelligence with the transportation industry. It is required to improve the infrastructure layout, three-dimensional interconnection, and modern high-quality comprehensive three-dimensional transportation network; develop technologies such as automatic driving and vehicle-road coordination; widely apply intelligent high-speed rail, intelligent roads, intelligent shipping, etc., and strive to achieve convenient, comfortable, and cost-effective transportation services. It is the top-level design and system planning of China's transportation development, and the top-level deployment of modern comprehensive transportation bodies has been completed (Xinhua net, 2019). In July 2019, the Ministry of Communications issued the Outline of the Digital Transportation Development Plan, which proposed the use of data-based and panoramic display methods to improve the support capabilities of comprehensive transportation operation monitoring and early warning, online monitoring and so on. With data as the key element, we empower transportation and related industries to improve the quality of travel and logistics services. That is, through the new generation of information technology such as digital twins, the information on road vehicle status and road surface condition is collected, and the operation trajectory is designed for logistics and transportation. Not only that, but possible emergencies (such as ambulances and fire trucks) can also be macro-regulated.

The Ministry of Industry and Information Technology, the Ministry of Science and Technology and other ministries and commissions have also issued policies to promote the digital transformation of the transportation field from multiple dimensions such as key technology research and development, technology application innovation, and industrial integration development. In December 2017, the Ministry of Industry and Information Technology issued the "Three-Year Action Plan for Promoting the Development of a New Generation of Artificial Intelligence Industry (2018-2020)", which applies artificial intelligence to the transportation system, focusing on bringing together a certain scale of industry application data in transportation and other fields to support entrepreneurship and innovation. In June 2019, the Ministry of Science and Technology took "integrated transportation and intelligent transportation" as a key special project, proposing intelligent coordination of means of transport, supervision and coordination of traffic operation, multi-mode integrated transportation integration, and comprehensive transportation safety risk prevention and control and emergency rescue four directions, and provide special scientific research funding support to strengthen the transformation and application of research results in the field of intelligent transportation (Ministry of Science and Technology of the People's Republic of China, 2019).

Translation and application of research results in the field of transportation (Ministry of Science and Technology of the People's Republic of China, 2019).

By 2020, the "Smart Transportation Makes Travel More Convenient Action Plan (2017-2020)" issued by the Ministry of Transport has reached its final year. In the past five years, the intelligence of intercity transportation travel has gradually become higher, first of all, to expand the market-oriented application of railway passenger transport information, the government has greatly supported the China Railway Corporation to study the conditions for the orderly opening of the railway ticket system, provided the society with passenger information such as train numbers, stops, remaining tickets, and train delays according to the principle of marketization, and carried out the construction of a road passenger transport network ticketing system. It has promoted comprehensive information on various modes of transportation in the relevant catering and service industries to make citizens' journeys smoother and more time-saving (Xinmin Evening News, 2020). To accelerate the expansion and application of expressway ETC, the government has also formulated the "Guiding Opinions on Promoting the Healthy Development of the Application of Expressway Electronic Non-Stop Toll (ETC) Systems". At the same time, the Chinese government has further regulated the development of shared bicycles and issued the "Guiding Opinions on Encouraging and Regulating the Development of Internet Rental Bicycles" to encourage and regulate the development of Internet rental bicycles. In addition, to accelerate the implementation of the "13th Five-Year Plan" Modern Comprehensive Transportation System Development Plan issued in 2017, the government has carried out customized passenger transportation and smart airport construction and other actions (China Communications News Network, 2019).

ADDED VALUE

Building virtual transportation systems in the digital space for simulation, experimentation, and trial and error can reduce the cost of physical ontologies and reduce waste. At the same time, it can provide a decision-making basis for urban transportation authorities to formulate emergency plans and provide technical support for management departments to optimize traffic management and scheduling. Marai et al. (2020) considered the physical vulnerability of current urban transportation and used new media to collect real-time data for assessment, which could help urban transportation decision-makers analyze the potential risks associated with data-driven infrastructure management during extreme weather events. Based on a virtual environment, it provides dynamic route planning for emergency vehicles such as ambulances and fire trucks to save lives. In addition, for emergencies, it can also be done quickly, three-dimensional presentation of geographical location, movement trajectory, and other information, and ultimately provide an important reference value for the planning and construction of the entire city (Deng et al., 2021).

In terms of safety, from a technical point of view, building a digital twin of a transportation system is conducive to increasing safety. Aksenov et al. (2022) mentioned that by constructing a physical space-filling information model based on the rail transit machine station system, using digital twin technology, coupled with the method explored by Krugilikov et al. (2013) to form a trajectory model in the context of collecting ballast prism state data through the "elastic wave" method, will facilitate the development of safety measures to reduce the risk of injury to maintenance employees during the introduction of new mechanized maintenance equipment and rail complexes. Pedersen et al. (2020) investigate an autonomous ship system based on autonomous navigation systems and DTs technology that combines sensors with new media technologies to generate test scenarios and assess the safety of ship navigation. Liu et al. (2021) propose that the maritime transport system based on the digital twin Internet of Things presents extraordinary transmission and security performance, achieving the safe transmission of data. Digital twins are also applicable in the field of drones. Wang et al. (2022) through research, it has been found that the construction of a drone digital twin system can significantly improve the safety performance of drones when flying in their airspace.

Figure 2. The structure of the digital twin system based on road-side sensing
Source: Guo et al., 2021

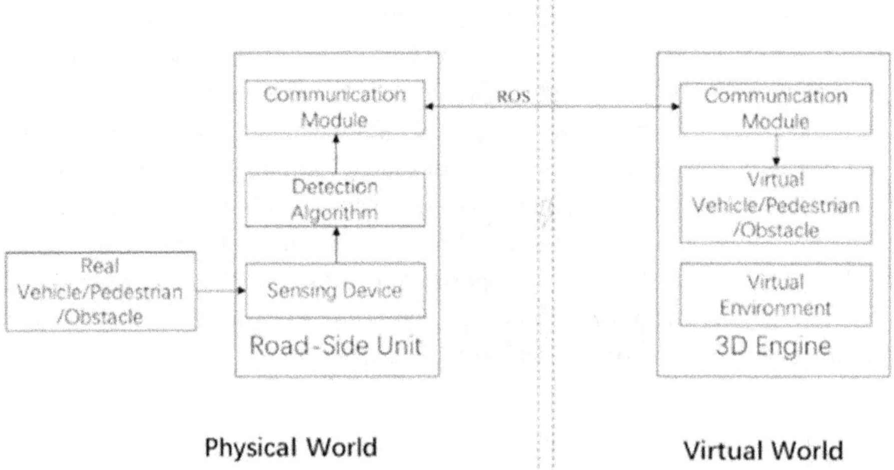

Digital twins can also be used in smart high-speed, vehicle-to-vehicle collaboration, and autonomous driving. The all-weather access system is one of the key applications of the current construction of smart high-speed-based digital twin technology (Tao et al., 2021). Through the sensors deployed at both ends of the road, the data information of vehicles and roads is collected in real-time and processed by digital twin technology, and the road information travelled by vehicles is uploaded to the digital twin visualization platform synchronously to help traffic managers make early warning judgments about the road environment (China Automotive News, 2021). Through data fusion docking, infrastructure cloud platform, big data center, vehicle-road collaborative business supervision and management, etc., the vehicle-road collaboration platform constructs a digital twin visualization and interaction system of traffic simulation "one map", reproduces the macro- and microscopic traffic flow operation process, provides reliable tools for traffic flow problems such as congestion traceability, and provides managers with a reliable decision-making basis (Zhao et al., 2021). Digital twins can also improve the accuracy of smart driving tests.

By building a real-world 1:1 virtual scene, it restores the operation rules of the physical world, meets the training needs of artificial intelligence algorithms in intelligent driving scenarios, and greatly improves training efficiency and safety. For example, by collecting laser point cloud data, establishing high-precision maps, and constructing a digital twin model of autonomous driving (The Paper, 2020). From pure models to semi-physical simulations, from closed sites to open sites, and finally to the process of road testing, and finally to achieve safe and reliable autonomous driving (Chen et al., 2022).

In the process of digitization, high-speed rail stations, airports and other places have introduced life brands, which are not only convenient for citizens' lives but also fast and efficient. Many high-speed rail stations and airport face recognition "paperless" ticket inspection methods save passengers the trouble of queuing up for formalities on site, and the journey process has become greener and more convenient. At the same time, many high-speed rail stations and airports have introduced affordable internet celebrity catering and lifestyle brands. In using these merchants to buy things can use the WeChat Mini Program

Figure 3. Functional architecture of autonomous driving systems
Source: Tsinghua University Suzhou Automotive Research Institute

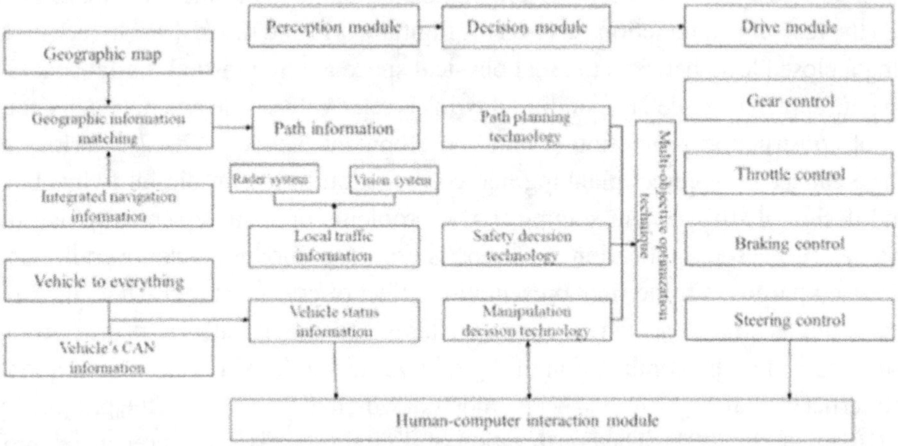

to order, eliminating the wasted time of on-site ordering, but also can better plan the time, in the same as the mall site can also be used to relieve the troubles caused by transportation delays.

With the support of digital twins and other technologies, China's modern comprehensive three-dimensional transportation network is becoming more and more perfect. China adheres to the leadership policy of transportation first and has built the world's largest high-speed railway network, highway network, world-class port cluster, aviation and navigation to the world, comprehensive transportation network exceeded 6 million kilometers, by the end of 2021, and the total railway operating mileage of the country has exceeded 150,000 kilometers, of which high-speed rail is more than 40,000 kilometers. The total length of highways in the country exceeds 5.2 million kilometers, and the expressway covers 98% of cities with a population of more than 200,000; about 2,660 berths of 10,000 tons or more in ports nationwide; the operating mileage of urban rail transit is 8,708 kilometers, and the rail transit of megacities is accelerated into the network; the number of nationally certified transportation airports has reached 248; and the proportion of rural poor areas that can send express delivery into them is more than 80%. This represents the continuous improvement of the terminal service system of the transportation industry; the construction of multi-level integrated comprehensive transportation hubs has been accelerated, "6 main axes, 7 corridors and 8 channels" The main skeleton space of the national comprehensive three-dimensional transportation network has initially taken shape (People's Daily, 2022).

EXISTING CHALLENGES

Although digital twins are at the forefront of smart transportation, there are pilot projects or actual projects in China based on digital twins such as vehicle-road collaboration, autonomous driving and smart high-speed, but there is still a certain gap between the digital twin transportation system of true global management, synchronous visualization, and virtual and real interaction (Communication Information News, 2021). According to the "Application Layer and Application Data Interaction Standard for Vehicle Communication Systems for Cooperative Intelligent Transportation Systems" (2020) issued by the Society of Automotive Engineers of China, most of China's digital twin projects still stay in offline "digital simulation" of traffic conditions and problems. There is no breakthrough in the collaborative opening of traffic data applications and system platforms, and the "digital twin" that has landed has not truly realized the real-time interaction between the digital world and the real world required by the digital twin, forming a closed loop between the real physical space and the virtual digital space.

Secondly, the application scenario of digital twins is relatively single and not deep enough, and lacks the purpose of construction promotion (China Development Network, 2022). Various transportation segments have carried out project planning and construction based on digital twins, but in the initial planning and design, it still stays in solving traffic problems in a single scenario, and the solution of traffic problems in specific application scenarios is not deep enough, lacks overall planning for road traffic, and has no purpose for the final presentation effect of construction (Wu et al., 2019).

In addition, the lack of clear construction standards and specifications in the current digital twin construction system, and the combination of digital twin technology and next-generation technologies such as the Internet of Things, have also been hindered, which in turn hinders the progress of digital twin technology (Bhatti et al., 2021). The construction of digital twins is a comprehensive project covering the entire industry field, but due to the real world, the fields are different, the demand responsible parties behind the project are not the same, and there are often duplicate constructions for the same area (CCTV, 2022). Although a series of documents such as the 14th Five-Year Plan for Digital Transportation has been introduced, the project construction standards and specifications for the application of data and system architecture have not been unified, and there may be a debate on unified standards in subsequent project collaborative processing and integration applications Issue.

Digital twin networks also have challenges in supporting autonomous driving control. It is now difficult to ensure that the operating status of smart vehicles and the perceived environmental information is transmitted to the digital twin network server in real-time, and there is a transmission delay. In addition, hackers may hack or tamper with communications between vehicles and digital twin network servers, thereby compromising the digital twin modeling structure and misleading vehicle driving control (Wu et al., 2021).

Key technologies for digital twins are in urgent need of innovative breakthroughs (China Development Network, 2022). Digital twins were born out of the explosive development of multi-advanced discipline technology and rely on the rapid development of multiple means of perception. Wu et al. (2021) when proposing digital twin data transmission, there are high requirements for privacy and security, such as the need for encryption during the transmission of personal information. However, many key technologies involved in the current digital twin, such as collaborative computing and simulation, have a low degree of development, and the integration of technologies needs to be strengthened. Therefore, the maturity of massive data loading technology, cloud-edge computing collaboration technology and multiple data source integration capabilities also needs to be improved, and the ability of artificial intelligence and edge computing to quickly analyze and process dynamic data is also insufficient; the maturity of twin scenarios and dynamic traffic reality interaction models is not high. The application of digital twins in the field of transportation is still at the primary level, and algorithms and models in various professional fields need to be further developed.

To realize the digitization of transportation infrastructure, it is necessary to integrate multiple data sources, combined with building information modeling (BIM), geographic information system (GIS) data, and traffic condition aerial photography data, to build a digital transportation infrastructure platform with large storage capacity, high storage efficiency, simple structure and stable operating environment (Sun et al., 2020). The form of data alone is no longer enough to meet the development of modern intelligent transportation systems. The combination of digital twin technologies needs to be improved, and quantitative results are lacking (Tu et al., 2022). However, with the successful use of more and more intelligent network control systems, urban rail clouds and other big data platforms, this problem is slowly being solved.

METHOD

CiteSpace Made a Data Visualization

This chapter first analyzed the timeline of visualization from 2017 to 2022 with data of WOS based on the keywords of "digital twin and digital transformation" (figure 2) and found that the research on it is mainly focused between 2018 and 2021 which shows that digital twin has already helped the digital transformation in many sectors like digital manufacturing in the fourth timeline and smart manufacturing in the fifth timeline. However, the keyword "transportation" isn't shown in the figure. That means digital twin technology applied in the transportation sector is not a hot spot before.

Figure 4. Timeline visualization with the keyword of "digital twin" and "digital transformation"
Source: Web of Science

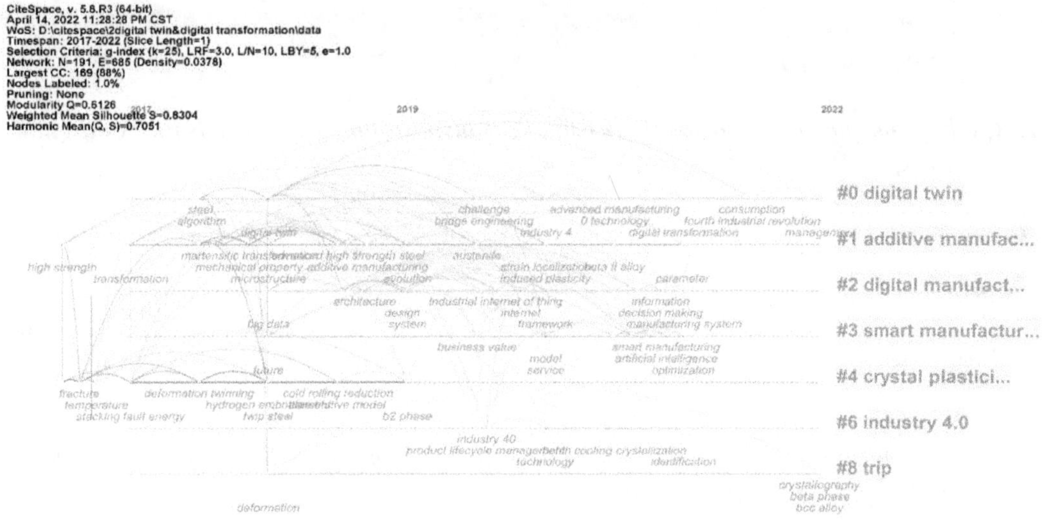

Next, we analyzed the keywords of "digital twin" and "transportation". It could visually see the fact that the digital twin was accompanied by technologies like cloud computing, 5g mobile communication, and modeling may help the digital transformation in the transportation sector.

Figure 5. Cluster visualization with the keyword of "digital twin" and "transportation"
Source: Web of Science

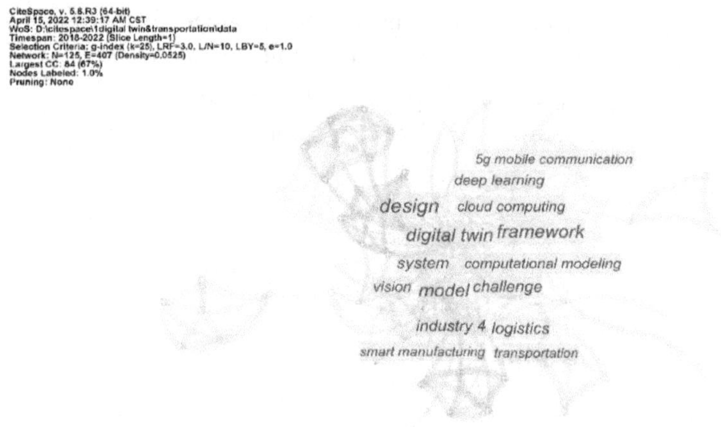

A Questionnaire

The authors published a survey for Chinese society entitled "Applying Digital Twin Technologies in China's Transportation Sector to Promote Its Digital Transformation", which was attended by 603 people from 25 provinces (municipalities). Most of them are from Jiangsu and Sichuan provinces. In the questionnaire, a whopping 93% of respondents believe that China needs digital transformation (Table 2). When it comes to "To what degree do you think the digital application in China's transportation sector

Table 2. Result of the question "Do you think China's transportation sector needs further digital transformation at this stage?"

Categories	Frequency	Percent	Cumulative Percent
Yes	586	97.18%	97.18%
No	17	2.82%	100%
Total	**603**	100%	

Table 3. Result of the question "To what degree do you think the digital application in China's transportation sector has proceeded at this stage?"

Categories	Frequency	Percent	Cumulative Percent
Very low	34	5.64%	5.64%
Low	138	22.89%	28.53%
Average	345	57.21%	85.74%
High	81	13.43%	99.17%
Very high	5	0.83%	100%
Total	**603**	100%	

has proceeded at this stage?" Nearly three-fifths of respondents believe that the digital application of Transportation in China is at an average level (Table 3).

About 60 percent of the 271 people who "know how to drive but think that digital transformation in transportation hasn't optimized their driving experience" are over 40 years old (Table 4). This should be because they may not use software such as electronic navigation, or even know what the specific behavior of digital transformation is. Of the 155 people who "know how to drive and think that digital transformation in the field of transportation has optimized their driving experience", when discussing how digital transformation in the field of transportation optimizes their driving experience, after removing 15 invalid answers, 56 people (about 31%) filled-in accurate answers about navigation systems and electronic maps, 11 people (about 8%) filled-in answers about automatic parking, and 9 people (about 6%) expressed the intuitive feeling of "convenience and speed". Others mention tools such as ETC in bits and pieces. It can be seen that although China has promoted the digitization of the field of transportation in all aspects, the scope of convenience that citizens feel from the digital process is still very narrow.

Table 4. Selected result with the condition of "know how to drive but think that digital transformation in transportation hasn't optimized their driving experience"

Categories	Frequency	Percent	Cumulative Percent
Under 20	20	7.38%	7.38%
20-30	38	14.02%	21.4%
31-40	47	17.34%	38.74%
41-50	126	46.49%	85.23%
Over 50	40	14.76%	100%
Total	**271**	100%	

Table 5. Result of the question "Technologies you have learned to use to accelerate digital transformation"

Categories	Frequency	Percent
I don't know any technologies that help accelerate digital transformation	41	6.8%
Other technologies	43	7.13%
Digital twin	79	13.1%
Simulation	94	15.59%
Cloud computing	290	48.09%
Big data	439	72.8%
Blockchain	221	36.65%
The internet of things	380	63.02%
Augmented reality	172	28.52%
Artificial intelligence	527	87.4%
Total	**603**	

When it comes to tools to accelerate the digitization of transportation in China (Table 5.), artificial intelligence, big data and the Internet of Things are in order of arrangement, the top three familiar tools for the public. Digital twins, simulation, and augmented reality are, in order of order, lesser-known tools. Only 13.1% of people know about digital twin technology, which shows that digital twin technology is a relatively unpopular technology in the eyes of ordinary citizens.

The questionnaire's questions about digital twins (Q10 to Q13) use the skill of skip pattern. Q10 is "How much do you know about digital twin technology," Q11 is "Do you think that digital twin technology can help accelerate the digital transformation in China's transportation sector," Q12 is "Do you have any suggestions for the digital transformation in China's transportation sector where digital twin technology is applied." Then Q13 is another question about digital transformation in transportation, as in Table 3 mentioned earlier. Its flow is as shown in Figure 4:

Figure 6. The respondent's understanding of digital twin technology
Source: The Questionnaire

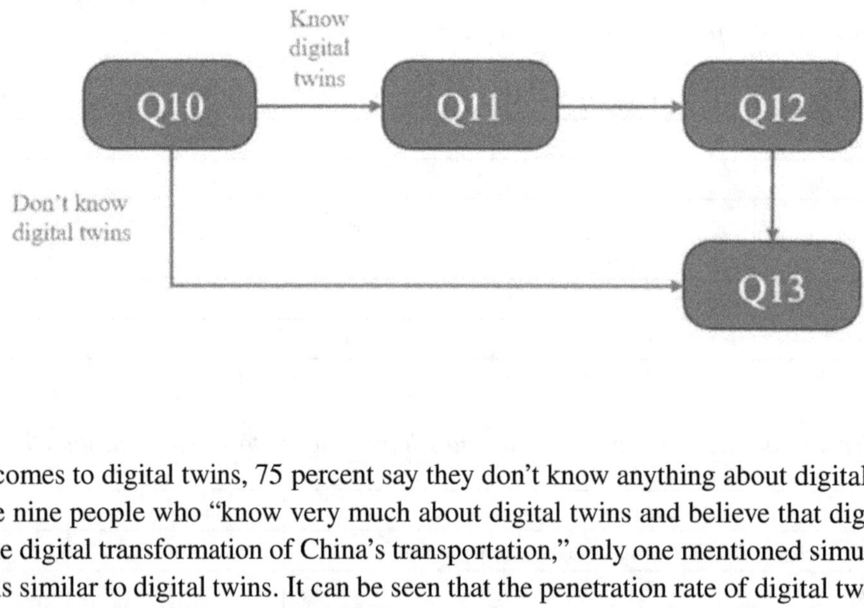

When it comes to digital twins, 75 percent say they don't know anything about digital twins. In addition, of the nine people who "know very much about digital twins and believe that digital twins can accelerate the digital transformation of China's transportation," only one mentioned simulation, a technology that is similar to digital twins. It can be seen that the penetration rate of digital twin technology among the Chinese masses is low, which may lead to more problems in the digital process of digital twins applied to transportation.

The masses believe that digital transformation is more successful in urban rail transit, high-speed railway and Aviation (Table 6), while the digital transformation process is slower in the three aspects of taxi, bus, and highway (Table 7). There are traces of this.

China's rail transit, high-speed rail and aviation fields have developed most rapidly in recent years, and a variety of new-generation information technologies have naturally been used when developing. Further, the units that take these means of transportation are usually a large part of the group, and after the use of a new generation of information technology, it becomes convenient to manage. This is why their digital transformation has been more successful. People have also experienced the convenience of "The large digital screens clearly show information on flights and trains.", "It is convenient and fast to go through paperless entrance gates.", "The introduction of life brands' scanning code reservation service saves time." and "The planning of travel times is more precise and reliably based on the train schedules and flight timetables." (Table 8). In contrast, the areas of taxis, bus systems and highways

Table 6. Result of the question "Which fields of China's transportation do you think have better digital transformation at this stage?"

Categories	Frequency	Percent
Taxi	147	24.38%
Bus	204	33.83%
Urban rail transit	441	73.13%
Highway	70	11.61%
Shipping	86	14.26%
High-speed railway	318	52.74%
Aviation	261	43.28%
Total	**603**	

Table 7. Result of the question "Which fields of China's transportation do you think have worse digital transformation at this stage?"

Categories	Frequency	Percent
Taxi	300	49.75%
Bus	288	47.76%
Urban rail transit	98	16.25%
Highway	261	43.28%
Shipping	105	17.41%
High-speed railway	47	7.79%
Aviation	49	8.13%
Total	**603**	

Table 8. Result of the question "What kind of experience have you gained from the digital transformation of airports, high-speed railway stations and other places?"

Categories	Frequency	Percent
The large digital screens clearly show information on flights and trains.	538	89.22%
It is convenient and fast to go through paperless	435	72.14%
The introduction of life brands' scanning code reservation service saves time.	285	47.26%
The planning of travel times is more precise and reliably based on the train schedules and flight timetables.	453	75.12%
Total	**603**	

have a long history of development, but the units that take these means of transportation are a group of people in smaller units, even very small units such as families or individuals. This has led to a slow process of digital development.

In the digital transformation of the transportation system, there are also many daily problems (Table 9). "Data between different departments is not connected", "Information is not updated timely," and "Not intelligent enough" are the questions most common among people. Of course, the problems of "Mainly displayed on large-screen" and "The construction period data cannot serve the operation period which will result in waste" also exist, but they exist in a narrower scope. These two different options expose two completely different problems. The range of people who came into contact with the "Mainly displayed on large-screen" option was relatively wide, indicating that the information displayed on the station's large screen was popular with the people. But in contrast, the "The construction period data cannot serve the operation period which will result in waste" option, most of the people who are exposed to this problem are people who have some understanding of digitalization in the transportation field or work in the transportation sector, so the contact area of the masses is narrow, and the number of people who choose this option is also small. When it comes to the institutional problems (Table 10), it is assumed that "Digital transformation has a greater impact on the traditional division of labor," "Lack of clear strategic goals and practical methods," "The foundation of digital transformation is weak, the motivation is insufficient, and there is a shortage of talents," "The information interface between different systems has certain limitations, resulting in poor data mobility," "Low information utilization rate and low digital governance level," and "The standard system is not perfect yet": these 6 options were almost voted on. This shows that people still have a more objective attitude toward the problem of investment in the digital transformation of the transportation system.

Table 9. Result of the question "What do you think are the main daily problems facing the digital transformation of China's transportation sector at this stage?"

Categories	Frequency	Percent
Information is not updated timely	321	53.23%
Data between different departments is not connected	378	62.69%
Mainly displayed on large-screen	156	25.87%
Not intelligent enough	291	48.26%
The construction period data cannot serve the operation period which will result in waste	104	17.25%
Other problems	34	5.64%
I have never experienced the application of digital transformation in the transportation sector in daily life, I can't answer this question.	34	5.64%
The application of digital transformation in the field of transportation in daily life has no problem.	28	4.64%
Total	**603**	

Table 10. Result of the question "What main institutional problems do you think China faces in the digital transformation of transportation at this stage?"

Categories	Frequency	Percent
Digital transformation has a greater impact on the traditional division of labor	242	40.13%
Lack of clear strategic goals and practical methods	202	33.5%
The foundation of digital transformation is weak, the motivation is insufficient, and there is a shortage of talents	256	42.45%
The information interface between different systems has certain limitations, resulting in poor data mobility	253	41.96%
Low information utilization rate and low digital governance level	157	26.04%
The standard system is not perfect yet	166	27.53%
Other problems	22	3.65%
I have never known about the system of digital transformation in the transportation sector, I can't answer this question	88	14.59%
The system of digital transformation of the transportation sector is perfect and there is no problem	16	2.65%
Total	**603**	

People also have great expectations for the digitization of transportation systems (Table 11), they want a unified public transportation platform or mobile application, unified development of a taxi dispatching platform, automatic driving, smart charging pile, Smart bus and smart service areas. These digital application scenarios should be added to China's transportation infrastructure. This means that the degree of digital transformation in these areas is still lacking. And the infrastructure that people need is the one that is useful and helps them, and it needs to be built the most. These aspects can serve as a direction for digital transformation.

Table 11. Result of the question "What digital application scenarios do you think need to be added for China's transportation infrastructure?"

Categories	Frequency	Percent
No other application scenarios need to be added	12	1.99%
Other application scenarios	32	5.31%
Smart service area	271	44.94%
Smart bus	265	43.95%
Smart charging pile	296	49.09%
Automatic driving	255	42.29%
Unified development of taxi dispatching platform	305	50.58%
Unified public transportation platform or mobile application	418	69.32%
Total	**603**	

Overall, there is still a lot of room for improvement in the digital transformation of China's transportation sector, and digital twin technologies need to be understood by residents and applied to the digital transformation of transportation systems. China not only needs to vigorously promote digital transformation, but also let digital transformation penetrate the lives of residents and bring real convenience to residents. Otherwise, digitalization is just an empty word at the policy level. At the same time, the transportation department can also introduce some digital transformation tools to the masses through social platforms or the development of specific software, so that the masses can enjoy the convenience brought by the digital process and have some understanding of the development of a new generation of information technology and technology.

Of course, the survey is aimed at mostly ordinary residents and school students, there is a lack of researchers engaged in digital twin research, and there is a shortage of staff in the transportation sector; the distribution areas are mostly limited to the more developed areas of China and do not circulate in China's poor cities. This is where this questionnaire is not rigorous. However, the problems reflected by residents and students in more developed areas also have certain reference significance in some aspects.

CASES IN CHINA

Through the application of digital twins by the Suzhou Institute of Automotive Research of Tsinghua University, the digital transformation of Shenzhen Airport and the "intelligent transportation cloud" in Guizhou Province, this paper explores the inextricable connection between digital twins and the digital transformation of transportation systems from point to point. Autonomous driving is an indispensable part of the digital transformation of transportation, and autonomous driving is not without challenges (Ge et al., 2020). The Suzhou Automotive Research Institute of Tsinghua University uses digital twin technology to create new solutions such as integrated decision-making and control frameworks for autonomous vehicles and applies them to the research of intelligent networked vehicles and autonomous driving, which is expected to promote the landing of autonomous driving. Shenzhen Bao'an International Airport has constantly tried and made mistakes in digital transformation, reflecting the distinctive characteristics of exploration and innovation in digital transformation. The digital transformation of transportation should take into account comprehensive urban transportation planning (Yang et al., 2021). The digital transformation process of Guizhou Province confirms the digital twin, integrating a new generation of information technology such as big data to carry out digital transformation according to the characteristics of urban transportation. It not only accelerates the digital transformation of the locality but even hopes to lead Guizhou Province to the forefront of national transportation data exchange.

Tsinghua University Suzhou Automotive Research Institute Applies Digital Twin Technology to the Research of Intelligent Networked Vehicles and Autonomous Driving

Digital twins are an integral part of the field of autonomous driving testing. How to test and evaluate autonomous vehicles efficiently and credibly has become the key to whether autonomous vehicles can be safely on the road, so it is necessary to fully test the overall performance of the controller and the whole vehicle (Chen et al., 2022). In the current self-driving test system based on the scene library, there are mainly Software in the Loop (SIL), Hardware in the Loop (HIL), Vehicle in the Loop (VIL), and real vehicle field testing 3. Several methods of real vehicle road testing. At present, the industry generally adopts HIL's test method to test the controller, but in the HIL test, only the controller under test is the real hardware, vehicle dynamics, roads, drivers, etc. are virtual models, and there is a deviation from reality, and the validity of the test results is difficult to guarantee (Suzhou Automotive Research Institute, Tsinghua University, 2020).

Virtual Reality in the Loop (VRIL) is a vehicle closed-loop test in which a real vehicle is driven in a real test site and mapped to a virtual test environment. In the virtual simulation system, the environment, road, traffic participants, the model of the test vehicle, the sensor model of its configuration are established, the target information detected by the virtual sensor in the simulation environment is sent to the test vehicle equipped with the automatic driving algorithm for information fusion and decision control, and the test vehicle is running in the test site at the same time. The motion state information of the test vehicle is collected and fed back to the virtual scene, to complete the synchronization of the virtual and real states, and realize the closed-loop real-time simulation test of the entire digital twin system (Ge et al., 2020). Combined with the scene library data, the test conditions that are close to the real traffic environment can be quickly set, and the efficiency and authenticity of the test can be effectively improved.

Figure 7. Static Path Planning Design Phases in Integrated Decision Control
Source: Tsinghua University Suzhou Automotive Research Institute

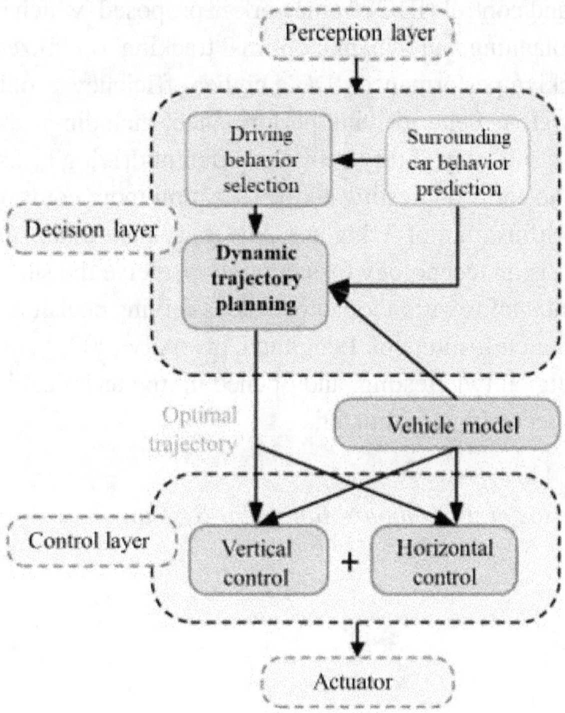

Figure 8. Dynamic Optimal Tracking Design Phases in Integrated Decision Control
Source: Tsinghua University Suzhou Automotive Research Institute

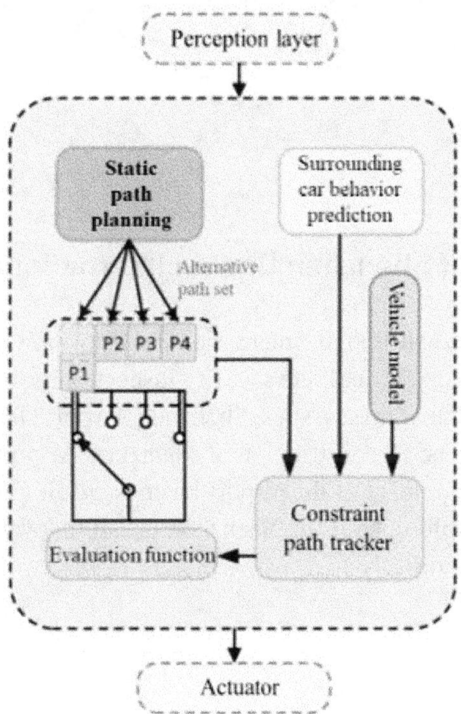

Guan et al. (2021), Tsinghua University, Professor Li Shengbo, Tsinghua University School of Vehicle and Transportation, and Suzhou Automotive Research Institute, Tsinghua University An autonomous vehicle integrated decision and control (IDC) framework is proposed, which decomposes driving tasks into hierarchical static path planning and dynamic optimal tracking, optimizes them separately and follows paths with optimal tracking performance. IDC's higher efficiency in online computing compared to benchmark methods, as well as better driving performance, including traffic efficiency and safety, produce great interpretability and adaptability between different driving tasks.

To help self-driving cars to carry out testing and verification more conveniently and efficiently, the Suzhou Automotive Research Institute of Tsinghua University also established an onboard binocular system in 2020, which uses vision technology to accurately perceive the surrounding environment and provide accurate environmental information for autonomous driving decision control. At the same time, the Suzhou Automotive Research Institute of Tsinghua University (2021) conducted in-depth research on the key technologies of digital twin testing, and opened up the technical link of digital twin testing based on the self-developed scene library "mirror".

Figure 9. Application Scenarios of the Onboard Binocular System
Source: Tsinghua University Suzhou Automotive Research Institute

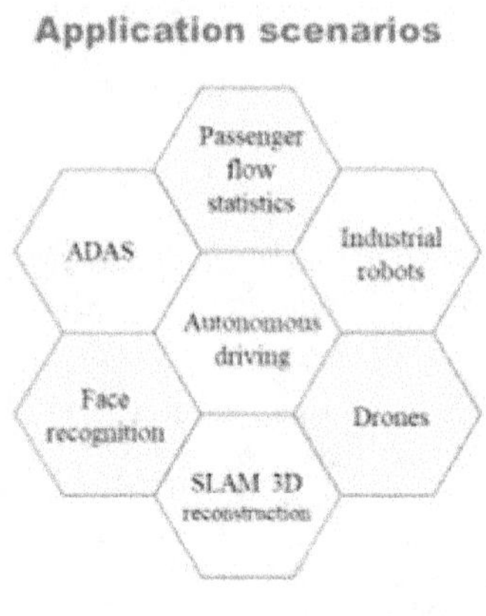

Digital Transformation of Shenzhen Bao'an International Airport

The congenital defects at Shenzhen Bao'an International Airport are very obvious: the boarding gate is too far and there is no diversion channel, the security inspection team is too long, the average airport stay time is too long, etc. To address these issues, Shenzhen Airport Group decided to advance the digital transformation of airports. The IT construction of Shenzhen Airport Group is still lagging in 2019, which does not match the development of the airport business itself (Yu, 2021). In addition, Shenzhen Airport, as a city window, should uphold the Shenzhen urban innovation gene and actively integrate into the construction of smart cities.

Shenzhen Bao'an International Airport has also encountered many difficulties in the process of digital transformation. Civil aviation is an industry that attaches great importance to safety and supervision, so the guidance of industry standards and the support of regulatory policies are particularly important (Guo, 2021). Airports are a traditional industry, and there is no standard template for airport digital transformation, and there is no mature experience to learn from.

Shenzhen Airport regards digital transformation as a major research topic of the Group's Board of Directors, and at the same time, in the mid-term adjustment of the Group's "13th Five-Year Plan" strategy, it regards digital transformation as one of the Group's five major development strategies (Yu, 2021). Next, I consulted Huawei's IT department and cultivated more than 100 digital transformation talents (Ye, 2020). The Group released the "White Paper" on Digital Transformation, thus forming a situation of joint participation and joint promotion.

In terms of overall planning, Shenzhen Airport has comprehensively sorted out the organizational structure, business areas, information systems and operating processes, and on this basis, it has systematically planned a blueprint for transformation (Shi & Ye, 2017). In terms of system architecture, Shenzhen Airport fully integrates the existing information system, breaks down internal information barriers, and realizes the integration and interoperability of data. In terms of governance system, we have established an IT governance system that meets the requirements of transformation through organizational adjustment and business transformation (China Informatization Weekly, 2019).

To put digital transformation in place, Shenzhen Baoan International Airport Airport News (2019) said that three initiatives have been taken: The first initiative is to implement IT organizational changes to provide organizational guarantees for transformation. The airport will integrate the IT organizations and personnel scattered in all corners of the group to form a group digital management center. The second initiative is to implement it in an orderly and phased manner according to the blueprint. By actively exploring the application innovation of cloud computing, big data and artificial intelligence in security, service and operation, several key information application projects have been newly built and upgraded, and the satellite hall has been put into operation as the goal, to build the airport into the "most experienced digital airport".

In this way, Shenzhen Bao'an International Airport has transformed into a "passenger differentiated security inspection model" relying on big data analysis, which certifies passengers' qualifications through the gates of the airport's domestic security inspection area, realizes the classification and diversion of passengers, and can remind passengers whether checked baggage needs to be opened for inspection. "Paperless" boarding eliminates the hassle of queuing up for procedures on site, and the boarding process has become greener and more convenient (State-owned Assets Supervision and Administration Commission of Shenzhen Municipal People's Government, 2021). At the same time, Shenzhen Bao'an International Airport has introduced affordable, net-red catering and lifestyle brands. In these merchants, items can be ordered with WeChat Mini Programs, not only saving time but also allowing better planning. At airports like malls, you can also schedule flight delays.

In 2021, Bao'an Daily said that Shenzhen Bao'an International Airport has begun a new round of digital processes for optimizing the three-dimensional transportation network. Strive to improve the airport's landside transportation connection facilities, optimize regional transportation connections, improve transportation organizations, actively develop air- and air-to-rail intermodal services, extend air transport capacity, and build a three-dimensional transportation network that combines air, land, rail, sea, and other modes of transportation (Bao'an Daily, 2021).

"Intelligent Transportation Cloud" Makes Transportation in Guizhou Province More Convenient

Highway construction in Guizhou Province is accelerating, and the large network with efficient information collection is becoming more and more intensive. Guizhou Province belongs to the underdeveloped areas of China, which is a heavily hilly area, with poor highway lines, relatively low technical standards, and is seriously affected by bad weather. To achieve today's achievements, it is inseparable from the deep integration of big data and big transportation, and it is also inseparable from the valuable experience provided by the digital twin pilot (Guizhou Provincial Big Data Bureau).

The first means of transportation to start digital transformation were coaches. The history of long-distance buses is much older than that of high-speed rail and airplanes, which means that the facilities of the bus station are relatively old, the management difficulty is relatively large, and the ride experience is relatively poor. Before the digital transformation, buying a bus ticket had to wait in line and get it from the staff at the window. This method of collecting tickets is a waste of time, especially when there is a long queue to buy tickets during the Spring Festival. On February 5, 2016, after the Guizhou Provincial Road Passenger Transport Network Ticketing System began to run online, it is only necessary to query or book and purchase bus tickets in the province at Internet terminals such as mobile phones and computers, and then click on the network ticket collection business at the self-service ticket machine, scan the ID card, enter the verification code, and in less than 1 minute, a brand new bus ticket will be printed out by itself (Guizhou Daily).

Figure 10. Guiyang, Guizhou Province, piloted the digital twin transportation system
Source: Guizhou Provincial Big Data Bureau

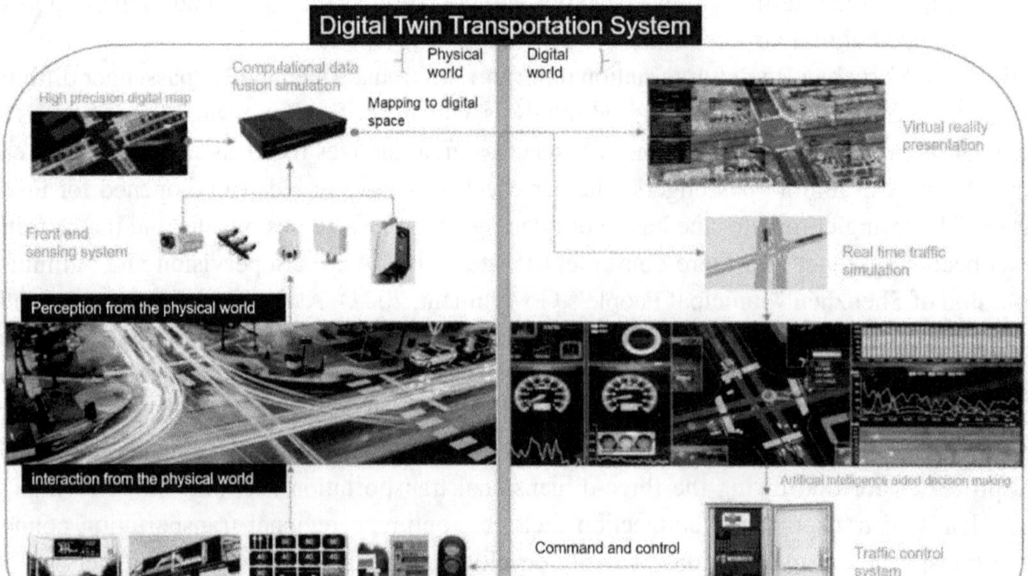

As the first batch of Kaili bus stations in Guizhou Province to open an online booking and ticket purchase business, 50% of the bus tickets for the line can be purchased online during the Spring Festival in 2016, and the vigorous online ticket sales allow the masses who arrive at the county seat and above to travel faster and more conveniently, which reduces the pressure of the window "long queue" to a certain extent. On July 30, 2019, Focus Guiyang said that the coverage rate of the road passenger network ticketing system in Guizhou Province has exceeded half.

Guiyang piloted the digital twin traffic system, mainly using the existing video surveillance resources at the current intersection, deeply integrating millimeter-wave radar, and fully integrating data through the holographic perception of network objects, including motor vehicles, non-motorized vehicles, pedestrians and other traffic elements, and introducing real-world information into the twin traffic simulation system (Intention Information Report, 2021). Combined with high-precision maps, real-time data is tracked for vehicle positioning and trajectory depiction, video profiling is analyzed, the significance of individual activities of the target is judged, the relationship between people, vehicles, roads and the environment is comprehensively studied, and traffic problems such as waste of traffic resources, rigid signal system functions, unpredictable traffic events and rapid response are finally solved (Communication Information News, 2021).

According to the "14th Five-Year Plan for digital transportation development in Guizhou Province" (2022), in the next step, Guizhou Province will improve the level of transportation informatization and intelligent management, accelerate the construction of digital transportation industry ecology, strive to achieve a breakthrough in traffic data aggregation, and strive to build a data aggregation center in all fields of transportation in Guizhou Province, a national transportation industry data disaster recovery center and a southwest regional transportation data exchange center.

CONCLUSION & RECOMMENDATIONS

In recent years, the urban transportation bureaus have also actively explored the use of artificial intelligence, big data, and other means to empower the fine management of urban transportation, gradually solving a series of problems. Digital twins are also slowly being applied to their digital transformation. Digital twins arose because of perception control technology and boomed with integrated technology integration and innovation. Combining the actual needs of the moment and finding the right application scenario is the key to the development of digital twins in the transportation industry. Digital twin related technologies cannot be limited to the creation of a virtual-twin space to carry out a new round of technology research and development, to achieve self-breakthrough, to meet the new goals and new needs of the development of the transportation industry in the context of digitalization in the transportation field. With the development of digital twins, digital twins are no longer just a technology, but also a new model and new driving force for development. This chapter uses a questionnaire entitled "Applying Digital Twin Technologies in China's Transportation Field to Promote Its Digital Transformation" and three examples from the Suzhou Automotive Research Institute of Tsinghua University, Shenzhen Bao'an International Airport and Guizhou "Intelligent Transportation Cloud" to briefly describe the current digital transformation process of China's transportation field and the penetration of digital transformation of the transportation system in the lives of the masses. It is not difficult to see that although China has some digital trends in the field of transportation, such as ETC on the common highways, etc., as far as the questionnaire is concerned, the degree of digitization is not enough, and residents have not received real convenience from the digitization process. The publication of this questionnaire is limited to Jiangsu and Sichuan provinces, which is the limitation of this questionnaire.

The digital twin traffic control system is an emerging technology path for the construction of a new intelligent transportation system based on urban traffic data standardization and digital accumulation from quantitative change to qualitative change, under the background of major breakthroughs in information technology such as perception modeling and artificial intelligence, which is a cutting-edge advanced mode of traffic intelligence and sustainable operation, and a key component of digital city construction. The transportation system needs to integrate inside and outside and establish a positive interaction feedback mechanism between transportation and urban land use, social and economic development and other systems. Such a digital transformation is healthy. At the same time, the public transport system gives full play to its differentiated advantages in terms of operating speed, reliability, and accessibility of services. Innovations in transport technology and mobility modes will also accelerate their digitalization. Although the government has issued many relevant laws, due to the large number of cities in China and the different management methods, it is necessary to improve the relevant laws on digital transformation in the transportation sector. In addition, a transportation platform that tries to provide users with one-stop high-quality travel services can be developed. Users enter the place of departure and destination in the platform, and the platform can automatically match the optimal route or even ticket or ticket. Through digital twin technology, a series of measures such as route optimization, operation and arrival time matching, and hub station construction are developed to give full play to the synergy effect between the two or more, and jointly enhance the attractiveness of public transportation services. Let citizens understand the true meaning of digital transformation in the convenience of travel. Finally, although a new generation of information technologies such as digital twin technologies has been applied to intelligent transportation, these technologies are not mature enough, and there is still room for further research. The above is the advice made in this chapter on the current digital transformation of China's transportation sector, and these suggestions are implemented in the lives of the masses. It is hoped that it can have a certain role in suggesting the future direction of digital transformation in the field of transportation. Finally, although a new generation of information technology such as digital twin technology has been applied to intelligent transportation, because most of China's digital twin project application scenarios are relatively single, they still stay in the offline "digital simulation" of traffic conditions and problems and lack clear construction standards and purposes. At the same time, there is a problem of transmission delay in digital twin networks, and these key technologies urgently need innovative breakthroughs, and there is room for further research.

ACKNOWLEDGMENT

The authors extend sincere gratitude to:

- Our colleagues from Soochow University, the Australian Studies Centre of Shanghai University, the European Business University of Luxembourg and Krirk University as well as the independent research colleagues who provided insight and expertise that greatly assisted the research, although they may not agree with all of the interpretations/conclusions of this chapter.
- China Knowledge for supporting our research.
- The Editor and the International Editorial Advisory Board (IEAB) of this book who initially desk reviewed, arranged a rigorous double/triple blind review process and conducted a thorough, minute and critical final review before accepting the chapter for publication.
- All anonymous reviewers who provided very constructive feedbacks for thorough revision, improvement and fine tuning of the chapter.

REFERENCES

Aksenov, V., Semochkin, A., Bendik, A., & Reviakin, A. (2022). Utilizing Digital Twin for Maintaining Safe Working Environment among Railway Track Tamping Brigade. *Transportation Research Procedia, 61*, 600–608. doi:10.1016/j.trpro.2022.01.097

Bao'an Daily. (2021, September 27). *Sea, Land, Air And Rail Three-Dimensional Traffic Reaches All Directions.* Bao'an Daily. http://barb.sznews.com/PC/content/202109/27/content_1099775.html

Bhatti, G., Mohan, H., & Singh, R. R. (2021). Towards The Future of Smart Electric Vehicles: Digital Twin Technology. *Renewable & Sustainable Energy Reviews, 141*, 110801. doi:10.1016/j.rser.2021.110801

Central Broadcasting Network. (2022, April 27). *Smart Transportation Observation: Solving Efficiency And Safety Problems Is The Core Of Intelligent Networking The Digital Twin Should Help Urban Traffic Governance*. CNR. http://tech.cnr.cn/techph/20220427/t20220427_525808782.shtml

Chen, G. (2022). Design Scheme and Application of Foundation Treatment for Non-stop Construction of New Connecting Roads in Shenzhen Airport. *Subgrade Engineering, 130-133*. doi:10.13379/j.issn.1003-8825.202101010

Chen, X., Jin, Z., Zhang, Q., Li, P., Zhang, S., Sun, J., Tian, X., Wang, Y., & Zhang, J. (2022). Research on Automatic Driving Simulation Test System Based on Digital Twin. *Journal of Physics: Conference Series, 2170*(1), 012039. doi:10.1088/1742-6596/2170/1/012039

China Automotive News. (2021, October 23). Digital Twin: "Black Technology" Incarnates As An Automotive Industry Booster. *China Automotive News*. http://www.cnautonews.com/

China Economic Network. (2016, February 5). Guizhou Provincial Road Passenger Transport Network Ticketing System was put into Trial Operation. *China Economic Network.* http://district.ce.cn/newarea/roll/201602/05/t20160205_8776229.shtml

China Informatization Weekly. (2019, November 16). Shenzhen Airport Digital Transformation and Exploration. *China Information Weekly.*

Chinese Academy of Information and Communications Technology. (2021, December 20). *Big Data White Paper (2021).* CAICT.

Communication Message News. (2021, June 4). How Can Digital Twins Empower Smart Transportation? *Communication Message News.* http://www.txxxb.com/yc/yc/2021/0604/247776.shtml

Deng, T., Zhang, K., & Shen, Z.-J. M. (2021). A Systematic Review of a Digital Twin City: A New Pattern of Urban Governance toward Smart Cities. *Journal of Management Science and Engineering*, *6*(2), 125–134. doi:10.1016/j.jmse.2021.03.003

Department of Transportation of Guizhou Province. (2017, February 14). *Promote The Development Of Intelligent "Transportation Cloud"*. Department of Transportation. https://www.guizhou.gov.cn/home/gzyw/202109/t20210913_70356107.html

Department of Transportation of Guizhou Province. (2022, January 27). *Guizhou Province "14th Five-Year" Digital Transportation Development Plan*. Department of Transportation. http://jt.guizhou.gov.cn/xxgkml/ztfl/zcfg/gfxwj/202201/t20220128_72435899.html

Dygalo, A., Keller, A., & Shcherbin, A. (2020). Principles of Application Of Virtual And Physical Simulation Technology In Production Of Digital Twin Of Active Vehicle Safety Systems. *Transportation Research Procedia*, *50*, 121–129. doi:10.1016/j.trpro.2020.10.015

Focus Guiyang. (2019, July 30). The Coverage Rate Of The Road Passenger Network Ticketing System In Guizhou Province Is More Than Half. *Focus Guiyang*. https://gy.focus.cn/zixun/06d41c6b1f08d13a.html

Ge, Y., Wang, Y., & Han, Q. (2020). Test Method of Connected and Automated Vehicles Based On Digital Twin. *ZTE Communications.*, *26*(1). doi:10.12142/ZTETJ.202001006

Guan, Y., Ren, Y., Sun, Q., Li, E. B., Ma, H., Duan, J., Dai, Y., & Cheng, B. (2022). Integrated Decision and Control: Toward Interpretable and Computationally Efficient Driving Intelligence. *IEEE Transactions on Cybernetics*. doi:1 doi:0.1109/TCYB.2022.3163816 PMID:35439160

Guangzhou Daily. (2021, November 30). Reveal the technological elements behind The First Insensible Toll Station in China. *Guangzhou Daily*. https://www.gzdaily.cn/amucsite/web/index.html#/detail/1717097 (gzdaily.cn).

Guizhou Daily. (2016) "Intelligent Transportation Cloud" Makes Guizhou More Convenient. *Guangzhou Daily*. http://www.gov.cn/xinwen/2016-03/14/content_5053045.htm

Guizhou Provincial Big Data Bureau. (2021, May 24). *Big Data Deep Mining Integration, Guiyang Pilot Digital Twin Transportation System.* GPBDB. https://www.guizhou.gov.cn/home/gzyw/202109/t20210913_70368737.html

Guo, X. (2021). Suggestions On The Transformation And Development Of Civil Airport Operation And Reflections On The Transformation To A Management Model: Take Shenzhen Airport As An Example. *Aviation Think Tank.* http://att.caacnews.com.cn/zsfw/jcgl/202111/t20211101_59856.html

Kruglikov, A. A., Lazorenko, G. I., Morozov, A. V., & Yavna, V. A. (2013). Designing Intelligent Systems And Monitoring Of Transport Infrastructure By Elastic Waves. *9th EAGE International Scientific and Practical Conference and Exhibition on Engineering and Mining Geophysics*, Gelendzhik, Russia.

Li, Y., Han, W., Shen, H., & Chen, H. L. (2021). Research on the Application of AR Visualization in Urban Traffic Management. *Journal of Transportation Engineering*, *21*(2), 57–61, 67.

Liang, C. (2020). Application of Bentley Digital Twin Technology in Highway Quality Engineering Construction. *China ITS Journal, 6.*

Liao, S., Wu, J., Bashir, A. K., Yang, W., Li, J., & Tariq, U. (2021). Digital Twin Consensus for Blockchain-Enabled Intelligent Transportation Systems in Smart Cities. *IEEE Transactions on Intelligent Transportation Systems*. doi:10.1109/TITS.2021.3122566

Liu, J., Li, C., Bai, J., Luo, Y., Lv, H., & Lv, Z. (2021). Security in IoT-Enabled Digital Twins of Maritime Transportation Systems. *IEEE Transactions on Intelligent Transportation Systems*, 1–9. doi:10.1109/TITS.2021.3122566

Lv, Z., Zhang, S., & Xiu, W. (2020). Solving the Security Problem of Intelligent Transportation System with Deep Learning. *IEEE Transactions on Intelligent Transportation Systems*, *22*(7), 4281–4290. doi:10.1109/TITS.2020.2980864

Lv, Z. H. & Xie, S.X. (2021). Artificial Intelligence in the Digital Twins: State Of The Art, Challenges, and Future Research Topics. *Digital Twin*, 1-12.

Marai, O. E., Taleb, T., & Song, J. S. (2020). Roads Infrastructure Digital Twins: A Step Toward Smarter Cities Realization. *IEEE Network*, (99), 1–8.

Menouar, H., Güvenc, I., Akkaya, K., Uluagac, A. S., Kadri, A., & Tuncer, A. (2017). UAV-Enabled Intelligent Transportation Systems for the Smart City: Applications and Challenges. *IEEE Communications Magazine*, *55*(3), 22–28. doi:10.1109/MCOM.2017.1600238CM

Ministry of Transport. (2017, September 26). *Smart Transportation Makes Mobility Easier Action Plan (2017-2020).* Ministry of Transport. https://xxgk.mot.gov.cn/2020/jigou/kjs/202006/t20200623_3317082.html

Ministry of Transport. (2019, July 25). *Outline of Digital Transport Development Plan.* Ministry of Transport. https://xxgk.mot.gov.cn/2020/jigou/zhghs/202006/t20200630_3321233.html

Pedersen, T. A., Glomsrud, J. A., Ruud, E. L., Simonsen, A., & Erikson, B. O. H. (2020). Towards Simulation-Based Verification of Autonomous Navigation Systems. *Safety Science*, 129.

People's Daily. (2022, May 2). The Transportation Industry Has Made Great Strides Towards A Transportation Power, And The Transportation Industry Has Achieved Leapfrog Development. *People's Daily.* http://paper.people.com.cn/rmrb/html/2022-05/02/nbs.D110000renmrb_06.htm

Rudskoy, A., Ilin, I., & Prokhorov, A. (2020). Digital Twins in the Intelligent Transport Systems. *Transportation Research Procedia 54* (2021), 927-935.

Shenzhen Bao'an International Airport Airport News. (2019, December 19). *Shenzhen Airport Builds The Most Experienced Digital Airport.* Shenzhen Bao'an International Airport. https://www.szairport.com/szairport/kgxw/201912/9360ae6092c94540bcc9518a9b685994.shtml

Shi, D., & Ye, D. (2017). Shenzhen Airport Will Build An International Aviation Hub Facing The Asia-Pacific Region And Radiating The World. *Air Transport and Business,* 36-37.

State-Owned Assets Supervision and Administration Commission of Shenzhen Municipal People's Government. (2021, November 29). *Shenzhen Airport Has Created A "Heart" For 30 Years And Promoted Digital Transformation.* GZW. http://gzw.sz.gov.cn/gkmlpt/content/9/9408/post_9408956.html#1904

Sun, Y., Kuai, R., Li, X., & Tang, W. (2020). Latency Performance Analysis For Safety-Related Information Broadcasting In VEMAC. *Transactions on Emerging Telecommunications Technologies, 31*(5). doi:10.1002/ett.3751

Tao, F., Liu, W., Liu, J. H., Liu, X., Liu, Q., Qu, T., Hu, T., Zhang, Z., Xiang, F., Xu, W., Wang, J., Zhang, Y., Liu, Z., Li, H., Cheng, J., Qi, Q., Zhang, M., Zhang, H., Sui, F., & Cheng, H. (2018). Digital Twin and Its Potential Application Exploration. *Jisuanji Jicheng Zhizao Xitong*, *24*(1), 1–18.

Tao, F., & Zhang, M. (2017). Digital Twin Shop-Floor: A New Shop-Floor Paradigm Towards Smart Manufacturing. *IEEE Access : Practical Innovations, Open Solutions*, *5*, 20418–20427. doi:10.1109/ACCESS.2017.2756069

The Central Committee of the Communist Party of China and the State Council. (2019, September 19). *Outline For the Construction of a Transportation Powerhouse.* CN. http://www.gov.cn/zhengce/2019-09/19/content_5431432.htm

The General Office of the Ministry of Transport. (2017, January 22). *Action Plan for Promoting the Development of Smart Transportation (2017-2020).* XXKG. https://xxgk.mot.gov.cn/2020/jigou/zhghs/202006/t20200630_3319779.html

The Ministry of Industry and Information Technology. (2017, November 14). *A Three-Year Action Plan to Promote the Development of a New Generation of Artificial Intelligence Industry (2018-2020).* MIIT. https://www.miit.gov.cn/jgsj/kjs/jscx/gjsfz/art/2020/art_291b5e6bc13f415494e84a0e9eac78f1.html

The Ministry of Science and Technology of the People's Republic of China. (July 12, 2019). Comprehensive Transportation and Intelligent Transportation. *Key Special Project Large-scale Networked Vehicle Collaborative Service Platform, Beijing.*

The Paper. (2020, August 18). China's Space Industry Trend Report: High-Precision Maps Reconstruct the World and Form A Digital Twin Closed Loop. *The Paper.* https://www.thepaper.cn/newsDetail_forward_8773349

The Paper. (2021, March 1). What Exactly Does The National Integrated Three-Dimensional Transportation Network Look Like? 4 Poles, 6 Axes, 7 Corridors, 8 Channels. *The Paper.* https://www.thepaper.cn/newsDetail_forward_11505409

Tsinghua University Suzhou Automotive Research Institute. (2020, October 23). *In-vehicle Binocular System.* TSARI. https://www.tsari.tsinghua.edu.cn/scientific/znwlCar/2020-10-23/348.html

Tu, Z., Qiao, L., Nowak, R., Lv, H., & Lv, Z. (2022). Digital Twins-Based Automated Pilot for Energy-Efficiency Assessment of Intelligent Transportation Infrastructure. *IEEE Transactions on Intelligent Transportation Systems, 23*(11), 22320–22330. Advance online publication. doi:10.1109/TITS.2022.3166585

Wang, G.T. (2015). Urban Transportation and Informatization. *Urban transport of China, 13*(3), 1-4.

Wang, W., Li, X., Xie, L., Lv, H., & Lv, Z. (2022). Unmanned Aircraft System Airspace Structure and Safety Measures Based on Spatial Digital Twins. *IEEE Transactions on Intelligent Transportation Systems, 23*(3), 2809–2818. doi:10.1109/TITS.2021.3108995

Wu, Y., Zhang, K., & Zhang, Y. (2021). Digital Twin Networks: A Survey. *IEEE Internet of Things Journal, 8*(18), 13789–13804. doi:10.1109/JIOT.2021.3079510

Wu, Z., Wu, X., & Wang, L. (2019). Prospect of Development Trend of Smart Transportation under the Background of Building China into a Country with Strong Transportation Network. *Transportation Research, 5*(4), 26–36. doi:10.16503/j.cnki.2095-9931.2019.04.003

Xinhua Net. (2019, September 24). *China's Construction Of A Transportation Power Has Opened A New Chapter.* Xinhua Net.

Xinhua News Agency. (2019, September 19). *Outline for the Construction of a Transportation Powerhouse.* State Council of People's Republic of China. http://www.gov.cn/zhengce/2019-09/19/content_5431432.htm

Xinhua News Agency. (2021, February 24). *Outline of the National Comprehensive Three-Dimensional Transportation Network Plan.* State Council of People's Republic of China. http://www.gov.cn/zhengce/2021-02/24/content_5588654.htm

Xinmin Evening News. (2020, October 17). Intelligent Transportation Facilitates People's Travel. *Xinmin Evening News.*

Yang, D., Guo, J., Gu, Y., Zhao, Y., Zhang, X., Ding, Q., Qian, L., Shao, D., Chen, X., Zhou, T., Bai, F., Cui, Y., Zhang, Y., & Wang, G. (2021). Digital Transformation of Urban Comprehensive Transportation Planning. *Urban Transport of China*, *19*(6), 107–113. doi:10.13813/j.cn11-5141/u.2021.0605

Ye, D. (2020). *Shenzhen Airport Group And Huawei Signed A Deepening Strategic Cooperation Agreement*. Civil Aviation Resource Network.

Yu, Z. (2021). Reflections and Practices of Digital Transformation at Shenzhen Airport. *Civil Aviation Resource Network*. http://news.carnoc.com/list/566/566097.html

Zhao, L., Liu, H., Zhang, X., & Wang, F. (2021). Research on Vehicle-Infrastructure Collaboration Virtual Simulation Platform Based on Digital Twin. *Mobile Communications*, *45*(6), 7–12.

APPENDIX: A QUESTIONNAIRE

A Survey of the Application of Digital Twin Technologies in the Field Of Transportation in China to Facilitate Its Digital Transformation

Background: On July 25, 2019, the Ministry of Transport issued the Outline of the Digital Transportation Development Plan. Its goal is to build a digital collection system, transmission system and application system to accelerate the development of transportation informatization to digitization, networking and intelligence.

Digital twin technology refers to the construction of a virtual system in information space that can map and characterize physical equipment to visualize data, thus reflecting the whole life cycle process of the corresponding entity. Digital twin technology plays an important role in the digital transformation of China's transportation sector.

*Note: The answer to this questionnaire is only used for academic research. There is no commercial use. Your personal information will be kept strictly confidential.

1. Your gender:
   A. Male
   B. Female
2. Your age:
   A. Under 20
   B. 20-30
   C. 31-40
   D. 41-50
   E. Over 50
3. Your city of residence:
4. Do you have a driving license?
   A. Yes
   B. No (If your answer is No, SKIP directly to Question 6)
5. Does the digital transformation in the transportation sector optimize your driving experience?
   A. Yes
   B. No
6. What kind of experience have you gained from the digital transformation of airports, high-speed railway stations and other places? (You can choose more than one answer)
   A. The large digital screens clearly show information on flights and trains
   B. It is convenient and fast to go through paperless entrance gates
   C. The introduction of life brands' scanning code reservation service saves time
   D. The planning of travel times is more precise and reliably based on the train schedules and flight timetables

7. What examples of digital transformation in the transportation sector have you experienced in your daily life? (You can choose more than one answer)
   A. Real-time update of traffic conditions
   B. Forecast of congested sections to provide the best route
   C. GPS positioning and electronic map
   D. Electronic Station Sign
   E. Parking space forecast
   F. Arrival reminder
   G. Other examples
   H. I have never experienced digital transformation in the transportation sector.
8. Do you think China's transportation sector needs further digital transformation at this stage?
   A. Yes
   B. No
9. Technologies you have learned to use to accelerate digital transformation: (You can choose more than one answer)
   A. Artificial intelligence
   B. Augmented reality
   C. The internet of things
   D. Blockchain
   E. Big data
   F. Cloud computing
   G. Simulation
   H. Digital twin
   I. Other technologies
   J. I don't know any technologies that help accelerating digital transformation.
10. How much do you know about digital twin technology?
   A. I know little about it (SKIP to Question 13)
   B. I know a little bit about it
   C. I know it well
11. Do you think that digital twin technology can help accelerate the digital transformation in China's transportation sector?
   A. Yes
   B. No
12. Do you have any suggestions for the digital transformation in China's transportation sector where digital twin technology is applied?
13. How well-developed do you think the digital application in China's transportation sector is at this stage?
   A. Very low
   B. Low
   C. Average
   D. High
   E. Very high

14. Which fields of China's transportation do you think have better digital transformation at this stage? (You can choose one to three choices)
    A. Taxi
    B. Bus
    C. Urban rail transit
    D. Highway
    E. Shipping
    F. High-speed railway
    G. Aviation
15. Which fields of China's transportation do you think have worse digital transformation at this stage? (You can choose one to three choices)
    A. Taxi
    B. Bus
    C. Urban rail transit
    D. Highway
    E. Shipping
    F. High-speed railway
    G. Aviation
16. What do you think are the main daily problems facing the digital transformation of China's transportation sector at this stage? (You can choose more than one answer)
    A. Information is not updated promptly
    B. Data between different departments is not connected
    C. Mainly displayed on large-screen
    D. Not intelligent enough
    E. The construction period data cannot serve the operation period which will result in waste)
    F. Other problems
    G. I have never experienced the application of digital transformation in the transportation sector in daily life, I can't answer this question
    H. The application of digital transformation in the field of transportation in daily life has no problem.
17. What main institutional problems do you think China faces in the digital transformation of transportation at this stage? (You can choose more than one answer)
    A. Digital transformation has a greater impact on the traditional division of labor
    B. Lack of clear strategic goals and practical methods
    C. The foundation of digital transformation is weak, the motivation is insufficient, and there is a shortage of talents
    D. The information interface between different systems has certain limitations, resulting in poor data mobility
    E. Low information utilization rate and low digital governance level
    F. The standard system is not perfect yet
    G. Other problems.
    H. I have never known about the system of the digital transformation of the transportation sector, I can't answer this question.

I. The system of digital transformation of the transportation sector is perfect and there is no problem.

18. What digital application scenarios do you think need to be added to China's transportation infrastructure? (You can choose more than one answer)
    A. Unified public transportation platform or mobile application
    B. Unified development of taxi dispatching platform
    C. Automatic driving
    D. Smart charging pile
    E. Smart bus
    F. Smart service area
    G. Other application scenarios
    H. No other application scenarios need to be added

Chapter 8
Big Data and Artificial Intelligence:
Creative Tools for Destination Competitiveness

Sandhya H.
CHRIST University (Deemed), India

Bindi Varghese
CHRIST University (Deemed), India

ABSTRACT

With the advancement of ICT, the tourism industry has undergone a digital transformation where management, marketing, and communication are largely using web based applications. Automation of processes like ticketing and reservation, online hotel booking, E visa processing, etc., indicates the significant reliance of the sector on technology and world wide web for its services. Big data and artificial intelligence are a fairly new and innovative approach to addressing this issue of managing and analyzing huge datasets collected from multiple sources. This chapter focuses on understanding the role and importance of big data and artificial intelligence in the tourism industry and its impact on improving the overall image and attractiveness of destinations.

INTRODUCTION

Every economy in the world is geared up towards implementing major structural and technological changes and advancements in the present world. Adapting to such disruptive changes has called for societies to take up new challenges and constantly evolve and progress with a steady pace. Townsend (2013) suggests that the ideal method to be adopted by societies is to transform themselves with the use of cutting-edge technologies and applications. Technological advancements are prevalent in every sector, but with respect to travel and tourism, the impact of technology is huge. Literature propagates the

DOI: 10.4018/978-1-6684-6366-6.ch008

importance of the development of smart communities to support and initiate urbanization, expansion and growth of the various industries functioning within the complex macro environment. The term 'smart destination' was coined in the early 1990s as a complex integration of technological infrastructure into the social and economic structure of an economy (Johnson, 2022). Later the acceptance of technology has contributed to the improvement and standardization of the various interconnected systems, subsystems and processes (Buhalis, 2015).

The need to evolve and imbibe technology into the operational structure and governance has been realized by the public and private bodies and this has led to the evolution of a smart and digital era that we are living in today. Adapting to the rapid changes in technology and advancing along at the same pace has created a wide range of opportunities for businesses and industries worldwide to explore, exploit and emerge as successful market leaders (Bolívar and Meijer, 2015). The role of Information and Communication Technologies (ICT) in enhancing customer experience and creating value for the products and services has been well researched in the past decades (Wang et al., 2013; Buhalis & Amaranggana, 2014; Gretzel et al., 2015).

Through smart economies, communities are well equipped with the infrastructure needed to create a connection between society and technology which leads to the creation of a digital ecosystem. Such a digital ecosystem helps to address the social, economic and environmental challenges caused by urbanism. McNaughton et al. (2016) highlighted the importance of integrating open data to boost the sustainability of the tourism sector in developed and developing countries. Since the tourism sector involves free flow of data and information about tourism products, services and destinations, it is essential to channelise this flow of data in the right means so that sustainable development and value creation can be achieved (Buhalis and Jun., 2011).

Smart technologies in tourism had been in existence for the last few decades with the introduction of Global distribution systems (GDS), Centralized Reservation Systems (CRS), Web based technologies, E tourism services, etc (Buhalis, 2003, Werthner and Ricci, 2004). The wide acceptance of Social media in marketing and promotion has created a platform for effective information dissemination, communication, and brand image building in tourism (Sigala et al, 2012). Due to the increasing reliance of people on smart technologies in the present day, ICT has contributed towards enriching customer experience and providing value added services to the customers (Wang et al, 2013). Xian et al, (2015) describes the transformation process of a destination to 'smart destination' by focusing on the extensive use of technology and offering accurate information, personalized services and valuable experiences to the customers. Boes et al, (2015) stated that increased dependence on IT can enable suppliers and customers to get actively involved in the process to co-create value and share the benefits arising out of such collaborations. Buhalis and Amaranggana (2014) propagate that smart tourism not only enriches experience but also improves the image and competitiveness of the destination.

Role of Destination Management Organizations (DMOs)

Destination Management Organizations (DMOs) facilitate the effective management, marketing and governance of tourism destinations by involving in the strategic planning, policy making and implementation processes (Buhalis, 2003). Destination Managers have realized the potential benefits of integrating smart technologies into the tourism ecosystem; like improving the quality of services, creating personalized and customized experiences, enhancing value and levels of satisfaction, competitive advantage in the global market, better visibility and destination image etc (Varghese, 2014). Fig 1.0 explains the role of DMOs in managing and marketing the destination to ensure that the destination is competitive, ensures effective stakeholder collaboration and focuses on continued growth and prosperity. The major functions of DMOs listed in the model include destination marketing and branding, building alliances and meaningful partnerships within the destination and within stakeholders, destination governance 0through strategic planning and policy building and careful governance of the destination through attraction and visitor management (Varghese, 2014). The strategies adopted by destinations to integrate technology into their governance and management includes introducing better broadband and Wi-Fi facilities at destination; creating digital information centers and virtual guides, making mobile applications available to the tourists, facilitating contactless booking and reservation facilities at destinations etc (Gretzel et al., 2015). For implementing such a complex system, it is essential that the destination and the service providers are well equipped to handle these applications and skilled to operate and use them (Mariani, M et al., 2021). Developing a skilled workforce by training and educating the local citizens has been prioritized by many destinations globally which has further led to the overall development and standardization of communities and societies (Varghese B, 2014). One of the key focus areas of DMOs has been the creation of an innovative network to encourage better involvement, participation and investment by the private players; build an efficient city resource management system and thereby bring together the various stakeholders directly or indirectly involved in the tourism business (Shafiee et al., 2019).

Sheehan et al., (2016) propagates the future role of DMOs as an intelligent agent, facilitator and executive capable of identifying, engaging and learning from internal and external stakeholders directly or indirectly involved in the destination. They function as intelligent knowledge sharing centers which facilitate information exchange to the tourists and other stakeholders at the destination (Varghese, 2014). Through such a system, destinations can quickly adapt themselves to the changing environments, identify and exploit market possibilities, assess the various environmental, social and economic risks involved and provide for effective planning and forecasting for the future. DMOs also function as facilitators and innovators as they collect the feedback about products and services from the customers and provide necessary inputs to the service providers on creating better and innovative products and services to enhance customer experiences (Ye et al., 2020). Tourism organizations work in a close-knit interconnected system that provides customers with real time information and simultaneously collect and optimize data and information for facilitating better strategic and operational management (Gretzel et al. 2015).

Destination Competitiveness Index: Factors, Determinants, and Indicators

Any destination must make sure that its overall "appeal" and the tourist experience delivered are superior to those of the alternative places available to potential visitors in order to attain competitive advantage for its tourism sector. Literature substantiates that destination competitiveness is associated with multiple factors and variables which may be objectively or subjectively verified. "Richness of culture and heritage," "quality of the tourism experience," and visitor numbers, market share, tourist expenditure, employment, and value added by the tourism industry are some of the most commonly linked variables to destination competitiveness. One of the earlier definitions for destination competitiveness was proposed by Go and Govers (1999) which measured a destination's competitive position in comparison to other destinations and that competitiveness can be measured by how well it performs across seven different criteria. The elements include amenities, accessibility, service quality, overall affordability, location image, climate and environment, and attractiveness. Ritchie and Crouch (2000) has some major contributions in the field of destination competitiveness and suggests that a destination's competitiveness can be improved

Figure 1. DMO Model for destination competency and managerial paradigm
Source: Bindi Varghese (2016)

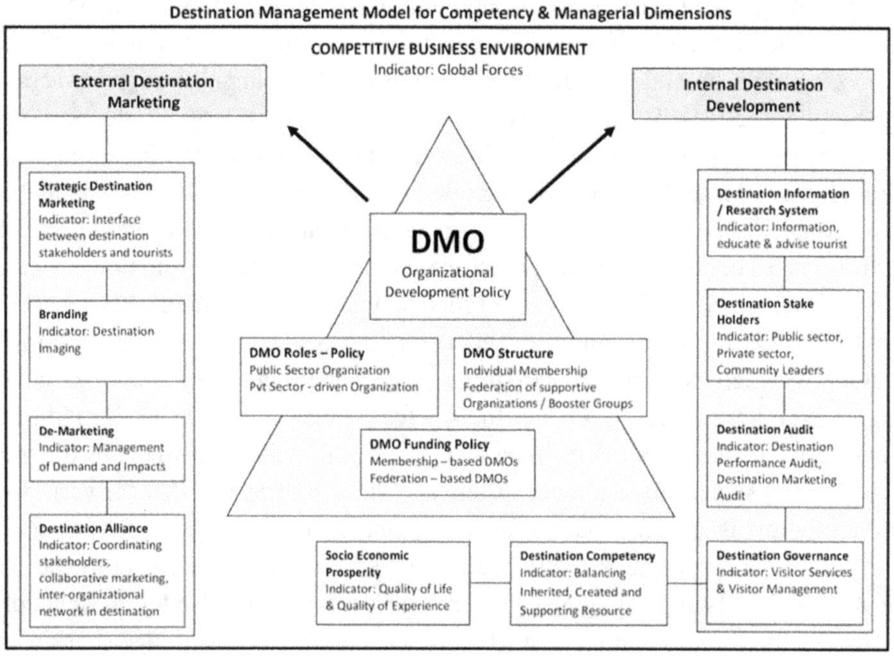

through innovation, risk taking, investments, specialization and productivity enhancement. Strategic alliances and partnerships also firm a base for developing meaningful relationships among key tourism stakeholders and thereby improving the overall attractiveness and image of the destination. A destination's competitive index can be measured using various micro and macro factors that exist in the internal and external environment. The major competitiveness factors, according to Ahn and Bessiere (2023), are Destination governance, Environmental regulations, Safety and Security concerns, Health and Hygiene factors, General and Physical Infrastructure, Technological Infrastructure, Stakeholder collaboration and Partnerships, and Human and Natural Resources at the destination.

DMOs as a Facilitator for Competitiveness

DMOs should ideally also act as the driving force behind the overall planning of the destination's growth. The goal in every situation is to make it possible for the diverse components of the tourism industry to collaborate and, as a result, compete more successfully. Inter-firm cooperation is encouraged by the interdependence of business and shared self-interest in the success of the destination, such as marketing partnerships, sectorial associations, and management structures.

The importance of destination planning in the long-term success of a destination has been widely researched throughout the globe. There is a need for high level strategic planning by incorporating the major social changes, the economic and political structure and the technological developments at the destination in order to ensure sustainable growth and balanced development. Ritchie and Crouch (1996) developed the Destination Vision Plan which includes the core values and principles of the destination, the long-term goals and objectives and the strategies and plans adopted by the destination for its long-term sustainability. Mintzberg (1987) highlighted that policy formulation and development of core strategies are an ongoing and continuous process for destination. This process essentially involves four phases i.e., Definitional Phase, Analytical Phase, Operational Phase, and Implementation phase. During the definitional phase, the destination system, core values and strategies and the long-term vision of the destination is defined. Analytical phase consists of internal and external analysis of the existing policies, tourist behavior at the micro and macro levels and impact assessment. During this stage, a destination undertakes a thorough analysis of the destination, its resources and capabilities through an extensive auditing process. The demand patterns and tourist behavior is assessed to understand the travel motivators and thereby improve the marketing and promotional strategies adopted by the destination planners. Once the analysis is completed, the next stage focuses on developing strategies and long-term plans based on the analytical conclusions drawn from the previous stage. The final phase is the implementation phase which involves the translation of the policies and strategies into reality. The implementation phase involves planning ahead and anticipating various contingencies, risks and unforeseen events that might occur in the future and preparing in advance how to deal with such contingencies.

Ritchie and Crouch (2003) evaluated the significance of the micro and macro environment factors and integration of policies and strategies in alignment to the resources at the destination to create a competitive destination. Ritchie and Crouch (2003) has developed the most detailed and comprehensive model for destination competitiveness by including 36 major determinants on competitiveness which are broadly classified under qualifying and amplifying determinants, destination policy, planning and development, destination management, core resources and attractors, supporting factors and resources etc. The location, safety and security considerations, destination image and carrying capacity are the qualifying determinants for a destination to be competitive. This is supported by the tourism system, the policies, strategies, long term vision and goal of the destination. Careful planning and management of the destination is essential to ensure that the destination is attractive and competitive. Effective destination management includes management of visitors, resources, finance and human capital. Crisis management also plays a significant role in destination management as destinations have to prepare in advance to face any contingencies that might occur in the future (Spencer et al., 2012). The core resources of the destination are the major attractors and these include the climate, culture, history, physiography, structures and superstructures, events, festivals, traditions and customs. These are supported by the marketing ties, infrastructure, hospitality, enterprise and the political and social structure of the destination. The concept of destination audit was developed by Ritchie and Crouch (2003) which essentially means keeping record of all the resources at the destination and ensuring careful upkeep and maintenance of these resources. Auditing of destinations helps to minimize wastage, encourage careful investment and upkeep of the resources, develop and maintain infrastructure and avoid unnecessary procedural delays in planning and implementation.

Transforming Cities to Smart Tourism Destinations

The tourism market is dynamic and constantly evolving which requires the optimization of information and resources to manage the demand. Due to the high degree of reliance of tourists on smart devices like mobile phones, social media platforms, network technologies and the internet, destinations have to keep up with the fast pace in order to survive the competition and grow. Buhalis (2019) suggests that destination management has to focus more on incorporating Information and Communication Technologies (ICT) to provide value added services and customized experiences to the customers and improve the overall efficiency of the destination. Incorporating new technologies also paves way for the internal and external stakeholders to create a smart and sustainable ecosystem facilitating mutual growth and development (Gretzel, 2015).

Smart experience creation focuses on enhancing and enriching the experience of tourists through personalization and real-time monitoring. Tourists play a major role in participating and co-creation of such experiences by sharing them on social media or word-of-mouth publicity. Smart businesses involve the complex business ecosystem that integrates Information technology and provides smart solutions to tourists visiting the destination (Gretzel, 2015). Such smart businesses can provide better and intelligent decision-making capabilities, greater potency, and therefore enhance the overall tourist experience experiences (Ye et al., 2020). A major concern of smart tourism development is the huge gap between the physical business worlds with the digital realm. The lack of awareness and readiness to accept technology is a major restricting factor in the implementation of smart business technologies. Social media and the internet are essential instruments facilitating a smooth transition to smart business models and creating a smart tourism ecosystem. As smart tourism relies on the free sharing of information and the enormous amount of data resources it can create better value propositions (Mehraliyev, 2019. Though there are some disadvantages like security and privacy concerns, affordability, etc., the digital footprints created by smart tourists have a high and major impact on destination marketing and promotional strategies.

The significance of the smart evolution of destinations is highlighted by defining the characteristics and features of smart tourism. The higher dependence of the present generation on smart applications like mobile phones, social media and the internet has created a huge impact on the strategic planning and marketing practices followed by destinations worldwide (Ye et al., 2020). Dorcic et al., (2019) researched the implications of smart mobile applications for the development of tourism destinations. Three major perspectives of smart tourism were deliberated in the paper; consumer perspective, technological perspective, and provider perspective. One of the research gaps identified is the information governance and privacy concerns faced by a smart tourist and how to tackle these issues at the organizational and management level. It was identified that there are still tourists who are not well informed of technology and are reluctant to use smartphones as information guides or service providers (Mehraliyev, 2019). Though many destinations are already in the process of transforming themselves into smart tourism destinations, there is still scope for research and development in terms of the implementation of cost-effective and affordable technologies and smart solutions to tourists (Buhalis and Amaranggana, 2015).

The smart transformation of a destination requires the development of a regionally integrated and connected system of stakeholders working together in the tourism supply chain (Ye et al., 2020). The need to include academia such as universities and other research organizations is also indicated in the research paper as these institutions are directly or indirectly involved in the tourism development process (Gretzel et al., 2015). Traditional businesses need further renewal and creation of value-added customized services to the tourists thus creating a more personalized experience. Implementation of a smart and intelligent system to monitor and forecast the tourism demand is also essential at this phase (Li et al., 2017). This is possible only through effective stakeholder engagement and collaboration and at the same time promoting social entrepreneurship in the tourism sector of the region. The role and functions of the regional DMOs have been redefined as they are not limited only to the creation of travel packages and upkeep of the tourist sites, there is a shift in the focus of DMO's functionality to being an information and knowledge sharing center of innovative tourism products and services to the tourists in the most efficient and effective manner (Varghese, 2014).

Mariani et al., (2021) researched on formulating Destination Business Intelligence Units (DBIU) and Big Data Analytics (BDA) to develop an integrated conceptual model for measuring Tourism Destination Competitive Productivity (TDCP) and implementing such a framework for sustainable development.

Tourists' Experiences With Smart Tourism Technology

Each traveler can have a unique experience thanks to the experiential character of the tourism sector through personal interactions with and perceptions of Smart Tourism Technologies. Despite the fact that all tourists may partake in the same activities at the same location, the memorability of their experiences varies, which influences how they will evaluate their trip (Jeong and Shin, 2019). The main elements that appeared to have an impact on travelers' memorable experiences are interactivity, customization, and informativeness (Mariani et al., 2021). The most frequently used technological applications include google maps, ride share apps, city guide and local tourism mobile apps, mobile payment gateways, smart parking facilities in cities, public wifi at destinations, smart traffic surveillance system, smart climate forecast systems, Internet of Things, Smart Street Lights, Virtual Reality and Augmented Reality (Ye et al., 2020).

Although there are many destination-specific aspects that influence tourists' behavior, how they experienced STTs in the highly digital tourism destination environment has a significant impact on their satisfaction with STTs and intention to return to the destination. It is obvious that tourism agencies in smart destinations need to focus more on enriching the overall experience of visitors with the help of technology enabled solutions.

Creative Tools for Competitiveness Through Smart Tourism Applications

Smart tourism consists of several different "smart" elements and layers, including: (1) Smart Destinations, which are unique examples of smart cities that have integrated ICTs into their physical infrastructure; (Gretzel, 2015), (2) Smart Experience, which focuses on technology-mediated traveler engagement through personalization, context awareness, and real-time monitoring (Tsaih et al., 2018); and (3) Smart Business, which refers to the intricate business ecosystem that fosters and supports the exchange of touristic information (Koo et al, 2019).

According to Lopez de Avila (2015), a smart tourism destination is one that is innovative, built on a foundation of cutting-edge technology, and accessible to all. It makes it easier for visitors to interact with and integrate into their surroundings, enhances the quality of their visit, and enhances the quality of life for locals. Smart Experience can be created using smart technologies like Artificial Intelligence which helps to anticipate tourist behavior and patterns. Through the collecting of large amounts of data, Artificial Intelligence /Machine Learning might learn visitor behavior to forecast habits and interests, recommend hotels, restaurants, and airlines, and promote social groups (Mehraliyev, 2019). Robotic chatbots could offer 24/7 online helpdesk assistance to tourists looking for travel-related information (Li et al., 2017). Smart technologies focus on Improving the on-site experiences of visitors by providing interactive, location-based, rich information services ; (Huang et al., 2017); Robotic and Chatbot offer travel destinations, information searches, and location promotion using GPS analysis and purchase data; To help travelers find trip information, Natural Language Processing (NLP) could offer instant language translation. It is important to provide a platform to share the traveler experiences for the destination to get better reach and visibility (Ye et al., 2020). Smart Technologies focus on enabling passengers to share their trip experiences so they can influence other travelers' decision-making, relive and reinforce their travel memories, and build their social media profiles and status; By automatically creating photo books and bookmarking photos with geographical data, AI/ML could improve the visitor experience (Jeong and Shin, 2019). Also, tourists could post pictures to social media groups to share their travel experiences with their loved ones (Buhalis, 2019).

Destination Competitiveness Through Big Data and Artificial Intelligence (BDAI)

Dameri and Cocchia (2013) states that a destination is transformed into a smart ecosystem through sustainable policies; enhanced customer value through better human capital investment, effective stakeholder collaboration and value co creation. Smart tourism involves a wide range of technology enabled devices and applications such as smartphones, sensors and beacons, software and applications,artificial intelligence, big data, virtual reality, augmented reality, internet of everything (IoT), cloud computing, autonomous vehicles, drone technology (Buhalis and Leung, 2018; Koo et al, 2019; Buhalis and Amaranggana, 2014).

Figure 2. Structure of smart tourism
Zhu, Zhang & Li (2014)

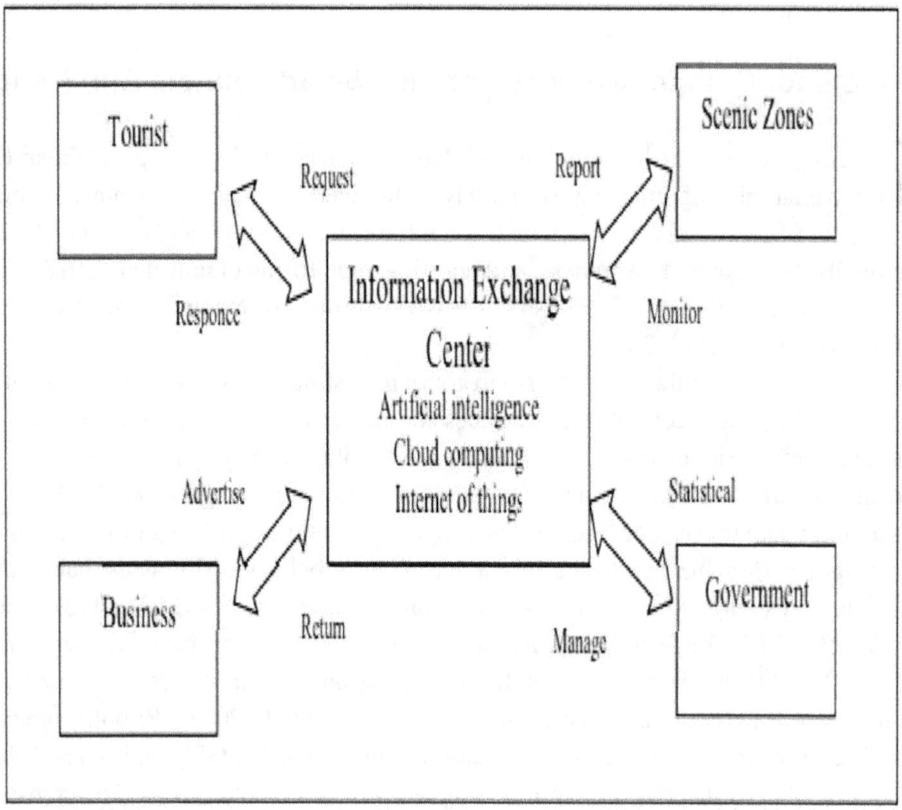

Results and Discussion

In the fast changing world, adapting and accepting technology is the need of the hour for destinations to maintain a competitive edge in the global market. Technology has become an inevitable part of our daily lives as it plays a major role in the day to day activities and plans of a common man. Tourism industry with its wide scope and multi-dimensional nature of operations, require effective integration and implementation of such technologies to ensure the smooth functioning and flow of tourists in and out of various destinations. From the literature reviews that formed the theoretical base for this study, it has been proved that smart tourism technologies play a major role in improving a destination's image and thereby make it more competitive. By developing a smart infrastructure, tourists can avail a wide range of services easily and in a hassle free manner with the help of their mobile devices and smart phones (Xiang and Fesenmaier, 2017; Ardito et al., 2019; Tsaih and Hsu, 2018). Tourists can be benefitted through ease in transactions, faster and quicker responses and easy access to data and information. From the management perspective, technology helps to easily market and manage the destinations resources, create a competitive edge for the destinations, plan and allocate the resources, understand tourist behavior patterns and create unique personalized experiences for the tourists (Križaj et al., 2021; Gretzel et al., 2015).

Implications and Conclusion

The chapter bears several implications for the tourism planners and policy makers in terms of efficient methods of exploitation of technology in the travel industry. Firstly, the policymakers have to adopt a more holistic approach to integrate tourism policies with the economic policies of the destination and include allied sectors like transportation, hospitality (Varghese, 2014), ICT etc. The second aspect involves the policymakers to be made aware of the digitization prospects of the destination and the various government supported smart tourism initiatives (Li et al., 2017).

DMOs should focus on the investment opportunities and the resources and funds available for implementing such technology enabled applications at the destination. Another key concern is the effective collaboration of stakeholders in managing and marketing the destination. The internal and external stakeholders also play a significant role in the digital transformation process of the destination (Gretzel et al., 2015).

Since there is an enormous amount of data that is being pumped into the market, the main concern of the destination planners is to make use of this data to create a competitive edge for the destination (Boes et al., 2015. The data has to be used to create value-added personalized services to the customers and provide real time information about destinations, tourist sites, and activities (Zhu et al., 2014). Data analytics and Business Intelligence units helps to measure the competitiveness and productivity of the destination and contribute further to improve the services and attractiveness of the destination, create new and innovative products, formulate better and faster communication process, create more global visibility through modern technology enabled marketing and promotional campaigns and contribute to the sustainable development of the destination.

REFERENCES

Ahn, Y. J., & Bessiere, J. (2023). The Relationships between Tourism Destination Competitiveness, Empowerment, and Supportive Actions for Tourism. *Sustainability*, *15*(1), 626. doi:10.3390u15010626

Ardito, L., Cerchione, R., Del Vecchio, P., & Raguseo, E. (2019). Big data in smart tourism: Challenges, issues and opportunities. *Current Issues in Tourism*, *22*(15), 1805–1809. doi:10.1080/13683500.2019.1612860

Boes, K., Buhalis, D., & Inversini, A. (2015). Conceptualising smart tourism destination dimensions. In I. Tussyadiah & A. Inversini (Eds.), *Information and Communication Technologies in Tourism 2015* (pp. 391–404). Springer. doi:10.1007/978-3-319-14343-9_29

Bolívar, M. P. R., & Meijer, A. J. (2016). Smart governance: Using a literature review and empirical analysis to build a research model. *Social Science Computer Review*, *34*(6), 673–692. doi:10.1177/0894439315611088

Buhalis, D. (2019). Technology in tourism-from information communication technologies to eTourism and smart tourism towards ambient intelligence tourism: A perspective article. *Tourism Review*.

Buhalis, D., & Amaranggana, A. (2013). Smart tourism destinations. In *Information and communication technologies in tourism 2014* (pp. 553–564). Springer. doi:10.1007/978-3-319-03973-2_40

Buhalis, D., & Jun, S. H. (2011). E-tourism. *Contemporary Tourism Reviews*, *1*, 2-38. www.goodfellowpublishers.com/free_files/Contemporary-Tourism-ReviewEtourism-66769a7ed0935d0765318203b843a64d.pdf

Buhalis, D., & Leung, R. (2018). Smart hospitality-Interconnectivity and interoperability towards an ecosystem. *International Journal of Hospitality Management*, *71*, 41–50. doi:10.1016/j.ijhm.2017.11.011

Dameri, R. P., & Cocchia, A. (2013, December). Smart city and digital city: twenty years of terminology evolution. In *X Conference of the Italian Chapter of AIS*. ITAIS.

Del Chiappa, G., & Baggio, R. (2015). Knowledge transfer in smart tourism destinations: Analyzing the effects of a network structure. *Journal of Destination Marketing & Management*, *4*(3), 145–150. doi:10.1016/j.jdmm.2015.02.001

Dorcic, J., Komsic, J., & Markovic, S. (2019). Mobile technologies and applications towards smart tourism – state of the art. *Tourism Review*, *74*(1), 82–103. doi:10.1108/TR-07-2017-0121

Gretzel, U., Sigala, M., Xiang, Z., & Koo, C. (2015). Smart tourism: Foundations and developments. *Electronic Markets*, *25*(3), 179–188. doi:10.100712525-015-0196-8

Gretzel, U., Werthner, H., Koo, C., & Lamsfus, C. (2015). Conceptual foundations for understanding smart tourism ecosystems. *Computers in Human Behavior*, *50*, 558–563. doi:10.1016/j.chb.2015.03.043

Huang, C. D., Goo, J., Nam, K., & Yoo, C. W. (2017). Smart tourism technologies in travel planning: The role of exploration and exploitation. *Information & Management*, *54*(6), 757–770. doi:10.1016/j.im.2016.11.010

Jeong, M., & Shin, H. H. (2019). Tourists' Experiences with Smart Tourism Technology at Smart Destinations and Their Behavior Intentions. *Journal of Travel Research, 004728751988303*. doi:10.1177/0047287519883034

Johnson, A.-G. (2022). Why are smart destinations not all technology-oriented? Examining the development of smart tourism initiatives based on path dependence. *Current Issues in Tourism*. doi:10.1080/13683500.2022.2053071

Komninos, N., Pallot, M., & Schaffers, H. (2013). Special issue on smart cities and the future internet in Europe. *Journal of the Knowledge Economy*, *4*(2), 119–134. doi:10.100713132-012-0083-x

Koo, C., Mendes-Filho, L., & Buhalis, D. (2019). Smart tourism and competitive advantage for stakeholders: Guest editorial. *Tourism Review*, *74*(1), 1–4. doi:10.1108/TR-02-2019-208

Križaj, D., Bratec, M., Kopić, P., & Rogelja, T. (2021). A technology-based innovation adoption and implementation analysis of European smart tourism projects: Towards a smart actionable classification model (SACM). *Sustainability*, *13*(18), 10279. doi:10.3390u131810279

Li, Y., Hu, C., Huang, C., & Duan, L. (2017). The concept of smart tourism in the context of tourism information services. *Tourism Management*, *58*, 293–300. doi:10.1016/j.tourman.2016.03.014

Mariani, M., Bresciani, S., & Dagnino, G. B. (2021). The competitive productivity (CP) of tourism destinations: An integrative conceptual framework and a reflection on big data and analytics. *International Journal of Contemporary Hospitality Management*, *33*(9), 2970–3002. doi:10.1108/IJCHM-09-2020-1102

McNaughton, M. L., McLeod, M. T., McNaughton, M., & Walcott, J. (2016). Open data as a catalyst for problem solving: empirical evidence from a small island developing states (SIDS) context. In 2016 *Open Data Research Symposium, Madrid, Spain*. https://drive.google.com/ open?id=0B4TpC6ecmrM7OEN6OVlIUXh1d1U

Mehraliyev, F., Choi, Y., & Köseoglu, M. A. (2019). Progress on smart tourism research. *Journal of Hospitality and Tourism Technology*, *10*(4), 522–538. doi:10.1108/JHTT-08-2018-0076

Mintzberg, H. (1987). The strategy concept I: Five Ps for strategy. *California Management Review*, *30*(1), 11–24. doi:10.2307/41165263

Ritchie, J. R., & Crouch, G. I. (2010). A model of destination competitiveness/sustainability: Brazilian perspectives. *Revista de Administração Pública*, *44*(5), 1049–1066. doi:10.1590/S0034-76122010000500003

Shafiee, S., Ghatari, A. R., Hasanzadeh, A., & Jahanyan, S. (2019). Developing a model for sustainable smart tourism destinations: A systematic review. *Tourism Management Perspectives*, *31*, 287–300. doi:10.1016/j.tmp.2019.06.002

Sheehan, L., Vargas-Sánchez, A., Presenza, A., & Abbate, T. (2016). The use of intelligence in tourism destination management: An emerging role for DMOs. *International Journal of Tourism Research*, *18*(6), 549–557. doi:10.1002/jtr.2072

Sigala, M., Christou, E., & Gretzel, U. (Eds.). (2012). *Social media in travel, tourism and hospitality: Theory, practice and cases*. Ashgate Publishing, Ltd.

Spencer, A., Buhalis, D., & Moital, D. (2012). A hierarchical model of technology adoption for small owner-managed travel firms: An organizational decision-making and leadership perspective. *Tourism Management, 33*(5), 1195–1208. doi:10.1016/j.tourman.2011.11.011

Townsend, A. M. (2013). *Smart cities: Big data, civic hackers, and the quest for a new utopia*. WW Norton & Company.

Tsaih, R.-H., & Hsu, C. C. (2018). *Artificial Intelligence in Smart Tourism: A Conceptual Framework*. ICEB 2018 Proceedings, Guilin, China. 89. https://aisel.aisnet.org/iceb2018/89

Varghese, B. (2016). A Strategic Evaluation on Competency of Karnataka Destinations through Destination Management Organizations. *American Journal of Industrial and Business Management, 06*(02), 102–108. doi:10.4236/ajibm.2016.62010

Wang, D., Li, X. R., & Li, Y. (2013). China's "smart tourism destination" initiative: A taste of the service-dominant logic. *Journal of Destination Marketing & Management, 2*(2), 59–61. doi:10.1016/j.jdmm.2013.05.004

Werthner, H., & Ricci, F. (2004). E-commerce and tourism. *Communications of the ACM, 47*(12), 101–105. doi:10.1145/1035134.1035141

Xiang, Z., & Fesenmaier, D. R. (2017). Big data analytics, tourism design and smart tourism. *Analytics in smart tourism design: concepts and methods*, 299-307. doi:10.1177/0047287514522883

Wang, A. D., O'Leary, J. T., & Fesenmaier, D. R. (2020). Adapting to the Internet: Trends in Traveller's Use of the Web for Trip Planning. *Journal of Travel Research, 54*(4), 511–527. doi:10.1177/0047287514522883

Ye, B. H., Ye, H., & Law, R. (2020). Systematic review of smart tourism research. *Sustainability, 12*(8), 3401. doi:10.3390u12083401

Young, A., & Verhulst, S. (2017). Jamaica's interactive community mapping: open data and crowdsourcing for tourism. In S. G. Verhulst & A. Young (Eds.), *Open Data in Developing Economies: Toward Building an Evidence Base on What Works and How* (pp. 206–223). African Minds. https://muse.jhu.edu/chapter/2062325

Zhu, W., Zhang, L., & Li, N. (2014). Challenges, function changing of government and enterprises in Chinese smart tourism. *Information and communication technologies in tourism, 10*, 553-564.

Chapter 9
Future of Public Sector Enterprises in the Metaverse

Richmond Anane-Simon
https://orcid.org/0000-0003-2512-7772
Independent Researcher, Ghana

Sulaiman Olusegun Atiku
https://orcid.org/0000-0001-9364-3774
Namibia University of Science and Technology, Namibia & Walter Sisulu University, South Africa

ABSTRACT

This chapter explores the future of public sector enterprises in the contemporary metaverse technological sphere. A literature review approach was adopted to explore customer expectations, sustainable governance forms, and innovations in the metaverse ecosystem. Findings indicate that most of the technologies that make up the metaverse's building blocks already exist or are in an advanced stage of development. Furthermore, there was a rise in the use of metaverse technology like virtual and augmented reality (AR), particularly in the healthcare sector during the pandemic. Digital currency adoption, including the use of cryptocurrencies and central bank digital currencies, was also noted. Public sector enterprises should rely on a variety of interdependent players rather than assuming they have a monopoly on knowledge or resources to govern. A process of co-evolution between non-government and governmental entities is a sine qua non for effective social innovation by public sector enterprises in the metaverse ecosystem.

INTRODUCTION

Public sector enterprises, or "public enterprises," are generally known as the business entities owned, run, and controlled by the federal, state, or local government. The term "public sector undertakings" also applies to these. Any business or industry that is owned and operated by the government with the goal of maximizing social welfare and upholding the public interest is considered a public sector enterprise. Public enterprises are state-owned institutions established for social welfare or to meet social needs. Examples include banks, life insurance companies, state-owned transport companies, hospitals,

DOI: 10.4018/978-1-6684-6366-6.ch009

etcetera. State-owned companies may be as old as the state itself. A number of waves contributed to the development of the state-owned enterprise sector, and the motivations behind their establishment are as diverse as the emerging trends in public entrepreneurship in the West. According to Obinger, Herbert, Schmitt, and Traub (2016), fiscal monopolies, political and economic modernization, the need for a strong military, Marxist ideology, and natural monopolies have historically been the five main forces behind state ownership. The evolution of public services (public utilities) and personal social services in European countries came in four stages: the pre-welfare state of the late nineteenth century; the advanced welfare state that peaked in the 1970s; the neo-liberal policy phase since the early 1980s; and the recent phase since the mid-2000s (Hellmut, 2018). Public sector enterprises can have a significant impact on a country's GDP. In the United States, for example, public sector enterprises accounted for 4.6% of the country's GDP in 2019, according to data from the Bureau of Economic Analysis. In terms of employment, data from the International Labour Organization (ILO) indicates that the public sector employed around 22% of the global workforce in 2021, with significant variations across regions and countries. In some countries, such as Sweden and Norway, the public sector employs more than 30% of the workforce. Public sector enterprises are often responsible for investing in infrastructure and other public goods. In 2020, global public investment was estimated to be around 20% of GDP, according to data from the Organization for Economic Co-operation and Development (OECD). Public sector enterprises can also drive innovation, particularly in areas such as healthcare and energy. In 2020, the US National Institutes of Health (NIH) funded over $41 billion in medical research and development, while the US Department of Energy invested over $5 billion in energy innovation (Bernard et al., 2021).

Undeniably, the world has gone through several stages of technological revolution and advancement in its quest to increase productivity. These advancements have changed the ways and manners in which businesses are conducted. Water and steam were used to mechanize production during the first Industrial Revolution. The second produced goods in large quantities using electric power. The third automated production used electronics and information technology. The boundaries between physical, digital, and biological spaces will be blurred by a fusion of technologies that will mark the beginning of the Fourth Industrial Revolution (Klaus, 2016). The production, management, and governance of not only private sector businesses but also public sector enterprises or businesses may change simultaneously as a result of the anticipated changes in the Fourth Industrial Revolution and the post-pandemic era.

The internet has evolved from "Web 1.0," or read-only Web, to "Web 2.0," or user-generated Web, and its anticipated revolutionary leap, "Web 3.0," shall be characterized by largely decentralized protocols (O'Dair & O'Dair, 2019). This new Internet version seeks to create a decentralized online network where users own and have control over the resources they produce. The result is the creation of the Metaverse, a virtual world that combines physically persistent virtual environments with enhanced physical reality created digitally (Buhalis et al., 2022; Tlili et al., 2022). Undoubtedly, all enterprises, including public-sector enterprises, will be impacted by these massive future technological innovations. The implications of these innovations on public sector governance, customer satisfaction, product development, social welfare, and protection of the public interest may be quite enormous if underestimated. To ensure sustainability in this new era of enterprise or business revolution, there have been calls for a review of management and governance approaches across the business divide.

BACKGROUND

In Neal Stevenson's science fiction book Snow Crash, which was released in 1992, the term "metaverse" was created and used for the first time (Stephenson, 2003). It represented a separate virtual reality (VR) world made from computer graphics that users from all over the world could connect to and access using goggles and earphones. A protocol known as "The Street" serves as the Metaverse's backbone, connecting various virtual communities and places in a manner similar to how the Information Superhighway functions. Users manifest as avatars, which are customizable digital bodies, in the metaverse. Even though Stephenson's "metaverse" is digital and artificial, experiences there can actually affect how one feels about their physical selves. The Matrix, a VR cyberspace featured in William Gibson's 1984 science fiction book Neuromancer, is considered a literary predecessor to the Metaverse (Dionisio, Burns, & Gilbert, 2013). The metaverse, which is still in its development stages, has various elements, including: digital currencies, marketplaces/digital commerce, online shopping, non-fungible tokens, workplace infrastructure, social media, device independence, gaming, digital assets, digital humans, natural language processing, concerts, social events, and entertainment events. The development of MetaEnterprises has a strong theoretical and technical foundation in artificial society, computing experiments, parallel execution theory (the ACP theory) Fei-Yue, (2004); Wang and Lansing, (2004); Wang, (2004), and cyber-physical-social systems (CPSSs) technology (Wang, 2010). As a result, any real business will soon have one or even more corresponding MetaEnterprises with various functions in the Metaverse, and these entities will be connected via CPSS (Wang, Qin, Wang, & Hu, 2022). Public sector enterprises are no exception to this phenomenon since their activities will be obstructed. There are core elements which will characterize the metaverse (Pietro & Cresci, 2021).

1) ***Activities:*** The metaverse is a virtual universe or substrate that can accommodate and link a wide range of applications. As a result, the metaverse offers users a variety of activities that mirror the variety of applications it contains (Pietro & Cresci, 2021). The metaverse offers unprecedented networking opportunities, which makes it especially convenient for social interactions. The metaverse will also support more conventional activities like chatting with users, making audio or video calls, and befriending other users. For example, these features will be made accessible by incorporating current messaging and video conferencing apps into the metaverse. The shared virtual spaces of the metaverse will also enable additional forms of social interaction, such as the interactions between 3D avatars that are typical of massively multiplayer online games (MMOs). Another significant group of metaverse activities will be gaming and other forms of entertainment, such as the opportunity to attend art exhibits and concerts. First off, metaverses share a number of traits with massively multiplayer online games. Furthermore, both the revenue and user bases of the gaming industry are constantly expanding. Games and entertainment in general will be among the most popular activities in the metaverse as a result of the interaction of these two elements. Notably, metaverse shows can be both entirely virtual, as in the case of the concerts held within the virtual worlds of online games like Fortnite, Minecraft, and Roblox, or entirely physical but still reachable via the metaverse, as in the case of a real-world concert that allows metaverse users to attend via VR (Cheng et al., 2022). Another category of activities that will profit from the cyber-physical integration made possible by the metaverse is sports and fitness. Wearable sensors and AR/VR in particular will enable realistic and immersive virtual sport simulations with previously unheard-of personalization and customization possibilities (Atiku, 2018). The same factors can be

taken into account for learning and other educational activities, which will greatly profit from the metaverse's immersion and 3D capabilities. In addition to being used for commerce, the metaverse will also be used for work and business. With regards to the former, rich, immersive meetings will be possible in the metaverse thanks to digital twins, VR, and the accessibility of embedded messaging and videoconferencing apps. Furthermore, one or more online marketplaces will serve as a support for both conventional and cutting-edge forms of commerce by offering both tangible and intangible products for sale. With regard to the latter, in particular, the marketplace will link independent content producers with their prospective clients (i.e., users of the metaverse), enabling business opportunities to scale to unprecedented heights.

2) ***Immersiveness***: The mobile Internet revolution witnessed over the last ten years has made it possible for us to access our online services and social networks from virtually anywhere (Kaisara et al., 2022). But at the moment, using screens and mobile devices limits our ability to enter and enjoy virtual environments. Instead, with the introduction of the metaverse, AR and VR technologies will also be able to access online virtual spaces. Indeed, the pervasiveness and immersiveness of the metaverse—achieved through an unprecedented fusion of the virtual and real worlds—are two of its key and distinguishing characteristics (Pietro & Cresci, 2021). The ways in which we access it, engage with it, and receive feedback from it are all part of how the metaverse is pervasive and immersive in many different ways. For example, the upcoming generation of AR-capable smart devices, such as compact smart glasses, will enable mobile, immersive access to the metaverse. Instead, portable and comfortable VR goggles will make access to the workplace or home possible. There will be a number of new possibilities that will increase immersiveness as we move from 2D interfaces to 3D virtual environments. First off, numerous 2D services and applications that we currently use on a daily basis, including Dropbox, Slack, Zoom, Facebook, Instagram, and many more, will be integrated into the metaverse as applications. Users will then inhabit the metaverse as avatars, transitioning from static 2D profile pictures to dynamic, customized 3D avatars. Users will be able to represent themselves with either photorealistic, cartoonish, or fully fictional avatars depending on the activity, application, or virtual space in use. In order to further close the gap between the virtual and physical dimensions, users will also have the option of making virtual versions of tangible objects (also known as "digital twins") and sharing them in the metaverse. Finally, the use of wearable sensors and devices will strengthen the connection between our real and virtual worlds by providing users with unheard-of sensory feedback and by supplying the metaverse with orders of magnitude more real-world data (Bermejo & Hui, 2021).

3) ***Interoperability:*** The metaverse can be thought of from an architectural perspective as a unifying framework or substrate that connects the plethora of applications and services that are embedded into it. Interoperability is a crucial aspect of the metaverse. As a result, users can experience it in many ways. For instance, they will be able to interact with several applications at once, just like we do on our desktop or mobile devices every day. Currently, this degree of app compatibility is normal and expected for general-purpose physical devices. For online virtual environments, it is, however, unprecedented. Consider MMOs, which are large virtual worlds where players can typically only engage in a select number of activities that are similar and related. By extending this idea, the metaverse will also have connected spaces and activities. The ability to switch between various virtual thematic spaces will actually be possible, as will the ability to stop one activity and begin another (for instance, pausing a game in a specific space to join a friend in another space). This interconnectedness will also include virtual items like avatar clothing. The interoperability of the

metaverse will enable users to purchase specific virtual items as Non-Fungible Tokens (NFTs) from an application's store and use them with their avatars in other applications, spaces, and throughout the entire metaverse in one of the possible metaverse evolutions where users instead of platforms will own the items (Pietro & Cresci, 2021).

Other elements include but are not limited to:

4) ***Sustainability***: The metaverse is said to be sustainable if a closed economic loop and a consistent value system are maintained with a high degree of independence. On one side, it should be open, i.e., continually igniting users' passion for creating digital content and open inventions (Allam, 2022). On the other hand, to avoid Single Point of Failure (SPoF) issues and avoid being ruled by a small number of strong entities, it should be based on a decentralized architecture if it is to remain enduring.
5) ***Scalability:*** The term "scalability" refers to the ability of the metaverse to continue operating effectively regardless of the quantity of avatars using it at any given time, the complexity of the scene, or the type, scope, and range of interactions between them (Dionisio & Gilbert, 2013).
6) **Heterogeneity**: The heterogeneity of the metaverse encompasses heterogeneous virtual spaces (e.g., with distinct implementations), heterogeneous physical devices (e.g., with distinct interfaces), heterogeneous data types (e.g., unstructured and structured), heterogeneous communication methods (e.g., mobile and satellite communications), as well as the diversity of human thinking (Wang et al., 2022).

Other elements may surface as the metaverse develops and becomes fully functional. The global economic downturns due to the impact of the pandemic are an indication that the global economy is not resilient enough, as businesses that were not operating in virtual environments could not withstand the economic shocks. This has raised concerns about the impacts of existing business models on enterprise sustainability. Even though some awareness of the need for sustainability in business organizations has been created over the past two decades, there are still uncertainties on how to achieve it. Early business models emerged at the end of the 20th century, motivated by the need to describe and analyze new forms of business. For example, e-business and virtual organizations (Alt & Zimmermann, 2014; Wirtz, Pistoia, Ullrich, & Gottel, 2015). The notion received attention as a general management concept through the early publications of Chesbrough and Rosenbloom (2002) and Magretta (2002), who linked business models to strategy and innovation. As more individuals spend enormous amounts of time in virtual spaces, public sector enterprises will be left with no option but to transition into this space or risk shrinking away. The Fourth Industrial Revolution comes with numerous complexities, undoubtedly disrupting business processes, organizational forms, and labor markets. The World Economic Forum highlights four main effects that the Fourth Industrial Revolution has on businesses: customer expectations, product enhancement, collaborative innovation, and organizational forms. It held that the emergence of global platforms and other new business models means that talent, culture, and organizational forms will have to be rethought (Klaus, 2016). Hence the chapter aims at exploring the future of public sector enterprises in an anticipated and much more technologically advanced ecosystem called the Metaverse.

Customer Expectations in the Metaverse

Public-sector enterprises are wholly or partially owned by the state, the larger society serves as their primary customer. Goods and services produced by these enterprises are expected to benefit the populace (customers). The customer, in turn, pays for the utility derived. As a result, the importance of populace (customer) satisfaction for the sustainability of public enterprises cannot be overstated (Hengboriboon et al., 2022). Modern technologies now place a strong emphasis on speed and convenience. Hence, customers prefer self-transactions. They seek speedy and simple transactions. They select customized goods and services for themselves. They prefer shopping expeditions that will result in memorable experiences, and they like to post about them on social platforms. Participants in the research conducted by Eman, Kathy, and Ibrahim (2013) identified several aspects of customer service in metaverse retailing that are similar to those in 2D e-retail service, including competence, courtesy, human contact, and responsiveness. A participant in the study maintained that "in Second Life (SL), you can just have instant interaction; if you need to ask about the product, if you need a demonstration, or if there is something that is not quite right about it, you can contact the person who created it and ask about it instantly, whereas with a 2D website, you have to email the person and then wait for a reply." Another respondent expressed similar sentiments: "I would hope for the maximum amount of real interaction between me and other avatars." As public sector enterprises are commercial in nature and usually self-funded, their economic sustainability lies on the line if they are unable to meet the technological expectations of their customers in the future space of business—the metaverse.

Digital Security and Safety Expectations

Despite the metaverse's positive signs, security and privacy concerns remain the main obstacles impeding its development. The management of huge data streams, widespread user profiling practices, unfair results of AI algorithms, and the safety of physical infrastructures and human bodies are only a few examples of the wide spectrum of security lapses and privacy invasions that may occur in the metaverse (Wang et al., 2022). Accordingly, as the metaverse uses a range of cutting-edge technologies and systems that are based on them, it is possible that it will inherit weaknesses and defects from those technologies. Emerging technology incidents have included the appropriation of wearable technology (Atiku, 2018) or cloud storage, the theft of virtual currencies, and the misuse of AI to create false news. Second, due to the blending of many technologies, the consequences of current risks can be exacerbated and made worse in virtual worlds, while new threats that do not exist in physical or digital places can emerge, including virtual stalking and virtual espionage (Leenes, 2008).

A digital replica of the real world can be created using more precise and pervasive personal data in the metaverse, which creates new opportunities for crimes involving private big data (Falchuk, Loeb, & Neff, 2018). For instance, users will unavoidably utilize wearable AR/VR gadgets with built-in sensors to gather extensive amounts of data on biometric traits, facial expressions, eye movements, hand movements, voice patterns, and brain wave patterns in order to create a virtual scene using AI algorithms. Additionally, since individuals in the metaverse must be uniquely recognized, headsets, VR glasses, and other gadgets may be unlawfully used to track users' real-world locations (Shang, Chen, Wu, & Yin, 2022). Last but not the least, hackers can leverage compromised devices and system flaws as entry points into real-world equipment like home appliances to endanger personal safety and even endanger key infrastructures like water supply systems, high-speed train systems, and power grid systems (Hu,

Li, Fu, Cansever, & Mohapatra, 2015). However, current security precautions may not always be efficient and may not be flexible enough for use with metaverse applications. The immersiveness, hyper-spatiotemporality, sustainability, interoperability, scalability, and heterogeneity inherent in the metaverse, in particular, may present a number of challenges for the provision of effective security (Allam, 2022; Pietro & Cresci 2021). These security challenges notwithstanding, there are emerging countermeasures to mitigate risks in the metaverse. Some key recommendations for data management and privacy countermeasures for the sustainability of the metaverse include key management for wearable devices, identity authentication for wearable devices, cross domain identity identification, etcetera (Wang et al., 2022). A public enterprise should brace itself to meet complex expectations from customers in the metaverse ecosystem, or else risk being out of business.

Sustainable Management Practices for the Metaverse

Due to digital security and safety concerns, it places enormous management and governance expectations on public sector enterprises and regulators in the metaverse, as well as enormous regulatory hurdles. There are three existing potential solutions (Wang, et al., 2022):

1. ***Artificial Intelligence (AI) Governance***: With the widespread convergence of perception, processing, and actuation, AI will play a significant role in enabling fully autonomous digital self-governance of people and society in the metaverse. AI-based techniques can be used to identify aberrant or Sybil accounts in the metaverse.
2. ***Decentralized Governance***: To maintain large-scale metaverse governance, centralized regulatory structures may run into a number of standard and technical challenges as well as issues with the compatibility of global laws. Avatar democracy can be promoted through collaborative governance, which prevents the concentration of regulatory rights. Blockchain technologies provide prospective decentralized options for cooperative governance in the metaverse, where smart contracts provide an easy-to-use method for decentralized governance that operates automatically.
3. ***Trusted Digital Forensics***: Digital forensics, which has been extensively studied in pictures and films, is a tool for accountability in the metaverse during conflicts. The development of a generic forensic method for digital camera photos is based on the observation that in-camera and post-camera image processing, which leaves a sequence of distinct fingerprint traces on the digital camera image (Swaminathan, Wu, & Liu, 2008). To determine whether a certain digital image is from a specific scanner, camera, or computer graphics application, the estimated post-camera fingerprints can be used to validate image authenticity.

Collaborative Innovations for Sustainability in the Metaverse

Virtual worlds (VWs), otherwise referred to as collaborative virtual environments (CVEs), offer a richer collaborative setting for interacting with others and fostering a sense of community. Collaboration within virtual worlds is becoming more prominent due to metaverse platforms. With the help of various social interactions, users of these platforms can create virtual worlds that can replicate real-life experiences. Effective commercialization of innovative ideas, including goods, services, processes, and business models, is a key driver of economic growth. Collaborative innovation is the next big idea that needs to shape up with actionable items, allowing players across the value chains to participate in the emergence of new collaborative business models (Atiku & Fields, 2017; Kattel, Lember, & Tnurist, 2020). Anchored in the solid foundations of intrapreneurship, collaborative innovation is the engine of modern, agile organizations capable of creating new capacity, which can pioneer radical new ideas while testing the limits of markets. The CocoVerse Framework is a true best friend for growth and collaboration in VWs (Greenwald, Corning, & Maes, 2017). The platform offered users the ability to modify items using hand-held tools in an immersive virtual environment. The Megacity VW was introduced by Sharma, Devreaux, Scribner, Grynovicki, and Grazaitis (2017) for emergency training. Megacity enables interaction between AI agents and digital avatars, which represent users' projections in the real world. It's adoption in agent security training programs was detailed by Passos, Silva, Abreu Mol, and Carvalho (2017). Autodesk 3Ds Max was used to generate the 3D models for the environment, while Unity 3D was utilized to make the VW. Key innovative concepts reviewed include the following:

MetaEducation

To adapt education to contemporary mediums, people have used a variety of technologies over time. The last few years have seen a huge increase in the popularity of online learning using PCs and mobile devices (Atiku & Boateng, 2020). Different client applications (desktop, mobile, and web-based) are available to access the content on modern learning platforms, along with sharing, monitoring, and personalized experiences (Atiku & Anane-Simon, 2020). Additionally, new platforms for virtual worlds (VWs) and metaverses, which we can use for online learning, are developing. Studies on education provided by VWs have been conducted over the years with promising outcomes. Students who require assistance in the development of their learning skills can find a variety of tools in virtual environments. For instance, students can complete assignments while learning about a new topic. One illustration is iSocial (Stichter, Laffey, Galyen, & Herzog, 2013), a desktop program for students with autism spectrum disorder (ASD) that aims to improve their social competencies by enabling social interaction in VWs. In contrast to the physical world, Stendal and Balandin (2015) assert that people with ASD prefer using VWs and feel more at ease communicating. Additionally, they emphasize how VWs give autistic individuals a way to participate in a virtual society without relying on social cues. A tool for helping students with ASD develop their social skills is the educational VW software that Fernandez-Herrero and Lorenzo (2019) introduced.

In tertiary education, a case study was conducted by Krajcovic et al., (2021) for the purpose of teaching lean management and drawing on VW's educational experience. The authors concluded that the VW experience is a compelling teaching strategy in a university setting, and the educational game in VW for the 5S methodology (sort, set, shine, standardize, and sustain) has a high acceptance rate. Also, multi-user (networking) VWs were regarded by Warburton as a new development in the area of e-learning that utilized the Second Life platform (Warburton, 2009). Along with the e-learning platform, Moodle, a VW was utilized to inspire students. Sloodle (Simulation Linked Object-Oriented Dynamic Learning), a learning system for virtual environments, served as an illustration of how Moodle could be integrated with the Second Life platform. Sloodle is an open-source solution that combines the Moodle learning management system with multi-user VRs, Second Life, and OpenSim (LMS). With the help of Sloodle, materials downloaded from Moodle can communicate with and interact with objects in Second Life. Thousands of teachers and students use Sloodle on a global scale. Research on Sloodle has been published by Nunes et al. (2017). Cruz-Benito et al. (2016) also highlighted an Usalpharma software architecture for learning inside VWs for the virtual learning environment via a high-speed network, while Zizza et al. (2018) suggested an alternative option. They discussed how several learning modules created with the high-fidelity VW platform were put into use.

Mystakidis (2020) studied the effects of gamification in social virtual worlds on student engagement, as well as the impact of VW involvement on skill development. The integration of quest-based techniques and virtual agent solutions into VWs allowed participants to obtain high levels of engagement based on the interactive VW experience. In addition, curricular gamification in social VR opens up new multidisciplinary collaborations that can enrich and differentiate, in contrast to the present online learning techniques. Furthermore, in order to facilitate remote access utilizing a VR interface (Zhang et al., 2018), creating a virtual laboratory (VL) with experimental hardware in the loop is required. Microsoft Kinect was utilized to monitor human motions and communicate information to the avatar in VL human-computer interaction, which uses cameras to make it possible for users to interact with the VW. Other advanced educational metaverse platforms are being developed, including the VoRtex software, which has a web platform, access control, and the MicroLesson feature as the main components (building blocks) of its architecture (Aleksandar & Aleksandar, 2022). Future VW efforts from academic institutions, individuals, and other stakeholders are to be anticipated. Public sector enterprises in the field of education intending to remain relevant in the metaverse era cannot overlook the adoption of VR technologies to facilitate teaching and learning.

MetaHealth

With the strain of long-term, chronic disease, growing expenses, aging populations, an inadequate health workforce, and finite resources, the health care system is unsustainable (Thomason, 2021). Finding methods that bring healthcare out of the hospital and into the home is crucial. The pharmaceutical and biotechnology industries have witnessed a significant upheaval because of the direct revolution in healthcare occasioned by digital health. Innovators and healthcare professionals were inspired by the COVID-19 outbreak to develop techniques that would make it possible to handle patients remotely and outside of hospitals. Smartphone adoption is rising, and wearable technology adoption is rising as well. The age of the metaverse is now rapidly approaching. The introduction of digital services will be one of the most important aspects in the transformation of health care over the next ten years (World Economic Forum, 2016).

Medical practice has traditionally involved direct contact with patients so that doctors can see both physical and emotional reactions. The pandemic has, nevertheless, prompted a sharp acceleration in remote care technology (Thomason, 2021). For instance, in 2020, 95% of healthcare facilities were able to offer telemedicine, up from 43% in the years prior to the pandemic (Demeke, 2021). The rising acceptance of electronic informed consent was reported by Tufts University to be the second-largest emerging trend, behind the use of telemedicine delivery, studying the effects of COVID-19 on clinical research (Le Breton, Lamberti, Dion, & Getz, 2020).

The potential use of Metaverse in health care is enormous. Real-time advice can be given in the surgeon's field of view through immersive experiences that replicate surgical procedures. Access to information within the operating room's sterile environment will be made possible by augmented reality (AR), improving surgical accuracy and flexibility. The Metaverse will enable coordinated medical procedures as well as simultaneous planning, training, and education (Thomason, 2021). When used in conjunction with AI, this can strengthen clinical judgment and guarantee more precise interventions that are customized for each patient. An illustration is Veyond Metaverse (https://www.veyondmetaverse.com), which is creating a future healthcare metaverse ecosystem. It intends to enhance training and education by providing a platform for simultaneous planning, training, and education as well as collaborative medical treatments. The Metaverse will initially be utilized for patient care management, surgical simulations, diagnostic imaging, rehabilitation, and health management (Sagentia Innovation, 2021). These technological advancements can speed up patients' education regarding their diseases or treatment options. At the point of care in a clinical context, AR and VR can assist care teams. When used in conjunction with radiology, AR can give medical professionals the ability to project images of the patient's body, such as CT (computed tomography) scans, directly onto the patient even as they move. This gives medical professionals better views of the patient's internal anatomy (Thomason, 2021). This can enhance the patient experience; for instance, intravenous injections may benefit from technology from AccuVein (https://www.accuvein.com), which can project a map of the patient's veins onto the skin. The real and virtual worlds will combine due to Zimmer Biomet's announcement of OptiVu™ Mixed Reality (Biomet, 2021) and the acquisition of Digital Surgery by Medtronic. Through data interconnection, avatars will simulate practical consultations, individualized care, treatment, and diagnosis. In addition to being utilized to treat addictions and phobias, extended reality headsets are also being used to influence users' psychological experiences (Slater, 2020). In the area of fitness and wellness, where AR may give better workouts with coaching from virtual instructors, gamification is a new approach for linking healthcare professionals and patients. "Move-to-Earn," another novel idea, encourages users to be active. As an illustration, Genopets (Hoogendoorn, 2021) uses data from wearables and smartphones. These technological innovations may be used by public-sector health care enterprises to provide advanced care to patients and customers.

Role of Insurtech and Fintech in Public Sector Enterprises' Participation in the Metaverse Ecosystem

As the metaverse ecosystem expands, many public-sector organizations are investigating the potential benefits of insurtech and fintech technologies. Current research and thinking on the subject, based on a variety of academic and industry sources, suggests the following.

1) Increased Efficiency and Cost Savings: Insurtech and fintech solutions can assist public sector enterprises in streamlining operations, lowering overhead costs, and increasing efficiency. Blockchain technology, for example, can help to simplify and automate administrative processes, whereas machine learning algorithms can improve risk assessment and underwriting. (Fernandez, 2020; Carr, 2021)
2) Improved Customer Experience: Insurtech and fintech solutions can also assist public sector enterprises in providing a better customer experience. Chatbots and other AI-powered tools, for example, can improve customer service and reduce response times, whereas mobile apps can make it easier for customers to access and manage their accounts. (Ruf, 2021; Sproul, 2019)
3) Increased Innovation and Flexibility: Insurtech and fintech solutions can also assist public sector enterprises in more quickly innovating and adapting to changing market conditions. The use of open APIs and data analytics tools, for example, can assist in identifying emerging trends and customer needs, while agile development methodologies can enable faster product development and deployment (Cohen, 2021; Novak, 2020)

4) Improved Risk Management: Insurtech and fintech solutions can also help public sector organizations manage risk more effectively. Predictive analytics and AI-powered underwriting tools, for example, can help identify high-risk customers and mitigate potential losses, while real-time monitoring and fraud detection tools can help prevent fraudulent activity (Fernandez & Sproul, 2020)
5) Greater Access and Inclusion: Insurtech and fintech solutions can also assist public sector enterprises in reaching underserved and marginalized populations, such as those who do not have access to traditional financial services. Mobile apps and other digital tools, for example, can make it easier for people in remote or low-income areas to access insurance and financial products, while alternative credit scoring models can help to extend credit to those with limited credit histories (Carr and Ruf, 2021).

Overall, there are a variety of potential benefits that insurtech and fintech solutions can provide to public sector enterprises operating within the metaverse ecosystem. These include increased efficiency, improved customer experience, increased innovation and flexibility, better risk management, and greater accessibility and inclusion. As the metaverse evolves, it will be critical for public sector enterprises to stay current on new technologies and trends in order to remain competitive and provide high-quality services to their customers.

Digital Currencies, Decentralized Finance, Decentralized Autonomous Organizations (DAO)

According to Outlier Ventures (Burke, 2021), a real Metaverse is one in which value can be earned, spent, lent, borrowed, or invested interchangeably in both the physical and virtual worlds, without the need for a central authority. Such a metaverse would have its own economy and local currencies. The Metaverse, which combines DeFi, NFTs, decentralized governance, decentralized cloud services, and self-sovereign identity, can facilitate the exchange of material, financial, and intellectual property. The financial services industry has evolved during the past ten years. Cryptocurrencies, in particular Bitcoin, have become increasingly significant. Although the cryptocurrency market has demonstrated a lot of potential, its wholly decentralized structure also has several downsides. Since there is no regulatory body to monitor and regulate the activities associated with such digital transactions, investing in cryptocurrencies has a high risk of loss. It is sustained by a small number of investors or market sentiment; cryptocurrency has a volatile quality that foreshadows the instability and weakness it can infuse into the economic system as a whole (Aysan, Khan, Topuz, & Tunali, 2021). Bitcoin is seen by financial professionals as the biggest threat to the financial system, and they believe that quick action is needed to maintain control (Aysan, Kayani, & Kayani, 2020).

To maintain control of the system and prevent any shifts, several central banks have launched national digital currencies (Auer, Cornerlli, & Frost, 2020). However, many governments continue to be averse to the idea of a digital currency, considering it unfeasible and asserting that doing so will only increase the financial system's exposure to cybersecurity risks like ransomware and threats to financial stability (Nicole, 2021; Parn & Edwards, 2019). That notwithstanding, the People's Bank of China (PBOC) has announced a goal of having more than a billion users before any other country can make substantial strides toward developing a central bank digital currency (CBDC) of its own (Hoffman et al., 2020). The year 2021 also saw the announcement by Ghana's central bank of plans to launch its own CBDC, which could be used for both local and international payments. At the moment, the eCEDI is just in the pilot phase. The Sand Dollar was issued by the Central Bank of The Bahamas in October 2020. It was the first CBD in the world to cover an entire country. Nigeria became the first African nation to create a CBDC in October 2021. The eNaira is kept in a digital wallet and can be used for both money transfers and contactless payments in retail establishments. Seven Eastern Caribbean Union countries created their own digital currency to facilitate faster transactions and to serve those who do not have bank accounts. The seven countries are Antigua and Barbuda, Dominica, Grenada, Montserrat, St. Kitts and Nevis, Saint Lucia, and St. Vincent and the Grenadines. The Indian government has announced the introduction of a "digital rupee" by 2022–23. The Bank of Russia, Russia's central bank, announced in February 2022 that it had completed initial trials of the CBDC, also known as the "Digital Ruble." The Central African Republic has also adopted Bitcoin as legal tender. Many economists think that the long-standing idea of money is under threat due to the introduction of digital currency. Another most talked-about emerging technological development in global finance is DeFi (decentralized finance), which is on par with FinTech (financial technology), RegTech (regulatory technology), cryptocurrencies, and digital assets. Decentralized financial service provision using a combination of infrastructure, markets, technology, techniques, and applications is what we consider DEFI to entail at its core. Decentralized financial service delivery refers to the provision of financial services by numerous actors, with end users dispersed across numerous geographical spaces.

Another important concept based on decentralized technology is the creation of decentralized autonomous organizations. The idea of DAOs is one particularly fascinating development that the rise of blockchain technology has brought. According to its creator, Satoshi Nakamoto (who goes by the mysterious or fictitious name), blockchain technology and Bitcoin were created to replace conventional financial governance structures (Nakamoto, 2008). Consequently, the DAO has two parts that make up its governance. The first is the internal governance element, which is characterized by non-hierarchical modes of government and has elements of a quasi-democracy. The idea of "one person, one vote" is not guaranteed. These elements are more akin to a form of democracy than democracy itself. For instance, an actor may receive more votes than others, tilting the voting scales in his favor. The weight of voting rights can be determined by a number of factors, depending on how they are designed. Voting weight within the DAO may be based on token ownership, capped at particular amounts (Quiniou, 2019). The financial services industry is being forced to reinvent itself, embrace opportunities, and take risks as a result of rapid technological development. Public-sector enterprises could also leverage on this decentralized technology to deliver value to the masses of customers they serve.

SUSTAINABLE ORGANIZATION FORMS IN THE METAVERSE

Public enterprises must be run in the public interest, which results to organizational and business problems. One of the issues is balancing the demands for sufficient management autonomy and close political control. An exclusive act of Parliament establishes the public corporation, which is widely used in Great Britain and widely imitated elsewhere in the world (Encyclopaedia Britannica, 2009). This act specifies the public corporation form's authority, organizational structure, and relationship with other governmental entities. It has a legal entity because it is a corporation. The Treasury meets its capital needs, but it is expected to cover its current costs through regular business operations. There are no civil servants among its staff, and the minister in charge frequently appoints the top management. The state company, which is merely a regular joint-stock company whose shares are owned entirely or in part by the state, is another administrative structure that is well-liked in some parts of the world. Public enterprises may be subjected to political restraints in their pricing policy that are at odds with their long-term financial sustainability goals. On the other hand, for social reasons, they might get secret subsidies or extra protection that is not offered to rivals. These elements frequently cause managerial disorientation and tend to distort the corporation or company's normal commercial operations. Public enterprises may appear to be extremely inefficient and, in times of challenging trading conditions, may be a drain on public resources, in part due to these non-commercial considerations. However, determining a public enterprise's efficiency is not a simple task. The typical commercial criterion of profit may be used to evaluate its performance when it produces a marketable good competes with other goods, such as steel or coal. Economists have created ideas like cost-benefit analysis as a performance measurement tool in the case of a utility with monopoly power. In the developed world, many state-owned businesses have recently received financial goals that balance their social and business obligations (Encyclopaedia Britannica, 2009).

Collaborative Governance

According to Ansell and Gash (2007), collaborative governance is a governing arrangement where one or more public agencies directly engage non-state stakeholders in a collective decision-making process that is formal, consensus-oriented, and deliberative and that aims to make or implement public policy or manage public programs or assets. The complexity of the metaverse favors collaborative governance as it comprises many actors. According to Borgonovi, Bianchi, and Rivenbark (2019), fragmentation is frequently the root of inconsistency in efforts to enhance community results. The dynamic and complex nature of "wicked" problems, which involve multi-level, multi-actor, and multi-sectoral concerns, prevents them from being grouped within a single organizational boundary (Head & Alford, 2013; Laegreid & Rykkja, 2014; Bianchi, 2015). In order to frame public value, its drivers, and the strategic resources required to have an impact on community outcomes, a public-sector institution collaborates with other community stakeholders (Ansell & Gash, 2007).

This collaboration may even be more essential in the metaverse ecosystem as public enterprises seem to have shrinking business space and tendencies. To pursue community resilience and sustainable socioeconomic development, this requires the co-design, co-production, and co-assessment of policies by community stakeholders (Bovaird, 2007; Osborne, 2021; Torfing & Ansell, 2017). This viewpoint "posits both a plural state, where numerous interdependent players contribute to the provision of public services, and a pluralist state, where numerous processes enrich the policy-making system." The terms new public governance (Osborne, 2010), policy networks (Klijn & Koppenjan, 2000; Rhodes, 1990), network governance (Rhodes, 2017), cross-sector collaboration (Bryson, Crosby, & Stone, 2006), and public value governance (Bryson, Crosby, & Bloomberg, 2014), participatory governance (Fung & Wright, 2001), holistic governance (Perry, Seltzer, & Stoker, 2002), integrated governance (Hood, 2005), and interactive governance (Torfing, Peters, Pierre, & Sørensen, 2012), meta-governance (Bekkers, et al.), are just a few more that apply to multi-actor collaboration, typically led by a public sector organization.

Andrej and Jilles (2020) proposed a more recent governance concept called decentralized network governance, which sees power as existing in particular and dynamic interactions. in contrast to traditional systems of governance that consider power to lie in identities or roles. Therefore, depending on the relationship between various actors requiring governance, power may lie with any actor, including individuals, businesses, or the state. Each of the innovative network powers can be possessed by several actors in various roles and interactions. State participation, of course, can occur in a variety of ways, from a regulator or central bank acquiring a stake in, co-managing, and publicly coordinating to a system or network being fully owned. The potential structures include anything from domestic real-time gross settlement (RTGS) systems to SWIFT-like systems, speedier payment systems, property registries, and central bank digital currencies. Given that full ownership would necessitate re-concentration and create a new single point of failure, eliminating important benefits of decentralization, the first variant of DEFI—a public-private partnership where a public authority assumes a node function (Auer & Boehme, 2020)—is probably the most prudent option in many situations. It is impossible to limit public sector enterprises to a centralizing, bureaucratic public administration model in a world of constant, complicated change and competing interests.

RECOMMENDATIONS

Although there are some obstacles standing in the way of achieving a fully developed metaverse, there are also many factors that give cause for optimism. Public sector businesses need to establish ways to identify their customers' shifting demands and develop strategies for meeting them because customer expectations are continuously changing. In terms of governance, Andrej and Jilles (2020) posit that it is critical to include third parties in the theoretical model of decentralized network governance since private (business) players own the key components of network power. For example, several different actors can carry out governance duties within a blockchain (on-chain governance) or outside of a blockchain system (off-chain governance). In some circumstances, these take on a particular function. In order to equalize the influence exerted by the dominating or deviant actors, alliances between them must be forged in order to effectively control power interactions. As decentralized systems portray a key feature of the metaverse, we propose and recommend a "cotonomous" governance structure for public sector enterprises intending to operate in the metaverse. By "cotonomity," we propose a governance framework that allows for interdependence rather than independence among the various (on-chain and off-chain) actors. This proposition will allow public sector enterprises to collaborate with key on-chain businesses or actors in their areas of business. This is in harmony with Zetzsche, Arner, and Buckley (2020), who proposed a public-private partnership where a public authority or corporation assumes a node function on the blockchain. The chapter also recommends that public-sector enterprises adopt and begin to familiarize themselves with key technological enablers of the metaverse ecosystem, including VR, AR, and artificial intelligence.

FUTURE AREAS OF RESEARCH

Future research should explore how the public sector enterprise can both exploit the amazing economic and social opportunities as well as reduce harm in the metaverse. It would also be fascinating to investigate stakeholders' involvement, trust, and conflict dynamics more qualitatively and in-depth in future studies. In order to protect users, more research is needed on how to address the security and privacy concerns.

CONCLUSION

Customers' expectations vary along with technology. Transactions simplicity, self-sufficiency, speed, tailored goods, and customer experiences are most anticipated. Public sector businesses need to grasp that the quick service, individualized goods and services, and enjoyable and non-monotonic understanding that customers demand, both in the present and in the future, cannot be achieved by conventional public sector methods. The venture capitalist and author Matthew Ball views the metaverse as the mobile Internet's successor state rather than a virtual realm. He views the metaverse as a framework for a life that is extremely connected and predicts that as various products and capacities mesh, the metaverse will gradually take shape. Due to globalization, increased technological specialization, and functional divergence, governance systems have evolved over time. However, in order to keep up with technological advancements in the digital space generally and the growing usage of blockchain technology in particular, governance must also adapt. Also, stakeholders should prepare for increasingly fluid dynamics within and surrounding digital networks if the future of governance is actually one of shifting roles and power coalitions. Finally, in light of profoundly democratized platforms, conventional command-and-control governance methods are not very useful.

REFERENCES

Abdelkafi, N., & Makhotin, S., & Posselt. (2013). Business Models Innovation for Electric Mobility- What can be learned from Existing Business Model Patterns. *International Journal of Management*, 1–41.

Aleksandar, J., & Aleksandar, M. (2022, January 20). VoRtex Metaverse Platform for Gamified collaborative learning. *Electronics (Basel)*, *11*(3), 2–20. doi:10.3390/electronics11030317

Allam, Z., Sharifi, A., Bibri, S. E., Jones, D. S., & Krogstie, J. (2022). The Metaverse as a Virtual Form of Smart Cities: Opportunities and Challenges for Environmental, Economic, and Social Sustainability in Urban Futures. *Smart Cities*, *5*(3), 771–801. doi:10.3390martcities5030040

Alt, A., & Zimmermann, H.-D. (2014). *Electronic Markets and Business Models.*

Andrej, Z., & Jilles, H. (2020). Decentralized Network Governance: Blockchain Technology and Future Regulation. *Frontiers in Blockchain,* 1-11.

Ansell, C., & Gash, A. (2007). Collaborative Governance in Theory and Practice. *Journal of Public Administration: Research and Theory*, *18*(4), 534–571. doi:10.1093/jopart/mum032

Atiku, S. O. (2018). Reshaping Human Capital Formation Through Digitalization. In P. Duhan, K. Singh, & R. Verma (Eds.), *Radical Reorganization of Existing Work Structures Through Digitalization* (pp. 52–73). IGI Global. doi:10.4018/978-1-5225-3191-3.ch004

Atiku, S. O., & Anane-simon, R. (2020). Leadership and Innovative Approaches in Higher Education. In N. Baporikar & M. Sony (Eds.), *Quality Management Principles and Policies in Higher Education* (pp. 83–100). IGI Global. doi:10.4018/978-1-7998-1017-9.ch005

Atiku, S. O., & Boateng, F. (2020). Rethinking Education System for the Fourth Industrial Revolution. In S. Atiku (Ed.), *Human Capital Formation for the Fourth Industrial Revolution* (pp. 1–17). IGI Global. doi:10.4018/978-1-5225-9810-7.ch001

Auer, R., & Boehme, R. (2020). The technology of retail central bank digital currency. *BIS Quarterly Review,* 1-16.

Auer, R. A., Cornerlli, G., & Frost, J. (2020). *Rise of the Central Bank Digital Currencies: Drivers.* Approaches and Technology.

Aysan, A., Kayani, F., & Kayani, U. N. (2020). The Chinese Inward RDI and Economic Prospects amid Covid 19 Crisis. *Pakistan Journal of Commerce and Social Science*, 1088-1105.

Aysan, A., Khan, A., Topuz, H., & Tunali, A. S. (2021). Survival of the Fittest: A Natural Experiment from Crypto Exchanges. *The Singapore Economic Review*, 1–20. doi:10.1142/S0217590821470020

Bekkers, V. J., Edelenbos, J., Nederhand, J., Steijn, A. J., Tummers, L. G., Voorberg, W. H., & Edelenbos, J. (n.d.). *The social innovation perspective in the public sector: Co-creation, self-organization, and meta-governance.*

Bermejo, C., & Hui, P. (2021). A survey on haptic technologies for mobile augmented reality. *ACM Computing Surveys*, *54*(9), 1–35. doi:10.1145/3465396

Bernard, M. A., Johnson, A. C., Hopkins-Laboy, T., & Tabak, L. A. (2021). The US National Institutes of Health approach to inclusive excellence. *Nature Medicine*, *27*(11), 1861–1864. doi:10.103841591-021-01532-1 PMID:34764481

Bianchi, C. (2015). Enhancing Joined-Up Government and Outcome-Based Performance Management through System Dynamics Modelling to Deal with Wicked Problems: The Case of Societal Ageing. *Systems Research and Behavioral Science*, *32*(4), 502–505. doi:10.1002res.2341

Biomet, Z. (2021, December 19). OptiVu™ mixed reality. ZimmerBioMet. https://www.zimmerbiomet.com/en/products-and-solutions/zb-edge/optivu.html

Borgonovi, E., Bianchi, C., & Rivenbark, W. C. (2019). Pursuing Community Resilience through Outcome-Based Public Policies: Challenges and Opportunities for the Design of Performance Management Systems. *Public Organization Review*, *19*(4), 153–158. doi:10.100711115-017-0395-1

Bovaird, T. (2007). Beyond Engagement and Participation: User and Community Coproduction of Public Services. *Public Administration Review*, *67*(5), 846–860. doi:10.1111/j.1540-6210.2007.00773.x

Breuer, H., & Ludeke-Freund, F. (2014). Normative Innovation fòr Sustainable business Models in Value Networks. In S. C. In K. Huizingh (Ed.), *The proceedings of XXV ISPIM Conference.* Dublin, Ireland: Lappeenranta: University of Technology Press. http://papers.ssrn.com/sol3/papers.cfm?abstract_

Bryson, J. M., Crosby, B. C., & Bloomberg, L. (2014). Public Value Governance: Moving beyond Traditional Public Administration and the New Public Management. *Public Administration Review*, *74*(4), 445–456. doi:10.1111/puar.12238

Bryson, J. M., Crosby, B. C., & Stone, M. M. (2006). The Design and Implementation of Cross-Sector Collaborations: Propositions from the Literature. *Public Administration Review*, *66*(s1), 44–55. doi:10.1111/j.1540-6210.2006.00665.x

Buhalis, D., Lin, M. S., & Leung, D. (2022). Metaverse as a driver for customer experience and value co-creation: Implications for hospitality and tourism management and marketing. *International Journal of Contemporary Hospitality Management*, *35*(2), 701–716. doi:10.1108/IJCHM-05-2022-0631

Burke, J. (2021). Reintroducing the open Metaverse OS paper. *Outlier Ventures*. https://outlierventures.io/research/the-open-Metaverse-os/

Carr, D. (2021). How Insurtech Is Transforming the Public Sector. *Insurance Journal*. https://www.insurancejournal.com/news/national/2021/01/26/597401.htm

Cheng, R., Wu, N., Chen, S., & Han, B. (2022). Will metaverse be nextg internet? vision, hype, and reality. *IEEE Network*, *36*(5), 197–204. doi:10.1109/MNET.117.2200055

Chesbrough, H., & Rosenbloom, R. (2002). The Role of the Business Model in Capturing Value from Innovation: Evidence from Xerox Corporation's Technology Spin-off Companies. *Industrial and Corporate Change*, *11*(11), 529–555. doi:10.1093/icc/11.3.529

Cohen, S. (2021). 5 Key Insurtech Trends to Watch in 2021. *Insurtech Insights*. https://www.insurtechinsights.com/5-key-insurtech-trends-to-watch-in-2021/

Cruz-Benito, J., Maderuelo, C., García-Peñalvo, F., Therón, R., Pérez, B., Jonás, S., & Martin, A. (2016, July 13). Usalpharma: A Software Architecture to Support Learning in Virtual Worlds. *IEEE Revista Iberoamarica de Technologias del Aprendizaje*, *11*(3), 194–204. doi:10.1109/RITA.2016.2589719

Demeke, H. B., Merali, S., Marks, S., Pao, L. Z., Romero, L., Sandhu, P., Clark, H., Clara, A., McDow, K. B., Tindall, E., Campbell, S., Bolton, J., Le, X., Skapik, J. L., Nwaise, I., Rose, M. A., Strona, F. V., Nelson, C., & Siza, C. (2021). Trends in use of telehealth among health centers during the COVID-19 pandemic— United States, June26–November 6, 2020. *Morbidity and Mortality Weekly Report*, *70*(7), 240–244. doi:10.15585/mmwr.mm7007a3 PMID:33600385

Dionisio, J., III, W., & Gilbert, R. (2013, July). 3D virtual worlds and the metaverse: Current status and future possibilities. *ACM Computing Surveys (CSUR)*, *45*(3), 1-38.

Dionisio, J. D., Burns, W. G., & Gilbert, R. (2013). 3D Virtual worlds and the metaverse. *ACM Computing Surveys*, *45*(3), 1–38. doi:10.1145/2480741.2480751

Eman, G., Kathy, K., & Ibrahim, A. (2013). Metaverse-retail service quality: A future framework for retail service quality in the 3D internet. *Journal of Marketing Management*, *29*(13-14), 1493–1517. doi:10.1080/0267257X.2013.835742

Encyclopaedia Britannica. (2009, February 9). *Home*. Britannica. https://www.britannica.com/topic/public-enterprise

Falchuk, B., Loeb, S., & Neff, R. (2018). The social metaverse: Battle for privacy. *IEEE Technology and Society Magazine*, *37*(2), 52–61. doi:10.1109/MTS.2018.2826060

Fei-Yue, F. (2004). Parallel system methods for management and control of complex systems. *Control Decis*, *19*(5), 485–489.

Fernandez, I. (2020). The Impact of Insurtech on the Public Sector. *Fintech Magazine*. https://www.fintechmagazine.com/insurance-and-protection/impact-insurtech-public-sector

Fernandez-Herrero, J., & Lorenzo, G. (2019). An immersive virtual reality educational intervention on people with autism spectrum disorders (ASD) for the development of communication skills and problem-solving. *Education and Information Technologies*, 1689–1722.

Fields, Z., & Atiku, S. O. (2017). Collective Green Creativity and Eco-Innovation as Key Drivers of Sustainable Business Solutions in Organizations. In Z. Fields (Ed.), *Collective Creativity for Responsible and Sustainable Business Practice* (pp. 1–25). IGI Global. doi:10.4018/978-1-5225-1823-5.ch001

Fung, A., & Wright, E. O. (2001). Deepening Democracy: Innovations in Empowered Participatory Governance. *Politics & Society*, *29*(1), 5–41. doi:10.1177/0032329201029001002

Greenwald, S. W., Corning, W., & Maes, P. (2017). Multi-User Framework for Collaboration and Co-Creation in Virtual Reality. *Proceedings of the 12th International Conference on Computer Supported Collaborative Learning (CSCL)*, (pp. 18-22). ACM.

Head, B., & Alford, J. (2013). Wicked Problems: Implications for Public Policy and Management. *Administration & Society*. doi:10.1177/0095399713481601

Hellmut, W. (2018). Public and Personal Social Services in European Countries from Public/Municipal to Private—And Back to Municipal and "Third Sector" Provision. *International Public Management Journal*, 413–431. http://www.tandfonline.com/action/showCitFormats?doi=10.1080/10967494.2018.1428255

Hengboriboon, L., Sayut, T., Srisathan, W. A., & Naruetharadhol, P. (2022). Strengthening a company–customer relationship from sustainable practices: A case study of petrotrade in Laos. *Cogent Social Sciences*, *8*(1), 2038355. doi:10.1080/23311886.2022.2038355

Hoffman, S., Garnaut, J., Izenman, K., Johnson, M., Pascoe, A., Ryan, F., & Thomas, E. (2020). *The Flip side of China's Central Bank Digital Currency.*

Hood, C. (2005). The Idea of Joined-Up Government. A Historical Perspective. In V. Bogdanor, Joined-Up Government. Oxford University Press: Oxford, British Academy.

Hoogendoorn, R. (2021, September 6). *Genopets combines physical activity with play-to earn gaming.* Play to Earn. https://www.playtoearn.online/2021/09/06/genopets-combines-physical-activity-with-play-to-earn-gaming/

Hu, P., Li, H., Fu, H., Cansever, D., & Mohapatra, P. (2015). Dynamic defense strategy against advanced persistent threat with insiders. *IEEE Conference on Computer Communications (INFOCOM)*, (pp. 747-755). IEEE. 10.1109/INFOCOM.2015.7218444

Innovation, S. (2021, November 4). What does the Metaverse hold for health care? *Sagentiainnocation.* https://www.sagentiainnovation.com/insights/what-does-the-metaverse-hold-for-healthcare/https://www.sagentiainnovation.com/insights/what-does-the-metaverse-hold-for-healthcare/

Kaisara, G., Atiku, S. O., & Bwalya, K. J. (2022). Structural Determinants of Mobile Learning Acceptance among Undergraduates in Higher Educational Institutions. *Sustainability*, *14*(21), 13934. doi:10.3390u142113934

Kattel, R., Lember, V., & Tõnurist, P. (2020). Collaborative innovation and human-machine networks. *Public Management Review*, *22*(11), 1652–1673. doi:10.1080/14719037.2019.1645873

Klaus, S. (2016, January 14). *Agenda*. World Economic Forum. http://www.weforum

Klijn, E. H., & Koppenjan, J. (2000). Public Management and Policy Networks: The Theoretical Foundation of the Network Approach to Governance. *Public Management*, *2*(2), 135–158. doi:10.1080/14719030000000007

Krajcovic, M., Gabajová, G., Furmannová, B., Vavrík, V., Gašo, M., & Matys, M. (2021). A Case Study of Educational Games in Virtual Reality as a Teaching Method of Lean Management. *Electronics (Basel)*, *10*(7), 838. doi:10.3390/electronics10070838

Laegreid, P., & Rykkja, L. (2014). Governance for Complexity – How to Organize for the Handling of wicked Issues. *Stein Rokkan Centre for Social Studies*. https://bora.uib.no/handle/1956/9384

Le Breton, S., Lamberti, M. J., Dion, A., & Getz, K. A. (2020). COVID-19 and Its impact on the future of clinical trial execution. *Applied Clinical Trials*. https://www.appliedclinicaltrialsonline.com/view/covid-19-and-its-impact-on-the-future-of-clinical-trialexecution

Leenes, R. (2008). Privacy in the Metaverse: Regulagting a Complex Social: Construct in a Virtual World. In The Future of Identity in the Information Society (pp. 95-112).

Lui, H., Bowman, M., Adams, R., Hurliman, J., & Lake, D. (2010). Scaling virtual worlds: Simulation requirements and challenges. *In Proceedings of the Winter Simulation Conference (WSC). The WSC Foundation*, (pp. 778-798). ACM.

Magretta, J. (2002). Why Business Models Matter. *Harvard Business Review*, (80(5)), 86–92. PMID:12024761

Mystakidis, S. (2020). *Distance Education Gamification in Social Virtual Reality: A Case Study on Student Engagement, 11th International Conference on Information, Intelligence, Systems and Applications*. Piraeus, Greece. 10.1109/IISA50023.2020.9284417

Nakamoto, S. (2008). Bitcoin: A Peer-to-Peer Electronic Cash System. *Bitcoin*. https://bitcoin.org/bitcoin

Nicole, S. (2021, August). Biden administration sanctions virtual currency exchange following spike in ransonware attacks. *CBS News*.

Novak, M. (2020). The Top 10 Fintech Trends in 2020. *Forbes*. https://www.forbes.com

Nunes, F., Herpich, F., Amaral, É., Voss, G., Zunguze, M., Medina, R., & Tarouco, L. (2017). A dynamic approach for teaching algorithms: Integrating immersive environments and virtual learning environments. *Computer Applications in Engineering Computing*, 1-20.

O'Dair, M., & O'Dair, M. (2019). Blockchain: the internet of value. *Distributed Creativity: How Blockchain Technology will Transform the Creative Economy*, 15-30.

Obinger, H. Schmitt, C., & Traub, S. (2016). The Emergence of Public Enterprises in Historical Perspective. In The Political Economy of Privatization in Rich Democracies (pp. 6-25). Oxford University Press. from doi:10.1093/acprof:oso/9780199669684.003.0002

Osborne, S. P. (2010). The (New) Public Governance: A Suitable Case for Treatment? In S. P. Osborne (Ed.), *In the New Public Governance? Emerging Perspectives on the Theory and Practice of Public Governance* (pp. 1–16). Routledge. doi:10.4324/9780203861684

Osborne, S. P. (2021). *Public Service Logic. Creating Value for Public Service Users, Citizens, and Society through Public Service Delivery.*

Parn, E., & Edwards, D. (2019). Cyber Threats confronting the digital built environment common data environment vunerabilities and blockchain deterence. *Engineering, Construction, and Architectural Management*, *26*(2), 245–266. doi:10.1108/ECAM-03-2018-0101

Passos, C., Silva, M. H., Abreu Mol, A. C., & Carvalho, P. V. (2017). Design of a collaborative virtual environment for training security agents in big events. *Cognition Technology and Work*, *19*(2-3), 315–328. doi:10.100710111-017-0407-5

Perry, L. D., Seltzer, K., & Stoker, G. (2002). *Towards Holistic Governance*. Palgrave Macmillan. doi:10.5040/9781350391246

Pietro, D. R., & Cresci, S. (2021, December 12-15). Metaverse: Security and Privacy Issues. *IEEE TPS*, 1-8.

Porter, M., & Krammer, M. (2011). Creating Shared Value. *Harvard Business Review*, (89(1/2)), 62–67.

Quiniou, M. (2019). Blockchain: The Advent of Disintermediation. In *Hoboken John*. Wiley. doi:10.1002/9781119629573

Rhodes, R. A. (1990). Policy Networks: A British Perspective. *Journal of Theoretical Politics*, *2*(3), 293–317. doi:10.1177/0951692890002003003

Rhodes, R. A. (2017). *Network Governance and the Differentiated Polity*.

Ritika, M., & Montu, B. (2018). Business Sustainability: Exploring the Meaning and Significance. *IMI Konnect*, 8-9.

Schaltegger, S., & Burritt, R. (2005). Corporate Sustainability. In H. Folmer & T. Tietenberg (Eds.), *International Yearbook of Environment and Resource Economics* (pp. 185–222). Edward Elgar.

Schaltegger, S., Hansen, E. G., & Ludeke-Freund, F. (2016). Business Models for Sustainability: Origins, Present Research and Future Avenues. *Organization & Environment*, *29*(1), 3–10. doi:10.1177/1086026615599806

Shang, J., Chen, S., Wu, J., & Yin, S. (2022, February). ARSpy: Breaking location-based multi-player augmented reality application for user location tracking. *IEEE Transactions on Mobile Computing*, *21*(2), 433–447. doi:10.1109/TMC.2020.3007740

Sharma, S., Devreaux, P., Scribner, D., Grynovicki, J., & Grazaitis, P. (2017). Megacity: A Collaborative Virtual Reality Environment for Emergency Response, Training, and Decision Making. *Electronic Imaging*, 70-77.

Slater, M. e. (2020). The ethics of realism in virtual and augmented reality. *Frontiers in Virtual Reality*. doi:10.3389/frvir.2020.00001

Stendal, K., & Balandin, S. (2015). Virtual worlds for people with autism spectrum disorder: A case study in Second Life. *Disability and Rehabilation,* 1-8.

Stephenson, N. (2003). *Snow Crash: A Novel.* Random House Publishing Group.

Stichter, J., Laffey, J., Galyen, K., & Herzog, M. (2013). iSocial: Delivering the Social Competence Intervention for Adolescents (SCI-A) in a 3D Virtual Learning Environment for Youth with High Functioning Autism. *Journal of Autism and Developmental Disorders*, 417–430. PMID:23812663

Swaminathan, A., Wu, M., & Liu, K. R. (2008, March). Digital image forensics via intrinsic fingerprints. *IEEE Transactions on Information Forensics and Security*, *3*(1), 101–117. doi:10.1109/TIFS.2007.916010

Thomason, J. (2021). MetaHealth - How will the Metaverse Change Health care? *Journal of Metaverse*, *1*(1), 13–16.

Tlili, A., Huang, R., Shehata, B., Liu, D., Zhao, J., Metwally, A. H. S., Wang, H., Denden, M., Bozkurt, A., Lee, L.-H., Beyoglu, D., Altinay, F., Sharma, R. C., Altinay, Z., Li, Z., Liu, J., Ahmad, F., Hu, Y., Salha, S., & Burgos, D. (2022). Is Metaverse in education a blessing or a curse: A combined content and bibliometric analysis. *Smart Learning Environments*, *9*(1), 1–31. doi:10.118640561-022-00205-x

Torfing, J., & Ansell, C. (2017). Strengthening Political Leadership and Policy Innovation through the Expansion of Collaborative Forms of Governance. *Public Management Review*, *19*(1), 37–54. doi:10.1080/14719037.2016.1200662

Torfing, J., Peters, B., Pierre, J., & Sørensen, E. (2012). *Interactive Governance: Advancing the Paradigm.*

United Nations Industrial Development Organization. (2013). *UNIDO ANNUAL REPORT 2012*. UN.

Wang, F. (2004). Computational theory and method on complex system. *China Basic Sci*, 5-12.

Wang, F. (2010, July/August). The emergence of intelligent enterprises: From CPS to CPSS. *IEEE Intelligent Systems*, *25*(4), 85–88. doi:10.1109/MIS.2010.104

Wang, F., & Lansing, J. (2004). From artificial life to artificial societies: New methods for studies of complex social systems. *Complex Sys. Complex. Sci*, *1*(1), 33–41.

Wang, F. Y. (2004). Computational experiments for behavior analysis and decision evaluation of complex systems. *Journal of Systems Simulation*, *16*(5), 893–987.

Wang, F. Y., Qin, R., Wang, X., & Hu, B. (2022). Metasocieties in metaverse: Metaeconomics and metamanagement for metaenterprises and metacities. *IEEE Transactions on Computational Social Systems*, *9*(1), 2–7. doi:10.1109/TCSS.2022.3145165

Wang, F.-Y., Qin, R., Wang, X., & Hu, B.IEEE. (2022). MetaSocieties in Metaverse: MetaEconomics and MetaManagement for MetaEnterprises and MetaCities. *IEEE Transactions on Computational Social Systems*, *9*(1), 2–5. doi:10.1109/TCSS.2022.3145165

Wang, Y., Zhou, S., Zhang, N., Xing, R., Lui, D., & Luan, T. H. (2022, August). *A Survey on Metaverse: Fundamentals, Security and Privacy*. doi:10.48550/arXiv.2203.02662

Warburton, S. (2009). Second Life in higher education: Assessing the potential for and the barriers to deploying virtual worlds in learning and teaching. *British Journal of Educational Technology*, *40*(3), 414–426. doi:10.1111/j.1467-8535.2009.00952.x

Whiteman, G., Walker, B., & Perego, P. (2013). Planetary boundaries: Ecological foundations for corporate sustainability. *Journal of Management Studies*, *50*(2), 307–336. doi:10.1111/j.1467-6486.2012.01073.x

Wirtz, B., Pistoia, A., Ullrich, S., & Gottel, V. (2015). Business Models: Origin, Development and Future Research Perspectives. *Long Range Planning*. doi:10.1016/j

World Business Council for Sustainable Development. (2012). Public Policy Options to Scale and Accelerate Business and Actions Towards Vision 2050. Geneva, Switzerland.

World Economic Forum. (2015, August). Collaborative Innovation: Transforming Business, Driving Growth. *Regional Agenda*, 1-44.

World Economic Forum. (2016). Building the healthcare system of the future. WEF. https://reports.weforum.org/digital-transformation/building-the-healthcare-system-of-the-future/

Zetzsche, D. A., Arner, D. W., & Buckley, R. P. (2020, September 30). Decentralized Finance. *Journal of Financial Regulation*, 172-203. doi:10.1093/jfr/fjaa010

Zhang, Z., Zhang, M., Chang, Y., Aziz, E., Esche, S., & Chassapis, C. (2018, April 27). Collaborative virtual laboratory environments with hardware in the loop. *Cyber-Physical Laboratories in Engineering and Science Education*, 363-402.

Zizza, C., Starr, A., Hudson, D., Nuguri, S., Calyam, P., & He, Z. (2018). Towards a Social Virtual Reality Learning Environment in High Fidelity. *Proceedings of the 15th IEEE Annual Consumer Communications & Networking Conference (CCNC)*. IEEE. 10.1109/CCNC.2018.8319187

Section 2
Green Collaboration, Creativity, and Innovation

Chapter 10
Holistic Creativity and Harmonious Mega-Innovation for Sustainable Development: Master Keys and Examples

Vadim Kotelnikov
Innompics, USA

ABSTRACT

This chapter focuses on systemic cross-functional and cross-border approaches to the creation of a harmonious mega-innovation (HarmInn) that fosters sustainable development globally. Harmonious mega-innovation is applied across several sectors of the economy. HarmInn uses a holistic development process that puts the well-being of people, organisations, and our planet at the centre of innovative strategies and entrepreneurial actions aimed at designing and building a better and sustainable future. Holistic mega-creativity and harmonious mega-innovation start from adopting a global mindset and seeing the Earth as a small planet to take care of, not a huge planet to survive on. The author argues that disruptive innovations may break obsolete man-made rules but should stay in harmony with eternal laws of the universe. The author uses the example of the global Innompic ecosystems and World Innompic Games to illustrate this approach.

HARMONIOUS MEGA-INNOVATION (HARMINN)

Harmonious Mega-Innovation (HarmInn) is an impact innovation that fosters sustainable development globally in a natural and lasting way. It is an innovation-driven large-scale complex enterprise that is transformational and impacts millions of people while staying in harmony with the laws of nature and fair social systems. All-inclusive love – passion for work, all people, and the whole World – is the main driving force behind harmonious mega-innovation (Kotelnikov, 2020).

DOI: 10.4018/978-1-6684-6366-6.ch010

Holistic innovation (HI) is a visionary big-picture systems approach to inventing, designing, combining, harmonizing, producing, and delivering complex innovative value. The process of holistic innovation integrates and synergises complementary activities such as visioning, inventing, designing, strategising, implementing, and marketing.

The metaphoric Fruit-Tree model helps to understand and implement holistic innovation projects easier (Kotelnikov, 2021).

Globally thinking innopreneurs are innovative entrepreneurs who initiate and implement harmonious mega-innovation projects. Impact crosspreneurs are passionate innovators who keep creating a great positive impact in various areas in an entrepreneurial fashion either as individuals or through their enterprises (BeC Glossary 2020).

Noble innovative entrepreneurs are actors of positive change, not spectators. They are innopreneurs for a better world. "They push the human race forward, and while some may see them as the crazy ones, we see genius, because the ones who are crazy enough to think that they can change the world, are the ones who do," said Steve Jobs about disruptive innovators and impact entrepreneurs (Jobs 1997).

Harmonious Mega-Innovation is applied across several sectors of the economy. HarmInn uses a holistic development process that puts the well-being of people, organisations and our planet at the centre of innovative strategies and entrepreneurial actions aimed at designing and building a better and sustainable future.

A harmonious mega-innovator thinks of the Earth as a small planet to take care of, not a big planet to survive on. Specific attitudes, competences, and skills of a harmonious mega-innovator include global mindset, passion for the World and sustainable-development, holistic mega-creativity, systems thinking, knowledge of the Universal Laws of Nature and knowing how to stay in harmony with them.

MOVING TO A HIGHER LEVEL

Below are some specific features of holistic creativity and harmonious mega-innovation:

- **Attitude:** Up from seeing the Earth as a big planet to survive on to seeing the Earth as a small planet to take care of.
- **Creativity:** Up from brainstorming to alternating brainstilling and brainstorming (Brainstilling is used to visualize the big picture, assess the impact of brainstormed ideas on the Earth as a whole, and re-brainstorm, if necessary).
- **Innovation:** Up from breaking rules to breaking rules while staying in harmony with the eternal Laws of the Universe (Tree model; harmonizing energies of Yin-Yang and 5 Basic Elements).
- **Design:** Up from product-focused Design for Environment (DfE) to Mega-Design for Sustainable Development (MD4SD).
- **Investment and Financing:** Up from project-focused Sustainable Finance and Impact Investment to Mega-Impact Finance and Investment (MIFI).

PASSION-DRIVEN CREATIVITY AND INNOVATION

"Love the whole world as if it were yourself; then you will truly care for all things." – Lao Tzu

Passion is the main driving force behind creativity and innovation. Passion makes creativity and innovation an easy and enjoyable task. There are no difficult tasks, only interesting ones if you love what you do. "Choose a job you love, and you will never have to work a day in your life," said Confucius.

Green creativity is inspired by the passion for the environment and a burning desire to preserve nature as well as to make the innovation at hand environment friendly.

EXAMPLE: INNOMPIC PLANET OF LOVING CREATORS AS HARMINN

Innompic Games is a harmonious mega-innovation that keeps creating increasingly greater well-being and brighter future for both humankind and our planet.

Global Innompic Ecosystem (Innompics, 2018) and annual World Innompic Games as its 'heartbeats' is a harmonious mega-innovation that keeps creating unique and increasingly greater value for well-being, sustainable development, and brighter future for both humankind and our planet.

Launched in 2017, annual World Innompic Games – intellectual Olympics for creative and entrepreneurial innovators – are designed to make a global positive impact for centuries ahead.

While being a civilizational breakthrough and a disruptive innovation, Innompic Games stay in harmony with eternal laws of the Universe.

HOLISTIC CREATIVITY

Holistic creativity and harmonious mega innovation require holistic thinking and holistic intelligence.

Holistic thinking is the ability to see things as a whole (or holistically), to understand and predict the many different types of relationships between the many elements in a complex system, and also perceive the whole picture through sensing its large-scale patterns (BeC Glossary, 2020).

In harmonious mega-innovation, effective invention of a creative business design, entrepreneurial or innovation strategies, requires both holistic and creative thinking.

- Brainstilling is a powerful tool of holistic thinking as well as a catalyst of subconscious mega-creativity and macro-creativity.
- Brainstorming is an effective tool of creative problem solving (CPS) primarily at the meso- and micro-levels.

The 3-step strategic-creativity process – brainstilling, brainstorming, brainstilling (3Bs) – is a proven way of inventing a harmonious business design. The second brainstilling phase is used to visualise the resulting big picture and assess the impact of the ideas brainstormed. The brainstorming session should be repeated if the resulting big picture is not good enough (Kotelnikov, 2013).

Micro-, Meso-, Macro-, and Mega-Creativity

Scale-wise and process-wise there are four levels of creativity: micro-, meso-, macro-, and mega-.

There are many different approaches to defining micro-, meso-, macro-, and mega-creativity. We shall focus on mega-creativity in this chapter.

Yet, processes at all four levels are involved in mega-creativity and are interdependent. For instance, micro-level creativity of 'Edison as inventor of genius.' was "influenced by the macro processes of his cultural context as well as meso processes in the work groups and organizations he was part of" (Fathall, 2016). MegaCreativity of Innompians addressing global challenges at World Innompic Games is influenced by the overall design of Innompics (macro-creativity) (Kotelnikov, 2019), creative teamwork of participants (meso-creativity), and individual creativity of team members (micro-creativity).

Let's start with relevant definitions of all these creativities.

MicroCreativity is about unleashing creativity on a daily basis or finding a creative solution to a simple problem. Micro-creativity can be both intra-personal and inter-personal. In turn, there are three levels of individual creativity: conscious, subconscious, and divine (Kotelnikov, 2011).

Some forms of micro-creativity include:

- Improvisation – a creative activity of saying, doing, or making something without specific preparation
- Spontaneous ideas that are relative to a discussion
- Serendipitous ideas
- Entertaining out-of-the-box sayings in real-life conversations and writings in social-network exchanges
- Group processes – both onsite and online – that can help group members have creative moments in which they generate new, useful ideas.

MicroCreativity training consists usually of quick exercises designed to break you out of your habitual thinking patterns and jumpstart your brain into thinking differently.

MesoCreativity can be situation- or sequence-focused. For instance, meso-innovation may refer to ideas of a user community that are implemented in the next generations of products. Meso-level creativity may also refer to innovation-focused work groups and organizations ('creative units'). Process-wise, meso-creativity could be about involving organizational and small group-level processes (Fathall, 2016).

MacroCreativity is about finding creative systemic solutions to complex problems. MacroCreativity may include holistic creativity and creative strategizing as sub-components.

MegaCreativity is about finding creative solutions to global problems or inventing mega-breakthroughs (Kotelnikov, 2017). At the intra-personal level, MegaCreativity is sometimes referred to as the ability to generate a lot of great ideas fast (Aleinikov, 2002). Light-speed subconscious creativity is a major vehicle of mega-creativity.

EXAMPLE: HOLISTIC BUSINESS DESIGN OF INNOMPICS

The Fruit Tree Model

The business-design of Innompic Games and the global Innompic Ecosystem is based on the Tree model. The roots-to-fruits model of a harmonious mega-innovation was invented and applied to Innompic Games to create a sustainable business design that would ensure natural growth of Innompic Games and the global Innompic ecosystem for centuries ahead.

Harmonising the Five Basic Elements

The 5 Basic Elements of the Universe are also harmonized in the business design of Innompic Games (Kotelnikov, 2019).

The Five Basic Elements are Fire, Earth, Water, Metal, and Wood. These elements are understood as different types of energy in a state of constant interaction and flux with one another. The Five Elements do not only mean Fire, Earth, Water, Metal, and Wood. They also mean Movement, Change, and Development. They are changing, moving, waning, and expanding all the time.

The most important of all is the balance of all five elements. The movements of five elements are stable and predictable when they are in balance, and vice versa. There are affinity and enmity relationships between the five elements. The affinity relationship means generating, supporting, helping, producing, etc. The enmity relationship means destroying, overcoming, etc. There are also two cycles of imbalance, an overacting cycle, and an insulting cycle.

Affinity Relationships

- Water can help tree (Wood) grow
- Wood can help Fire to burn
- Fire can help to produce dust (Earth)
- Earth can help mineral (Metal) to form
- Metal can hold Water

Enmity Relationships

- Water quenches Fire
- Fire melts Metal
- Metal chops Wood
- Wood parts Earth
- Earth absorbs Water

Understanding of these relationships helped the business architect of Innompic Games to design a harmonious and sustainable business model (Kotelnikov, 2016). Thanks to its harmonious design, annual low-cost World Innompic Games managed to attract and engage aspiring innovators from over 40 countries within five years.

Table 1. Affinity relationships among the five basic elements of Innompic Games

Fire can help to produce dust (Earth)	Passionate IG Leaders inspired and guided by the Innompic vision and mission attract and engage aspiring world-changers, design the harmonious innovative value that Innompic Games are to create, and establish the values-based Innompic social system and culture.
Earth can help mineral (Metal) to form	Innompic mission, values and culture inspire Innompians to invent new things, to design mega-innovations holistically and harmoniously, and to perform at Innompic Games' creation show enthusiastically and innovatively.
Metal can hold Water	Innompic challenged-based constructive competition contests inspire entrepreneurial creativity, accelerated learning, help to test processes, and produce actionable feedback.
Water can help tree (Wood) grow	Feedback and suggestions received from participants help the business architect to develop the design of Innompic Games further creatively and to invent new inspiring and stretching tasks for contestants.
Wood can help Fire to burn	Innovations help achieve new milestones on the way to the Innompic vision of turning the Earth to the Planet of Loving Creators.

Sustainable Growth Strategies

Strategic development of the global Innompic Ecosystem is a continuous innovation process that is driven by holistic creativity and harmonises proactive (Yang) and adaptive (Yin) strategies.

In the state of peace, good fortune, and success (I-Ching, Hexagram 11), Yin, the Receptive, which moves downward, stands above; Yang, the Creative, which moves upward, is below. Hence their influences meet and are in harmony, so that all living things bloom and prosper.

EXAMPLE: BUSINESS DESIGN OF INNOMPIC GAMES

Yin is above: The business design of the Innompic Planet of Loving Creators and Innompic Games stays in harmony with the laws of the Universe such as balance of the Five Basic Elements and a Tree-shaped business system.

Yang in below: Proactive trend-setting innovations break outdated man-made laws and practices but stay in harmony with universal laws.

EXAMPLE: DESIGNING AND IMPLEMENTING A LASING-PEACE MODEL

Sustainable development is impossible without lasting peace. World peace, harmony and wellbeing come naturally when every individual loves the World and all people and loves to create wonderful value for others and make the World better.

The Innompic Planet of Loving Creators is a leading peace platform with its united- nationals preached-and-practiced approaches, all-win joyful creativity and constructive competition contests, and collaborative creative activities. The Global Innompic Ecosystem and World Innompic Games as its peak events nurture peace, unite people, help them grow as innovators, and turn the Earth to the Planet of Loving Creators by organising the right flow of Yin-Yang energies and balancing them dynamically (Innompics, 2021).

Helping People Grow as Loving Creators

Most importantly, Innompic Games help people grow as loving creators who see the Earth as a small planet to take care of. The vision of Innompics is to turn the Earth to the Planet of Loving Creators. Having engaged people from 40+ countries since 2017, the global Innompic Ecosystem grows rapidly towards its vision.

EXAMPLE: INNOMPUS AS HARMINN

At the 1st World Innompic Games 2027 held in Pune, India the teams were to address the creative challenge 'How To Create World's Best Innovation City'. The Russia team who won the 'Best Innovation Team' award focused on a creative design of InnoSphere of INNOMPUS, a harmonized all-inclusive multi-functional complex – the heart of the Word's Best Innovation City (Innompics-Ru, 2017).

The mega-innovators chose the Indian version of the Basic Elements to harmonise their mega-invention:

1. **Earth:** Name, Mission; Value Mantra; Slogan; Thinking Tools; Float tanks.
2. **Water:** Circulation of Expertise & Opportunities; Intellectual cross-pollination; InnoBall.
3. **Fire:** Vision; Yang rooms; Micro e-courses; Fun; Celebrations.
4. **Air:** Creativity contests; Experiments; Transfer of successful practices.
5. **Space:** Yin rooms; Intellectual cross-pollination; Creative chaos; Suggestion system.

Learning Activities for Kids at INNOMPUS are also harmonized through interaction of the 5 Basic Elements:

1. **Earth:** Strong vs. weak foundation; healthy vs. unhealthy roots; the core reason.
2. **Water:** Nurturing life; flowing around obstacles; managing currents; relaxing sounds.
3. **Fire:** Nurturing burning desire; fighting "fires"; think-fast games; think-better games.
4. **Air:** Sports; learning from feedback; change-creating games; survival games.
5. **Space:** Open creativity exercises; intelligence raising challenges; communication games.

Inventive Thinking Techniques for Sustainable Development

We all have the potential to be mega-creative, but researchers say that most people use less than 10 percent of their brain.

Innovators and serendipitous people see things differently than others. Innovators find entrepreneurial opportunities; serendipitous people make accidental discoveries where most people see just what is before them. Some people read a book to gain knowledge, while others read the same book with "lights going off in their brain". Aspirations, attitude, and habits make the difference.

Those who want to grow as sustainable-development leaders and mega-innovators tend to start by visualising the desired future they want to make a reality. Having done so, they cultivate a burning belief in their abilities to do the impossible in order to achieve their vision. "The first step is to establish that something is possible; then probability will occur," advises Elon Musk (Musk, 2022).

Inventive Questions

Inventive questions are an effective tool for making discoveries and inventions. In particular, "Why? What If?" questions help discover hidden problems and break rules in order to invent unusual solutions.

Below are sample inventive questions for green creativity and harmonious innovation:

- Is my mind open to outside-the-box discoveries?
- What does all this information about unsustainable development really mean?
- What opportunities for greener development are hidden in this information?
- How could this information be reframed to inspire my green creativity?
- Can I display this information in a way which would cast more light on my core sustainability question?
- How could I make greater sense of all my apparently unrelated green ideas?
- Why do we do this in such an environmentally unfriendly way? What assumptions should be challenged?
- What would be a cleaner way to produce it?
- How else could we utilise these wastes?
- What is still missing in my green design?
- What if I morph all these green ideas, adopt a beginner's mind, and look at the situation with new eyes to discover new possibilities for sustainable development?

Examples From World Innompic Games 2020

The following creative concepts and implementation strategies were developed by the participants of the 'Team Leader' contest:

- **Clean City.** Noble Goal: To inspire citizens to keep the city clean (Dennis, 2020).
- **Clean Energy for 1 Million People.** Noble Goal: To provide clean energy to One Million People in Uzumba Marambapfungwe District in Zimbabwe by the year 2025 (Chiropa, 2020).

Holistic Thinking and Holistic Intelligence

Holistic thinking is a cognitive style characterized by the ability to see things as a whole in the context of a situation. It is the ability to understand and predict the many different types of relationships between the many elements in a complex system. Holistic thinking is also about perceiving the whole picture through sensing its large-scale patterns and the relationships between its disparate elements.

Holistic Intelligence is the ability to see the 'big picture', comprehend it, and to think and act systemically.

Holistic Intelligence is a must-have competence and a great source of sustainable competitive advantage in today's rapidly changing world and increasingly complex business systems caused by the IT revolution and the following shift from linear to systemic innovation.

Global Mindset

Global citizens are people who have a global mindset, understand the wider world, and their impactful role in it. They are citizens of the world who take care of it. They take an active role in various communities and work both individually and with others to make our planet more sustainable and peaceful (World Intelligence Glossary, 2022).

The Internet revolution made global expansion easy. Today, global expansion is the ultimate goal for many mature companies, startup teams, and digital solopreneurs. Yet, global expansion requires a global mindset and world intelligence. That's why global mindset has emerged as both a key leadership competence and a source of competitive advantage of corporations in the 21st century.

Global mindset is the set of individual qualities, communication skills, and actionable knowledge that enhances the information processing capabilities of decision makers and their cross-cultural intelligence. Global mindset empowers those in leadership roles and leads to more effective strategic business design and managerial action in a global context.

"As such, global mindset involves three complementary aspects: (1) an openness to and awareness of multiple spheres of meaning and action; (2) complex representation and articulation of cultural and strategic dynamics; and (3) mediation and integration of ideals and actions oriented both to the global and the local" (Levy et al., 2014).

CASES: NOURISHING GLOBAL MINDSET

World Innompic Games

The vision of Innompic Games is to turn the Earth to a Planet of Loving Creators.

Being challenge-based constructive competition events, annual World Innompic Games inspire participants to address various global challenges creatively and thus build their global mindset and capabilities for mega-innovation. Examples of global challenges addressed by participants of World Innompic Games include:

- Design a virtual global innovation accelerator for green startups
- Design a virtual sustainable-development platform that would attract and engage one billion people
- How to inculcate breakthrough creativity and innovative culture in universities
- Design a sustainability-focused social system
- Design a global network of cultural and touristic ecosystems for digital solopreneurs
- Design outside-the-box sustainable-development strategies.

In addition, the World Innompic Games harness diversity, inspire global cross-pollination of ideas among international participants, and help Innompians grow as Loving Creators. This is very important because a positive self-concept results in high self-confidence and impacts self-success (Nasrul, 2021).

GE

In the GE Leadership Effectiveness Survey (LES), 'Global Mind-set' is one of the 10 Characteristics to be assessed. (Slater, 2003).

The following Performance Criteria are used:

1. Demonstrates global awareness / sensitivity and is comfortable building diverse / global teams.
2. Values and promotes full utilization of global and work force diversity.
3. Considers the global consequences of every decision. Proactively seeks global knowledge.
4. Treats everyone with dignity, trust, and respect.

The ranking scale is from 1 (Significant Development Needed) to 5 (Outstanding Strength).

IBM

IBM provides another example of a corporate approach to developing a global mindset. IBM sustained its global competitiveness by creating a globally integrated enterprise. The firm uses a number of structures and processes, such as Global Enablement Teams (GETs), to develop a global mindset among its employees and increase their competence in cultural adaptability. IBM also uses social media platforms that help to bridge cultural differences and social gaps among its employees in 100+ countries (IBM, 2012).

Singapore Airshow

The commercial aviation industry is responsible for around 2.1% of total carbon dioxide emissions and 12% of emissions from transport (Singapore Airshow, 2022).

To speed up invention of creative solutions to emission reduction, the organisers of the Singapore Airshow invited the Founder of Innompics World to demonstrate the Innompic approaches to holistic green creativity and Harmonious Innovation A-Z/360 (Kotelnikov, 2015), as well as to serves as a jury at the 'What's Next' startup contest.

- At the Singapore Airshow 2018, demo Innompic Games were presented to showcase Innompic approaches to holistic creativity and harmonious innovation.
- At the Singapore Airshow 2022, the 'Innompics Air' concept was presented, and the most promising startups were selected and awarded.

In 2022, the Singapore Airshow eNews published a report on some developments in the field of Sustainable Aviation that took place within the year (Singapore Airshow 2022).

Impact Innovators and Impact Entrepreneurs

Virtuous entrepreneurs innovate and create innovative value to improve the World (Kotelnikov, 2008).

Impact Innovators are those who innovate to make a great positive change in the world. Impact Innovators develop new harmonious solutions in pursuit of a more enjoyable, valuable, and healthier future for all.

Impact Entrepreneurship refers to the process favoring the creation and growth of enterprises that are ethical, transparent, and have a meaningful impact on our lives and society. Unlike non-profits, they focus on creating social impact through market strategies. Impact entrepreneurs focus on the triple-bottom-line: People-Planet-Profit, to achieve the appropriate balance between each dimension. (EMMIE, 2022)

Several universities around the world launched programmes that help their students to develop competencies related to entrepreneurship with a specific slant toward impactful projects. For instance, Erasmus Mundus Master in Impact Entrepreneurship (EMMIE) has courses on:

- New ventures with impact
- Strategies for impact of entrepreneurial projects
- Business modelling and impact entrepreneurship
- Financing entrepreneurial impact opportunities (EMMIE, 2022).

In India, virtual accelerator WEneurs Forum launched their Aadyaa Programme to promote impact entrepreneurship. This Launchpad is the 1st step towards the "Womenpreneur" journey. It is a specially curated entrepreneurship program for girl students from Standard 8th to 12th and is open to school students from across the country. The Programme serves as a catalyst and help germinate & accelerate the entrepreneurial journey of the young aspirational young minds in creating a new innovative India (WEneurs, 2022).

Impact Startup

"You never change things by fighting the existing reality. To change something, build a new model that makes the existing model obsolete." – Buckminster Fuller

Impact startups (sometimes called 'impactful startup') are early-stage companies that show innovation in following the United Nations Sustainable Development Goals (UN SDG, 2015) – 17 mega-objectives in the fight for a better world.

Impact entrepreneurs building impactful startups want to make a positive impact on society, the economy, the environment, and the world.

Many startups are created with the intention of solving a social or environmental or sustainable development problem. For example, there are startups working on energy efficiency, developing clean energy sources, improving access to education, and providing clean water to communities in need. By addressing these types of problems, startups can have a positive impact on society and the environment (FosterCapital, 2022).

Web Summit (Web Summit, 2022) defines Impact startups based on four factors:

1. Pursuing one or more of the SDGs through their operations.
2. Adapting enterprise objectives to be more sustainable.
3. Solving industry problems that could have a far-reaching effect.
4. Prioritizing social collaboration and idea-sharing over competition and profit.

In order to build a successful impactful startup, it is essential to have the right resources in place. This includes everything from an effective entrepreneurial team to sufficient financial resources.

For instance, systemic support of impactful startups under the Startup India Initiative unlocked people's entrepreneurial potential in a conducive environment and helped the Indian economy grow rapidly and turned India to the third largest startup ecosystem globally (DST 2022).

Keys to a Successful Impact Startup

The four critical components of a successful impact startup are:

1. Impact Entrepreneur
2. Impact Innovation
3. Impact Financing
4. Impact Marketing

Impact Entrepreneurs are visionaries who strive to contribute to sustainable development and the creation of an inclusive circular economy. They create human, ecological, and economic value through a holistic entrepreneurial approach by bringing innovative ideas into reality.

Impact Innovation is an entrepreneurial way to contribute to sustainable development locally or globally through development of new products and/or services. The impact-innovation process involves all stakeholders with the aim of cross-pollinating knowledge, perceptions, and ideas in order to develop effective all-win solutions (Innovarsity, 2022).

Impact Financing refers to various forms of raising funds for positive-impact projects. These forms include impact investing, grants, sponsorship, and donations. Impact investing is an investment strategy that aims to generate specific beneficial global, social, or environmental effects in addition to financial gains. Impact investing, or sustainable investing, covers a range of activities, from putting cash into green energy projects to investing in green businesses or companies that demonstrate environmental, social and governance (ESG) values such as social inclusion, environmental responsibility, or good governance.

Impact Marketing is a purpose-led framework that includes social and sustainable considerations and benefits into the overall marketing strategy. Impact marketing aims at both growing sales and making a positive impact in the world while doing it. Effective forms of impact marketing are white marketing and experiential marketing. White marketing appeals to higher-self and noble desires of prospective customers. Experiential marketing lets a prospect experience the positive impact of the innovation. In digital marketing, inspirational content marketing can help attract followers, partners, and supporters.

SEVEN LESSONS LEARNED FROM SUCCESSFUL IMPACT ENTREPRENEURS

Below is the summary of the lessons learned from impact entrepreneurs who grew their impact startups successfully.

1. Formulate an inspiring vision of the desired future you want to create and get clear on your mission from the start.

2. Build a passionate, synergistic, and committed entrepreneurial team of mission-mates and treat them as partners.
3. Focus on your vision and be passionate about your customers if you want to keep creating outstanding innovative value for them relentlessly. Revenues will follow.
4. Play entrepreneurial simulation games with your most promising business ideas to strengthen both your business model and entrepreneurial capabilities of your team.
5. Take daring entrepreneurial actions; be prepared to face a lot of challenges on your journey towards your vision; make it a habit to treat challenges as entrepreneurial opportunities and address them creatively.
6. Entrepreneurial persistence is a key to great achievements. Stay focused on your vision but be flexible on your ways to get there; balance inside-out innovations with outside-in adaptations.
7. Start marketing your world-changing innovation to enthusiasts, others will follow.

Virtuous Spiral of Innovation

The route from an invention to market success is never a straight line. It is a virtuous spiral process – a recurring cycle of ideas inspired by simulation, experimentation and market learning.

"There are really two things that have to occur in order for a new technology to be affordable to the mass market. One is you need economies of scale. The other is you need to iterate on the design. You need to go through a few versions," advises Elon Musk (Musk, 2022).

The virtuous spiral illustrates the evolving experience, knowledge, strategies, and processes reflecting the response to the discoveries made thanks to simulation and testing exercises. Various tools such as learning SWOT questions and assessments provide insights into the current state and help gain increased awareness of how to proceed further towards the desired state.

In Yin-Yang terms, the virtuous-spiral innovation journey can be presented as a sequence of proactive inside-out (Yang) and adaptive outside-in (Yin) phases.

Each upgraded set of strategies is developed as a response to learning SWOT questions asked at the end of the previous curve.

Learning SWOT Questions

"Winners never quit and quitters never win… We would accomplish many more things if we didn't think of them as impossible," said Vince Lombardi, a legendary football coach and motivator.

This is true for mega-innovation and creative transformation as well.

KoRe learning SWOT questions are designed to analyse feedback and help innovators to come out as winners from every step forward into a *terra incognita* on their daring innovation journey (Kotelnikov, 2009).

While classic Strengths, Weaknesses, Opportunities, Threats (SWOT) analysis is done before strategy formulation exercise, Learning SWOT Questions are asked after a beta-test, or a simulation move.

Strengths

- What went well and why?
- What should be done the same way next time?

Weaknesses

- What underperformed and why?
- What should be done better next time?

Opportunities

- What went unexpectedly well and why?
- What new directions should be explored?

Threats

- What failed dramatically and why?
- What assumptions should be checked?

White Marketing and Selling

"The only way you can conquer me is through love, and there I am gladly conquered." – Lord Krishna

White marketing is about passion-driven appeal to noble desires of prospective customers and their higher self (Kotelnikov, 2014). White Marketing and Impact Marketing arts and skills are essential for making a harmonious mega-innovation a success.

White marketers are helpers, not sellers. They don't push, they pull. They listen to prospective partners, investors, or customers, ask empathetic questions, and strive to figure out what kind of partnership would benefit them most.

Case: Teaching Harmonious Mega-Innovation at Innompirsity

Innompic University (Innompirsity) integrates proprietary and classic approaches to nourishing entrepreneurial creativity, systemic innovation, and achievements.

In particular, Innompisity helps aspiring innovators to prepare to win. Innoball (Innovation Brainball) entrepreneurial simulation game (Kotelnikov, 2012) is a trademark prepare-to-win tool that students master at Innompirsity.

To illustrate, the Founder of Innompirsity serves on Advisory Boards of several pre-accelerators and accelerators across the globe, including Change 90 Inititative (Canada) and WEneurs Forum (India), to help their aspiring and first-time entrepreneurs master Innompirsity trademark approaches and tools boosting holistic creativity and harmonious innovation.

Integrating NLP and InnoBall Simulation Game

Below are the five NLP steps for creating inevitable success (McDermott, 2001):

1. Create and appreciate a compelling stretch goal
2. Set your brain on the path toward achieving your goal so that it's working on it all day long
3. Vividly imagine that you have already achieved your goal, then walk back and examine the pathway toward your goal

4. Notice specific steps on the pathway you took to get there, including all those elements that led toward your goal
5. Go back to the present with a new appreciation for the steps on the path to your stretch goal and take action.

Storytelling: Preparation and Simulation of Implementation

"If you want to change the world with skill, start with stories. Stories build awareness, mindshare, and desire," advises JD Meier, a former Microsoft top manager and Mr. Innovation World award winner (Meier, 2022).

Example: HealthBiotics Startup Success Story

Below is the HealthBiotics startup success story created by Kseniya, Ms. Innovation World award winner while designing the world-changing journey of a real-world startup (Kseniya, 2020). The story is divided into two parts – preparation and simulation of implementation of the business plan. InnoBall (Innovation Brainball) entrepreneurial simulation game (Kotelnikov, 2012) was used during the simulation phase.

Impact Entrepreneurship Success Story: HealthBiotics

Healthbiotics is a startup that produces a natural agent that cleans environment, increases yield of agriproduce, and heals diseases.

THE STORY BEGINS

Marfa, a young biologist and healthcare specialist, wanted to make people healthier. The best way to achieve this is to get people closer to Mother Nature.

The famous ancient doctor Hyppocrates taught 'Doctors treat diseases, but it is Nature who cures them". Marfa wanted to do something that would help people to live in a cleaner, more natural environment and eat healthy natural food.

Marfa started searching for best solutions and found scientists who invented Healthbiotics, an all-natural and highly effective victorious agent that can clean environment, improve and increase yield of agriproduce, and prevent and heal diseases naturally.

Yet, the scientists were not entrepreneurial enough to commercialise their invention and make it applied globally. Marfa established the 'HealthBiotics' startup with the scientists and obtained a small startup loan.

InnoBall (Innovation Brainball) Entrepreneurial Simulation Game

CHALLENGE #1

'Healthbiotics' needed success stories to be able to persuade target customers to switch from chemicals to this all-natural and more cost-effective agent. Yet, prospective clients didn't want to switch to a new and better process from an old one that was good enough for them.

SOLUTION

Marfa found three organisations that faced a severe crisis:

1. A municipal sewage treatment organisation whose bad performance caused civilian protest actions.
2. A poultry farm that was about to file for bankruptcy because their major corporate customers stopped buying their antibiotics-full chickens.
3. A hospital where the mortality rate among the patients was too high.

Facing a life-of-death dilemma, these three organisations agreed to replace chemicals with heatlhbiotics and achieved amazing positive results within 9 months.

These three success stories helped 'HealthBiotics' to start acquiring new customers.

CHALLENGE #2

A chemical giant who saw the growing 'HealthBiotics' business as a threat to their profits, launched a campaign aimed at killing the impact startup.

SOLUTION

Marfa partnered with 3 popular environment-concerned bloggers who interviewed her about the great benefits of healthbiotics for people and also about how the chemical giant tried to kill her noble business.

Social networkers raised a campaign against the chemical giant and forced the corporation to discontinue their attempts to kill 'HealthBiotics'.

CHALLENGE #3

'HealthBiotics' was doing well locally, but had no funds to expand globally.

SOLUTION

Marfa diversified into more profitable areas such skin care, baby care, and shrimps farming. She also managed to win a big contract for cleaning a subway from pathogens. Increased revenue streams made it possible for 'HealthBiotics' to start expanding globally.

Marfa's social entrepreneurship big dream came true!

END OF STORY

From Simulations to Real Impact Innovations

Innompic Games and InnoBall simulations helped hundreds of Innompians to launch and grow high-impact startups since 2017. Examples include Healthbiotics, FreeFloat, and Change 90 Initiative.

Since its invention in 2010, InnoBall has helped thousands of mature companies as well to create breakthrough innovations more effectively and efficiently. Examples include Leadership Vision Centre (LVC), IPE Lab, NPD Co., and Vorobyev DroneTech.

Sustainable Finance and Impact Investing

Harmonious mega-innovation requires Impact investing that favours profitable companies, organisations, and funds that generate a measurable, beneficial social or environmental impact.

In turn, the financial sector holds enormous power in funding and bringing awareness to issues of sustainability.

Sustainable finance refers to the process of taking environmental, social and governance (ESG) considerations into account when making investment decisions in the financial sector, leading to more long-term investments in sustainable economic activities and projects.

Several financial institutions developed e-learning services to facilitate access to sustainable finance. For instance, the e-Learning platform on Sustainable Finance and Corporate Governance developed by Innovativkonzept GmbH, in cooperation with ESG experts of DEG - Deutsche Investitions- und Entwicklungsgesellschaft mbH and KfW's Development Bank offers several e-learnings on sustainability and corporate governance for financial institutions and private companies. Complex content is simplified and delivered from a risk perspective, the opportunity perspective and from a motivational point of view (Janischewski, 2022).

Measuring Mega-Innovation for Sustainable Development

"What gets measured gets done," advised Peter Drucker. So, we must measure what we value. In this chapter, we shall focus on measuring contribution of green creativity and mega-innovation to sustainable development.

Best metrics

1. are simple and easy to understand and apply
2. tell users what's going on and why
3. help users discover ways to make things better.

Assessing Capabilities of Innovators

Innovation is primarily about people. Assessors need to see an entrepreneurial team in action to be able to assess its entrepreneurial capabilities, Entrepreneurial simulation games make it possible. For instance, invented in 2010 InnoBall (Innovation Brainball) entrepreneurial simulation game is a 3-in-1 tool. It helps:

1. Develop entrepreneurial smartness
2. Develop an effective business model and entrepreneurial strategies
3. Assess the entrepreneurial capabilities of the team and its individual members (Kotelnikov, 2012).

InnoBall helps build a rainbow of entrepreneurial thinking skills: entrepreneurial creativity, strategizing, business model innovation, anticipation, creative problem solving, turning problems to opportunities, intellectual teamwork, quick assessment of ideas, and entrepreneurial communication.

According to their self-assessment before and after an InnoBall game, entrepreneurial smartness of players grows remarkably within a matter of hours.

Venture capital investors use InnoBall for quick assessment of entrepreneurial capabilities of startup teams. Venture capitalists say, “Ideas are a dime a dozen, only execution skills count.” According to surveys, venture capital investors invest primarily in entrepreneurial teams and their capability to build a highly profitable startup.

Earlier, it took months for prospective venture capital investors to assess the execution skills of a startup team. InnoBall-based assessment helps venture capital investors to make their investment decisions better and faster. Invented in 2010, InnoBall (Innovation Brainball) entrepreneurial simulation game made it possible for venture capital investors to see entrepreneurial teams in action and assess their execution skills within a matter of hours.

Sustainable Scoring Platform for Employing Institutions

At the institutional level, sustainable development and climate change challenges can be addressed systemically across three dimensions – Economy, Society, and Environment – to inspire and help people and businesses to behave sustainably.

According to Mirjam Gawellek, the author of Sustainable Scoring Platform for Employing Institutions (Gawellek, 2021), sustainability must be understood by the general population, and sustainable business must become global. Mirjam Gawellek suggests inventing and developing an international employer rating platform. In this platform employing institutions (such as companies, organisations and public administration) would be rated according to easy-to-understand social, environmental and economic metrics. The nine (3x3) meaningful metrics would give everyone a picture of the state of sustainable action or management. The key figures would be chosen in a way that they could be understood by all employees and be in balance in the three dimensions mentioned above. All employees would rate their (past and) current employers. The ratings would be published together with the employer’s presentation on a platform for all interested stakeholders (especially applicants, partners and competitors). In this way, it would quickly become clear which companies, organizations and public administrations act sustainably, and who needs to become more sustainable in a specific dimension.

LEARNING HOLISTIC CREATIVITY AND HARMONIOUS MEGA-INNOVATION

Gamification 10+

Gamification of learning is the addition of game elements such as a competitive aspect, awards, and fun to the learning process:

Gamification 10+ (BeC Glossary, 2020) creates much greater user value. In addition to accelerated learning benefits, Gamification 10+ provides multiple personal, social, and business benefits that include better team creativity, better business design, better value creation, better strategizing, higher profits, and an effective team assessment (Kotelnikov, 2022).

InnoBall (Innovation Brainball) entrepreneurial simulation game and Innompic Games are examples of Gamification 10+ master tools that focus on development of disruptive projects and impact innopreneurs respectively.

Similarly, Gamification 10+ can enhance green creativity and radically innovative sustainable development projects dramatically by increasing their impact and reducing failure risks.

Table 2. Benefits of gamification 10+

Business Benefits	InnoBall	Innompic Games
Accelerated Learning	Very High	Very High
Creativity Development	Very High	Very High
Entrepreneurial Thinking	Very High	High
Leadership Development	High	Very High
Nourishing Creative Teamwork	High	Very High
Connecting People	High	Very High
Winning Business Design	High	Very High
Idea Implementation	Very High	High
Value Creation	Very High	High
Creating Innovation	Very High	Very High
Focus on Profitability	Very High	High
Sustainable Growth	Very High	Very High

Being a registered partner of the United Nations Sustainable Development Goals (UN SDG), World Innompic Games inspire green creativity in participants and give special awards for outstanding contribution to UN SDG.

Innompirsity

Innompirsity (Innompic University) is a virtual global Innompic university for World-changers that teaches Innompiology, disruptive innopreneurship, how to create holistic harmonious mega-innovations and civilizational breakthroughs. Launched in 2017 by the Founder of Innompic Games and the global Innompic Ecosystem, Innompirsity focuses on its trademark techniques and teaches by examples to a much larger extent than most other universities (Innompirsity, 2019).

ChEduFuntion is a trademark style of education at the Innompic University (Innompirsity). ChEduFuntion stands for Challenge-based Education + Fun + Action. This is the Innompic way of inspiring accelerated learning of innopreneurial arts and skills A-Z/360.

Fun. Benefits of fun-based education are countless. Incorporating fun into educational activities helps students process, apply, and memorize information more effectively. ChEduFuntion students have fun all the time – while addressing creative challenges, while showcasing their creative solutions in an artful way, while watching other students showcasing their inventions, while playing simulation games, and so on.

Action. Learning by doing is the most effective way of learning. Confucius said, "I hear, and I forget. I see, and I remember. I do, and I understand." ChEduFuntion students learn by doing while inventing, during spoken innovation contests, while playing InnoBall simulation game, and during many other specially designed activities.

Topics taught at the Inomopic University include:

- First Impessionism
- Holistic Thinking

- Holistic Creativity
- Strategic Creativity
- Outside-the-Box Intelligence
- Advanced Ideation Techniques
- Subconscious Ideation
- Entrepreneurial Serendipity
- Trend-Setting Intelligence
- Futures Thinking
- Harmonious Mega-Innovation
- Master of Business Synergies (MBS)
- Spoken Innovation
- Innovation Architect
- Jazz-like Innovation Process A-Z/360
- White Marketing
- How To Market Radically Innovative Products and Services

Students showcase what they learned at Innompic Games conducted at a global, national or corporate level.

Example: Futures Thinking Contest

At the Innompic University, teachers teach by example.

Here is the "Education 2030 Forecast: Online & Fun" by Dr. Michael Zelin, a teacher at the Innompic University and the Founder of Innompics USA Inc.

Online schools, proctored online testing, online tutoring, e-textbooks, educational gaming: gamification of education has been accelerating and will further accelerate with development of new technologies, increase in speed of transfer information, and storage capabilities.

Edutainment, future of education, forecast

- *Efficiency of teaching will further increase: one instructor can teach thousands of students utilizing the Massive Online Open Course model.*
- *Efficiency of learning will also increase. Currently, most of the work students do is discarded. New systems leveraging value of students' work beyond academic grading will allow students to post their work online and collaborate (Zelin 2020).*

Innompiology

Innompiology is a young social science that examines and explains how to create harmonious mega-innovations and civilizational breakthroughs (Innompiology, 2020).

Civilizational breakthrough is a result of a World-changing radical, holistic, systemic, harmonious, and synergised mega-innovations that create a lasting civil impact and cause a global paradigm shift.

Innompiology was launched on 10th May 2020 to study how the Innompic civilizational breakthrough and some other mega-breakthroughs were achieved and to develop recommendations for creation of harmonious mega-innovations. Some of these findings and recommendations have been presented in this chapter. Innompiology-based mega-thinking and mega-implementations approaches, such as mega-creativity, mega-innovation, and harmonious innovation, are taught at the Innompic University.

Takeaways: Practical Suggestions

- Harmonious mega-innovation is about all-inclusive love – love what you do, love all people, and love the whole World.
- See yourself as a Loving Creator – see the Earth as a small planet to take care of, not as a huge planet to survive on.
- While creating harmonious mega-innovations, you may break man-made rules but should stay in harmony with Universal laws.
- Alternate brainstilling and brainstorming sessions to make sure that your proactive innovation contributes to sustainable development in the best possible way.
- Know how to engage your light-speed subconscious mind to be able to master holistic creativity and harmonious mega-innovation.
- Harness the power of diversity. Cross-pollinate knowledge and ideas with people from diverse cultures to achieve harmonious creative breakthroughs.
- Play entrepreneurial simulation games, like InnoBall, with high-promising ideas to increase your chances of succeeding.
- Ask learning SWOT questions after every major experiment to enhance your business model and implementation strategies.
- Target enthusiasts first in your marketing strategy – early adopters and the vast majority will follow.
- Entrepreneurial persistence is a key to great achievements. Stay focused on your vision but be flexible on your ways to get there; balance inside-out innovations with outside-in adaptations.

REFERENCES

Aleinikov. (2002). *MegaCreativity: 5 Steps to Thinking Like a Genius*. Walking Stick Press.

BeC Glossary. (2020). *Glossary of Thinking, Entrepreneurship, Business, Finance, Sustainable Development, and Environment Terms*. https://1000ventures.com/doc/glossary.html

Chiropa, T. (2020), *Clean Energy for 1 Million People*. http://www.innompics.com/events/ig2020/contests_sbm_tl_zimbabwe_tc_energy.html

DST. (2022). *75 Impactful Startups: DST Incubation Programme*. Vigyan Prasar.

EMMIE. (2022). *Erasmus Mundus Master in Impact Entrepreneurship Web Summit (2022), Impact Startups at Web Summit 2023*. EMMIE.

FosterCapital. (2022). *Use Resources To Build an Impactful Startup.* https://fastercapital.com/content/Use-resources-to-build-an-impactful-startup.html

Gawellek, M. (2021). *Sustainable Scoring Platform for Employing Institutions.* http://www.innompics.com/events/ig2021/sbm-socsys_sustainable-scoring_mg.html

GlossaryW. I. (2012). http://1world1way.com/coach/glossary.html

Hakim, N. (2021). *Social Networking Site (SNS) for Students.* http://innompics.com/events/ig2021/sbm-socsys_socnet_nasrul.html

IBM. (2012). *Developing global leadership: how IBM engages the workforce of a globally integrated enterprise.* https://www.ibm.com/downloads/cas/K7EWX39G

Innompics. (2018). *Global Innompic Ecosystem as a Harmonious Mega-Innovation.* http://innompics.com/org/ecosystem.html

Innompics. (2021). *Innompic Planet of Loving Creators as the World's Leading Peace Platform.* http://innompics.com/games/ig-way_peace-nobel.html

Innompiology. (2020). *INNOMPIOLOGY - The social science that examines and explains how to create harmonious mega-innovations and civilizational breakthroughs.* http://innompics.com/org/innompiology.html

Innompirsity. (2019). *Inompic University – growing disruptive innopreneurs and mega-innovators,* http://innompics.com/org/innompirsity.html

Innoteam-Ru. (2017). *INNOMPUS – harmonized all-inclusive multi-functional complex.* Russia Team, 1st World Innompic Games. http://innompics.com/innompics1/contests_bc2_russia-innompus.html

Innovarsity. (2022). *Glossary of Innovation Terms.* Innovation University (Innovarsity). http://www.innovarsity.com/coach/glossary.html

Janishewski, J. (2022). *e-Learning on Sustainable Finance and Corporate Governance.* https://www.1000ventures.com/doc/glossary-finance-sustainable_elearning.html

Jobs. (1997). *Here's to the crazy ones.* Steve Jobs speech. http://innompics.com/coach/innopreneur-crazy_sj.html

Kotelnikov, D. (2020). *My Clean City Team.* http://www.innompics.com/events/ig2020/contests_sbm_tl_ru-in_denko.html

Kotelnikov, V. (2008). *Virtuous Entrepreneur: 6+6 Engines.* http://kotelnikov.biz/coach/entrepreneur_12drivers.html

Kotelnikov. (2009). *Learning SWOT Questions.* 1000ventures.com

Kotelnikov, V. (2010). *KoRe 10 Innovative Thinking Tools (10 KITT).* http://www.kotelnikov.biz/coach/creativity_10magictools.html

Kotelnikov, V. (2011). *Three Levels of Individual Creativity: Conscious, Subconscious, and Divine.* http://www.kotelnikov.biz/coach/creativity_3levels.html

Kotelnikov, V. (2012). *Create Breakthrough by Playing INNOBALL Simulation Game*. http://www.innoball.com / http://kotelnikov.biz/coach/innogames.html

Kotelnikov, V. (2013). *3Bs of Strategic Creativity: Brainstilling, Brainstorming, Brainstilling*. http://kotelnikov.biz/coach/creativity_strategic_3b.html

Kotelnikov, V. (2014). *White Marketing – Enjoy Both Nobel Joy and High Revenues*. http://kotelnikov.biz/coach/marketing_white.html

Kotelnikov, V. (2016). *Business Design of Innompic Games: Harmonising the Five Basic Elements*. IG Way. http://innompics.com/games/ig_about_5be.html

Kotelnikov, V. (2017). *Innompic Games as a Cilizational Breakthrough*. http://innompics.com/games/innompics_civbreak.html

Kotelnikov, V. (2018). *How To Develop Serendipity: KoRe 10 Tips*. http://www.kotelnikov.biz/coach/discovery_serendipity.html

Kotelnikov, V. (2019). *Business Design of Innompic Games*. http://www.innompics.com/org/ig_biz-design.html

Kotelnikov, V. (2020). *Harmonious Mega-Innovation*. http://innompics.com/coach/innovation-mega-harmonious.html

Kotelnikov, V. (2021). *Holistic Innovation (HI): What, Why, and How*. http://innompics.com/coach/innovation-holistic.html

Kotelnikov. (2022). *Gamification 10+*. Innompirsity.

Kotelnikova, K. (2020). *HealthBiotics Startup Success Story*. http://innompics.com/events/ig2020/contests_sbm_msiw_ru-in_ksu.html

Levy, O., Beechler, S., Taylor, S., & Boyacigiller, N. A. (2014). *Global Mindset. In Wiley Encyclopedia of Management*. John Wiley & Sons, Ltd.

Lewis, H., Gertsakis, J., Grant, T., Morelli, N., & Sweatman, A. (2001). *Design + Environment, A Global Guide to Designing Greener Goods*. Routledge.

McDermott & Jadd. (2001). *NLP Coach. A Comprehensive Guide to Personal Well-being and Professional Success*. Judy Platkus (Publishers) Limited.

Meier, J. D. (2022). *Innovation Explained – The Big Ideas of Innovation All In One Place*. http://innompics.com/people/meier-jd.html

Moghaddam, F. M., & Covalucci, L. (2016). Macro, Meso, and Micro Creativity: The Role of Cultural Carriers. In The Palgrave Handbook of Creativity and Culture Research (pp. 721-741). Palgrave.

Musk. (2022). *Lessons from Elon Musk: Be a Genius Innovator*. Innovarsity. http://www.innovarsity.com/coach/quotes_a_musk.html

Singapore Airshow. (2022), *A Year in Review: Developments in Sustainable Aviation*. Singapore Airshow 2022 Report.

Slater, R. (2003). *Jack Welch and the GE Way*. Tata McGraw-Hill Edition.

UN SDG. (2015). *United Nations Sustainable Development Goals*. United Nations Department of Economic and Social Affairs.

WEneurs. (2022). *WEneurs Forum, Annual Newsletter 2022*. Author.

Zelin, M. (2020). *Education 2030 Forecast: Online & Fun*. Innompics USA. http://innompics.com/events/ig2020/contests_sbm_futh_zelin.html

Chapter 11
Embedding Sustainability in Project Management:
A Comprehensive Overview

Ramakrishnan Vivek
https://orcid.org/0000-0001-5691-6825
Sri Lanka Technological Campus (SLTC), Sri Lanka

Rohit Bansal
https://orcid.org/0000-0001-7072-5005
Vaish College of Engineering, Rohtak, India

Nishita Pruthi
https://orcid.org/0000-0002-6094-0972
Maharshi Dayanand University, India

ABSTRACT

The issue of sustainability is the most pressing issue facing the world today. Companies are incorporating sustainability concepts into their marketing, corporate communication, annual reports, and other aspects of their operations. Projects are critical in the implementation of more sustainable business practices, and the notion of sustainability has increasingly been associated with project management in recent years. The growing body of research on this subject offers compelling evidence that taking sustainability into account has an influence on project management procedures and practices. The requirements for project management, on the other hand, do not take into consideration the sustainability agenda. The purpose of this chapter is to give a comprehensive overview of the relationship between sustainability and project management. Following the publication of these results, it will be possible to further enhance project management techniques and standards in order to address the role that projects play in the creation of sustainable development.

DOI: 10.4018/978-1-6684-6366-6.ch011

INTRODUCTION

The sustainability principles we discuss in today's project management was in its inception when McKinlay, (2008) stated for project managers to "take responsibility for sustainability" as the opening line in his keynote address in 2008 World Congress's of the International Project Management Association (IPMA). Several authors, including Brent and Labuschagne (2006), Labuschagne and Brent (2005, 2007, 2008), and Pade, Mallinson, and Sewry (2006, 2008), were conducting research on the subject, and the Association for Project Management recognised that "the planet earth is in peril due to a variety of fundamental sustainability threats" and that "Project and programme managers are uniquely positioned to contribute to sustainable management practises" (Association for Project Management, 2006, pp. 1, 7).

However, according to a paper published in 2009, sustainable development in temporary organisations such as initiatives and programmes is seldom investigated (Gareis et al., 2009, p. 1). Furthermore, it has been said that project management has failed to address the sustainability issue in a significant manner (Eid, 2009, p. 288). On the issue of sustainability, more recent advancements have claimed that the relationship between sustainability and project management is gaining momentum and that much of the research into project sustainability has been conducted in the last few years (Silvius & Tharp, 2013, p. xix).

While the rising focus on sustainability in project management is a positive development, it also offers some challenges, since the concept of sustainability is intuitively understood but difficult to define in clear, practical terms (Briassoulis, 2001). The content and meaning of corporate sustainability might change depending on the circumstances (van Marrewijk, 2003).

Projects, on the other hand, raise challenges for the local community and government, one of which is the question of sustainable development. The question of how institutions such as governments, corporations, and other organisations may develop and execute ideas without risking the lives and prosperity of future generations is crucial to consider.

DISCUSSION

Sustainability in Projects

The majority of meanings of the word "sustainability" are concerned with the relationship that exists between people and the resources that they use (Voinov, 2007). As defined by Wimberley (1993, p. 1), being "sustainably resourced" implies supplying food, fibre, and other natural and social resources essential for a group's survival while doing so in a manner that ensures critical resources are preserved for current and future generations, among other things. This is somewhat close to the widely used Brundtland Commission idea, which characterised it as progress that meets present needs without compromising the ability of future generations to meet their own (WCED, 1987).

However, although alternative definitions of sustainability have been given, the vast majority of scholars working in the subject feel that sustainability highlights the need of achieving a harmonious balance between social, environmental, and economic goals at the same time. These goals are referred to as the "three pillars of sustainable development" (Azapagic and Perdan, 2000; Labuschagne and Brent, 2005; Sillanpää, 1999) since they are intertwined with one another.

Sustainability principles and aims are increasingly being incorporated into business policies and operations as a result of increased public demand (Labuschagne and Brent, 2005). Implementing sustainability principles may be driven by both ethical and financial motives. Hart (1997) established a framework for sustainable value, which defines four fundamental components of sustainability strategy, with the focus on the possibility for producing win-win circumstances that assist both shareholders and the environment.

Bossink (2002) used the scenario of the Netherlands to demonstrate how the nation's innovative and sustainable policy influenced the way in which distinctive economic expansion and design strategies were created and implemented by Dutch construction firms, according to the author. As an example of how standards and prizes may be used to promote engagement of local stakeholders in low-income housing projects in South Africa, Ross et al. (2010) demonstrated how standards and rewards can be used to encourage participation of local stakeholders.

More specifically, organisational and management scholars have investigated the role of firms in contributing to sustainable development (Shrivastava, 1995), the function of stakeholders in the determination of sustainable development strategies (Sharma and Henriques, 2005), with the principles of corporate social responsibility and corporate long-term viability (Van Marrewijk, 2003). Project management methodologies are not immune from the obligation to include sustainability into their planning and execution. Existing project management concepts, as per some, do not effectively handle social and environmental problems, and as a result, they should be revised (Labuschagne and Brent, 2005).

To do this, the goal of this study is to characterise the important sustainability discourses in the field of project management. Our goal is to determine if project organisations or hosts actively use independent sustainability practices, that is, methods targeted at accomplishing their sustainability-related goals in the face of uncertainties, in order to achieve their sustainability-related objectives. The word "strategy" refers to the process of choosing different actions that will result in value being created (Porter, 1996).

However, sustainability strategies, is from the other side, are often characterised as plans and routes for success (Artto et al., 2008), and they are designed to handle the specific obstacles and opportunities connected with sustainability.

On the basis of a review of papers published in journals, during their investigation, the researchers came across two divergent viewpoints on project sustainability. The first examines the situation from the perspective of the project organisation, which is in charge of delivering the investment asset, whereas the second examines the situation from the perspective of the host organisation, which is the regulatory authority that has authorised the project and is in charge of administering the region in which the project is located.

In contrast, we discovered eight separate sustainability strategies that were used by either the program or its sponsor, or by either of these organisations, throughout the course of the study. Using an instructive empirical scenario involving the deployment of an innovative system, we compare and contrast our findings, and we explain the consequences of our findings for both project research and practical application.

Sustainable development, according to a widely recognised definition, is "development that meets the needs of the present without threatening the ability of future generations to satisfy their own requirements."

Significantly more, even if about half of the articles published in the sample do not explicitly contain an interpretation of sustainable development, the results of the study are still significant., 28 percent of the articles published correspond to the World Commission on Environment and Development's definition of sustainability (Bruntland, 1987) as a theoretical first step.

Following the study's findings, the 'triple bottom line' or 'Triple-P (People, Planet, Profit)' definition of the concept of sustainability is the one that is most frequently used (86 percent) by the periodicals in the sample in their definitions of sustainability, according to the findings of the study (Elkington, 1997).

Despite the fact that the articles cover a wide range of subjects, the articles differ in their handling of the numerous 'P's. The economic component of their themes is covered in 96 percent of the articles, the social aspect is covered in 89 percent, and the environmental side is covered in 86 percent of the articles. Sustainability-related projects and project management articles tend to emphasise a combination of financial and ecological components, while publications on sustainable development projects are more likely to focus on the social elements of the projects. It is evident from the following list of publications that the sample comprises articles that investigate other dimensions or "principles" of sustainability that are significant to project management.

Using the example of Gareis et al. (2011b), sustainability is defined in terms of the following fundamentals: the economic, social, and ecological orientations; the short-, medium- and long-term orientations; the local, regional, and global orientations; and the value orientations of people and organisations.

Sustainability is alluded to as a moral concept in the final aspect, value orientation, which calls for the formation of specific values as the basis for people's attitudes and behaviours (Eid,2009; Eskerod and Huemann, 2013; Schieg, 2013). There have been several discussions about:

- About "risk reduction," which was presented by Gareis et al. (2010), Huemann et al. (2010), and Goedknegt, (2012).
- About "transparency," which was presented by Khalfan (2006) and Silvius et al. (2012).
- About "performance," which was presented by Khalfan (2006) and Silvius et al. (2012).
- About "other sustainability features and principles," which was presented by Gareis et al. (2010) and Goedknegt (2012).

However, there is a serious lack of literature that fills the gap among sustainability and project management literature (Marcelino-Sádaba et al., 2015;). Further, according to Brones et al. (2014), key project management systems, such as the Project Management Body of Knowledge (PMBoK), the International Construction Bureau (ICB), ISO 21500:2012, and Prince2, have failed to include environmental sustainability across the project life cycle.. In support of this conclusion, Silvius and Schipper (2014) suggest that current PM guidelines fail to address the problem of environmental sustainability. However, there are many unsolved questions and the integration of sustainable project management is hard (Marcelino-Sádaba et al., 2015; Silvius and Schipper, 2014).

Sustainable project management may test the system limitations of project management (Silvius and Schipper, 2014). Sustainable development in the realm of project management may be seen from a number of different viewpoints. A sustainable viewpoint on PM exists, according to Carvalho and Rabechini (2011), both within and outside. Throughout the project life cycle, the internal perspective is intertwined with the project management process and the project management areas of responsibility. Taking an external perspective is linked to sustainable development in a broad sense, and it takes into account the project's social and environmental ramifications, as well as its economic and environmental advantages.

There has been special interest in the linkages among PM and Ecodesign, or the environmental aspect of sustainability from an environmental perspective, or the interaction among PM and the environment (Johansson and Magnusson 2006). Those employed in the public sector and the construction business, in particular, are concerned about the social component of sustainability, which is becoming more significant (Campbell et al., 2008). The Triple Bottom Line (TBL), which considers environmental, social, and economic factors, has been the subject of more recent study and development (Pulaski and Horman, 2005).

According to Marcelino-Sádaba et al. (2015), a four-dimensional framework for project management has been established that incorporates sustainability from the perspective of the triple bottom line (economic, social, and environmental). According to Martens and Carvalho (2016, 2017a), the most critical concerns affect both strategic and tactical viewpoints in four areas: the sustainable innovation business model, stakeholder management, economic and competitive advantage, environmental legislation, and resource conservation and conservation, and environmental policies and resource conservation and conservation for starters. it's vital to think about how to align project management ideas with sustainable development principles in order to integrate sustainable product and project lifecycle management (Labuschagne and Brent, 2005, 2008).

A product and/or a project may be considered sustainable when the term is employed in the context of project management, similar to how the terms scope and quality are used to different aspects of project management (Carvalho and Rabechini, 2011). During the project scope management process, the term "Product scope" may be used to refer to a specific product. In addition to project scope, it is important to consider the qualities and functions that differentiate a product, service, or result. Project management, according to the Project Management Institute (2013), is "the systematic use of best practises to produce outcomes."

The phrase "project sustainability management" is used in this study to refer to both the product and the project, similarly to how the word "product sustainability management" is used. The word "sustainability" may refer to both goods and processes, according to Silvius and Schipper (2015), and can be used to both. Aside from that, when we speak to the product, we are referring to the service and other intangibles that are created as a result of the project. It is critical to note out that the recommended model did not take the viewpoint of organisational sustainability into mind.

SUSTAINABILITY IN PRODUCT PERSPECTIVE

Life cycle analysis literature (such as the ISO 14062 and ISO 14006 standards) has reached a mature stage, according to Labuschagne and Brent (2005). In order to establish a more sustainable PM environment, this literature may be utilised to include environmental considerations into the product viewpoint (Marcelino-Sádaba et al., 2015).

However, the area of project management has not yet been successful in integrating this content into its practices. By integrating Design for Environment (DfE) and Project Management (PM), it is feasible to promote environmental sustainability engagement from the start of the process; connect stakeholders' aspirations with both the project's intent on sustainability; and define success criteria linked to the project's environmental effects from the initiation stage (Brones et al., 2014; Brones and Carvalho, 2015).

Ecology-friendly design is defined as "a product development process that considers the complete life cycle of a product and considers environmental aspects at all stages of the process," and it includes eco-efficiency, remanufacturing and recycling as well as resource conservation and waste minimization throughout the product's lifetime (Glavi and Lukman, 2007). Originally focused on physical products, ecodesign has now expanded to encompass services-oriented life cycle design (Bhamra and Evans,

1999), which is meant to prevent "Impact transmission and result in environmentally-sound services" (Bonvoisina et al., 2014).

In accordance with Kuehr (2007), environmental technologies (ET) can be divided into four categories, each of which has a different level of eco-effectiveness: "measuring technologies on the environment, cleansing technologies or end-of-pipe approach [cleaning technologies or end-of-pipe approach], cleaner technological approaches, clean technological approaches, or zero impact technological approaches [zero impact technological approaches]."

Sustainable Project Management is the management of project organised change in policies, assets or organisations, with consideration of the economical, social and environmental impact of the project, its result and its effect, for now and future generations.

Figure 1. Sustainability in project management context

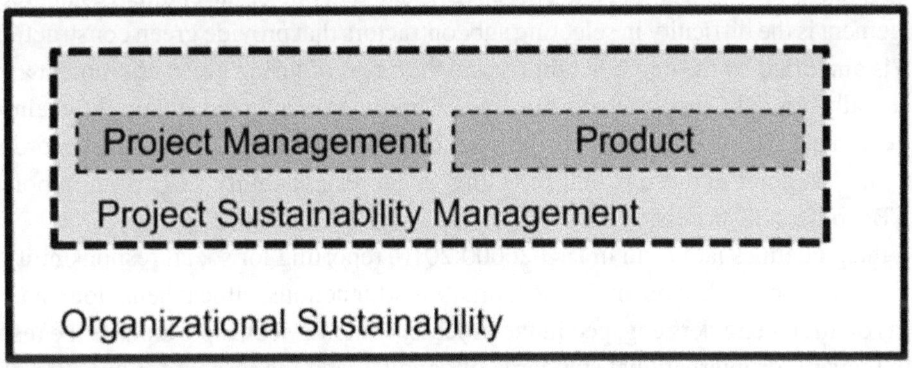

SUSTAINABILITY IN PROJECT PERSPECTIVE

According to Carvalho and Rabechini (2015), the TBL perspective should be utilised to resolve challenges related to sustainability in the PM Process and Knowledge sectors of the organisation (PMfS).

Stakeholder communication and human resource management should be included within the project management scope (soft side), procurement, and risk management processes. A number of possible opportunities for integrating sustainability concerns into all stages of the project management lifecycle are also identified by Silvius and Schipper (2014).

As stated by Marcelino-Sádaba et al. (2015), The elements affecting the management of sustainable projects include the use of sustainable goods and processes, as well as an organization's commitment to sustainability, and staff who are taught and knowledgeable about sustainability issues (among other things).

Examples of sustainable project goods include policies emerging from Corporate Social Responsibility (CSR), ecodesign, and economic products that are aligned with ISO criteria in the social and environmental spheres (ISO 26000, ISO 14000 among others).

There are a variety of obstacles to overcome, ranging from a lack of enthusiasm on the part of key stakeholders to a lack of capabilities in the field of sustainability management (Carvalho and Rabechini, 2011).

Among the most significant parts of incorporating environmental considerations into project management, Brones et al., (2014) emphasised the relevance of the supply chain, quality, timeliness, and risk as the most critical aspects of their study. According to Silvius and Schipper (2014), strengthening five fundamental skills is required to bridge the sustainability skill gap between project managers: systems theory abilities, predictive characteristics, ethical competencies, conceptual engagement, and interpersonal competency.

When it comes to incorporating sustainability into project management, the soft side is critical in order to get stakeholder buy-in and support (Carvalho and Rabechini, 2015). Stakeholder management, as a connection across traditional project management, and social and ethical issues (Marcelino-Sádaba et al., 2015), has the potential to improve stakeholders' participation and coordination, opening the path for more sustainable project management (Marcelino-Sádaba, 2015). The project's supply chain is also a crucial factor to take into account (Silvius and Schipper, 2014).

The green procurement and procurement (GP&P) programme in the area of project management (PM) is still in its early stages. Per the Hwang and Ng (2013), state that one of the challenges in green building project management is the difficulty in selecting subcontractors that provide green construction services. This problem is amplified by the unpredictability and high cost of hiring green equipment and supplies.

Environmentally friendly supply chain practices were widely adopted in supply chain operations as a result of external influences such as customer expectations and legal restraints, according to Kuei et al. (2015). The last point to make is that fulfilling social responsibility (SR) is an important part of meeting the TBL pillars of success.

Following the guidelines laid forth in ISO 26000:2010, reporting for social responsibility is defined by the following basic principles: openness, accountability for actions, ethical behaviour, and awareness of the concerns of all stakeholders engaged in the process. Other key concepts include the respect for the rule of law, the regard for international standards of conduct, and the respect for human rights, among others (International Standard, 2010).

All of these concepts may be utilised in the context of project management, among other things. The concept of social responsibility, according to Labuschagne and Brent (2005), comprises internal human resource management, external population management, stakeholder involvement, and macro-social performance. Further to our earlier debate, we recommend that project sustainability management (PSM) take into consideration both the product and the project's viewpoints. In the product viewpoint, there are two constructions that are used to refer to sustainability:

- Design for Environment (DfE)
- Environmental Technologies (ET)

Whereas the project perspective encompasses the project management process and knowledge areas focusing on sustainability (PMfS), the green procurement and partnership (GPP), and social responsibility in the project(SR), according to the literature on product sustainability management (PSM).

Project Success (PS)

PS is a notion with many different aspects. Traditionally, project success is measured by the project's ability to adhere to the project's scope, schedule, and cost goals (Wit, 1988). It has been suggested that project success may be divided into many additional strategic aspects, according to Shenhar and Dvir

(2007) such as project efficiency, effect on the team, impact on the client, business or direct success, and future preparation.

An element of environmental and social sustainability has recently been added to the equation (Carvalho and Rabechini, 2015; Martens and Carvalho, 2017b). A model for sustainable construction was established by Pulaski and Horman (2005), which includes sustainable goals into project management practises, resulting in an eight-category rating system: four categories for project performance assessment (such as cost estimation, quality assurance, schedule accuracy, and process efficiency), and four categories for sustainable building objectives (such as safety and health, maintainability, resource efficiency, and Leadership in Energy and Environmental Design (LEED) credit.

Establishing the project's context as the starting point for all aspects of the project and its management is critical to its success. A complete evaluation of the project's surrounding environment is required in order to include sustainability issues into the project management process (Silvius et al., 2012; Tharp, 2013). When it comes to addressing sustainability, the boundaries of the setting are stretched in terms of both time and geography. A substantial impact on the framework of the project is exerted by the sustainability aspects "short and long term," and "local and global."

Due to the fact that members of the project team may be based in developing nations such as India or China, and suppliers or customers may be located all over the world, a growing number of projects are addressing geoeconomic concerns in an increasingly globalised business environment. It goes without saying that a globalising business environment necessitates the implementation of global initiatives and the management of global projects.

A visual illustration of the scope of sustainable project management by Van den Brink (2009) illustrates the larger context of 'Sustainable Project Management' in contrast to traditional or current project management by mentioning local and global society as well as an enlarged time frame. In sustainable project management, the context of the project is addressed in connection with the organization's strategy, but it is also addressed in relation to society as a whole.

A project's internal or external environmental factors that surround or influence the success of the project are specified in the PMBOK Guide (Project Management Institute, 2013). However there is no mention of probable social or environmental effect as a component to consider in the PMBOK Guide. Many processes in PRINCE2 address the project environment, particularly during the start-up and initiation stages, but there is little mention of the project's larger social context for the rest of the project lifecycle (Office of Government Commerce, 2009). PRINCE2 pays ample attention to the management and development of the project team. It does mention the activity 'Design and appoint the project management team', but in later stages no reference is made.

Those who support the project's sustainability components, specifically those who support the concepts of "balancing or harmonising social, environmental, and economic interests"; "both short and long term"; and "both local and global," are likely to increase the number of stakeholders who support the project (Eskerod & Huemann, 2013; Tharp, 2013).

Environmental advocacy groups, human rights organisations, and non-governmental organisations (NGOs) are examples of sustainability stakeholders' (Silvius et al., 2012). In contrast to conventional sustainability stakeholders, the PMBOK Guide makes no mention of these organisations (Silvius, 2013). During the project's first stages, the identification of stakeholders is handled in detail across the various processes of PRINCE2 (Office of Government Commerce, 2009).

Another issue is that there is no particular recognition of potential stakeholders who might be affected by the project's environmental and/or social consequences. It is anticipated that incorporating sustainability concepts will have an influence on the project's specifications and requirements for its deliverable outputs, as well as the criteria for project excellence (Eid, 2009; Maltzman & Shirley, 2010; Taylor, 2010).

The incorporation of environmental or social issues into the project's overall purpose, projected output, and result are only a few of examples (Eid, 2009; Maltzman & Shirley, 2010; Taylor, 2010). With current project management standards in place, quality is first and foremost associated with achieving or surpassing the needs of the project sponsor, the project's client, or the ultimate user of the project's product (Silvius, 2013).

The needs and interests of other parties are taken into account to the extent that they may conflict with the expectations of the sponsors (Eskerod & Huemann, 2013). Implementing sustainable project management requires that the project's content, intended output/outcome, and quality standards be produced in partnership with a varied set of stakeholders and that the project be seen as a whole (Gareis et al., 2013). In order to do so, it is necessary to consider sustainable viewpoints such as "economic, environmental, and social," as well as "short-term and long-term," and "local and global" and developed together with a broad group of stakeholders. (as well as other perspectives) (Eskerod & Huemann, 2013).

Additionally, the influence of sustainability principles on the project's content must be recognised in the project's reasoning in order for it to be considered successful (Silvius & Schipper, 2012). It may be necessary to widen the scope of the project's costs, benefits, and business case in order to include non-financial considerations such as social or environmental considerations (Gareis et al., 2011a, 2011b). When considering the business case for a project, the PMBOK guide points out that environmental repercussions and social needs are also potential benefits to consider (Project Management Institute, 2013).

The business case in PRINCE2 is more generic in nature, rather than addressing specific social or environmental benefits in particular (Office of Government Commerce, 2009). In PRINCE2, the 'Lessons log' and the 'Lessons report' explicitly capture the lessons learned in a project. These lessons are explicitly addressed in the starting up stage of a project, in the process 'Capture previous lessons' None of the guidelines make any mention of the environmental or social ramifications of their recommendations. When considering the concept of sustainability in project management, it is recommended that the business case for a project take into account the triple bottom line of economic, social, and environmental benefits as well.

Investing decisions are made based on a multi-criteria approach that considers both quantitative and qualitative aspects of a situation (Silvius & Schipper, 2012). The components of project success are logically connected to the criteria for project excellence that have already been discussed in detail.

This indicates that the project's success is judged by the length of its life cycle and the outcome of the project (Craddock, 2013; Pade et al., 2008). Today's project management standards reflect a more constrained concept of project success than was previously the case. According to the Project Management Body of Knowledge, "success is defined as completing the project within the constraints of scope, time, money, quality, resources, and risk" (Project Management Institute, 2013, p. 35). Furthermore, while project success is often stated in a holistic manner, this broader set of criteria has minimal impact on the way projects are handled in the real world (Thomas & Fernandez, 2007).

The economic feasibility of a project is constantly emphasised by the constraints that must be overcome (Silvius et al., 2012). The project management methodology PRINCE2 includes six performance criteria. Despite the fact that these characteristics do not directly relate to sustainability, they may be included in performance measurements such as quality and benefits.

CONCLUSION

After conducting a research into the several elements of sustainability discussed in the literature on sustainability in project management, we decided that the following aspects of sustainability proved critical for comprehending the influence of sustainability on project management: To contribute positively to sustainable development, a business must adhere to the three sustainability 'pillars': social, environmental, and economic sustainability.

Each of the dimensions is intricately related to the others, which means that they have an influence on one another in various ways. Furthermore, as per Gareis et al., (2011b) a sustainable firm should consider both the short- and long-term consequences of its operations, rather than concentrating just on the short-term benefits.

When the dimension of 'short- and long-term orientation' is present, it draws attention to the whole of the issue under consideration. Aside from that, the increasing globalisation of economies has an impact on the geographical location in which businesses operate. The actions of foreign stakeholders, such as competitors, suppliers, and (potential) customers, have an influence on many organisations, whether they do so consciously or unconsciously. As a result, the behaviour and actions of organisations have an impact on economic, social, and environmental challenges on both a local and global scale, depending on their size.

As Gareis et al., (2011) elaborates to effectively address these nested and interconnected processes, sustainable development must be a joint effort spanning several levels, from global to regional to local. According to Robinson (2004) and Martens (2006), sustainable development is intrinsically a normative construct, reflecting the values and ethical concerns of society.

We as professionals, business leaders, and consumers will need to alter our implicit or explicit set of values, which will impact or guide our behaviour. This will be part of the transition necessary for more sustainable growth. When it comes to being transparent, a company must be forthright about its policies, decisions, and activities, as well as the environmental and social ramifications of those decisions and actions.

To make this possible, organisations must give stakeholders with current, accurate, and relevant information that enables them to analyse the organization's activities and rectify any potential issues. An organisation is responsible for its policies, decisions, and acts, as well as the consequences of those decisions and activities on the environment and society as a whole.

Stakeholder engagement requires a process of discussion and, ultimately, consensus building among all stakeholders as partners who collaboratively define difficulties, devise possible solutions, collaborate to put them into action, and monitor and analyse the results. As a result of the complexity, irreversibility, irreversibility, nonlinearity, and irreversibility of interactions between the environment and society, the precautionary principle recognises that avoiding damage is more effective than alleviating it.

A new round of discussion concerning the applicability of financial risk management systems to social and environmental challenges has erupted after the recent Deepwater Horizon oil disaster. There are seven types of waste: overproduction; waiting; transferring; inefficient processing; unnecessarily large inventories; superfluous or excessive motion; and faults. Additionally, the idea of waste removal underlies the cradle-to-cradle concept (McDonough & Braungart, 2002), which is built on the assumption that waste equals food.

Furthermore, the concept might be applied to social ideas as well. Additionally, organisations should avoid physically or psychologically straining their personnel in order to 'deplete' their ability to generate or develop labour or knowledge, since this will reduce their productivity. Businesses must manage not just their economic capital, but also their social and environmental capital if they are to stay viable in the long run.

Recommendation

As mentioned in the article's opening, the definition and meaning of sustainability vary according to circumstance. As a consequence, future research should concentrate on categorising our findings according to their context, such as industry. There was a strong presence of the building and construction industry in the publications included in the sampling. Other businesses, such as information technology, that rely largely on projects, such as the construction industry, got limited coverage in the periodicals.

In these conditions, we recommend that future study expand on the issue of sustainability in project management, which is now under investigation. Another topic that will be investigated in the future is the domain of sustainability and project management. We discovered that case studies constituted a disproportionately large part of empirical research. Case studies are well-suited for inquiry and the formulation of hypotheses, but they are less well-suited for hypothesis testing and knowledge generalisation than other methods of research. Therefore, we recommend that future research expand on current findings by evaluating them in order to develop knowledge that can be used across a wide range of situations.

Originality

This study has concentrated on an innovative pattern, a new perspective, sustainable project management, which is driving sustainability into project management processes. However, the pace at which this trend is evolving varies depending on the industrial setting in which it is occurring. The findings of this research have ramifications for both academics and business leaders. For those interested in the theoretical implications, the book serves as a synthesis of existing knowledge on the relatively new issue of sustainable project management, stressing the most significant features and connecting them via a conceptual framework. For researchers who are interested in the issue of sustainable project management, this book might serve as a useful reference source.

REFERENCES

Artto, K., Kujala, J., Dietrich, P., & Martinsuo, M. (2008). What is project strategy? *International Journal of Project Management*, *26*(1), 4–12. doi:10.1016/j.ijproman.2007.07.006

Azapagic, A., & Perdan, S. (2000). Indicators of sustainable development for industry: A general framework. *Process Safety and Environmental Protection*, *78*(4), 243–261. doi:10.1205/095758200530763

Bhamra, T. A., & Evans, S. (1999). The next step in ecodesign: service-oriented life cycle design. *IEEE International Symposium on Electronics and the Environment-ISEE*, 263–267. . 765887.10.1109/ISEE.1999.765887

Bonvoisina, J., & Lelaha, A. (2014). An integrated method for environmental assessment and ecodesign of ICT-based optimization services. *Journal of Cleaner Production, 68*, 144–154. doi:10.1016/j.jclepro.2014.01.003

Bossink, B. A. G. (2002). A Dutch public-private strategy for innovation in sustainable construction. *Construction Management and Economics, 20*(7), 633–642. doi:10.1080/01446190210163534

Brent, A. C., Heuberger, R., & Manzini, D. (2005). Evaluating projects that are potentially eligible for Clean Development Mechanism (CDM) funding in the South African context: A case study to establish weighting values for sustainable development criteria. *Environment and Development Economics, 10*(5), 631–649. doi:10.1017/S1355770X05002366

Brent, A. C., & Labuschagne, C. (2007). An appraisal of social aspects in project and technology life cycle management in the process industry. *Management of Environmental Quality, 18*(4), 413–426. doi:10.1108/14777830710753811

Brent, A. C., & Petrick, W. (2007). Environmental Impact Assessment (EIA) during project execution phases: Towards a stage-gate project management model for the raw materials processing industry of the energy sector. *Impact Assessment and Project Appraisal, 25*(2), 111–122. doi:10.3152/146155107X205832

Briassoulis, H. (2001). Sustainable Development and its Indicators: Through a (Planner's) Glass Darkly. *Journal of Environmental Planning and Management, 44*(3), 409–427. doi:10.1080/09640560120046142

British Standards Institute. (2007). *Occupational Health and Safety Management Systems — Requirements*. BSI Global.

Brones, F., Carvalho, M. M., & Zancul, E. S. (2017). Reviews, action and learning on change management for ecodesign transition. *Journal of Cleaner Production, 142*, 8–22. doi:10.1016/j.jclepro.2016.09.009

Bruntland, G. H. (1987). *Our Common Future: World Commission on Environment and Development*. Oxford University Press.

Campbell, C., Gibbs, A., Maimane, S., & Yugi, N. (2008). Hearing community voices: Grassroots perceptions of an intervention to support health volunteers in South Africa. *Journal of Social Aspects of HIV/AIDS Research Alliance, 5*(4), 162–177. doi:10.1080/17290376.2008.9724916 PMID:19194598

Carvalho, M. M., Patah, L. A., & Bido, D. S. (2015). Project management and its effects on project success: Cross-country and cross-industry comparisons. *International Journal of Project Management, 33*(7), 1509–1522. doi:10.1016/j.ijproman.2015.04.004

Carvalho, M. M., & Rabechini, R. (2015). Impact of risk management on project performance: The importance of soft skills. *International Journal of Production Research, 53*(2), 321–340. doi:10.1080/00207543.2014.919423

Construction Industry Institute (CII). (2006). *Leading indicators during project execution. Resarch Summary 220-1*. The University of Texas in Austin.

Corder, G. D., McLellan, B. C., & Green, S. (2010). Incorporating sustainable development principles into minerals processing design and operation: SUSOP (R). *Minerals Engineering, 23*(3), 175–181. doi:10.1016/j.mineng.2009.12.003

Craddock, W. T. (2013). How Business Excellence Models Contribute to Project Sustainability and Project Success. In A. J. G. Silvius & J. Tharp (Eds.), *Sustainability Integration for Effective Project Management* (pp. 1–19). IGI Global Publishing. doi:10.4018/978-1-4666-4177-8.ch001

Crawford, L. (2013). Leading Sustainability through Projects. In A. J. G. Silvius & J. Tharp (Eds.), *Sustainability Integration for Effective Project Management* (pp. 235–244). IGI Global Publishing. doi:10.4018/978-1-4666-4177-8.ch014

Cronbach, L. J., & Meehl, P. E. (1955). Construct validity in psychological tests. *Psychological Bulletin*, *52*(4), 281–302. doi:10.1037/h0040957 PMID:13245896

Deland, D. (2009). Sustainability Through Project Management and Net Impact. In *PMI Global Congress North America*. Philadelphia PA: Project Management Institute.

Dell, R. M., & Rand, D. A. J. (2001). Energy storage—A key technology for global energy sustainability. *Journal of Power Sources*, *100*(1), 2–17. doi:10.1016/S0378-7753(01)00894-1

Diamantopoulos, A., & Winklhofe, H. M. (2001). Index construction with formative indicators: An alternative to scale development. *JMR, Journal of Marketing Research*, *38*(2), 269–277. doi:10.1509/jmkr.38.2.269.18845

Dincer, I. (2007). Environmental and sustainability aspects of hydrogen and fuel cell systems. *International Journal of Energy Research*, *31*(1), 29–55. doi:10.1002/er.1226

Ding, G.K.C. (2008). Sustainable construction - The role of environmental assessment tools. *Journal of Environmental Management*, *86*(3), 451-464. doi: .jenvman.2006.12.025 doi:10.1016/j

Doran, J. W., Sarrantonio, M., & Liebig, M. (1996). *Soil health and sustainability*. Adv. Agron. doi:10.1016/S0065-2113(08)60178-9

Downe-Wamboldt, B. (1992). Content analysis: Method, applications, and issues. *Health Care for Women International*, *13*(3), 313–321. doi:10.1080/07399339209516006 PMID:1399871

Ebbesen, J.B., & Hope, A.J. (2013). Re-imagining the Iron Triangle: Embedding Sustainability into Project Constraints. *PM World Journal*, *2*(3).

Edum-Fotwe, F. T., & Price, A. D. F. (2009). A Social Ontology for Appraising Sustainability of Construction Projects and Developments. *International Journal of Project Management*, *27*(4), 313–322. doi:10.1016/j.ijproman.2008.04.003

Eid, M. (2009). *Sustainable Development & Project Management*. Lambert Academic.

Elkington, J. (1997). *Cannibals with Forks: the Triple Bottom Line of 21st Century Business*. Capstone Publishing.

Ellatar, S. M. S. (2009). Towards developing an improved methodology for evaluating performance and achieving success in construction projects. *Scientific Research and Essays*, *4*, 549–554.

Eriksson, P. E., Olander, S., Szentes, H., & Widén, K. (2013). Managing short-term efficiency and long-term development through industrialized construction. *Construction Management and Economics*, *32*(1–2), 97–108. doi:10.1080/01446193.2013.814920

Eskerod, P., & Huemann, M. (2013). Sustainable development and project stakeholder management: What standards say. *International Journal of Managing Projects in Business*, *6*(1), 36–50. doi:10.1108/17538371311291017

Falagas, M.E., Pitsouni, E.I., Malietzis, G.A., & Pappas, G. (2008). Comparison of PubMed, Scopus, web of science, and Google scholar: strengths and weaknesses. *The FASEB Journal, 22*(2), 338-342. doi:10.1096/fj.07-9492LSF

Falk, R. F., & Miller, N. B. (1992). *A Primer for Soft Modeling*. The University of Akron Press.

Faul, F., Erdfelder, E., Lang, A. G., & Buchner, A. (2007). G*Power 3: A flexible statistical power analysis program for the social, behavioral, and biomedical sciences. *Behavior Research Methods*, *39*(2), 175–191. doi:10.3758/BF03193146 PMID:17695343

Fellows, R., & Liu, A. (2008). Impact of participants' values on construction sustainability. *Proceedings of the Institution of Civil Engineers. Engineering Sustainability*, *161*(4), 219–227. doi:10.1680/ensu.2008.161.4.219

Fernandez-Sanchez, G., & Rodriguez-Lopez, F. (2010). A methodology to identify sustainability indicators in construction project management-application to infrastructure projects in Spain. *Ecological Indicators*, *10*(6), 1193–1201. doi:10.1016/j.ecolind.2010.04.009

Flynn, B. B., Kakibara, S. S., Schroeder, R. G., Bates, K. A., & Flynn, E. J. (1990). Empirical research methods in operations management. *Journal of Operations Management*, *9*(2), 250–284. doi:10.1016/0272-6963(90)90098-X

Fornell, C., & Lacker, D. F. (1981). Evaluating structural equation models with unobservable variables and measurement errors. *JMR, Journal of Marketing Research*, *18*(1), 39–50. doi:10.1177/002224378101800104

Forza, C. (2002). Survey research in operations management: A process-based perspective. *International Journal of Operations & Production Management*, *22*(2), 152–194. doi:10.1108/01443570210414310

Fthenakis, V. (2009). Sustainability of photovoltaics: The case for thin-film solar cells. *Renewable & Sustainable Energy Reviews*, *13*(9), 2746–2750. doi:10.1016/j.rser.2009.05.001

Gao, C., Hou, H., Zhang, J., Zhang, H., & Gong, W. (2006). Education for regional sustainable development: Experiences from the education framework of HHCEPZ project. *Journal of Cleaner Production*, *14*(9–11), 994–1002. doi:10.1016/j.jclepro.2005.11.043

Gareis, R., Huemann, M., & Martinuzzi, R.-A. (2011). What can project management learn from considering sustainability principles? *Project Perspectives*, *33*, 60–65.

Gareis, R., Huemann, M., Martinuzzi, R.-A., Sedlacko, M., & Weninger, C. (2011b). *The SustPM Matrix: Relating sustainability principles to project assignment and project management. EURAM11*. Talinn.

Gareis Atkinson, R. (1999). Project management: Cost, time and quality, two best guesses and a phenomenon, its time to accept other success criteria. *International Journal of Project Management*, *17*(6), 337–342. doi:10.1016/S0263-7863(98)00069-6

Genus, A., & Theobald, K. (2015). Roles for university researchers in urban sustainability initiatives: The UK Newcastle Low Carbon Neighbourhoods project. *Journal of Cleaner Production*, *106*, 119–126. doi:10.1016/j.jclepro.2014.08.063

Gilbert, R., Stevenson, D., Girardet, H., & Stern, R. (1996). *Making Cities Work: The Role of Local Authorities in the Urban Environment*. Earthscan Publications Ltd.

Glavič, P., & Lukman, R. (2007). Review of sustainability terms and their definitions. *Journal of Cleaner Production*, *15*(18), 1875–1885. doi:10.1016/j.jclepro.2006.12.006

Goedknegt, D. (2013). Sustainability in Project Management: Perceptions of Responsibility. In A. J. G. Silvius & J. Tharp (Eds.), *Sustainability Integration for Effective Project Management* (pp. 279–287). IGI Global Publishing. doi:10.4018/978-1-4666-4177-8.ch017

Goldemberg, J., Coelho, S. T., & Guardabassi, P. (2008). The sustainability of ethanol production from sugarcane. *Energy Policy*, *36*(6), 2086–2097. doi:10.1016/j.enpol.2008.02.028

Gössling, S., Hansson, C. B., Hörstmeier, O., & Saggel, S. (2002). Ecological footprint analysis as a tool to assess tourism sustainability. *Ecological Economics*, *43*(2), 199–211. doi:10.1016/S0921-8009(02)00211-2

Gregersen, H.M., Lundgren, A.L., & White, T.A. (1994). *Improving project management for sustainable development*. Midwest Universities Consortium for International Activities, Inc. (MUCIA), Policy Brief No. 7.

Gudergan, S. P., Ringle, C. M., Wende, S., & Will, A. (2008). Confirmatory tetrad analysis in PLS path modeling. *Journal of Business Research*, *61*(12), 1238–1249. doi:10.1016/j.jbusres.2008.01.012

Hair, J. F., Anderson, R. E., Tatham, R. L., & Black, W. C. (2005). *Multivariate Data Analysis* (4th ed.). Prentice Hall.

Hair, J. F., Hult, G. T. M., Ringle, C. M., & Sarstedt, M. (2014). *A Primer on Partial Least Squares Structural Equation Modeling (PLS-SEM)*. Sage.

Hart, S. L. (1997). Beyond greening: Strategies for a sustainable world. *Harvard Business Review*, *75*(1), 66.

Haugan, G. (2012). *The New Triple Constraints for Sustainable Projects, Programs, and ortfolios*. CRC Press.

Henderson, J. (2005). Google Scholar: A source for clinicians? *Canadian Medical Association Journal*, *172*(12), 1549–1550. doi:10.1503/cmaj.050404 PMID:15939908

Henseler, J., Ringle, C. M., & Sinkovics, R. R. (2009). The use of partial least squares path modeling in international marketing. *Adv. Int. Mark.*, *20*, 277–319. doi:10.1108/S1474-7979(2009)0000020014

Heravi, G., Fathi, M., & Faeghi, S. (2015). Evaluation of sustainability indicators of industrial buildings focused on petrochemical projects. *Journal of Cleaner Production*, *109*, 92–107. doi:10.1016/j.jclepro.2015.06.133

Herazo, B., Lizarralde, G., & Paquin, R. (2012). Sustainable development in the building sector: A Canadian case study on the alignment of strategic and tactical management. *Project Management Journal, 43*(2), 84–100. doi:10.1002/pmj.21258

Hill, R. C., & Bowen, P. A. (1997). Sustainable construction: Principles and a framework for attainment. *Construction Management and Economics, 15*(3), 223–239. doi:10.1080/014461997372971

Hsieh, H.-F., & Shannon, S. E. (2005). Three Approaches to Qualitative Content Analysis. *Qualitative Health Research, 15*(9), 1277–1288. doi:10.1177/1049732305276687 PMID:16204405

Hwang, B.-G., & Ng, W. J. (2013). Project management knowledge and skills for green construction: Overcoming challenges. *International Journal of Project Management, 31*(2), 272–284. doi:10.1016/j.ijproman.2012.05.004

Ibbs, C. W., & Kwak, Y. H. (2000). Assessing Project Management Maturity. *Project Management Journal, 31*(1), 32–43. doi:10.1177/875697280003100106

International Standard. (2010). ISO 26000:2010. Guidance on Social Responsibility. ISO.

International Standard. (2016). *ISO/CD 45001. Occupational Health and Safety Management Systems — Requirements*. Draft.

International Standards Organization. (2010). *ISO 26000 Guidance on Social Responsibility*. ISO.

IPMA. (2006). International Competency Baseline (3rd ed.). IPMA.

Jaillon, L., & Chi-Sun, P. (2010). Design issues of using prefabrication in Hong Kong building construction. *Construction Management and Economics, 28*(10), 1025–1042. doi:10.1080/01446193.2010.498481

Jarvis, C. B., MacKenzie, S. B., & Podsakoff, P. M. (2003). A critical review of construct indicators and measurement model misspecification in marketing and consumer research. *The Journal of Consumer Research, 30*(2), 199–218. doi:10.1086/376806

Johansson, G., & Magnusson, T. (2006). Organising for environmental considerations in complex product development projects: Implications from introducing a "green" sub-project. *Journal of Cleaner Production, 14*(15–16), 1368–1376. doi:10.1016/j.jclepro.2005.11.014

Johnson, C., Lizarralde, G., & Davidson, C. H. (2006). A systems view of temporary housing projects in post-disaster reconstruction. *Construction Management and Economics, 24*(4), 367–378. doi:10.1080/01446190600567977

Keeble, J. J., Topiol, S., & Berkeley, S. (2003). Using Indicators to Measure Sustainability Performance at a Corporate and Project Level. *Journal of Business Ethics, 44*(2-3), 149–158. doi:10.1023/A:1023343614973

Keeys, L.A. (2012). Emerging Sustainable Development Strategy in Projects: A Theoretical Framework. *PM World Journal, 1*(2).

Kennedy, A., & Smith, K. (1995). Soil microbial diversity and the sustainability of agricultural soils. *Plant and Soil, 170*(1), 75–86. doi:10.1007/BF02183056

Khalfan, M. M. A. (2006). Managing Sustainability within Construction Projects. *Journal of Environmental Assessment Policy and Management, 8*(1), 41–60. doi:10.1142/S1464333206002359

Khalili-Damghani, K., & Tavana, M. (2014). A comprehensive framework for sustainable project portfolio selection based on structural equation modeling. *Project Management Journal, 45*(2), 83–97. doi:10.1002/pmj.21404

Klotz, L., & Horman, M. (2010). Counterfactual analysis of sustainable project delivery processes. *Journal of Construction Engineering and Management, 136*(5), 595–605. doi:10.1061/(ASCE)CO.1943-7862.0000148

Knight, P., & Jenkins, J. O. (2009). Adopting and applying eco-design techniques: A practitioners perspective. *Journal of Cleaner Production, 17*(5), 549–558. doi:10.1016/j.jclepro.2008.10.002

Knoepfel, H. (2010). *Survival and Sustainability as Challenges for Projects*. International Project Management Association.

Kometa, S., Olomolaiye, P., & Harris, F. (1995). An evaluation of clients' needs and responsibilities in the construction process. *Engineering, Construction, and Architectural Management, 2*(1), 57–76. doi:10.1108/eb021003

Korkmaz, S., Riley, D., & Horman, M. (2010). Piloting evaluation metrics for sustainable high-performance building project delivery. *Journal of Construction Engineering and Management, 136*(8), 877–885. doi:10.1061/(ASCE)CO.1943-7862.0000195

Kuehr, R. (2007). Environmental technologies: From a misleading interpretations to an operational categorization and definition. *Journal of Cleaner Production, 15*(13–14), 1316–1320. doi:10.1016/j.jclepro.2006.07.015

Kuei, C., Madu, C. N., Chow, W. S., & Chen, Y. (2015). Determinants and associated performance improvement of green supply chain management in China. *Journal of Cleaner Production, 95*, 163–173. doi:10.1016/j.jclepro.2015.02.030

Kumaraswamy, M. M., & Thorpe, A. (1996). Systematizing construction project evaluations. *Journal of Management Engineering, 12*(1), 34–39. doi:10.1061/(ASCE)0742-597X(1996)12:1(34)

Kuper, M., Dionnet, M., Hammani, A., Bekka, Y., Garin, P., & Bluemling, B. (2009). Supporting the shift from state water to community water: Lessons from a social learning approach to designing joint irrigation projects in Morocco. *Ecology and Society, 14*(1), art19. doi:10.5751/ES-02755-140119

Labuschagne, C., & Brent, A. C. (2008). An industry perspective of the completeness and relevance of a social assessment framework for project and technology management in the manufacturing sector. *Journal of Cleaner Production, 16*(3), 253–262. doi:10.1016/j.jclepro.2006.07.028

Lam, E. W. M., Chan, A. P. C., & Chan, D. W. M. (2007). Benchmarking the performance of design-build projects: Development of project success index. *BIJ, 14*(5), 624–638. doi:10.1108/14635770710819290

Laws, D., & Loeber, A. (2011). Sustainable development and professional practice. *Proceedings of the Institution of Civil Engineers. Engineering Sustainability, 164*(1), 25–33. doi:10.1680/ensu.2011.164.1.25

Leal-Rodríguez, A. L., Roldán, J. L., Ariza-Montes, J. A., & Leal-Millán, A. (2014). From potential absorptive capacity to innovation outcomes in project teams through integrated contracts: Experiences with inclusiveness in Dutch infrastructure projects. *International Journal of Project Management*, *31*(4), 615–627.

Leal-Rodríguez, A. L., Roldán, J. L., Ariza-Montes, J. A., & Leal-Millán, A. (2014, August). the conditional mediating role of the realized absorptive capacity in a relational learning context. *International Journal of Project Management*, *32*(6), 894–907. doi:10.1016/j.ijproman.2014.01.005

Leurs, M. T. W., Mur-Veeman, I. M., Sar, R., Schaalma, H. P., & Vries, N. K. (2008). Diagnosis of sustainable collaboration in health promotion: A case study. *BMC Public Health*, *8*(1), 382. doi:10.1186/1471-2458-8-382 PMID:18992132

Lim, C., & Mohamed, M. Z. (1999). Criteria of project success: An exploratory reexamination. *International Journal of Project Management*, *17*(4), 243–248. doi:10.1016/S0263-7863(98)00040-4

Liu, C. H., Zhang, K., & Zhang, J. M. (2010). Sustainable utilization of regional water resources: Experiences from the Hai Hua ecological industry pilot zone (HHEIPZ) project in China. *Journal of Cleaner Production*, *18*(5), 447–453. doi:10.1016/j.jclepro.2009.11.011

Ma, U. (2011). *No Waste: Managing Sustainability in Construction*. Gower Publishing.

Madden, P. B., & Morawski, J. D. (2011). The future of the Canadian oil stands: Engineering and project management advances. *Energy & Environment*, *22*(5), 579–596. doi:10.1260/0958-305X.22.5.579

Maltzman, R., & Shirley, D. (2013). Project Manager as a Pivot Point for Implementing Sustainability in an Enterprise. In A. J. G. Silvius & J. Tharp (Eds.), *Sustainability Integration for Effective Project Management* (pp. 262–278). IGI Global Publishing. doi:10.4018/978-1-4666-4177-8.ch016

Marcelino-Sádaba, S., González-Jaen, L. F., & Pérez-Ezcurdia, A. (2015). Using project management as a way to sustainability. From a comprehensive review to a framework definition. *Journal of Cleaner Production*, *99*, 1–16. doi:10.1016/j.jclepro.2015.03.020

Martens, M. M., & Carvalho, M. M. (2017). Key factors of sustainability in project management context: A survey exploring the project managers' perspective. *International Journal of Project Management*, *35*(6), 1084–1102. doi:10.1016/j.ijproman.2016.04.004

Martens, P. (2006). Sustainability: science or fiction? Sustainability: Science, Practice, &. *Policy*, *2*(1), 36–41.

Mathur, V. N., Price, A. D. F., & Austin, S. (2008). Conceptualizing stakeholder engagement in the context of sustainability and its assessment. *Construction Management and Economics*, *26*(6), 601–609. doi:10.1080/01446190802061233

McDonough, W., & Braungart, M. (2002). *Cradle To Cradle: Remaking The Way We Make Things*. North Point Press.

McKinlay, M. (2008). *Where is Project Management running to …?* Keynote address delivered at the 22nd World Congress of the International Project Management Association, Rome, Italy.

Meech, J. A., McPhie, M., Clausen, K., Simpson, Y., Lang, B., Campbell, E., Johnstone, S., & Condon, P. (2006). Transformation of a derelict mine site into a sustainable community: The Britannia project. *Journal of Cleaner Production, 14*(3–4), 349–365. doi:10.1016/j.jclepro.2004.08.009

Mishra, P., Dangayach, G. S., & Mittal, M. L. (2011). An Ethical approach towards sustainable project Success. *Procedia - Social and Behavioral Sciences, 25*, 338-344. 10.1016/j.sbspro.2011.10.552

Mochal, T., & Krasnoff, A. (2013). GreenPM®: The Basic Principles for Applying an Environmental Dimension to Project Management. In A. J. G. Silvius & J. Tharp (Eds.), *Sustainability Integration for Effective Project Management* (pp. 39–57). IGI Global Publishing. doi:10.4018/978-1-4666-4177-8.ch003

Mokhlesian, S., & Holmén, M. (2012). Business model changes and green construction processes. *Construction Management and Economics, 30*(9), 761–775. doi:10.1080/01446193.2012.694457

Morfaw, J. N. (2012). *Fundamentals of Project Sustainability: Strategies, Processes and Plans*. CreateSpace Independent Publishing Platform.

MoriokaS. N.CarvalhoM. M. 2015. Sustainability and management of projects: a bibliometric study. *Production*. doi:10.1590/0103-6513.058912

Morris, P. W. (2013). *Reconstructing Project Management*. John Wiley & Sons. doi:10.1002/9781118536698

Mulder, J., & Brent, A. C. (2006). Selection of Sustainable Rural Agriculture Projects in South Africa: Case Studies in the LandCare Programme. *Journal of Sustainable Agriculture, 28*(2), 55–84. doi:10.1300/J064v28n02_06

Müller, R., Pemsel, S., & Shao, J. (2014). Organizational enablers for governance and governmentality of projects: A literature review. *International Journal of Project Management, 32*(8), 1309–1320. doi:10.1016/j.ijproman.2014.03.007

Müller-Pelzer, F. (2009). *Sustainability Management in CDM Project Activities: How to demonstrate and assess the contribution to sustainable development of Clean Development Mechanism (CDM) project activities*. SVH-Verlag.

Netemeyer, R. G., Bearden, W. O., & Sharma, S. (2003). *Scaling Procedures: Issues and Applications*. Sage Publications. doi:10.4135/9781412985772

Ning, C., Zhang, S., & Li, L. (2009). Sustainable Project Management: A Balance Analysis Model of Effect. *International Conference on Management and Service Science*, Wuhan, China. 10.1109/ICMSS.2009.5302357

Ny, H., Hallstedt, S., Robèrt, K.-H., & Broman, G. (2008). Introducing templates for sustainable product development: A case study of televisions at the Matsushita Electric Group. *Journal of Industrial Ecology, 12*(4), 600–623. doi:10.1111/j.1530-9290.2008.00061.x

Ochoa, J. J. (2014). Reducing plan variations in delivering sustainable building projects. *Journal of Cleaner Production, 85*, 276–288. doi:10.1016/j.jclepro.2014.01.024

Office of Government Commerce. (2010). *Management of Risk: Guidance for Practitioners*. HMSO.

Olander, S., & Landin, A. (2005). Evaluation of stakeholder influence in the implementation of construction projects. *International Journal of Project Management, 23*(4), 321–328. doi:10.1016/j.ijproman.2005.02.002

Orr, R. J., & Scott, W. R. (2008). Institutional exceptions on global projects: A process model. *Journal of International Business Studies, 39*(4), 562–588. doi:10.1057/palgrave.jibs.8400370

Ortiz, O., Castells, F., & Sonnemann, G. (2009). Sustainability in the construction industry: A review of recent developments based on LCA. *Construction & Building Materials, 23*(1), 28–39. doi:10.1016/j.conbuildmat.2007.11.012

Ostrom, E. (2009). A general framework for analyzing sustainability of socialecological systems. *Science, 325*(5939), 419–422. doi:10.1126cience.1172133 PMID:19628857

Pade, C., Mallinson, B., & Sewry, D. (2008). An Elaboration of Critical Success Factors for Rural ICT Project Sustainability in Developing Countries: Exploring the Dwesa Case. *The Journal of Information Technology Case and Application, 10*(4), 32–55. doi:10.1080/15228053.2008.10856146

Pade, C. I., Mallinson, B., & Sewry, D. (2006). An exploration of the categories associated with ICT project sustainability in rural areas of developing countries: a case study of the Dwesa project. *Proceedings of the 2006 annual research conference of the South African institute of computer scientists and information technologists on IT research in developing countries (SAICSIT),* 100-106. 10.1145/1216262.1216273

Pade-Khene, C.I., Mallinson, B., & Sewry, D. (2011). Sustainable rural ICT project management practice for developing countries: investigating the Dwesa and RUMEP projects. *Information Technology for Development, 17*(3), 187-212. doi:10.1080/02681102.2011.568222

Patanakul, P., & Shenhar, A.J. (2012). *What Project Strategy Really Is: the Fundamental Building Block in Strategic Project Management.* doi:10.1002/pmj

Pauly, D., Christensen, V., Guénette, S., Pitcher, T. J., Sumaila, U. R., Walters, C. J., Watson, R., & Zeller, D. (2002). Towards sustainability in world fisheries. *Nature, 418*(6898), 689–695. doi:10.1038/nature01017 PMID:12167876

Pearce, A. R. (2008). Sustainable capital projects: Leapfrogging the first cost barrier. *Civil Engineering and Environmental Systems, 25*(4), 291–300. doi:10.1080/10286600802002973

Peng, D. X., & Lai, F. (2012). Using partial least squares in operations management research: A practical guideline and summary of past research. *Journal of Operations Management, 30*(6), 467–480. doi:10.1016/j.jom.2012.06.002

Pennypacker, J. S., & Grant, K. P. (2003). Project management maturity: An industry benchmark. [March]. *Project Management Journal, 34*(1), 4–9. doi:10.1177/875697280303400102

Perrini, F., & Tencati, A. (2006). Sustainability and Stakeholder Management: The Need for New Corporate Performance Evaluation and Reporting Systems. *Business Strategy and the Environment, 15*(5), 286–308. doi:10.1002/bse.538

Pilbeam, C. (2013). Coordinating temporary organizations in international development through social and temporal embeddedness. *International Journal of Project Management*, *31*(2), 190–199. doi:10.1016/j.ijproman.2012.06.004

PMI. (2013). *A Guide to the Project Management Body of Knowledge (PMBOK® Guide)*. Project Management Institute, Incorporated.

Porter, M.E. (1996). What is strategy? *Harvard Business Review*, *74*(6), 61–78.

Pocock, J. B., Hyun, C. T., Liu, L. Y., & Kim, M. K. (1996). Relationship between project interaction and performance indicators. *Journal of Construction Engineering and Management*, *122*(2), 165–176. doi:10.1061/(ASCE)0733-9364(1996)122:2(165)

Prasad, S., Tata, J., Herlache, L., & McCarthy, E. (2013). Developmental project management in emerging countries. *Oper. Manag. Res.*, *6*(1-2), 53–73. doi:10.100712063-013-0078-1

Prieto, B. (2011). *Sustainability on Large, Complex Engineering & Construction Programs Utilizing a Program Management Approach. PMWorldToday, 13(7).*

Project Management Institute. (2013). *A Guide to Project Management Body of Knowledge (PMBOK® Guide)* (5th ed.). Project Management Institute Publishing.

Pulaski, M. H., & Horman, M. J. (2005). Continuous value enhancement process. *Journal of Construction Engineering and Management*, *131*(12), 1274–1282. doi:10.1061/(ASCE)0733-9364(2005)131:12(1274)

Raven, R., Jolivet, E., Mourik, R. M., & Feenstra, C. F. J. (2009). ESTEEM: Managing societal acceptance in new energy projects a toolbox method for project managers. *Technological Forecasting and Social Change*, *76*(7), 963–977. doi:10.1016/j.techfore.2009.02.005

Raz, T., Shenhar, A. J., & Dvir, D. (2002). Risk Management, Project Success, and Technological Uncertainty. *R & D Management*, *32*(2), 101–109. doi:10.1111/1467-9310.00243

Ringle, C. M., Wende, S., & Becker, J.-M. (2015). *SmartPLS 3*. SmartPLS GmbH. http://www.smartpls.com

Robichaud, L. R., & Anantatmula, V. S. (2011). Greening Project Management Practices for Sustainable Construction. *Journal of Management Engineering*, *27*(148), 48–57. doi:10.1061/(ASCE)ME.1943-5479.0000030

Robinson, J. (2004). Squaring the circle? Some thoughts on the idea of sustainable development. *Ecological Economics*, *48*(4), 369–384. doi:10.1016/j.ecolecon.2003.10.017

Rosen, M. A., Dincer, I., & Kanoglu, M. (2008). Role of exergy in increasing efficiency and sustainability and reducing environmental impact. *Energy Policy*, *36*(1), 128–137. doi:10.1016/j.enpol.2007.09.006

Ross, N., Bowen, P. A., & Lincoln, D. (2010). Sustainable housing for low-income communities: Lessons for South Africa in local and other developing world cases. *Construction Management and Economics*, *28*(5), 433–449. doi:10.1080/01446190903450079

Russell, J. (2008). Corporate social responsibility: what it means for the project manager. In *PMI Global Congress EMEA*. Project Management Institute.

Sanchez, M. A. (2015). Integrating sustainability issues into project management. *Journal of Cleaner Production, 96*, 319–330. doi:10.1016/j.jclepro.2013.12.087

Sandoval, M. C., Veiga, M. M., Hinton, J., & Sandner, S. (2006). Application of sustainable development concepts to an alluvial mineral extraction project in lower Caroni River, Venezuela. *Journal of Cleaner Production, 14*(3–4), 415–426. doi:10.1016/j.jclepro.2004.10.007

Saunders, M., Lewis, P., & Thornhill, A. (2012). *Research Methods for Business Students* (6th ed.). Pearson Education.

Scanlon, J., & Davis, A. (2011). The role of sustainability advisers in developing sustainability outcomes for an infrastructure project: Lessons from the Australian urban rail sector. *Impact Assessment and Project Appraisal, 29*(2), 121–133. doi:10.3152/146155111X12913679730836

Schaltegger, S., Freund, F. L., & Hansen, E. G. (2012). Business cases for sustainability: The role of business model innovation for corporate sustainability. *International Journal of Innovation and Sustainable Development, 6*(2), 95–119. doi:10.1504/IJISD.2012.046944

Schieg, M. (2009). The model of corporate social responsibility in project management. *Business: Theory and Practice, 10*(4), 315–321. doi:10.3846/1648-0627.2009.10.315-321

Seglen, P. O. (1994). Causal Relationship between Article Citedness and Journal Impact. *Journal of the American Society for Information Science, 45*(1), 1–11. doi:10.1002/(SICI)1097-4571(199401)45:1<1::AID-ASI1>3.0.CO;2-Y

Sharma, S., & Henriques, I. (2005). Stakeholder influences on sustainability practices in the Canadian forest products industry. *Strategic Management Journal, 26*(2), 159–180. doi:10.1002mj.439

Shen, L. Y., Tam, V. W. Y., Tam, L., & Ji, Y. B. (2010). Project feasibility study: The key to successful implementation of sustainable and socially responsible construction management practice. *Journal of Cleaner Production, 18*(3), 254–259. doi:10.1016/j.jclepro.2009.10.014

Shenhar, A. J., & Dvir, D. (2007). *Reinventing Project Management: The Diamond Approach to Successful Growth and Innovation*. Harvard Business School Press.

Shi, Q., Zuo, J., & Zillante, G. (2012). Exploring the management of sustainable construction at the programme level: A Chinese case study. *Construction Management and Economics, 30*(6), 425–440. doi:10.1080/01446193.2012.683200

Shiferaw, A. T., & Klakegg, O. J. (2012). Linking policies to projects: The key to identifying the right public investment projects. *Project Management Journal, 43*(4), 14–26. doi:10.1002/pmj.21279

Shiferaw, A. T., Klakegg, O. J., & Haavaldsen, T. (2012). Governance of public investment projects in Ethiopia. *Project Management Journal, 43*(4), 52–69. doi:10.1002/pmj.21280

Shrivastava, P. (1995). The role of corporations in achieving ecological sustainability. *Academy of Management Review, 20*(4), 936–960. doi:10.2307/258961

Siggelkow, N. (2007). Persuasion with case studies. *Academy of Management Journal, 50*(1), 20–24. doi:10.5465/amj.2007.24160882

Sillanpää, M. (1999). A new deal for sustainable development in business. In M. Bennet & P. James (Eds.), *Sustainable Measures*. Greenleaf Publishing.

Silvius, A. J. G., & Schipper, R. (2014). Sustainability in project management competencies: Analyzing the competence gap of project managers. *Journal of Human Resource and Sustainability Studies*, *2*(02), 40–58. doi:10.4236/jhrss.2014.22005

Singh, R. K., Murty, H. R., Gupta, S. K., & Dikshit, A. K. (2012). An overview of sustainability assessment methodologies. *Ecological Indicators*, *15*(1), 281–299. doi:10.1016/j.ecolind.2011.01.007

Tabassi, A. A., Roufechaei, K. M., Ramli, M., Abu Bakar, A., Ismail, R., & Pakir, A. H. K. (2016). Leadership competences of sustainable construction project managers. *Journal of Cleaner Production*, *124*, 339–349. doi:10.1016/j.jclepro.2016.02.076

Talbot, R., & Venkataraman, R. (2013). Managing Sustainability on Projects Using Indicators. In A. J. G. Silvius & J. Tharp (Eds.), *Sustainability Integration for Effective Project Management* (pp. 194–211). IGI Global Publishing. doi:10.4018/978-1-4666-4177-8.ch012

Tam, G. (2010). The program management process with sustainability considerations. Journal of Project. *Program & Portfolio Management*, *1*(1), 17–27. doi:10.5130/pppm.v1i1.1574

Taylor, T. (2010). *Sustainability Interventions - for Managers of Projects and Programmes. Centre for Education in the Built Environment*. The Higher Education Academy.

Tenenhaus, M., Esposito Vinzi, V., Chatelin, Y.-M., & Lauro, C. (2005). PLS path modeling. *Computational Statistics & Data Analysis*, *48*(1), 159–205. doi:10.1016/j.csda.2004.03.005

Tharp, J. (2013). Sustainability in Project Management: Practical Applications. In A. J. G. Silvius & J. Tharp (Eds.), *Sustainability Integration for Effective Project Management* (pp. 182–193). IGI Global Publishing. doi:10.4018/978-1-4666-4177-8.ch011

Thomas, G., & Fernandez, W. (2007). The Elusive Target of IT Project Success. In *Proceedings of the 2nd International Research Workshop on IT Project Management (IRWITPM)* (pp. 93-102). Association of Information Systems, Special Interest Group for Information Technology Project Management.

Thomson, C. S., El-Haram, M. A., & Emmanuel, R. (2011). Mapping sustainability assessment with the project life cycle. *Proceedings of the Institution of Civil Engineers. Engineering Sustainability*, *164*(2), 143–157. doi:10.1680/ensu.2011.164.2.143

Thoumy, M., & Vachon, S. (2012). Environmental projects and financial performance: Exploring the impact of project characteristics. *International Journal of Production Economics*, *140*(1), 28–34. doi:10.1016/j.ijpe.2012.01.014

Tilman, D., Cassman, K. G., Matson, P. A., Naylor, R., & Polasky, S. (2002). Agricultural sustainability and intensive production practices. *Nature*, *418*(6898), 671–677. doi:10.1038/nature01014 PMID:12167873

Tingström, J., & Karlsson, R. (2006). The relationship between environmental analyses and the dialogue process in product development. *Journal of Cleaner Production*, *14*(15–16), 1409–1419. doi:10.1016/j.jclepro.2005.11.012

Tiron-Tudor, A., & Dragu, I.-M. (2013). Project Success by Integrating Sustainability in Project Management. In A. J. G. Silvius & J. Tharp (Eds.), *Sustainability Integration for Effective Project Management* (pp. 106–127). IGI Global Publishing.

Toor, S., & Ogunlana, S. O. (2010). Beyond the "iron triangle": Stakeholder perception of key performance indicators (KPIs) for large-scale public sector development projects. *International Journal of Project Management, 28*(3), 228–236. doi:10.1016/j.ijproman.2009.05.005

Tranfield, D., Denyer, D., & Smart, P. (2003). Towards a Methodology for Developing Evidence-Informed Management Knowledge by Means of Systematic Review. *British Journal of Management, 14*(3), 207–222. doi:10.1111/1467-8551.00375

Turner, J. R. (2010). Responsibilities for Sustainable Development in Project and Program Management. In H. Knoepfel (Ed.), *Survival and Sustainability as Challenges for Projects Zurich*. International Project Management Association.

van den Brink, J. (2009). Duurzaam projectmanagement: verder kijken dan je project lang is [Sustainable project management: looking beyond the project]. *Projectie, 4*.

van Marrewijk, M. (2003). Concepts and definitions of CSR and corporate sustainability: Between agency and communion. *Journal of Business Ethics, 44*(2-3), 95–105. doi:10.1023/A:1023331212247

van Pelt, M. J. F. (1993). *Ecological Sustainability and Project Appraisal: Case Studies in Developing Countries*. Avebury.

Venkatraman, N., & Ramanujam, V. (1986). Measurement of business performance in strategy esearch: A comparison of approaches. *Academy of Management Review, 11*(4), 801–814. doi:10.2307/258398

Verrier, B., Rose, B., & Caillaud, E. (2016). Lean and Green strategy: The Lean and Green House and maturity deployment model. *Journal of Cleaner Production, 116*, 150–156. doi:10.1016/j.jclepro.2015.12.022

Vezzoli, C., & Sciama, D. (2006). Life cycle design: from general methods to product type specific guidelines and checklists: a method adopted to develop a set of guidelines/checklist handbook for the eco-efficient design of NECTA vending machines. *Journal of Cleaner Production, 14*(15–16), 1319–1325. doi:10.1016/j.jclepro.2005.11.011

Voinov, A. (2007). Understanding and communicating sustainability: Global versus regional perspectives. *Environment, Development and Sustainability, 10*(4), 487–501. doi:10.100710668-006-9076-x

WCED. (1987). Our Common Future. Oxford University Press.

Weninger, C., Huemann, M., Oliveira, J. C., Barros Filho, L. F. M., & Weitlaner, E. (2013). Experimenting with project stakeholder analysis: a case study. In A. J. G. Silvius & J. Thap (Eds.), *Sustainability Integration for Effective Project Management*. IGI Global. doi:10.4018/978-1-4666-4177-8.ch023

White, D., & Fortune, J. (2002). Current practice in project management— An empirical study. *International Journal of Project Management, 20*(1), 1–11. doi:10.1016/S0263-7863(00)00029-6

Wimberley, R. C. (1993). Policy perspectives on social, agricultural, and rural sustainability. *Rural. Sociol., 58*(1), 1–29. doi:10.1111/j.1549-0831.1993.tb00480.x

Winnall, J.-L. (2013). Social Sustainability to Social Benefit: Creating Positive Outcomes through a Social Risk-based approach. In A. J. G. Silvius & J. Tharp (Eds.), *Sustainability Integration for Effective Project Management* (pp. 95–105). IGI Global Publishing. doi:10.4018/978-1-4666-4177-8.ch006

Wit, A. (1988). Measurement of Project Success. *International Journal of Project Management, 6*(3), 164–170. doi:10.1016/0263-7863(88)90043-9

Wood, M., Mathieux, F., Brissaud, D., & Evrard, D. (2010). Results of the first adapted design for sustainability project in a South Pacific small island developing state: Fiji. *Journal of Cleaner Production, 18*(18), 1775–1786. doi:10.1016/j.jclepro.2010.07.027

Yunus, R., & Yang, J. (2014). Improving ecological performance of industrialized building systems in Malaysia. *Construction Management and Economics, 32*(1/2), 183–195. doi:10.1080/01446193.2013.825373

Zeng, S. X., Ma, H. Y., Lin, H., Zeng, R. C., & Tam, V. W. Y. (2015). Social responsibility of major infrastructure projects in China. *International Journal of Project Management, 33*(3), 537–548. doi:10.1016/j.ijproman.2014.07.007

Zhang, X., Wu, Y., Skitmore, M., & Jiang, S. (2015). Sustainable infrastructure projects in balancing urban–rural development: towards the goal of efficiency and equity. *J. Clean. Prod., 107*, 445–454. .2014.09.068 doi:10.1016/j.jclepro

Zhong, Y., & Wu, P. (2015). Economic sustainability, environmental sustainability and constructability indicators related to concrete- and steel-projects. *Journal of Cleaner Production, 108*, 748–756. doi:10.1016/j.jclepro.2015.05.095

Zuo, K., Potangaroa, R., Wilkinson, S., & Rotimi, J. O. (2009). A project management prospective in achieving a sustainable supply chain for timber procurement in Banda Aceh, Indonesia. *International Journal of Managing Projects in Business, 2*(3), 386–400. doi:10.1108/17538370910971045

Zwikael, O., & Ahn, M. (2011). The Effectiveness of Risk Management: An Analysis of Project Risk Planning across Industries and Countries. *Risk Analysis, 31*(1), 25–37. doi:10.1111/j.1539-6924.2010.01470.x PMID:20723146

Chapter 12
Sustainable Technologies Development:
An Analysis of the Brazilian Green Patents Program

Luan Carlos Santos Silva
https://orcid.org/0000-0002-8846-2511
Federal University of Grande Dourados (UFGD), Brazil

Carla Schwengber Ten Caten
Federal University of Rio Grande do Sul (UFGRS), Brazil

ABSTRACT

The objective of this study was to describe the Brazilian green patents pilot program from the National Institute of Industrial Property (INPI) by mapping the national scenario in concerning to the technological development of green patents. The methodology consisted in prospecting the green patents deposit per the technological sector, highlighting that only documents deposited in Brazil were considered. Besides that, the solid waste management area has greater documental recording showing a propensity to develop more technologies; therefore, the Brazilian government should create public policies aimed at strengthening these technologies, thereby stimulating and developing them in both universities and industries, granting governmental incentives for the manufacturing of products and processes provision, by creating specific lines of financing through development agencies.

INTRODUCTION

Green technologies have taken a very important position in the development of global sustainability. In accordance with several countries' recognition of the importance of developing these technologies to combat global climate change, governments have begun to recognize the relevance of patent-granting procedures as a mechanism to stimulate green technologies.

DOI: 10.4018/978-1-6684-6366-6.ch012

The innovation model based on the government-university-industry relationship is of fundamental importance for the development of green technologies. Through the interaction of these three actors, it is possible to create a system of sustainable and lasting innovation in the era of the knowledge economy.

Green technologies are important for promoting economic growth, as they create new opportunities for business and employment. They also help to reduce energy costs and improve energy efficiency, which can have a significant impact on the global economy. In addition, green technologies can promote social inclusion, as they can help to provide access to clean energy and improve the quality of life for people living in poverty.

The development of green technologies requires significant investment and collaboration between the public and private sectors. Governments have an important role to play in supporting the development of green technologies, by providing funding and incentives for research and development. The private sector also has a key role to play, by investing in research and development, and by bringing new green technologies to market.

There is a growing body of research on the importance of the UN Sustainable Development Goals for the development of green technologies. For example, a study by Boulanger et al. (2020) found that the SDGs are driving innovation in the green technology sector and that companies that align their business strategies with the SDGs are more likely to succeed. Another study by Atadana et al. (2020) found that the SDGs provide a framework for the development of green technologies and that companies that focus on the SDGs are more likely to be successful in the long term.

The United Nations Sustainable Development Goals (SDGs) are a set of 17 global objectives that aim to promote economic growth, social inclusion, and environmental sustainability. One of the key areas addressed by the SDGs is the development of green technologies. Green technologies are technologies that are environmentally friendly and sustainable, and they play a crucial role in addressing the challenges of climate change and environmental degradation. This paper explores the importance of the UN Sustainable Development Goals for the development of green technologies.

The SDGs aim to create a sustainable future for the planet, and green technologies are an important part of this vision. Green technologies are technologies that use renewable energy sources, reduce greenhouse gas emissions, and minimize the use of non-renewable resources. These technologies play a crucial role in addressing the challenges of climate change and environmental degradation. They are essential for achieving the SDGs, particularly those related to environmental sustainability, such as SDG 7 (Affordable and Clean Energy), SDG 11 (Sustainable Cities and Communities), and SDG 13 (Climate Action).

The UN Sustainable Development Goals are essential for the development of green technologies. Green technologies play a crucial role in addressing the challenges of climate change and environmental degradation, and they are important for promoting economic growth and social inclusion. The development of green technologies requires significant investment and collaboration between the public and private sectors, and the SDGs provide a framework for this collaboration. Governments and the private sector have an important role to play in supporting the development of green technologies, and the research by Boulanger et al. and Atadana et al. supports this assertion.

In 2009, the national patent offices in Japan, Israel, South Korea, the United Kingdom, the United States, Australia, and Canada created pilot programs to expedite the examination of patent applications for green technologies, initially focused on specific areas that aim to reduce the impact of climate change by emitting less or removing CO2 from the atmosphere. In addition to following these principles, the inventions should be related to waste management, alternative energy, agriculture, transportation, and energy conservation.

In Brazil, this pilot program started in 2012 to accelerate the examination of patent applications related to green technologies, aiming to reduce the time taken for patent examination by two years and contribute to the fight against global climate change. Currently, the Brazilian program is in its third phase, which covers the time period from April 2014 to April 2016.

The objective of this research was to describe the green patent program from the Brazilian National Institute of Industrial Property (INPI) by mapping the Brazilian technological development scenario regarding green patents. The methodology consisted of defining the monitoring period, the databases used, and the technology sector. The period established for document searching was from January 2011 to September 2015, considering only documents deposited in Brazil related to phases I to III of the pilot program.

HISTORICAL CONTEXTUALIZATION OF GREEN TECHNOLOGIES

The first discussions on sustainable green technologies began in 1972 at the United Nations Conference on the Human Environment held in Stockholm, Sweden, a global event discussing environmental sustainability issues stimulated by economic growth and industrial pollution. During the same period, the United Nations Environment Program was developed.

Nine years later, Brazil established a National Environmental Policy. This was the first important environmental landmark in the country, aiming to preserve, advance, and recover environmental quality to guarantee conditions for socioeconomic development, the interests of national security, and the protection of human dignity, as established in Article 4, item IV, of Law 6.938/81, with the objective of developing national research and technologies oriented to the rational use of environmental resources.

Six years later, in 1987, the Brundtland Report discussed sustainable development issues for the long-term protection of the environment, taking into account the economic development of world nations. The document, entitled "Our Common Future," was prepared by the World Commission on Environment and Development.

In 1992, for the first time, Brazil hosted an important event that gave continuity to previous discussions about the environment and green technologies. Rio de Janeiro hosted the United Nations Rio Conference, known as ECO 92. An action plan for sustainable development was discussed, which established strategies and a program of unified measures set to eliminate and reverse the developmental effects in all nations. This plan, known as Agenda 21, covered economic, social, and cultural issues regarding environmental protection and had, at the first moment, 150 signatory countries.

As a result of this series of events on environmental sustainability, the "Kyoto Protocol" was signed in 1997, consisting of an international treaty with more stringent commitments aimed at reducing greenhouse gas emissions. However, the protocol entered into force only in 2005.

Three years later, in 2008, there was pressure from the World Intellectual Property Organization (WIPO) by the UN General Secretary on the environment, technology, and patent issues. In the following year, 2009, a global framework for the patent system development with regard to green technologies was started. Countries such as Japan, Israel, South Korea, the United Kingdom, the United States, Australia, and Canada created their first pilot programs for granting green patents. The objective was to identify technologies and prioritize the process of patent analysis and granting, reducing the analysis time to up to two years.

In 2010, following the initiative of these countries to prioritize green patents, WIPO developed a virtual tool linked to the International Patent Classification System known as the WIPO Green Inventory. The tool has two clear objectives: (i) to facilitate the search and identification of green technologies, and (ii) to cooperate between universities and industries to invest in green technologies research and development resources.

Only four years later, in 2012, Brazil created its first green patents pilot program, with a renewal in 2013, advancing to its second phase, and in April 2014, the third phase of the program came into effect.

After this series of events, discussions, treatises, and programs that have established a global framework on sustainability and green technologies, many researchers have begun to understand and research the field of green technologies in the patent system. Green patents are essential to driving sustainable development innovation and effectiveness (Nitta, 2003; Nitta, 2005a; Nitta, 2005b; Nitta, 2005c). Thus, the granting of green technologies through the patent system will allow future generations to achieve more sustainable development.

With this opening for discussions on green technologies, in both national and international literature, a diversity of nomenclatures on this subject emerged, which are better explained in Table 1.

Table 1. Terminologies related to green technologies

Terminology	Authors
Advanced natural technologies	(Olson, 1991)
Environmental-friendly technologies	(Olson, 1991)
Clean technologies	(Mazon, 1992)
Cleaner technologies	(ECO, 1992)
Environmentally sensitive technologies	(Mazon, 1992)
Environmentally-friendly solutions	(Martinsons et all, 1997)
Environmental health technological innovations	(Barbieri, 1997)
Greener technologies	(Conway e Steward, 1998)
Technologies not aggressive to the environment	(Donaire, 1999)
Alternative environmental technologies	(Kolar, 2000)
Ecotechnology	(Smith, 2001)
Environmentally interesting technologies	(Unep, 2002)
Environmentally sustainable technologies	(Hall e Vrendenburg, 2003)
Environmentally friendly technologies	(Barbieri, 2004)
Environmentally beneficial technologies	(Jaffe et all., 2005)
Green technologies	(Kivimaa e Mickwitz, 2006)

Source: The authors.

The importance of green technologies has been widely acknowledged by researchers and scholars in the last few years. These technologies are essential for achieving Sustainable Development Goals, reducing greenhouse gas emissions, and mitigating the effects of climate change.

One of the key authors in this field is E. Van der Voet, who has published several papers on the importance of green technologies. In a 2019 paper, Van der Voet and his colleagues argued that green technologies are crucial for achieving a sustainable future and that a transition to green technologies is both necessary and feasible. They also emphasized the need for collaboration between the public and private sectors to accelerate the development and adoption of green technologies.

Another author who has written on green technologies is F. Boccia. In a 2018 paper, Boccia argued that green technologies are essential for reducing greenhouse gas emissions and mitigating the effects of climate change. Boccia also emphasized the importance of policy interventions to support the development and adoption of green technologies and highlighted the role of public-private partnerships in this process.

An author who has contributed to the literature on green technologies is J. Schipper. In a 2017 paper, Schipper emphasized the need for a holistic approach to the development of green technologies, which takes into account not only the technical aspects but also the social, economic, and political dimensions of the transition to a green economy. Schipper also highlighted the importance of international collaboration and cooperation to support the development and diffusion of green technologies.

For Roy et al. (2020), green technologies are critical for achieving Sustainable Development Goals, and a transition to a green economy is essential for addressing the challenges of climate change and environmental degradation. Roy et al. also emphasized the need for policy interventions to support the development and adoption of green technologies and highlighted the role of innovation and entrepreneurship in this process.

The importance of green technologies has been widely acknowledged by researchers and scholars in the last few years. Authors such as Van der Voet, Boccia, Schipper, and Roy and their respective publications have highlighted the critical role of green technologies in achieving a sustainable future and mitigating the effects of climate change. These authors have emphasized the need for collaboration between the public and private sectors, policy interventions, and international cooperation to accelerate the development and adoption of green technologies. The literature on green technologies provides a roadmap for the transition to a more sustainable and resilient future.

BRAZILIAN GREEN PATENTS PROGRAM

The Green Patents pilot program aims to accelerate the patent examination process and identify strategic green technologies in Brazil (INPI, 2012).

The first phase of the program began on April 17, 2012, and lasted for one year, granting 500 applications restricted to applications filed in the INPI since January 2011. The program was extended for another year until April 16, 2015. This period was the second phase of the program, restricted to applications deposited in the INPI and countries signatories of the CUP (Convention of the Paris Union), including utility models.

The third phase of the program started on 04/17/2014 and was scheduled to end by 04/16/2016. At this stage, the participation of applications filed via PCT (Patent Cooperation Treaty) was extended.

Applications submitted and granted in the program have a decision process that takes an average of two years, but some concession requests are granted in less than six months. Currently, 56 patents have been granted through the program. Figure 1 illustrates a flowchart regarding the Brazilian green patents program, which has a cleaner flow, and aims to reduce internal bureaucracy.

Figure 1. Flowchart of the Brazilian green patents program
Source: INPI (2017).

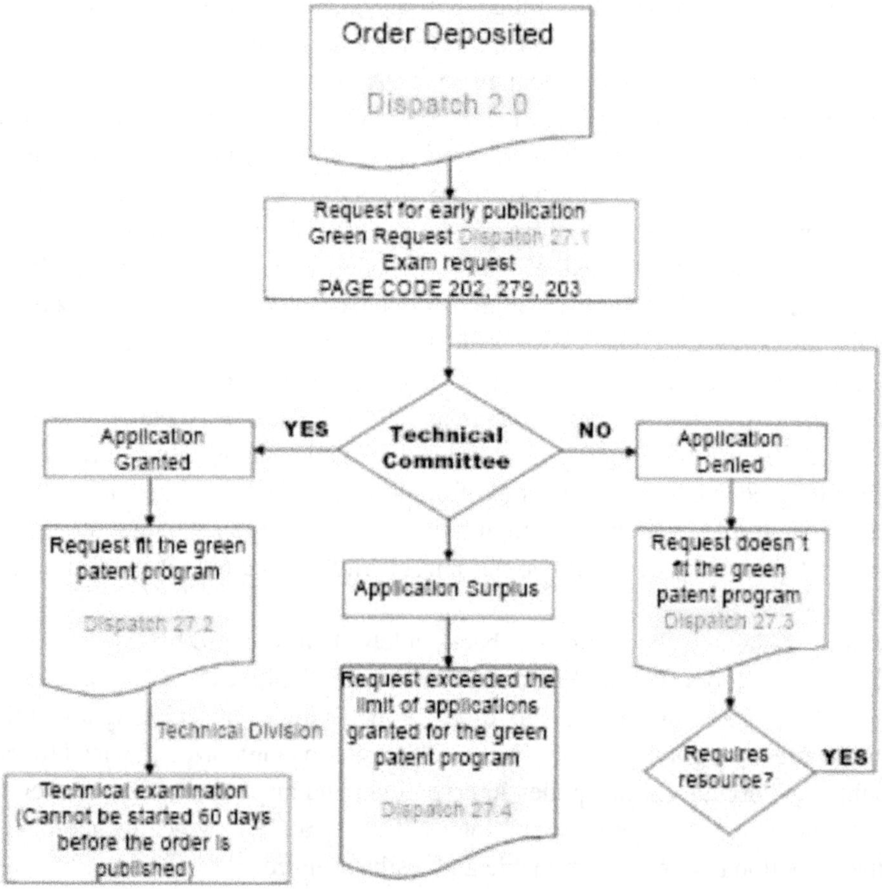

METHODOLOGICAL PROCEDURES

To establish the prospecting of patent documents, it was necessary to define the monitoring period, the databases to be used, and the technology sector. Data collection was carried out in the Directorate of Contracts, Geographical Indications, and Records (DICIG) from the National Institute of Industrial Property (INPI), Brazil. The period established for document searching was from January 2011 to September 2015, concerning phases I and III of the pilot program.

In the first step, we sought to analyze all the deposits designated for the green patents program. In the second stage, we sought to analyze only the patents granted, and thus the number of documents found was 56.

After carefully reading the titles and abstracts of patent documents foreseen by the searching strategy in the second stage, we identified only those referring to the green patent program applied to the following areas: waste management, alternative energy, agriculture, transport, and energy conservation. Subsequently, the data was detailed analyzed and tabulated using the Microsoft Excel 2007 program to generate the graphs. Additionally, the Wordle online software was used for the elaboration of word clouds.

RESULTS AND DISCUSSION

1. Distribution of Patent Applications

Figure 2 shows the distribution of the number of requests made in the Brazilian Green Patents Program regarding the areas of alternative energy, transport, waste management, energy conservation, and agriculture. The data analyzed correspond to the years 2013 and 2015 related to Phases I to III of the program.

Figure 2. Number of green patent applications
Source: The authors.

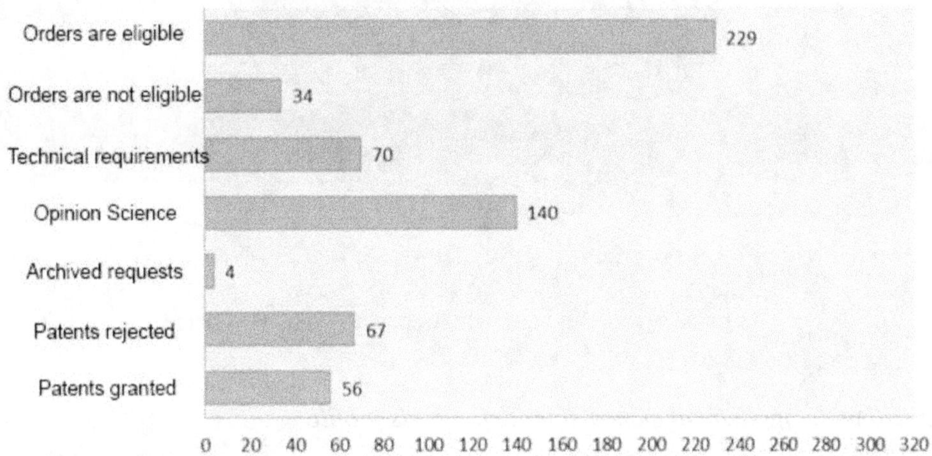

Until September 2015, a total of 56 patents had been granted as part of the program, which is a small number compared to the average number of patents granted in Brazil over the last ten years, corresponding to 3,000 patents per year. A total of 70 patent applications failed due to technical requirements, 229 applications were valid, and only 34 applications did not fit the program. Additionally, 4 applications have been archived.

2. Distribution of Depositor's Profile

Figure 3 shows the distribution of green patents deposited per country. As can be seen, Brazil has the highest percentage of documents deposited, accounting for 85%, followed by the United States with 10%, and the other countries accounted for 5% of deposits.

Figure 4 shows the distribution of patent documents per depositor's profile. The largest percentage comes from inventors, accounting for 45%. This number can be explained by the fact that independent inventors do not have the research and development infrastructure that companies and universities have.

In second place are the companies, accounting for 38%. And in third place are the universities, which obtained the lowest percentage of 17%. This number takes into account the investments made in research and development by universities.

Figure 3. Distribution of countries depositing green patents
Source: The authors.

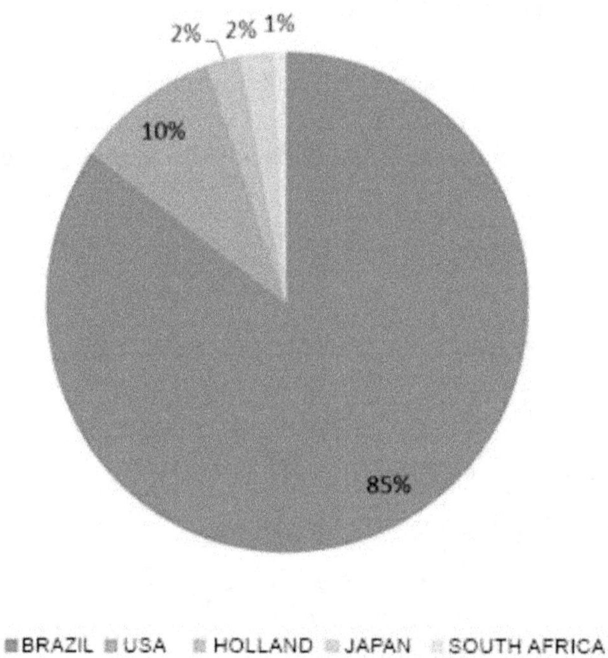

Figure 4. Distribution of patent documents per depositor's profile
Source: The authors.

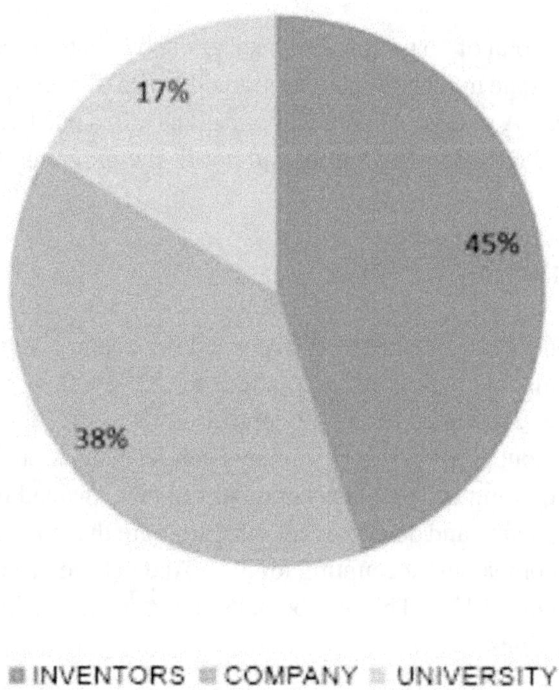

Figure 5 shows the distribution of patent documents granted per sector. The numbers within the circles represent the number of patent documents granted, and at the intersections are the documents deposited in partnership (co-ownership).

Figure 5. Distribution of documents granted per sector
Source: The authors.

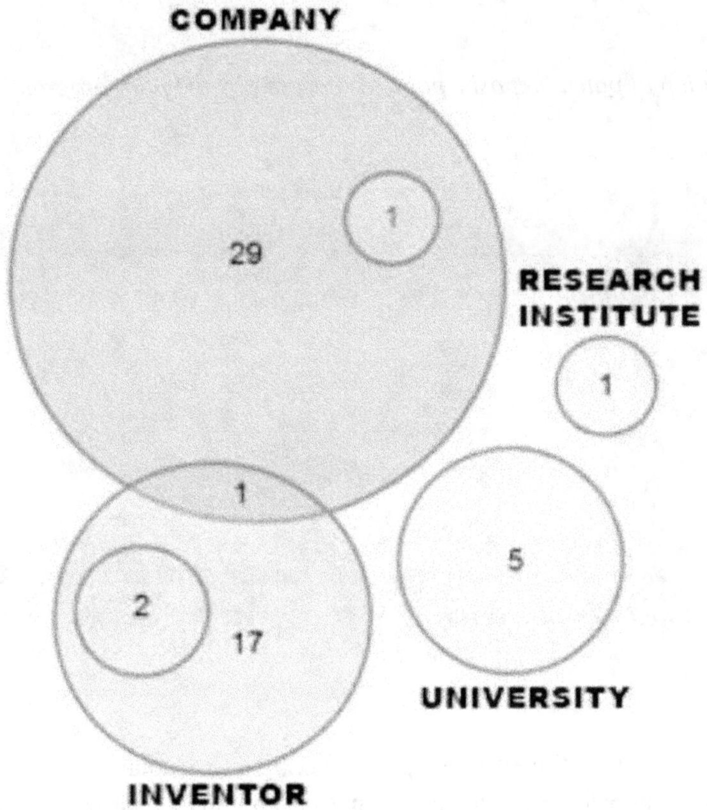

As can be seen, companies had a greater number of patents granted, followed by independent inventors and universities. Only one patent was granted to a research institute.

The number of patents developed in co-ownership is still small, with few partnerships occurring. One was between company-company, one between company-inventor, and two between the inventor-inventor. However, universities and research institutes did not establish any partnerships.

3. Distribution of Patent Deposits per Area and Time of Analysis

Figure 5 shows the distribution of documents deposited in the five areas covered by the green patents program. The solid waste management area had the highest percentage of contributions, accounting for 52.2% of deposits made. In 2012, the first patent document reviewed and granted (phase I of the program) belonged to the solid waste management area. The technology applied comprises a solid waste treatment process based on a gradient composed of two distinct thermal sources, patent number PI1104219.

The alternative energy and agriculture areas obtained similar positions, accounting for 25.4% each. However, the transport area had a low percentage of deposits, despite being one of the most polluting areas in Brazil. According to information obtained from the report of annual estimates of greenhouse gas emissions in Brazil, the transport area provided the highest rates of energy consumption over the last ten years (4.42% per year from 2002 to 2012). Additionally, CO2 emissions passed from 84 million tons in 1990 to 204 million tons in 2012 (Mobilize, 2014).

Figure 6. Distribution of patent deposits per area covered by the pilot program
Source: The authors.

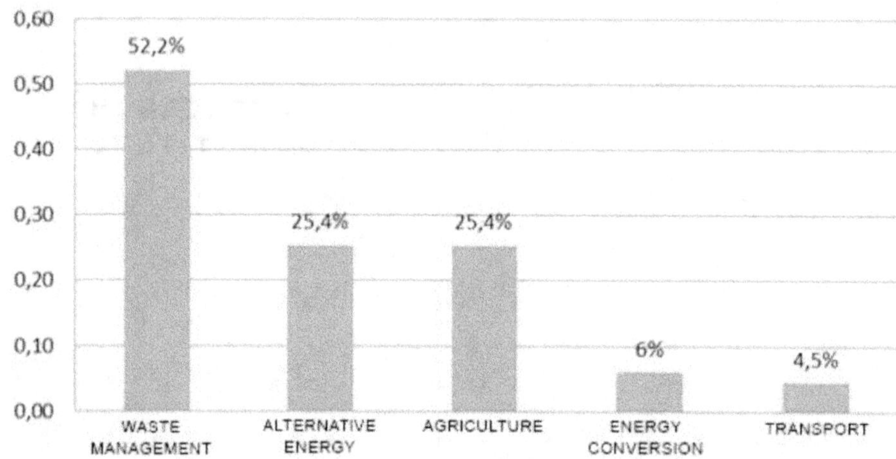

Figure 7 shows the distribution of patent grants, taking into account the analysis time, separated into six-month periods. Only patents granted between 2013 and 2015 (phases I to III of the program) were considered.

It is worth mentioning that the pilot program aims to significantly reduce the time required for patent grants. Currently, in Brazil, the average time between documental analysis and the patent grant is eight years.

Of the patents analyzed, four were granted within six months, and 92.6% of patents were granted within the expected period of 24 months. Only 7.4% of documents exceeded the expected analysis time.

CONCLUSION

The Brazilian green patents development scenario is in the early stage, with only 56 patents granted since the first phase of the program started in 2012. The lack of cooperation for the development of technologies in co-ownership among universities, industry, and inventors was identified.

Figure 7. Estimated time is taken for patent analysis
Source: The authors.

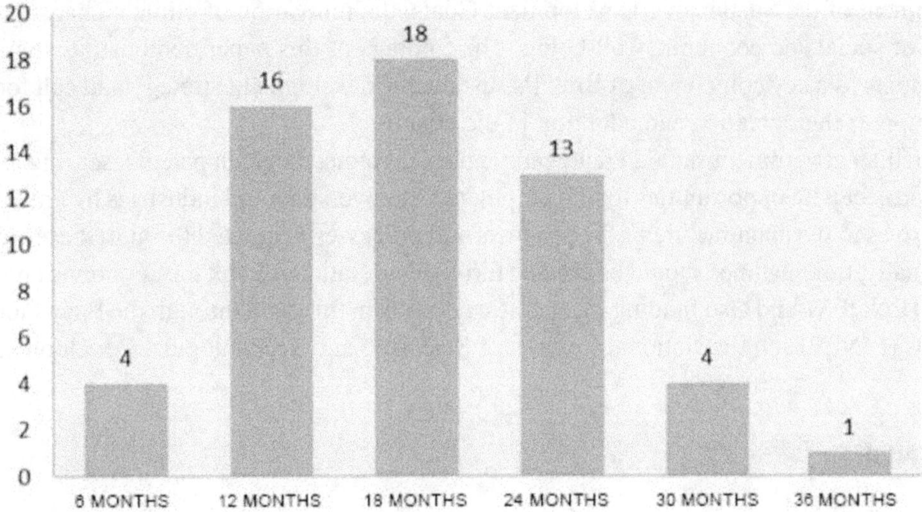

The solid waste management area has a high propensity to increase technologies due to the number of patents already registered.

The program created can benefit universities, industries, and society in general. Since a reduction in the documental analysis time and patent granting is its main objective, in some cases showing concessions made in less than six months, a shorter time in comparison to the Brazilian average time, which takes 16 months.

The INPI does not yet have a documental database exclusively designated for green patent analysis. Its creation will enable rapid access to the knowledge of these technologies by society, thus stimulating their licensing and encouraging innovation and development in the country.

Green patents are an essential tool for promoting innovation and sustainable development in Brazil. Although Brazil has made progress in recent years in promoting the development and adoption of green technologies, there is still significant room for growth in this area. The implementation of policies that incentivize the creation and use of green patents can provide a powerful tool for promoting sustainable development in Brazil. By encouraging research and development in green technologies and providing incentives for their adoption, green patents can contribute to the reduction of greenhouse gas emissions, the preservation of natural resources, and the improvement of social and economic well-being.

Brazil has significant potential to be a leader in green technologies, due to its abundant natural resources and its commitment to sustainable development. The implementation of policies that encourage the creation and use of green patents can help to realize this potential, by promoting the development and adoption of innovative technologies that contribute to a more sustainable and resilient future. As highlighted in this paper, the implementation of green patent policies in Brazil can also have important economic and social benefits, including the creation of new jobs and the improvement of access to sustainable products and services.

Overall, green patents represent a powerful tool for promoting sustainable development in Brazil. The implementation of policies that incentivize the creation and use of green patents can contribute to the achievement of the Sustainable Development Goals, the mitigation of climate change, and the improvement of social and economic well-being. The findings of this paper demonstrate the importance of green patents as a key component of Brazil's sustainable development strategy and call for continued efforts to support their creation and adoption in the country.

The Brazilian government should create public policies aimed at green patents, serving as a basis for a stimulus project with opportunities for development for universities and industries by granting government incentives to the manufacture of the products and processes generated through these technologies.

Additionally, funding lines should be created through the National Bank for Economic and Social Development (BNDES), and also funding for specific research in this field through the Financier of Studies and Projects (FINEP) and the National Council of Scientific and Technological Development (CNPq).

REFERENCES

Atadana, L., Borbor, S., & Rabbani, M. (2020). Sustainable Development Goals and Clean Energy: A Review. *International Journal of Energy Economics and Policy, 10*(5), 445–452.

Barbieri, J. C. (1997). Políticas públicas indutoras de inovações tecnológicas ambientalmente saudáveis nas empresas. *Revista Brasileira de Administração Pública, 31*(2), 135–152.

Barbieri, J. C. (2004). *Gestão ambiental empresarial: conceitos, modelos e instrumentos*. Saraiva.

Boccia, F. (2018). The importance of green technologies in the fight against climate change. *Energy Policy, 115*, 634–639.

Boulanger, P., Girod, B., Houdret, A., & Lashermes, B. (2020). Aligning Business Strategies with the Sustainable Development Goals: Insights from the Green Technology Sector. *Sustainability, 12*(14), 5844.

Brasil. (1981). *Política Nacional do Meio Ambiente, Lei nº 6.938, de 31 de agosto de 1981*. Retrieved June 25, 2015, from: http://www.planalto.gov.br/ccivil_03/Leis/L6938.htm

Conway, S., & Steward, F. (1998). Networks and interfaces in environmental innovation: A comparative study in the UK and Germany. *The Journal of High Technology Management Research, 9*(2), 239–253. doi:10.1016/S1047-8310(98)90006-2

Donaire, D. (1999). *Gestão ambiental na empresa*. Atlas.

ECO. (1992). Cleaner technologies. *Environmental Conservation, 19*(4), 361-369.

Hall, J. K., & Vrendenburg, H. (2003). Environmentally sustainable technologies. *Journal of Cleaner Production, 11*(2), 139–149.

Instituto Nacional de Propriedade Industrial (INPI). (n.d.). Retrieved September 19, 2015, from www.inpi.gov.br

International Patent Classification (IPC). (2006). *WIPO* (8th ed.). IPC.

Jaffe, A. B., Newell, R. G., & Stavins, R. N. (2005). A tale of two market failures: Technology and environmental policy. *Ecological Economics*, *54*(2-3), 164–174. doi:10.1016/j.ecolecon.2004.12.027

Kivimaa, P., & Mickwitz, P. (2006). The challenge of greening technologies: Environmental policy integration in Finnish technology policies. *Research Policy*, *35*(5), 729–744. doi:10.1016/j.respol.2006.03.006

Kolar, J. L. (2000). Alternative energy technologies. *Environmental Quality Management Journal*, *10*(3), 45–53.

Martinsons, M. G., So, S. K. K., Tin, C., & Wong, D. (1997). Hong Kong and China: Emerging markets for environmental products and technologies. *Long Range Planning*, *30*(2), 277–290. doi:10.1016/S0024-6301(96)00119-7

Mazon, R. (1992). Em direção a um novo paradigma de gestão ambiental: Tecnologias limpas ou prevenção da poluição. *Revista de Administração de Empresas*, *32*(2), 78–98. doi:10.1590/S0034-75901992000200009

Mobilize. (2014). *No Brasil, transporte é o vilão das emissões de CO2*. Retrieved from http://www.mobilize.org.br/noticias/7345/transporte-e-o-vilao-das-emissoes-de-co2.html

Nitta, I. (2003, Winter). Green Patent System: An invention for inventions toward sustainable development. *GIP Progress*.

Nitta, I. (2005a). Proposal for a green patent system: Implications for sustainable development and climate change. *Sustainable Development Law and Policy*, *5*, 61–65.

Nitta, I. (2005b). Green Intellectual Property: A tool for greening a society. *Ecological Economics*.

Nitta, I. (2005c). Patents and essential medicines: An application of the green intellectual property project, on the Submission site of the Commission of Intellectual Property Rights, Innovation and Public Health. WHO.

Olson, R. L. (1991). The greening of high tech. *The Futurist*, *25*(3), 28–34.

Organização das Nações Unidas (ONU). (1992). *Agenda 21*. Rio de Janeiro: Author.

Roy, R., Chattopadhyay, S., Mitra, S. K., & Ghosh, S. (2020). Green technologies for sustainable development: A review. *Journal of Cleaner Production*, *255*, 120247.

Schipper, J. (2017). The importance of a holistic approach to the development of green technologies. *Technological Forecasting and Social Change*, *124*, 215–222.

Smith, V. (2001). Ecotechnology. *Environmental Engineering Science*, *18*(5), 295–305.

UNEP. (2002). *Environmentally interesting technologies*. United Nations Environment Programme. Retrieved from https://www.unep.org/

Van der Voet, E., Van Oers, L. F., & De Haan, P. (2019). The importance of green technologies for a sustainable future. *Sustainability*, *11*(19), 5351.

Chapter 13
Green Logistics and Transport Processes:
Mitigating the Sixth Extinction

Enock Gideon Musau
Department of Transport and Supply Chain Management, University of Johannesburg, South Africa

ABSTRACT

The sixth mass extinction's emerging threat is sad, considering that humans have caused it. But, the growing world population leaves humans with no option in their desire to survive. This situation where humans need to survive and yet protect the habitat and many other similar scenarios complicate the process of seeking solutions to the looming disaster and extinction. Yet, with collaborations, humans can develop novel and innovative ideas to save them from the threats posed by the sixth mass extinction. This chapter examines how supply chain partners can collaborate to ensure green-sensitive and eco-innovative supply chain practices to help prevent this extinction.

INTRODUCTION

Supply chains plays vital roles in the development of national economies by expediting functions such as operations, distribution, customer service, marketing, transportation among others. Nevertheless, with these obligations, supply chains system is causing harm to the environmental protection. The transport supply chain, in particular, is threatening the environment through the emission of greenhouse gases, noise pollution, water pollution and pollutants due to exponential growth of commuter activities (Verma et al., 2016). Therefore, the emerging and growth of motorized mobility has resulted to negative environmental impacts (Zhao & Zhang, 2018).

The global supply chain has emerged as an effective way of investment, production, and trade in national economies. Through the supply chain, nations, particularly developing ones, have realized social and economic development by creating employment opportunities (Abbasi, 2017). Like other countries

DOI: 10.4018/978-1-6684-6366-6.ch013

globally, Kenya is staring at the reality of the sixth mass extinction, with scientific knowledge showing a decline in biodiversity and ecosystem services. According to the Intergovernmental Platform on Biodiversity and Ecosystem Services (IPBES), close to one million animals and plant species face the threat of extinction, weakening livelihoods, health, food security, and reduction in the quality of life (IPBES, 2019). One sector that contributes significantly to the degradation and loss of biodiversity is manufacturing which defines modern economies. Statistics show that the manufacturing industry accounts for 23% of employment worldwide and close to 15% of the global Gross Domestic Product (GDP) (World Bank, 2018). Yet, research also points to manufacturing as one of the main sources of greenhouse gas emissions (Golasa et al., 2021).

Exploring and stopping the threat of the sixth extinction requires eco-innovations upon which to prototype collective green creativity. One such eco-innovation is green manufacturing. Green manufacturing targets operations that recycle and reuse materials, reduce waste and pollution, moderate emissions, and use fewer resources. Therefore, green manufacturing is the establishment of environmentally friendly and renewable product processes in manufacturing (Dornfeld et al., 2013). Such a process brings positive change and limits the negative impact of manufacturing. The sixth mass extinction is emerging threat is sad, considering that humans have caused it.But,the growing world population leaves humans with no option in their desire to survive. This situation where humans need to survive and yet protect the habitat and many other similar scenarios complicate the process of seeking solutions to the looming disaster and extinction. Yet, with collaborations, humans can develop novel and innovative ideas to save them from the threats posed by the sixth mass extinction. This chapter endeavors to examine how supply chain partners can collaborate to ensure green-sensitive and eco-innovative supply chain practices that can help prevent this extinction.

BACKGROUND

Extinction is a part of life and involves humans, animals, and plants. However, in recent times, the human population has grown exponentially, putting a lot of pressure on the planet. This pressure has caused changes that threaten humankind and the natural world. Evidence indicates that human impacts are exacerbating the extinction of species by at least 1,000 times faster than ever before and that since 1970 the population of wild animals has decreased by more than two-thirds. According to scientists, what is happening today amounts to the sixth mass extinction following the fifth mass extinction that saw dinosaurs wiped out. According to new research published in the Journal Proceedings of the National Academy of Sciences of the United States of America (PNAS), this sixth mass extinction promises to be among the most severe environmental threats to the persistence of civilization. Yet, humans have caused this extinction, whose impact is more immediate (Hu et al., 2021).

It is argued that due to the increasing population, people continue to strain the planet in several ways. For instance, people are destroying the habitat by infringing on the natural environment; polluting the environment due to increased household, agricultural, and industrial waste; depleting the soil by intensifying agriculture; endangering aquatic species by chocking them with plastic waste; increasing climate emissions such as greenhouse gases including carbon dioxide and methane, putting the planet on the verge of a climate crisis; and relentless consumption of resources such as timber, oil, and minerals leading to the destruction of natural habitats globally.

A close examination of these human activities associated with habitat destruction identifies the supply chain that focuses on the core activities within organizations required to convert raw materials or parts to finished products or services as a significant player in the sixth mass extinction. Therefore, this chapter contributes to the efforts to mitigate human impacts and stop this extinction from a supply chain perspective. The chapter focuses on green logistics and transportation processes as creative eco-innovative supply chain practices that can be employed in addressing the sixth extinction that is already here with us

Concerns

Humankind still grapples with the reality of the destruction of the sixth extinction. Yet the solution to this sixth mass extinction resides within the humans themselves. According to Cowie et al. (2022), humans are the only species able to manipulate the biosphere on a large scale. Therefore, people can collaborate in setting things right proactively. Although supply chains play a critical role in national economies through functions such as product development, marketing, operations, distribution, finance, and customer service, supply chains remain of concern in the quest to mitigate the sixth extinction. With the growing awareness of the need to keep the earth sustainable, the supply chain has become a topic of interest. Supply chains contribute to environmental degradation in several ways: increased toxic waste, loss of biodiversity, water pollution, deforestation, hazardous air emissions, long-term damage to ecosystems, and generation of greenhouse gas emissions, all of which contribute to the threat of the sixth mass extinction. Never before has there been an urgent need for collaboration among supply chain stakeholders to identify a novel and innovative solution to such environmental degradation as there is today.

One supply chain concept emerging in modern society associated with lessening environmental degradation is green logistics which subsumes modern transportation process technology (Larina, 2021). The concern is that cargo turnover is increasing in modern society, almost constantly raising environmental safety issues. For instance, research demonstrates that 25% of CO_2 emissions and 23% of total energy consumption could be attributed to transport systems (Shabani & Shahnazi, 2019; Ulewicz et al., 2021). Such high CO_2 emissions and energy consumption proportions have resulted in vulnerability due to high fuel prices and damage to non-renewable fossil fuels. Therefore, green logistics is a concept often associated with sustainable development that targets the environment, the economy, and society. The chapter provides evidence to show that green logistics and transportation processes could be the panacea to mitigating the sixth extinction.

The Missions

The chapter is guided by the mission to comprehensively explore the concept of supply chain management from an environmental sustainability perspective. Consequently, the chapter examines the four supply chain logistics and their role in the rising levels of environmental degradation and biodiversity destruction. First, the four supply chain logistics components are identified, and each component is assessed for its contribution to environmental degradation and green practices under the component. Second, the chapter will enumerate supply chains' environmental impacts, including toxic waste, biodiversity loss, water pollution, deforestation, hazardous air emissions, long-term damage to ecosystems, and greenhouse emissions. It will further examine how these impacts are potential causes of the sixth mass extinction. Finally, the chapter demonstrates how the supply chain could be the key to eliminate the sixth extinction through greening.

The Supply Chain and Environmental Degradation

Although consensus is yet to be reached on the precise definition of the supply chain, it is perceived to subsume businesses and activities employed to design, produce, deliver, and consume products and services (Hugo, 2018). In essence, a supply chain brings together all parties directly or indirectly associated with satisfying customer requests (Chopra & Meindl, 2010). These parties include manufacturers, suppliers, customers, retailers, warehouses, and transporters. Therefore, supply chain management (SCM), perceived from a goods flow perspective, relates to avenues used to integrate these supply chain players to facilitate service level requirements at minimum system-wide costs (Simchi-Levi et al., 2008). The increasingly competitive business environment has made customers gain the upper hand. They not only expect goods to be delivered on time but also seek lesser prices. Therefore, manufacturers should be keen on SCM logistics components of procurement, production, transportation, and warehousing if they have to remain competitive. Yet, evidence shows that these logistical components are the major sources of environmental degradation and pollution in the supply chain (Golinska & Romano, 2012).

Procurement Logistics

Procurement logistics encompasses sourcing and procurement. These are two distinct functions despite being confused as synonymous among scholars. The two functions are complementary in their supply chain roles. Sourcing is a process through which suppliers are vetted, selected, and managed. Therefore it involves researching, strategizing, outlining quantity and quality metrics, and contracting suppliers who meet the quantity and quality criteria (Kandil et al., 2020). Through sourcing, organizations' supply chains are maintained and guarantee accessibility to required tools for realizing objectives.

On the contrary, procurement is viewed as acquiring the services and goods needed by organizations. In essence, procurement leverages existing supply chains to sustain a steady flow of supplies and inputs in organizations. According to Pereira et al. (2014), procurement aims at achieving resilience in the supply chain. Whichever way one defines supply chain sourcing and procurement, evidence shows that sourcing and procurement teams can significantly impact environmental sustainability by using eco-friendly practices (Disclosure Insight Action, 2021). Sourcing and procurement are supply chain logistic functions that pose top environmental risks, including recklessness in sourcing raw materials. It is argued that sourcing and extracting raw materials are vital cogs in corporate social responsibility (Bellow-Bravo et al., 2022). Reckless sourcing can be an avenue for environmental degradation and gravitation towards the sixth mass extinction.

For instance, in recent years, products like palm oil have been linked to practices that destroy biodiversity and global ecosystems, such as deforestation (Van der Ven et al., 2018). Reckless sourcing does not create room for due diligence understanding of suppliers and the source of their products. This lack of room may lead to sourcing and procuring products that consume fossil fuels, resulting in greenhouse emissions. Moreover, such recklessness increases the chances of sourcing and procuring products/services that produce non-recyclable waste, lead to air pollution, or impact native flora and fauna (Saporovskaya et al., 2020).

Mitigating the looming sixth mass extinction requires green creativity and eco-innovation in supply chain procurement logistics. Profound changes are experienced in planet earth's essential life-sustaining functions due to trends in environmental degradation associated with human activity (Ripple et al., 2019; Steffen et al., 2015). It is high time for supply chain sourcing and procurement to rid the environment of products that are potential causes of worldwide growth in affluence and green gas emissions. Never before has there been an urgent drive for green procurement than now when the planet is in real danger of extinction.

Greening Procurement Logistics

Green procurement also referred to as green purchasing, is conceived as environmentally preferable purchasing (EPP) that seeks procurement of services and products with minimum effects on human health and the environment (Terman & Smith, 2018). According to Srinivas (2022), green procurement relates to purchasing eco-friendly products and eco-friendly services and setting environmental targets in supplier contracts. Research shows that the procurement process leveraged on the criteria of ecological preference reduces waste, and costs and, more importantly, mitigates environmental impacts (Yang et al., 2019).

Green procurement is anchored on principles and activities geared towards pollution prevention. Consequently, green procurement assesses product cost, quantity contract, environmental impact, and compliance with multilateral requirements, including the Rotterdam Convention and the Kyoto Protocol (Srinivas, 2022). Therefore, it is apparent that ethics, social regeneration, environmental, and sixth mass extinction concerns posed by supply chain sourcing and procurement can be addressed through green purchasing, that considers environmentally preferable criteria.

Several strategies are frequently used to bring greening in procurement logistics and optimize procurement, including procurement automation, inventory optimization, and sourcing from multiple suppliers and distribution sites. For instance, in addition to handling heavy data processing, automation software also allows insight into inventory levels, risk assessments, and material compliance to eliminate excess waste. Meanwhile, optimizing inventory and reducing carrying costs also eliminates obsolete inventory that can lead to the emission of greenhouse gases. Sourcing from multiple suppliers and regions eliminates materials that may not be eco-friendly.

The critical role that green procurement logistics plays in supply chain management has been demonstrated empirically. Galeazzo et al. (2021) showed that by moderating tourists' behaviour toward green purchasing, then green procurement impacted financial performance positively. Hazaea et al. (2022), on the other hand, analyzed studies that had previously examined green purchasing. They determined that green purchasing had a positive effect on sustainable development.

Production Logistics

Production logistics is a supply chain function that subsumes materials management, distribution in factories, product management, and shipping. Michlowicz (2013) states that production logistics is primarily a move to expand capabilities to make reliable deliveries with the least logistical and production cost. Production logistics, synonymously used with internal logistics, aims to reduce the production lead time.

Under production logistics, several activities are undertaken. First, procured parts and materials must be managed and distributed within the factory. Besides, products need to be managed by packaging and shipping to warehouses. These activities though necessary, are avenues for environmental degradation that may contribute to the biodiversity crisis and the threat of mass extinction being witnessed. For instance, Albert et al. (2021) have used the Abuja context in Nigeria to show that poor management of procured materials considerably impacts waste generation. As noted in this chapter, waste generation is responsible for the increased emission of greenhouse gases. Kulkarni et al. (2017) define materials management as the provision of the right materials in correct quantities on time to cut costs. Consequently, poor material management leads to excess waste generation, degrading the environment and raising production costs.

Another example of the negative environmental impacts of poor management of materials is highlighted in the findings by Kim et al. (2012) in the road construction perspective. Their findings indicated that poor management of onsight equipment gave room to greenhouse gas emissions. The distribution and resource use within the factory is an activity that accounts for a large proportion of environmental degradation.

It has been documented that the distribution of materials within the factory involves using raw materials that have high chances of emitting pollutants (Kolorzek et al., 2018). An example is the uptake of land for mining activities contributing to biodiversity loss, the production of mine tailings being a source of water acidification, and the combustion of fuel during mining generating greenhouse gases (Obasi & Akudinobi, 2020). Such impacts on the environment contribute negatively to ecosystems and human health. Mining and production of biotic substances such as natural rubber and pulp lead to diffuse pollution at the mining facilities, whose impacts on the environment continue even in the post-closure phase (Kumar et al., 2021, Mishra & Maiti, 2019).

The United Nations Environmental Programme (UNEP, 2019) observes that intrinsic features of the commodity under production, such as techniques, technologies, conditions for socioeconomic and environmental frameworks, and practices in place to manage production determine the intensity and type of impacts on the environment. UNEP (2019), for instance, points out that the impact of the mining facility on the water is dependent upon specific water requirements in processing and extraction of the mineral, local water availability, and level of water depuration and reuse at the facility.

Production logistics involving product packaging relates to producing product containers based on design and evaluation. Packaging is identified as a sure way of moving products from the point of source to the point of consumption (Meherishi et al., 2019). Product packaging is deemed a strategic marketing tool to introduce a product to consumers (Rundh, 2016). Product packaging is important in multiple ways that include securing lasting brand loyalty. Primarily, product packaging plays the role of protecting the product during shipment between manufacturers and retailers. For instance, Wikistrom et al. (2019) argue that the increasing spatial distance of consumers from farms requires packaging systems that guarantee the protection and storage of food.

Product packaging is also associated with the marketing function of displaying and promoting the product. For instance, product packaging, especially in the food industry, often displays vital nutritional and ingredient information. The essence is to arouse customer curiosity and satisfaction (Chukwuma et al., 2018). Besides playing the promotional role, product packaging also attracts buyers. This attraction is maximized through well-designed packages that appeal to consumers. A core purpose of product packaging is product differentiation. Shreay et al. (2016) found that canned tuna differentiated across quantity surcharges to cater to the heterogenic consumer tastes. Chou and Wang (2012) had earlier shown that differentiation of bottle packaging by properties of the commodity, bottle shape design, and label design enabled consumers to differentiate between them on hypermarkets shelves.

In as much as product packaging critically protects, promotes, and differentiates products, this logistic production function has contributed significantly to environmental pollution and biodiversity loss. Product packaging, especially food packaging, is not designed for reuse or recycling and is often disposed of (Bodamer, 2016). Indeed, in the US, materials from food and food packaging are estimated to be approximately half of the solid waste from the municipal (US Environmental Protection Agency, 2014). The various forms of packaging used in food consume resources like energy, chemicals, water, wood, minerals, fibers, and petroleum. Consequently, the production of the packages leads to emissions like heavy metals and particles, greenhouse gases, sludge with toxic contaminants, and wastewater (World Bank Group, 1998).

Glass is one form of packaging usually used in food packaging. Although glass provides the advantage of being nonporous and impermeable, that maintains the flavor of beverages and food, how it is manufactured is what concerns biodiversity. According to the World Bank Group (1998), fossil fuels like liquefied petroleum gas, heavy and light fuel oils, and natural gas are burned to melt the feedstock material. The combustion of these fuels results in emissions such as nitrogen oxides, sulphur oxides, and greenhouse gases. Meanwhile, Aluminum packaging is also common for foods. Yet, evidence shows that aluminum production requires a process that is energy-intensive and uses plenty of water (Roppenheimer, 2014). Therefore, the process leads to emissions that include sulphur dioxide, greenhouse gases, wastewater, and polyclic aromatic hydrocarbons that threaten biodiversity.

Paper and plastic packaging have commonly been used in food packaging. However, plastic packaging has been found to have a huge impact on the environment. The ocean conservancy clean-up in 2018 shows that plastic comprised 70 percent of the top ten items collected (Carter n.d.). Paper and paperboard production creates waste water and air emissions like carbon monoxide, nitrogen oxides, and sulfur dioxide. Meanwhile, over 70 percent of plastic products associated with plastic polymers package food and are sourced from fossil fuels. The manufacturing of plastics contributes significant amounts of greenhouse gases to the environment. This contribution of greenhouse gases adds to the contribution of other emissions such as sulfur hexafluoride, nitrous oxides, perfluorocarbons, and hydrofluorocarbons (Natural Resources Canada, 2018).

Plastic packaging remains a human activity in supply chain production logistics that contributes significantly to Anthropocene extinction. Plastics permeate all corners of the planet and threaten human, avian, and marine life. The problem of plastic pollution has been so acute in oceans leading to its declaration by the United Nations as a planetary crisis (Harrabin, 2017). Yet, plastic pollution has not only been experienced in oceans but also in soils, where it causes microplastics that have detrimental effects. Plastics of all forms clog waterways and float in oceans exposing harm to animals and birds. Indeed, evidence shows that a sample of birds, turtles, and fish from seafood markets revealed the presence of plastics in 100 percent of sea turtles, 59 percent of sea birds, and 25 percent of fish (Ocean Conservancy, 2018).

Shipping constitutes another critical function in production logistics. Shipping relates to cargo movement from source to warehouses and consumers using ships. According to Bichon (2014), shipping desires the optimization of resources. Therefore, it involves planning and selecting the most profitable routes, the volume of cargo across these routes, number of ships available for the routes, ports to dock in, transit days for round trips, and costs (operational, port, fuel, and human resources). Although shipping is important to supply chain sustainability and most supply chain stakeholders' economic growth, environmental protection for eco-innovative sustainability remains a big challenge. Svanberg et al. (2018) determined that shipping was among the main global contributors to Co_2 emissions, accounting for between 2% and 3%, with fears that it could hit 17% by 2050 if not checked.

Environmental impacts of shipping include air pollution. To be energized, commercial ships burn fuel. This burnt fuel emits several air pollution by-products, including nitrogen oxides, carbon dioxide, and sulphur oxide. Carbon dioxide disrupts ocean circulation and the ecosystem by changing the ocean chemistry, increasing acidity, and threatening coral reefs (Uption & Folger, 2018). Nitrogen oxide, on the other hand, pollutes the environment leading to smog or ground-level ozone and respiratory issues in humans. Noise pollution is another environmental impact of shipping.

The danger posed by the increasing noise pollution is that it is likely to harm marine species that communicate, find orientation, and feed using sound. Research demonstrates that long-term sound transmission under water due to shipping noises induces stress response among the Nile Tilapia (Kusku, 2020). Shipping also impacts the environment through vessel discharges. There has been a tremendous decline in the number of accidental oil spills. Yet, research shows that oil spills persist and harm marine ecosystems enormously (Zhang et al., 2019). However, other than accidental oil spills, water discharges also threaten marine life. Cargo ships notably discharge grey water from ships' accommodation areas, oily bilge water, and black water containing urine and feaces (Kotrikla et al., 2021). These water discharges contaminate water and impact the marine environment negatively.

The contribution of production logistics to environmental degradation and the destruction of biodiversity is enormous. Therefore, production logistics is a supply chain function that holds a lot of promise for preventing the sixth mass extinction by leveraging green creativity, eco-innovation, and collaboration across the supply chain.

Environmental production logistics impacts can be mitigated through green production. Supply chain stakeholders must underpin production logistics on ethical and environmental compliance. Considering the population growth that increasingly strains the planet's ecosystems, green production is increasingly important in the supply chain (Baines et al., 2012). Consequently, production logistics in the supply chain should entrench green practices in material management, packaging, and shipping functions identified in this chapter as the primary production logistic contributors to environmental degradation and biodiversity destruction. For instance, the research identifies green material management as an approach that not only focuses on material waste but also promotes sustainable use of materials geared toward minimizing the negative impacts of poor material management (Khorasanizadeh et al., 2018).

Therefore, supply chain stakeholders need to look towards eco-innovation in material management. This way, products and services that comply with minimal resource use and less environmental impact can be procured and distributed within companies. Moreover, production logistics should take cognizance of sustainable consumption and production (SCP). The SCP, first defined in 1994 during the Oslo Symposium and adopted by the World Summit on Sustainable Development (WSSD) in 2002 (Wapner, 2003), seeks to strike a balance between material production and consumption. The essence is that global sustainable development lies in the way societies produce and consume goods. The hypothesis is that sustainable material management is a function of the balance in material procurement and processing. Indeed Lorek and Fuchs (2013) agree that sustainable consumption focuses on product efficiency while optimizing services and products.

Although packaging, especially of the single-use nature, is identified as a threat to environmental sustainability, consumers are becoming eco-conscious and are pushing for change. Regulatory and public concerns are revolutionalizing consumer packaging. For instance, on 28 August 2017, Kenya set what was then ranked as the world's strictest plastic bag ban (Mbugua, 2020). Although the country still struggles to reduce waste from single-use plastics, the commitment shown was quite positive towards environmental sustainability. Apart from Kenya, other countries around the world are putting in place regulations to minimize and manage packaging waste. These countries include Australia which is targeting packaging that is 100% recyclable, compostable, or reusable; China which proposes to ban single-use plastic by 2022; India that is pushing for proactive awareness campaigns and collection points; Canada is in the process of implementing the zero plastic waste strategy passed in 2018; and the European Union that is implementing a ban on selected single-use plastics (Berg et al, 2020).

Besides regulatory endeavors, perhaps production logistics in packaging should use green practices such as sharing best practices in material disposal and recycling, employing recyclable packaging materials, using edible or plant-oriented packaging and looking for plastic alternatives that are compostable and biodegradable, minimizing packaging throughout the supply chain.

In the case of shipping, air pollution should be addressed through emerging innovative technologies. For instance, sensors with real-time knowledge and smart alerts should be installed in ports and ships to monitor toxic levels in emissions, identify areas with ecological dangers, and minimize environmental impact. In the case of noise pollution, aerial and underwater acoustics modules should be employed to monitor noise pollution in real-time and limit environmental impacts. Additionally, innovative compliance management and monitoring tools can report water quality parameters like oxygen levels, total microalgae content, organic matter, salinity, and temperature.

Supply chain production logistics provide an avenue for escalating the biodiversity crisis through several pollutions of the environment. Despite this, production logistics provides a good example of what collaborations, creativity, and eco-innovations can achieve towards saving the planet. From regulatory undertakings requiring cooperation from all stakeholders through green practices such as reproduction, recycling, and reuse, to eco-innovations such as aerial and underwater acoustics, production logistics showcases the power inherent in creativity eco-innovations and collaboration in mitigating the threat of the sixth mass extinction.

Sales Logistics

Sales logistics is the supply chain function that encompasses delivery from warehouses to wholesalers, retailers, and consumers. Consequently, logistics management as a supply chain component meets customer demands through planning, controlling, and implementing the movement and storage of services and goods from source to destination. This way, companies minimize costs and maximize customer service (Harrison et al., 2019). The main three functions under supply chain sales logistics are inventory management, warehouse logistics, and transport logistics.

Supply Chain Inventory Management

Supply Chain Inventory Management is another supply chain function with great potential in creating a more sustainable world. Inventory is a sales logistic component that subsumes raw materials, goods-in-process, finished goods, or merchandise and remains a crucial asset for companies handling tangible goods (Zhong, 2021). Essentially, inventory management is a the Just-in-case (JIC), first-in-first-out (FIFO), and economic order quantity (EOQ) techniques, among others, to certify adequate availability of resources in the supply chain, including manufacturing facilities, warehouses, and last point of sale (Singh & Verma, 2018). Galea-Pace (2020) delineates three core inventory management steps, including purchasing inventory which involves buying raw materials and delivering them to warehouses; storing inventory, which involves storing inventory until required; and profiting from inventory where finished goods are pulled to satisfy orders.

Despite the critical role inventory management plays in ensuring sustainable supply chains, poor inventory management has been recognized as a recipe for environmental degradation and contributing to the biodiversity crisis (Cardoso et al., 2016). A critical consequence of poor inventory management is overstocking or excess inventory. Excess inventory relates to ordering products in excess of projected consumer demands, perhaps due to rising tariffs, over-purchasing, poor demand forecasting, or cancelled orders (Matinise, 2019). Indeed, in the recent past, Covid-19 has driven manufacturing shutdowns, shipping delays, and stores' temporary closures. These events have resulted in excess inventory across many brands, likely damaging the environment when destroyed or disposed of. Environmental impacts of excess inventory have been documented across diverse industries.

From the food industry, research shows that globally 65kg of food is wasted annually per person in terms of vegetable waste (25%), Cereals (24%), and fruits (12%). The environmental footprints occasioned by these wastes include 124g Co_2 (Chen et al., 2020). According to Cooper (2020), the United States ranks as the leading global contributor to food waste, where on average, a family of four wastes approximately $1,500 worth of food annually. Excess food waste does not auger well with the desire for a sustainable world. Evidence shows that overstocking in the food industry accounts for a third of the one-quarter of greenhouse emissions reportedly come from the food industry (Parker et al., 2018). Meanwhile, excess inventory in the beauty industry has raised concern about the excess expired products packaged in unrecyclable plastic that damages biodiversity (Bocken & Short, 2021). The fashion industry has also been flagged as contributing to greenhouse gas emissions worth 1.2billion tons annually through decomposing clothing due to overstocking (Huang et al., 2017), making it the second-largest global polluter.

Therefore, the conversation regarding supply chain inventory management, green creativity, and eco-innovation need not be ignored. Managing the impact of excess inventory requires that proper inventory technology that leverages green inventory practices is employed to reduce waste and the overall carbon footprint. Green inventory management is perceived as the desire to complement the cost focus in inventory with environmental considerations (Marklund & Berling, 2017). Scholars contend that going green in inventory management holds human inventory activities accountable and benefits the environment (Civelek, 2017). These scholars highlight that green creativity and eco-innovation in inventory management remain functions of the three principles of reduce, reuse, and recycle.

Consequently, to contribute to the prevention of the sixth extinction, supply chain stakeholders should focus on the three Rs in inventory management. This focus will not only lessen greenhouse gas emissions, prevent pollution, save energy, reduce waste, and allow maximization of value from products, but is also ultimately likely to sustain the environment for generations to come.

Warehouse Logistics

A warehouse in the supply chain is a central location fitted with enough facilities to receive shipments from production, break them down, reassemble them in particular orders, and ship them out to customers (Ten Himpel & Schmid, 2008). According to Rose (2004), a warehouse is a planned space where goods can be stored safely and efficiently until required for consumption. Consequently, warehouse logistics involve humans, processes, and programmes to keep these goods moving around and through the warehouse. The rationale for such a distribution network in the supply chain is overall cost reduction in the distribution function and competitive advantage in the market (MBA Knowledge Base, 2021). A warehouse specifically performs tasks in the supply chain, including receiving goods, sorting them, dispatching goods to storage, holding goods, packing goods, marshalling goods, dispatching goods to customers, and preparing records.

Despite the important roles warehousing through these functions play in sustaining the supply chain, warehousing and these activities produce indoor and outdoor pollutants that negatively impact humans and the environment (Shearston et al., 2020). Among the common indoor pollutants experienced in warehouses are volatile organic compounds from stored materials, forklifts used in loading and unloading materials, and inhouse material handling; carbon monoxide, nitrogen oxides, and hydrocarbons from forklifts and loading docks are also experienced. Similarly, toxic mould due to high humidity and poor ventilation is also an environmental issue (Ries et al., 2017). Meanwhile, outdoor pollutants from warehouse logistics include carbon dioxide emissions and particulates from diesel trucks used in warehouse logistics. Besides, the expansion of warehousing worldwide means that they consume a large amount of energy leading to an increase in carbon dioxide emissions (Ries et al., 2017).

The negative impacts of air pollution arising from warehouse logistics are well documented. For instance, research demonstrates that inhaling particulates emitted by diesel trucks that serve warehouse ranks among the top contributors to lung cancer (McEntee & Ogneva-Himmelberger, 2008). Other than lung cancer, warehousing has also been associated with air pollution causing asthma, coronary heart disorder, and chronic bronchitis (Horne et al., 2018; Shearston et al., 2020). Besides air pollution due to diesel trucks, research shows that other functions of warehouse logistics, like material handling, account for close to 13 per cent of overall supply chain emissions (World Economic Forum, 2009). In the UK, warehouses have been attributed to nearly 2 million tones of Co_2 (United Kingdom Warehouse Association, 2010).

Considering that warehouses play a crucial role in sustaining supply chains and in determining business success and operational efficiency, green warehousing practices that leverage creativity, eco-innovation, and collaboration can mitigate the warehousing air pollution threat to the environment. Green warehousing is the identification that, unlike traditional warehousing, modern warehousing can do more than create environmental pollution. In essence, green warehousing relates to sustainable warehousing functions and processes (Bartolini et al., 2019). Indeed the recognition that green warehousing reduces the carbon footprint in warehouses has led to green warehousing being recognized among supply chain processes that are environmentally sustainable (Kumar et al., 2015; Rostamzadeh et al., 2015).

Therefore, it is prudent to postulate that warehouse logistics belongs to the list of activities comprising humans' contribution to Anthropolene extinction that can be mitigated through green creativity and eco-innovation in the supply chain. The dynamic supply chains in today's business environment require that alternative lighting sources such as renewable energy and natural light be harnessed to cut costs and emissions from energy use. Similarly, using rain water capture and installing mechanisms to reduce water flow can optimize water use. Environmental sustainability in warehouse logistics can also be maximized using warehousing products such as a warehouse management system (WMS), order-picking technology, Barcoding, and Radio frequency identification (RFID). WMS, for instance, enables stores, distribution centres, and warehouses to be coordinated concurrently with shipping and transportation. This green warehousing practice allows the ideal product quantity in the chain, maximizing space and minimizing excess inventory that could lead to waste. On the other hand, order-picking technology reduces paper consumption and increases speed and accuracy in warehouse orders. Green processes such as Barcoding and RFID increase supply chain visibility and reduce paper use.

Transport Logistics

Transport logistics encompasses the complete logistic process needed to conduct a transport. Transport logistics distributes and provides goods at minimum cost in production logistics. The essence is to optimize transport in unloading, loading, capacity utilization, handover, and identification (Topolsek et al., 2018). Transport logistics examine interactions among several variables, including administrative variables such as personnel management and vehicle management; planning variables like transport control and transport strategies and operational variables such as transport technology and data transmission technology.

Ruziyer and Bakhriddinova (2022) delineate five modes of transportation, including truck freight, marine transportation, rail transportation, air transportation, and intermodal transportation. Truck freight, also known as road transportation, is ideal among supply chains that directly engage quick and small shipments to the warehouse or consumers (Ruziyer & Bakhriddinova, 2022). They posit that truck freight is cheaper than ship and air transport, faces fewer restrictions, is more accessible, allows for door-to-door shipment, and provides more options.

Marine transportation reportedly accounts for over 90% of the global economy, with the US relying on marine transportation for 70% of all its international merchandise trade (Spalding, 2016). Ship transport is ideal for heavy loads, accommodates more space and weight, costs less than air transport, and enhances shipment safety. Meanwhile, rail transportation is ideal for fast, scheduled ground freight. Its benefits are more carrying capacity, reduced delays, and accessibility. Although air transportation is relatively new compared to the other three (Ruziyer & Bakhriddinora, 2022), it is the best choice for fast uncompromising delivery. It is ideal for shipments covering long distances and requiring quick movement. Therefore, besides allowing for speedy deliveries, it offers enhanced security. On the contrary, some shipment requires two or more types of transportation. These shipments are therefore ferried through intermodal transportation.

The five types of transportation logistics are responsible for sustaining the supply chain and attract a lot of interest from supply chain stakeholders. Yet, they are the main sources of negative impacts of transport logistics on the environment. The negative impacts are often due to fuel use, accident, land use, noise, and congestion.

EMISSIONS DUE TO FUEL CONSUMPTION

The five modes of transportation consume fuels depending on the distribution of commodities, shipment size, and distance to be covered. These fuels are in the form of gasoline and diesel. Burning these fossil fuels results in greenhouse gases, of which the most important include carbon dioxide (Co_2), Nitrous oxide (N_2O), and Methane (CH_4). The build of these gases has affected global warming leading to climate change.

Another category of emissions due to fossil fuel burning are criteria air contaminants, including carbon monoxide, hydrocarbons (HC), Nitrogen Oxides (NOx), Sulphur Oxides (SOx), and particulate matter (PM) (Shahraeeni et al., 2015). Empirical evidence shows that these fuel consumption emissions have negatively impacted human health and crop production and have damaged the ecosystem encompassing biosphere, water, and soil (Maubach et al., 20008).

Environmental Impacts Due to Accidents

Accidents, depending on operationalization, have also been associated with environmental impacts. For instance, pipeline and ship oil spills have negatively impacted ecosystems and wildlife (Chilvers et al., 2021). Similarly, train derailments, truck accidents, or accidents in the gas pipeline can negatively affect the ecosystem. Land transportation systems have also been associated with habitat fragmentation and disruption of wildlife habitats by replacing them with roads, rails, and others (Adhikari & Hausen, 2018). Indeed, species diversity in an ecosystem is a function of the uninterrupted habitat's total size. Thus species diversity could be reduced by half if a road divides an area. Environmental impacts also result from noise. As already demonstrated during the discourse on warehouse logistics, noise impacts are more severe in marine transport, where marine species that depend on sound for communication and food are mostly affected.

The discourse on transport logistics confirms that transport logistics alongside warehouse logistics contributes largely to the destruction of the planet due to energy consumption. Evidence shows that in consuming over 30% of total energy globally, the transport industry lies second in consumption, just after the industrial sector (Giannakis et al., 2020). Considering that most of this energy is from fossil fuels, the supply chain should go green in transport logistics (Chien et al., 2021) to mitigate greenhouse emissions and the criteria air containments associated with transportation. Green or sustainable transport logistics have improved the quality of life (Abdelwahed Ahmed & Abd El Monem, 2020).

Several green transportation practices have been identified, including green trains, electric vehicles, service and freight vehicles, and monorails. Green trains use innovative technology and do not emit greenhouse gases. Electric vehicles, on the other hand, use electricity tapped from renewable energy, thus stopping the emission of harmful gases. Meanwhile, service and freight vehicles utilize electricity instead of fossil fuels to eliminate greenhouse gas emissions. Monorails are hauled by an electric locomotive and run on a single rail and therefore protect the environment from noise and air pollution. Such systems protect the environment from harmful emissions, air and noise pollution, and congestion (Abdel Wahid Ahmed & Abd El Monem, 2020). Therefore, adopting sustainable and green transportation in the supply chain will contribute positively to stemming the sixth mass extinction threat. Besides eliminating harmful greenhouse emissions, green transport practices will contribute to sustainable economies and improve human health.

Reverse Logistics

Reverse logistics forms the final essential logistical function in the supply chain. Reverse logistics relates to the movement of materials, goods or products from the consumer and back to the production point (Rajagopal et al., 2015). According to the Council of Logistics Management, reverse logistics is the process that oversees the implementation, controlling and planning in the supply chain to achieve cash-effectiveness in the flow of raw materials, finished goods, and in-process inventory. Often the flow is reversed from the consumption point to the origin–either for proper disposal or value recapture. Reverse logistics includes but is not limited to customers' return of goods, distribution partners' return of unsold goods in line with the contract, reuse of packaging, repairs and maintenance, refurbishment, recycling and disposal and selling to a secondary market.

Unlike the other supply chain logistics, reverse logistics compels the supply chain to take care of wastes, repairs, sourcing alternative markets, repackage, recycling and disposal, thereby ridding the environment of harmful materials. For instance, Oliveira et al. (2018) used batteries collected in Brazil to explore environmental advantages that may accrue from reverse logistics. They determined that reverse logistics had enabled the collection of 4,304,465 post-consumed batteries. Collecting these batteries rid the environment of 25,751 kilograms of chemical waste and 150,671 kilograms of solid waste. On the other hand, Alnoor et al. (2018) used the multinational oil company context in Iraq to show that reverse logistics impacted sustainable manufacturing positively.

Although reverse logistics may prove expensive, preventing the sixth extinction requires that the environment is devoid of any harmful substances to the ecosystem. Therefore, reverse logistics is a supply chain function that could be used to ensure that materials are recycled and disposed of properly, alternative markets are sourced for unused materials to avoid excessive inventory, and repackaging is undertaken when necessary.

Supply Chains and Greenhouse Gas Emissions

From the discussion and findings of the discourse on the four supply chain logistics, it is apparent that the supply chain, particularly the agri-food and industrial supply chains, are culpable for the present state that faces human's kind regarding the Anthropocene extinction. Although the Supply Chain is arguable the lever of modern business, its impact on the environment cannot be overlooked, particularly in such an era when the planet is staring at mass extinction. The supply chain is associated with the environmental impact experienced in the consumer sector. According to Bore and Swartz (2016), over 90% of impacts affecting the consumer sector's land, soil and air arise from the supply chain. Similarly, over 80% of greenhouse gas emissions in most consumer goods come from the supply chain. Consumer packaged goods supply chains have been at the forefront of greenhouse gas emissions. According to Mckinsey's analysis, greenhouse gas emissions from consumer-packaged-goods companies accounted for 33 Gigatons of Co_2e^2 (bore & Swartz, 2016). Therefore consumer packaged goods supply chains must be able to cut greenhouse gas emissions by more than half by 2050 to meet targets.

Greenhouse gas emissions experienced in the shipping process and international travel through water are two supply chain aspects that concern environmentalists. In 2018, it was estimated that total shipping emitted 1,056 million tonnes of CO_2, which amounts to 2.89% of global anthropogenic CO_2 emissions for the year (Fourth IMO GHG study, 2020). Besides sulphur oxide and nitrogen oxide have also attracted interest in marine emissions.

Taking cognizance of the understanding that protection of the planet requires sustainable supply chain practices, various practices that border on creativity, eco-innovativeness, and collaboration have been advocated for supply chains to advance green practices in their logistics. Products are now required to undergo an analysis of total environmental effects before production and distribution (Yadar et al., 2021). Supply chain efficiency is gaining more leverage on biofuels and renewable energy targeting economic growth and a sustainable environment. It is argued that renewable energies and biofuels reduce CO_2 emissions (Babatunde et al., 2020).

Supply chains are becoming more flexible with the need to adapt efficiently and fast to environmental changes and other unpredictable changes like competition and customer demands. Flexibility aligns with the green supply chain narrative that leverages environmental thinking. Several scholars have acknowledged the continuing progress in integrating green thinking into the supply chain (Ahi & Seacy, 2013; Alexander et al., 2014; Shan & Wang, 2018).

CONCLUSION

The supply chain is a significant contributor to improving the efficiency of plants, warehouses, and transportation. Through logistical functions such as procurement, production, sales and reverse logistics, the supply chain increases cash flow by ensuring timely delivery of products and services to customers. Supply chain management leverages the various supply chain functions to oversee the quick and efficient movement of items from source to destination. Among the direct benefits that accrue from supply chain management are reduced operational costs and improved customer service.

Yet, concerns are directed at the knowledge that supply chains cause 90% of companies' environmental impacts. This high proportion of reported environmental impacts puts the supply chain at the center of human activities responsible for the biodiversity crisis that is facing the world. Through the various functions, including procurement, packaging, shipping, warehousing, and transportation, the supply chain contributes environmental footprints such as carbon dioxide and other greenhouse gases from excess inventory, poor materials management, plastic packaging and shipping. Besides air pollution, shipping also emits toxic waste and noise that interfere with marine life. Meanwhile, transport logistics mainly employ fossil fuels that contribute emissions such as carbon dioxide, nitrogen oxide, sulphur oxide and particulate matter when they burn.

Luckily, companies, especially big multinational ones, are realizing the weight of the supply chains and the important roles they play and have since emphasized eco-friendly operations. Companies are looking to improve the various industries through networks of collaborations with suppliers, civil society, or intermediates to enhance creativity and eco-innovation. Such networks have been created in the case of palm oil or the better cotton initiative on cotton. Time is already here when similar networks can transcend to packaging for which suppliers can supply biodegradable or recycled materials. Sustainability in supply chains is also gaining support from reverse logistics. Receiving back packaging previously supplied provides an opportunity to reuse or recycle, which is an excellent strategy to address the demand for new materials and waste.

Although the prevention of Anthropocene extinction is a wicked problem, green creativity, eco-innovation and collaboration remain the only hope through which the supply chain can offer wicked solutions. The supply chain responsible for such a wicked problem is in the correct position to find solutions proactively. This proactivity is happening with technological innovations and collaborative international regulations governing environmental sustainability being put in place. Despite the threat of the sixth mass extinction being real, it is not insurmountable. Unlike the curse of the Cassandra, supply chain stakeholders are showing commitment to entrenching green supply chain practices that can protect the planet. In this case, supply chain stakeholders have chosen to be proactive to protect the supply chain from future harm. This proactivity is true to the adage that "when we make it dirty, we must clean it." Together we extinguish the threat of the sixth extinction through the supply chain.

RECOMMENDATIONS AND FUTURE AREAS

The governmental insititutions should put in place a green strategies and implementation plans for 2022-2030 so that various industries could take advantage of formulating policies guidelines from the laid down strategies which will establish standards,requirements and enforcement of environmental protection.The measures will foster for the mantainance of ecological balance in the natural habitat and resources conservation with the aim of supporting the wellness of the existing and posterity.The book chapter focused on theme discussions, this could have been limiting since it has been established that inattention to directional causality may lead to serious consequences (Hanafiah, 2020).Therefore future studies should consider formative constructs across the diversified supply chains to determine the contribution towards the sixth mass extinction and offer better insights.

REFERENCES

Abbasi, M. (2017). Towards socially sustainable supply chains–themes and challenges. *European Business Review*, *29*(3), 261–303. doi:10.1108/EBR-03-2016-0045

Abdel Wahed Ahmed, M. M., & Abd El Monem, N. (2020). Sustainable and green transportation for better quality of life case study greater Cairo–Egypt. *HBRC Journal*, *16*(1), 17–37. doi:10.1080/16874048.2020.1719340

Adhikari, A., & Hansen, A. J. (2018). Land use change and habitat fragmentation of wildland ecosystems of the North Central United States. *Landscape and Urban Planning*, *177*, 196–216. doi:10.1016/j.landurbplan.2018.04.014

Ahi, P., & Searcy, C. (2013). A comparative literature analysis of definitions for green and sustainable supply chain management. *Journal of Cleaner Production*, *52*, 329–341. doi:10.1016/j.jclepro.2013.02.018

Albert, I., Shakantu, W., & Ibrahim, S. (2021). The effect of poor materials management in the construction industry: A case study of Abuja, Nigeria. *Acta Structilia*, *28*(1), 142–167.

Alexander, A., Walker, H., & Naim, M. (2014). Decision theory in sustainable supply chain management: A literature review. *Supply Chain Manag: An International Journal*, *19*(5/6), 504–522. doi:10.1108/SCM-01-2014-0007

Alnoor, A., Eneizan, B., Makhamreh, H. Z., & Rahoma, I. A. (2018). The effect of reverse logistics on sustainable manufacturing. *International Journal of Academic Research in Accounting, Finance and Management Sciences*, *9*(1), 71–79.

Babatunde, O. M., Munda, J. L., & Hamam, Y. (2020). A comprehensive state-of-the-art survey on hybrid renewable energy system operations and planning. *IEEE Access : Practical Innovations, Open Solutions*, *8*, 75313–75346. doi:10.1109/ACCESS.2020.2988397

Baines, T., Brown, S., Benedettini, O., & Ball, P. D. (2012). Examining green production and its role within the competitive strategy of manufacturers. *Journal of Industrial Engineering and Management*, *5*(1), 53–87. doi:10.3926/jiem.405

Bartolini, M., Bottani, E., & Grosse, E. H. (2019). Green warehousing: Systematic literature review and bibliometric analysis. *Journal of Cleaner Production, 226*, 242–258. doi:10.1016/j.jclepro.2019.04.055

Batista, L., Bourlakis, M., Smart, P., & Maull, R. (2018). In search of a circular supply chain archetype–a content-analysis-based literature review. *Production Planning and Control, 29*(6), 438–451. doi:10.1080/09537287.2017.1343502

Berg, P., Feber, D., Granskog, A., Nordigarden, D., & Ponkshe, S. (2020). *The drive toward sustainability in packaging—beyond the quick wins*. Mckinsey & Company.

Bichou, K. (2014). *Port operations, planning and logistics*. CRC Press. doi:10.4324/9781315850443

Bocken, N. M., & Short, S. W. (2021). Unsustainable business models–Recognising and resolving institutionalised social and environmental harm. *Journal of Cleaner Production, 312*, 127828. doi:10.1016/j.jclepro.2021.127828

Bodamer, D. (2016, November 16). *14 Charts from the EPA's Latest MSW Estimates*. Waste460, Retrieved June 8, 2022, from https://www.waste360.com/waste-reduction/14-charts-epa-s-latest-msw-estimates

Bové, A. T., & Swartz, S. (2016). Starting at the source: Sustainability in supply chains. *McKinsey on Sustainability and Resource Productivity, 4*, 36–43.

Byrnes, T. A., & Dunn, R. J. (2020). Boating-and shipping-related environmental impacts and example management measures: A review. *Journal of Marine Science and Engineering, 8*(11), 908. doi:10.3390/jmse8110908

Cardoso, P., Carvalho, J. C., Crespo, L. C., & Arnedo, M. A. (2016). Optimal inventorying and monitoring of taxon, phylogenetic and functional diversity. *Biorxiv*, 060400. doi:10.1101/060400

Carter, B. (n.d.). *The Impact of Packaging on the Environment: Is Plastic the Only Demon?* Eco & Beyond.

Chen, C., Chaudhary, A., & Mathys, A. (2020). Nutritional and environmental losses embedded in global food waste. *Resources, Conservation and Recycling, 160*, 104912. doi:10.1016/j.resconrec.2020.104912

Chilvers, B. L., Morgan, K. J., & White, B. J. (2021). Sources and reporting of oil spills and impacts on wildlife 1970–2018. *Environmental Science and Pollution Research International, 28*(1), 754–762. doi:10.100711356-020-10538-0 PMID:32822011

Chopra, S. M. (2010). *Supply chain management: Strategy, planning and operations*. Academic Press.

Chou, M. C., & Wang, R. W. (2012). Displayability: An assessment of differentiation design for the findability of bottle packaging. *Displays, 33*(3), 146–156. doi:10.1016/j.displa.2012.06.003

Chukwuma, A. I., Ezenyilimba, E., & Agbara, N. O. (2018). Effect of product packaging on the sales volume of small and medium scale bakery firms in South East Nigeria. *International Journal of Academic Research in Business & Social Sciences, 8*(6), 988–1001.

Civelek, I. (2017). Sustainability in inventory management. In Intelligence, Sustainability, and Strategic Issues in Management (pp. 43-56). Routledge. doi:10.4324/9780203788394-3

Conservancy, O. (2018). *Fighting for Trash Free Seas: Ending the Flow of Trash at the Source*. Ocean Conservancy. Retrieved March 7, 2019, from https://oceanconservancy.org/trash-free-seas/

Conserve Energy Future. (n.d.). *What is green transportation?* Available from: https://www.conserve-energy-future.com/modes-and-benefits-of-green-transportation.php

Cooper, R. (2020, August 25). *Food Waste in America: Facts and Statistics*. RUBICON. https://www.rubicon.com/blog/food-waste-facts/

Cowie, R. H., Bouchet, P., & Fontaine, B. (2022). The Sixth Mass Extinction: Fact, fiction or speculation? *Biological Reviews of the Cambridge Philosophical Society*, *97*(2), 640–663. doi:10.1111/brv.12816 PMID:35014169

Disclosure Insight Action. (2021, February 9). *Environmental supply chain risks to cost companies $120 billion by 2026*. https://www.cdp.net/en/articles/supply-chain/environmental-supply-chain-risks-to-cost-companies-120-billion-by-2026

Galea-Pace, S. (2020, May 17). *Why is inventory management in the supply chain important?* https://supplychaindigital.com/digital-supply-chain/why-inventory-management-supply-chain-important

Galeazzo, A., Ortiz-de-Mandojana, N., & Delgado-Ceballos, J. (2021). Green procurement and financial performance in the tourism industry: The moderating role of tourists' green purchasing behaviour. *Current Issues in Tourism*, *24*(5), 700–716. doi:10.1080/13683500.2020.1734546

Giannakis, E., Serghides, D., Dimitriou, S., & Zittis, G. (2020). Land transport CO2 emissions and climate change: Evidence from Cyprus. *International Journal of Sustainable Energy*, *39*(7), 634–647. doi:10.1080/14786451.2020.1743704

Harrabin, R. (2017). *Ocean plastic a 'planetary crisis' – UN*. BBC. Retrieved March 7, 2019, from https://www.bbc.com/news/science-environment-42225915

Harrison, A., Skipworth, H., van Hoek, R. I., & Aitken, J. (2019). Logistics management and strategy: competing through the supply chain. Academic Press.

Hazaea, S. A., Al-Matari, E. M., Zedan, K., Khatib, S. F., Zhu, J., & Al Amosh, H. (2022). Green Purchasing: Past, Present and Future. *Sustainability*, *14*(9), 5008. doi:10.3390u14095008

Heal, G. (2020). Reflections—What would it take to reduce US greenhouse gas emissions 80 percent by 2050? *Review of Environmental Economics and Policy*.

Horne, B. D., Joy, E. A., Hofmann, M. G., Gesteland, P. H., Cannon, J. B., Lefler, J. S., Blagev, D. P., Korgenski, E. K., Torosyan, N., Hansen, G. I., Kartchner, D., & Pope, C. A. III. (2018). Short-term elevation of fine particulate matter air pollution and acute lower respiratory infection. *American Journal of Respiratory and Critical Care Medicine*, *198*(6), 759–766. doi:10.1164/rccm.201709-1883OC PMID:29652174

Hu, B., Huang, S., & Yin, L. (2021). The cytokine storm and COVID-19. *Journal of Medical Virology*, *93*(1), 250–256. doi:10.1002/jmv.26232 PMID:32592501

Huang, B., Zhao, J., Geng, Y., Tian, Y., & Jiang, P. (2017). Energy-related GHG emissions of the textile industry in China. *Resources, Conservation and Recycling*, *119*, 69–77. doi:10.1016/j.resconrec.2016.06.013

Hugos, M. H. (2018). *Essentials of supply chain management*. John Wiley & Sons. doi:10.1002/9781119464495

Kandil, N., Battaïa, O., & Hammami, R. (2020). Globalisation vs. Slowbalisation: A literature review of analytical models for sourcing decisions in supply chain management. *Annual Reviews in Control*, *49*, 277–287. doi:10.1016/j.arcontrol.2020.04.004

Khorasanizadeh, M., Bazargan, A., & McKay, G. (2018). *An Introduction to Sustainable Materials Management*. Academic Press.

Kim, B., Lee, H., Park, H., & Kim, H. (2012). Greenhouse gas emissions from onsite equipment usage in road construction. *Journal of Construction Engineering and Management*, *138*(8), 982–990. doi:10.1061/(ASCE)CO.1943-7862.0000515

Knowledge Base, M. B. A. (2021). *Warehousing Function of Logistics*. https://www.mbaknol.com/logistics-management/warehousing-function-of-logistics/

Kolotzek, C., Helbig, C., Thorenz, A., Reller, A., & Tuma, A. (2018). A company-oriented model for the assessment of raw material supply risks, environmental impact and social implications. *Journal of Cleaner Production*, *176*, 566–580. doi:10.1016/j.jclepro.2017.12.162

Kotrikla, A. M., Zavantias, A., & Kaloupi, M. (2021). Waste generation and management onboard a cruise ship: A case study. *Ocean and Coastal Management*, *212*, 105850. doi:10.1016/j.ocecoaman.2021.105850

Kumar, A., Jigyasu, D. K., Subrahmanyam, G., Mondal, R., Shabnam, A. A., Cabral-Pinto, M. M. S., ... Bhatia, A. (2021). Nickel in terrestrial biota: Comprehensive review on contamination, toxicity, tolerance and its remediation approaches. *Chemosphere*, *275*, 129996. doi:10.1016/j.chemosphere.2021.129996 PMID:33647680

Kumar, N., Agrahari, R. P., & Roy, D. (2015). Review of green supply chain processes. *IFAC-PapersOnLine*, *48*(3), 374–381. doi:10.1016/j.ifacol.2015.06.110

Kusku, H. (2020). Acoustic sound–induced stress response of Nile tilapia (Oreochromis niloticus) to long-term underwater sound transmissions of urban and shipping noises. *Environmental Science and Pollution Research International*, *27*(29), 36857–36864. doi:10.100711356-020-09699-9 PMID:32577967

Larina, E., Bottjer, D. J., Corsetti, F. A., Thibodeau, A. M., Berelson, W. M., West, A. J., & Yager, J. A. (2021). Ecosystem change and carbon cycle perturbation preceded the end-Triassic mass extinction. *Earth and Planetary Science Letters*, *576*, 117180. doi:10.1016/j.epsl.2021.117180

Lee, J., Bazilian, M., Sovacool, B., & Greene, S. (2020). Responsible or reckless? A critical review of the environmental and climate assessments of mineral supply chains. *Environmental Research Letters*, *15*(10), 103009. doi:10.1088/1748-9326/ab9f8c

Lorek, S., & Fuchs, D. (2013). Strong sustainable consumption governance–precondition for a degrowth path? *Journal of Cleaner Production*, *38*, 36–43. doi:10.1016/j.jclepro.2011.08.008

Maibach, M. (2008). *Handbook on estimation of external costs in the transport sector - Produced within the study Internalisation Measures and Policies for All external Cost of Transport.* IMPACT.

Marklund, J., & Berling, P. (2017). Green inventory management. In *Sustainable supply chains* (pp. 189–218). Springer. doi:10.1007/978-3-319-29791-0_8

Matinise, S. N. (2019). *Understanding waste management practices in the commercial food service sector* [Doctoral dissertation]. North-West University.

Mbugua, S. (2020, Jan 23). *2 Years Ago, Kenya Set The World's Strictest Plastic Bag Ban. Did It Work?* https://www.huffpost.com/entry/plastic-bag-ban-works-kenya_n_5e272713c5b63211761a4698

McEntee, J. C., & Ogneva-Himmelberger, Y. (2008). Diesel particulate matter, lung cancer, and asthma incidences along major traffic corridors in MA, USA: A GIS analysis. *Health & Place*, *14*(4), 817–828. doi:10.1016/j.healthplace.2008.01.002 PMID:18280198

Meherishi, L., Narayana, S. A., & Ranjani, K. S. (2019). Sustainable packaging for supply chain management in the circular economy: A review. *Journal of Cleaner Production*, *237*, 117582. doi:10.1016/j.jclepro.2019.07.057

Michlowicz, E. (2013). Logistics in production processes. *Journal of Machine Engineering*, *13*(4), 5–17.

Mishra, S., & Maiti, A. (2019). Applicability of enzymes produced from different biotic species for biodegradation of textile dyes. *Clean Technologies and Environmental Policy*, *21*(4), 763–781. doi:10.100710098-019-01681-5

Natural Resources Canada. (2018). *Greenhouse Gas Emissions from the Plastics Processing Industry.* Government of Canada. Retrieved March 7, 2019, from https://www.nrcan.gc.ca/energy/efficiency/industry/technical-info/benchmarking/plastics/5211

Oliveira Neto, G. C. D., Ruiz, M. S., Correia, A. J. C., & Mendes, H. M. R. (2018). Environmental advantages of the reverse logistics: A case study in the batteries collection in Brazil. *Production*, *28*(0), 28. doi:10.1590/0103-6513.20170098

Pereira, C. R., Christopher, M., & Da Silva, A. L. (2014). Achieving supply chain resilience: The role of procurement. *Supply Chain Management.*

Poppenheimer, L. (2014, July 7). *Aluminum Beverage Cans – Environmental Impact.* Green Groundswell. Retrieved March 7, 2019, from https://greengroundswell.com/aluminum-beverage-cans-environmental-impact/2014/07/17/

Rajagopal, P., Kaliani Sundram, V. P., & Maniam Naidu, B. (2015). Future directions of reverse logistics in gaining competitive advantages: A review of literature. *International Journal of Supply Chain Management, 4*(1), 39-48.

Ries, J. M., Grosse, E. H., & Fichtinger, J. (2017). Environmental impact of warehousing: A scenario analysis for the United States. *International Journal of Production Research*, *55*(21), 6485–6499. doi:10.1080/00207543.2016.1211342

Ripple, W. J., Wolf, C., Newsome, T. M., Barnard, P., & Moomaw, W. R. (2019). World scientists' warning of a climate emergency. *Bioscience*, biz088. Advance online publication. doi:10.1093/biosci/biz088

Ross, D. F. (2004). Warehousing. In *Distribution Planning and Control* (pp. 535–608). Springer. doi:10.1007/978-1-4419-8939-0_11

Rostamzadeh, R., Govindan, K., Esmaeili, A., & Sabaghi, M. (2015). Application of fuzzy VIKOR for evaluation of green supply chain management practices. *Ecological Indicators*, *49*, 188–203. doi:10.1016/j.ecolind.2014.09.045

Rundh, B. (2016). The role of packaging within marketing and value creation. *British Food Journal*, *118*(10), 2491–2511. doi:10.1108/BFJ-10-2015-0390

Ruziyev, B., & Bakhriddinova, Y. (2022). Logistics: Types of transport. *Science Progress*, *3*(2), 456–462.

Saporovskaya, T. Y., Prohorov, S. V., & Timakov, E. A. (2020, July). Composite materials based on non-recyclable polyethylene. *IOP Conference Series. Materials Science and Engineering*, *896*(1), 012078. doi:10.1088/1757-899X/896/1/012078

Shabani, Z. D., & Shahnazi, R. (2019). Energy consumption, carbon dioxide emissions, information and communications technology, and gross domestic product in Iranian economic sectors: A panel causality analysis. *Energy*, *169*, 1064–1078. doi:10.1016/j.energy.2018.11.062

Shahraeeni, M., Ahmed, S., Malek, K., Van Drimmelen, B., & Kjeang, E. (2015). Life cycle emissions and cost of transportation systems: Case study on diesel and natural gas for light duty trucks in municipal fleet operations. *Journal of Natural Gas Science and Engineering*, *24*, 26–34. doi:10.1016/j.jngse.2015.03.009

Shan, W., & Wang, J. (2018). Mapping the landscape and evolutions of green supply chain management. *Sustainability*, *10*(3), 597. doi:10.3390u10030597

Shearston, J. A., Johnson, A. M., Domingo-Relloso, A., Kioumourtzoglou, M. A., Hernández, D., Ross, J., Chillrud, S. N., & Hilpert, M. (2020). Opening a large delivery service warehouse in the South Bronx: Impacts on traffic, air pollution, and noise. *International Journal of Environmental Research and Public Health*, *17*(9), 3208. doi:10.3390/ijerph17093208 PMID:32380726

Shreay, S., Chouinard, H. H., & McCluskey, J. J. (2016). Product differentiation by package size. *Agribusiness*, *32*(1), 3–15. doi:10.1002/agr.21425

Simchi-Levi, D., Kaminsky, P., Simchi-Levi, E., & Shankar, R. (2008). *Designing and managing the supply chain: concepts, strategies and case studies*. Tata McGraw-Hill Education.

Singh, D., & Verma, A. (2018). Inventory management in supply chain. *Materials Today: Proceedings*, *5*(2), 3867–3872. doi:10.1016/j.matpr.2017.11.641

Spalding, M. J. (2016). The new blue economy: The future of sustainability. *Journal of Ocean and Coastal Economics*, *2*(2), 8. doi:10.15351/2373-8456.1052

Srinivas, H. (2022). *Sustainable Development: Concepts*. GDRC Reseaarch Output E-008. Global Development Research Center. Retrieved from https://www.gdrc.org/sustdev/concepts.html

Steffen, W., Richardson, K., Rockström, J., Cornell, S. E., Fetzer, I., Bennett, E. M., Biggs, R., Carpenter, S. R., de Vries, W., de Wit, C. A., Folke, C., Gerten, D., Heinke, J., Mace, G. M., Persson, L. M., Ramanathan, V., Reyers, B., & Sörlin, S. (2015). Planetary boundaries: Guiding human development on a changing planet. *Science*, *347*(6223), 1259855. doi:10.1126cience.1259855 PMID:25592418

Svanberg, M., Ellis, J., Lundgren, J., & Landälv, I. (2018). Renewable methanol as a fuel for the shipping industry. *Renewable & Sustainable Energy Reviews*, *94*, 1217–1228. doi:10.1016/j.rser.2018.06.058

Terman, J., & Smith, C. (2018). Putting your money where your mouth is: green procurement as a form of sustainability. *Journal of Public Procurement.*

Topolšek, D., Čižiūnienė, K., & Ojsteršek, T. C. (2018). Defining transport logistics: A literature review and practitioner opinion based approach. *Transport*, *33*(5), 1196–1203. doi:10.3846/transport.2018.6965

Ulewicz, R., Siwiec, D., Pacana, A., Tutak, M., & Brodny, J. (2021). Multi-criteria method for the selection of renewable energy sources in the polish industrial sector. *Energies*, *14*(9), 2386. doi:10.3390/en14092386

UNEP. (2019). *The emissions gap report 2019.* United Nations Environment Programme.

Upton, H. F., & Folger, P. F. (2018). *Ocean Acidification.* Congressional Research Service.

US Environmental Protection Agency. (2014). *Reducing Wasted Food & Packaging: A Guide for Food Services and Restaurants.* EPA. Retrieved March 7, 2019, from https://www.epa.gov/sites/production/files/201508/documents/reducing_wasted_food_pkg_tool.pdf

Verma, M., & Manoj, M., & Verma. (2016). Analysis of the influences of attitudinal factors on car ownership decisions among urban young adults in developing Country like India,Transportation Research Part F:Traffic psychology and Behaviour. *Vulume*, *42*(1), 90–103.

Wapner, P. (2003). World Summit on Sustainable Development: Toward a post-Jo'burg environmentalism. *Global Environmental Politics*, *3*(1), 1–10. doi:10.1162/152638003763336356

Wikström, F., Verghese, K., Auras, R., Olsson, A., Williams, H., Wever, R., Grönman, K., Kvalvåg Pettersen, M., Møller, H., & Soukka, R. (2019). Packaging strategies that save food: A research agenda for 2030. *Journal of Industrial Ecology*, *23*(3), 532–540. doi:10.1111/jiec.12769

World Economic Forum. (2009). *Supply Chain Decarbonization.* World Economic Forum.

Yadav, P., Singh, J., Srivastava, D. K., & Mishra, V. (2021). Environmental pollution and sustainability. In *Environmental Sustainability and Economy* (pp. 111–120). Elsevier. doi:10.1016/B978-0-12-822188-4.00015-4

Yang, S., Su, Y., Wang, W., & Hua, K. (2019). Research on developers' green procurement behavior based on the theory of planned behavior. *Sustainability*, *11*(10), 2949. doi:10.3390u11102949

Yarbrough, Q. (2021, November 16). *Production Planning in Manufacturing: Best Practices for Production Plans.* Planning, Project Management. https://www.projectmanager.com/blog/production-planning

Zhang, B., Matchinski, E. J., Chen, B., Ye, X., Jing, L., & Lee, K. (2019). Marine oil spills—oil pollution, sources and effects. In *World seas: An environmental evaluation* (pp. 391–406). Academic Press. doi:10.1016/B978-0-12-805052-1.00024-3

Zhao, P., & Zhang, Y. (2018). Travel behaviour and life course:Examining changes in car use after residential relocation in Beijing. *Journal of Transport Geography*, *7*, 41–53. doi:10.1016/j.jtrangeo.2018.10.003

Zhong, A. (2021, September 24). *4 Steps In Successful Supply Chain Inventory Management*. https://www.gep.com/blog/technology/4-steps-in-successful-supply-chain-inventory-management

KEY TERMS AND DEFINITIONS

Environmental Sustainability: The effort of meeting the expectations of both the existing and succeeding generations without jeopardizing the wellbeing of the biological community furnishing them.

Green Logistics: It is the effort of public or private entity to establish the appropriate strategies of supporting environmental sustainability through eco efficient activities with the focus of ultimate consumer satisfaction.

Sixth Mass Extinction: This is a phenomena attributed to the human destructive activities such as deforestation for food production which reshapes the natural habitation thus affecting the population growth of the animals .This contributes to the diminishing breeding capability rendering them risk of vanishing.

Supply Chain: This is the composition of single parties or organizational entities depending on the number of participants involved through the upstream and downstream flow of goods, works or services with distinct activities such as procuring,production and distribution from the source to the consumer.

Transportation: Transportation concept involves the movement of people or goods between the point of origin to the point of destination through infrastructural networks such as road,air or marine whose objective is to meet the market expectation.

Chapter 14
Does User Feedback Matter for Product Development?

Aslı Diyadin Lenger
İstanbul Gelişim University, Turkey

Aykan Candemir
Ege University, Turkey

ABSTRACT

This chapter investigates to what extent the feelings and thoughts of consumers are effective in the process of new product development and improvement. Inspired by the user innovation theory, the study analyzes the online feedback of the users of a smartphone brand in Turkey. This analysis covers a sentiment analysis performed using the support vector machines, the random forest, and the recurrent neural networks algorithms. By studying 2005 reviews, the chapter concludes that two strategies are proposed for the firms. A first strategy is a hybrid approach: Given the imported input-dependency, we know that the cost of imported inputs matters for firms. The second strategy is repositioning the brand in the long term. This chapter attempts to contribute to the literature by providing new evidence in a developing country case whether user comments are effective on the modified versions of a high-tech product.

1. INTRODUCTION

New Product Development (NPD) and improvement process can be considered as a precondition for survival in competitive markets (Tripathy and Katyayn, 2021) and also a key for success for the growth of firms. A firm should allocate both its time and financial resources to this process which should have the right idea at the background to begin with. Perhaps, the foremost challenge is the satisfaction of the customer needs which is a key factor for marketing. Consumer feedback appears to be very critical for NPD. The inclusion of customers in NPD and improvement will increase the efficiency of the process (von Hippel 1978, 1988). For a successful NPD, strategies about the process should keep pace with customer needs (Hsu, 2018). So, customers' feedback is like a gold mine for the company in that they can improve the product.The feedback reduces the risks and the money they will spend.

DOI: 10.4018/978-1-6684-6366-6.ch014

This study tries to explore whether consumer emotions and thoughts have an impact on NPD and improvement process. The studies proposing that understanding consumers matter for NPD generally analyze the issue from the perspective of firms based on case studies (Goffin and New, 2001; Fundin and Bergman, 2003; Tzoka et al., 2004; Fang et al., 2008; Joshi and Sharma, 2004; Alli, 2018). We adopt sentiment analysis to analyze the issue from the perspective of customers in the framework of social networks based on User Innovation Theory (Eric von Hippel, 1978) Today, there are 4.5 billion internet users, 5.2 billion mobile phone users, and 3.8 billion active social media users in the World (We are social, 2020) and social media provides accessible user feedback of any products easily. The relevant data were obtained from two social network platforms for this study; i.e., Twitter, and a local e-comment website. Each social media platform possesses huge data of many different products. Therefore, we confined ourselves to a specific technology-intensive product, a smartphone brand produced in Turkey, as a case for which the market is highly competitive and NPD is inevitable. The NPD is perhaps more important in high-tech products than in other types of products because the product life cycle is shorter and more complex (Oh et al., 2015). 2005 user comments out of 4954 raw data related to these versions was collected starting from 2014, in which the smartphone was launched, up to 2019. We also chronologically ordered all versions of the product to sketch the product improvement line.

We utilized the text mining method using customer feedback in the case of a high-tech product for the analysis. The most widely used method in recent years is text mining to analyze the user comments in written text. There are few studies using sentiment analysis with NPD. Rathore et al., (2018) have built a new way of conversational pattern using sentiment analysis with the case of three models of smartphones (Rathore et al., 2018). Wu et al., (2019) have developed an extended social media analytics framework. Rathore and Ilavarasan (2020) investigated the emotion change pre- and post-launch period for three products. Finally, Giannakis et al., (2020) use social media as an information tool for every stage of NPD. Unlike other studies using this method sufficing with only acquiring accuracy rate, the paper connects the results of the sentiment analysis with the NPD process. We used python as the programming software.

The plan of the paper is as follows: the next section reviews the theoretical and empirical literature. The third section explains the methodology used in the study. The fourth section presents the findings. The last section is devoted to the discussion and conclusion.

2. THEORETICAL FRAMEWORK

The NPD is a vital process for companies (Schilling and Hill, 1998) aiming at being up-to-date. It is a hotly debated topic in the literature due to the developing technologies, changing firm models, and the change in the profile of the consumer. Some of the studies below define the steps of the NPD process in detail. The models created for the NPD process aimed at making the process more efficient and applicable to the firms. Booz et al. (1982)'s model has seven stages (Bhuiyan, 2011). The authors focused on more than 700 businesses in the Fortune 1000 list between 1981 and 1986 and found that new products contribute 30% to the operating profit; however, the authors had predicted that this rate would be over 40% in technology businesses (Cooper and Kleinschmidt, 1987). In the model, the strategy development phase is before the phase of idea generation. The ideas obtained during the idea generation stage are subject to monitoring and evaluation in the next step whose purpose is to get a product idea that can be produced. After the selected idea has gone through the business analysis, the design stage starts in which the first step of turning the idea into a product is taken. Before a new product can go into mass production, it has to go through the testing phase. The problems that may arise in the production line are eliminated at this stage. The product passing all stages successfully is finally put on the market.

The seven-stage model of Marshall et al., (2008) starts directly with the phase of idea generation. The collected ideas are analyzed, marketing strategies are determined and their suitability for the business is reviewed, respectively, in the model. The product completing these stages is ready for production before the commercialization step starts. One of the most widely used models is the progressive model of Kotler and Armstrong (2010). The model consisting of eight stages starts with the idea generation and ends up with commercialization. The model is quite similar to that of Marshall et al. (2008) model except for the number of stages. The most compact model of the NPD is the three-stage model of Otto (2003). The process starting with the understanding opportunities continues with the stages of concept development and implementation.

Cooper (1988) observed that a NPD following a process increases the success rate. Cooper (1990) observed the deficiencies in the staged models and constructed a stage-gate model consisting of stages and five doors. Gate holders possessing the authority and experience are in a position of deciding "Go/Kill/Hold/Recycle" at the gates in the model (Cooper, 1988; 1990). Any stage must be successful to pass to the next gate in this model. The process management is very rigorous in the Stage-Gate model which is more comprehensive than others.

The essence of the literature above makes it clear that each step of the NPD process should be meticulously constructed and applied. Therefore, this issue has been increasingly attracting the attention of researchers and practitioners.

The studies on the NPD deal with the process in terms of its different dimensions. The marketing capability viewpoint points out the relationship between the NPD and marketing capability. For example, research using the data on high-tech products in China proposes that marketing capability is less influential on the NPD when market uncertainty and technological turbulence are high (Ju et al., 2018). Mu (2015) has shown that the marketing capability of the organizations has a positive effect on the process in China and America. Morgan et al., (2015) state that the entrepreneurial orientation strategy has a positive effect on the NPD process compared to the market-oriented strategy according to the analysis conducted with data on 206 mid-sized manufacturing firms in Sweden. The psychological effect in the NPD is another dimension. Sozo and Ogliari (2019) shown that the stimuli developed based on human emotions provide positive feedback on creativity in terms of quantity, quality, diversity, and innovation in a study focusing on designers coming up with ideas for a product portfolio.

The NPD process in business-to-business markets is an issue that needs to be discussed as much as it is in business-to-consumer markets. In business-to-business markets, crowdsourcing improves product line expansion more than the NPD process (Zahay et al., 2018). Fang (2008) proposes that high downstream customer network connectivity negatively affects innovation but positively affects the speed-to-market if customers are used as an information source in the business-to-business market based on an analysis of 143 customer-component manufacturer dyads. Low downstream customer network connectivity does not affect speed-to-market but positively influences innovation (Fang, 2008).

The NPD process requires a substantial amount of resources (Cooper, 1996), e.g., time, money, and information. Building a successful process, in the beginning, is very essential and it should be based on quality information (Zahay et al., 2004) not to lose these resources. Stakeholders (Jay Polonsky and Ottman, 1998), competitors (Cooper and Kleinschmidt, 1995) are some of the information resources, which firms can use to achieve their goals.

However, customers appear to be the most prominent information resource among them since they are making the final decision to buy. This point is initially very well developed by Eric von Hippel (1978) in the user innovation theory. This theory maintains that the inclusion of users in the NPD process is an efficient and smart attitude for firms (Riggs and von Hippel, 1994; von Hippel, 2001; von Hippel and Baker, 2006; de Jong and von Hippel, 2009; von Hippel, 2009). User innovation can increase the efficiency of the process while reducing the information symmetries between the users and the producers (von Hippel, 2006). Sometimes reaching out to all users might be cumbersome. So, the theory also develops the idea of *lead users* (von Hippel, 1986) who are aware of a specific need, and sometimes, those individuals may modify the product for that need. Before going into a larger mass of consumers, the idea of lead users can be used by firms. As a result, those people act as an intermediary in the spreading of this innovation without any payment (von Hippel and Baker, 2006; von Hippel, 2009). It seems that the lead users matter in the innovation process, even if business firms are not fully aware of their function (Von Hippel and Baker, 2006). Lilien et al., (2002) reports that lead users improve the innovation capability based on a natural experiment on a firm.

Many studies have also mentioned the positive effects of customer engagement. A study on 81 high-tech companies in China states that customer involvement as an information provider will reduce conflict between companies and customers, while their participation as co-developer will increase conflict (Wang et al., 2020). Quality market information has a positive effect on predictability and new product performance. Another study on 287 new product projects of medium and high technology companies in Poland shows that the quality of market information collected from customers is higher than that of competitors (Dabrowski, 2018). The information obtained from sources such as suppliers other than customers and competitors does not affect market information quality (Dabrowski, 2018). Customer participation also affects the financial profile of firms. Chang and Taylor (2016) indicate that the participation of customers in the initial stages of product development, i.e., the introduction of an idea or product, also affects financial performance positively. In the development phase, on the other hand, customer engagement has the opposite effect on the financial outcome. The benefit of customer participation varies according to different factors. For example, customer participation in developing countries or low-tech products positively affects the performance of the NPD process (Chang and Taylor, 2016). Surprisingly, Lin et al. (2013) reports a negative effect of customer participation on the NPD as the innovativeness degree increases in a study on 196 companies from 14 different industries in Taiwan.

Using consumers as an information source can be troublesome. So, generally lead users were used for innovation using consumer feedback. However, the recent technological developments facilitated reaching out to the great majority of consumers as an innovation source. For example, social media can be used as a tool for customer participation in the NPD process. Although lead users still matter for innovation more, today we can collect, sort, and analyze feedback from almost all users.

Social media has recently been the most extensive mass media enabling individuals to express their ideas, comments, thus it is an important data source. There are a total of 3,8 billion active social media users in the world today. This figure corresponds to 49% of the entire population in the world. The daily average social media usage of the population is 2 hours and 24 minutes (We are social, 2020). For example, Twitter, an online platform where users can express their feelings and thoughts anonymously, has become one of the most important data sources and this platform has nearly 353 million monthly active users in January 2021 (www.statista.com). Providing real-time information during many major cases (simultaneously with things that are happening) is another reason why Twitter is widely used as a data source (Krishnamurthy et al., 2008). Many studies used this social media platform because of these advantageous (Agarwal et al., 2011; Ghiassi et al., 2016; Ruhwinaningsih and Djatna, 2016; Lahuerta-Otero and Cordero-Gutiérrez, 2016; Estévez-Ortiz et al., 2016; Salas-Zárate, 2017; Jain and Jain, 2017; Gabrovšek et al., 2017; Paulose et al., 2018). Rautela et al. (2020) analyze the data obtained from the participation of 213 young social media users in India and show that customer participation in the process is important in the idea and commercialization stages; and therefore, affects the financial performance of businesses. The participation of customers through social media in the product improvement process ensures the production of products that meet their needs and therefore increases customer loyalty (Hidayanti et al., 2018). Social media has transformed the relationship between manufacturers and customers into an active, creative, and social cooperation in terms of co-creation in NPD (Piller et al., 2012). Carr et al.; (2015) suggest that social media should be used in order to reduce the costs of product development and to obtain a more practical approach based on an analysis conducted with the data obtained from a social media search engine on coffee freshness. Rathore and Ilavarasan (2020) collected consumer comments from Twitter before and after the launch of three new products: a car, pizza, and a smartphone. The findings of their research indicate that consumers' emotions changed from positive to negative for pizza whereas comments stayed positive both for the car and the phone. Social media use together with the firm technological capabilities supports the effect of supplier involvement on NPD performance (Cheng and Krumwiede, 2018). Cheng and Krumwiede (2018) found that supplier involvement has an inverted-U shape, i.e., both negative and positive effects on the NPD. In research using the 367 companies from 7 main manufacturing industries, the authors conclude that social media use increases the positive effects whereas it reduces the negative effect on the inverted-U function. Peltola and Mäkinen (2014) noted that the absorptive capacity of social media has increased; the amount of information accessed and the number of ideas increased accordingly in social media, which support the NPD process (Peltola and Mäkinen, 2014). Research on 384 participants working in the field of Research and Development social media together with e-trust are effective in the NPD process of high-tech products (Seyyedamiri and Tajrobehkar, 2019). Rathore et al. (2018) report that the conversation patterns created on social media data have shown that social media is an important tool in NPD in a sentiment analysis based on the data obtained from Twitter regarding the Samsung Galaxy S6, and S6 Edge.

Social media appears as an information source for the NPD. Comments on Twitter for a new car product were found to be effective at all stages of the NPD confirming that social media can be used as an information source and to discover new product ideas (Giannakis et al., 2020). Rakshit et al. (2021) showed, in a study based on data from SMEs in India, that the use of social media as a source of information during Covid-19 is an organized component in the NPD process for SMEs. If a business has used social media to gather market insights, open innovation has been shown to improve customer focus (Du et al., 2016). Social media can be used as an informal source of information in new product projects for multinational companies around the world (Bashir et al., 2017).

Social media also endorses learning mechanisms. It is used to identify organizational learning mechanisms and that it has a facilitating role in the relationship between these mechanisms and NPD (Jiao et al., 2021).

In sum, the key message of the literature on the relationship between social media and NPD is that the former positively affects the latter. However, Roberts and Piller (2016) concluded that business firms have low social media usage rates for the NPD in an analysis using data from 209 northern European businesses and the Product Development and Management Association's (PDMA) 2012 Global Comparative Performance Assessment Study. The authors interpret that firms do not know how to use social media effectively, or it may negatively affect innovation performance in some businesses (Roberts and Piller, 2016).

3. METHODOLOGY

Text mining, machine, and deep learning algorithms are of limited use in the literature, in social sciences in particular. The studies in computer and software engineering suffice with the technical stages such as model development or library building. These algorithms, as a scientific method, can be used to solve a social science problem. This is what this study attempts to.

The sentiment analysis, one of the data mining methods, is an important mining method used to analyze social media data (Li et al., 2015; Fersini et al., 2017). The sentiment analysis emerged in the early 2000s and is used extensively to learn the opinion of the society (Salas-Zárate et al., 2017; Jain and Jain; 2017). Sentiment analysis is the process of classifying opinions to identify emotions in a text, such as positive, negative, or neutral (Rani and Kumar, 2017). In many fields, this method is used (Li et al., 2016; Yu et al., 2016; Fersini et al., 2017) and some of which improved it (Hassan Khan et al., 2015; Grosse, 2015; Athanasiou and Maragoudakis, 2017; Amplayo and Song, 2017; Araque et al., 2017; Salas-Zárate et al., 2017; Jain and Jain; 2017). The sentiment analysis consists of the following steps:

3.1 The Collection of Data Sentiment Analysis

In the data collection phase, all the models of the smartphone are listed first. To reach big data, comments on all websites of these models were tried to be collected one by one. The data collection process is summarized as follows.

Figure 1. The data collection process

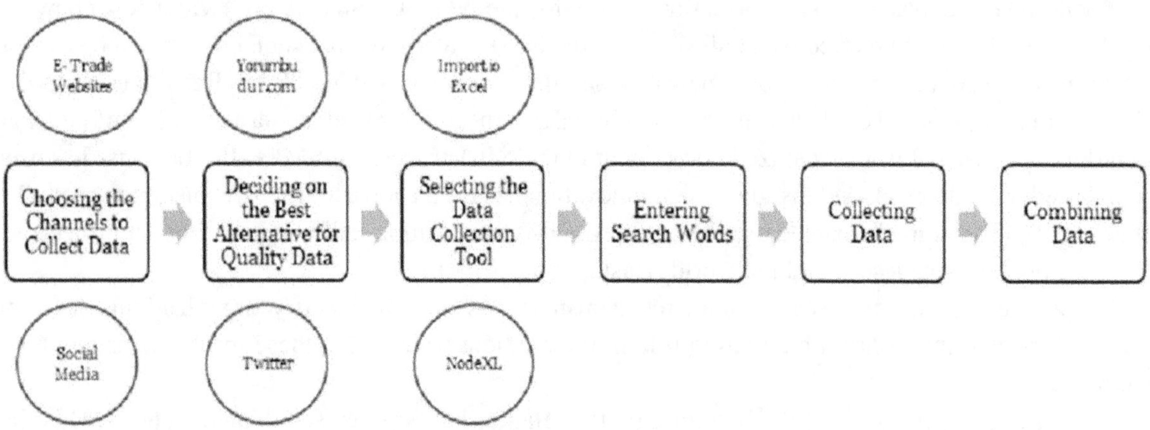

The data were collected within the constraints of requesting high data quality and the legal permission of sharing. The Twitter data from social networks were used for accessing data. The data from Facebook and Instagram does not meet the quality requirement. The *yorumbudur* the comments about the products are gathered from multiple e-commerce sites. Besides, we used the comments from a specific e-trade platform, *yorumbudur,* which involves only the comments on the product itself. The user comments on other e-commerce sites involve the comments both on the product and the stores selling the smartphone, which could not be separated. So, we used only the comments obtained from the *yorumbudur* in addition to Twitter. In addition to the search words above, all uppercase and lowercase letter combinations have been tried to prevent data loss. A total of 4954 data were obtained. All available data from 2014 to 2019 were tried to be collected.

3.2 Clearing the Data

We obtained 4954 comments in total, some of which are on the other this brand's products. Firstly, we removed the comments irrelevant to the smartphone product. Figure 2 outlines the data cleaning process.

Figure 2. Data cleaning process

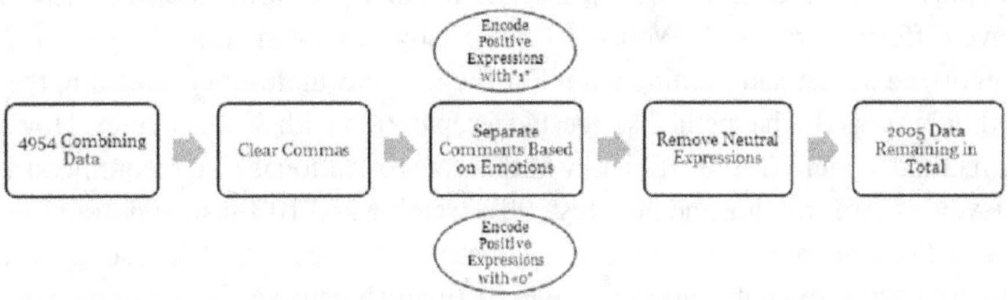

In the first step, we also excluded neutral statements because the neutral ones affect the quality of the data negatively. Positive comments were coded as 1 while the negative ones were coded as 0 manually. We aimed at increasing the data quality by coding all the comments manually. Because there is no comprehensive Turkish lexicon for the sentiment analysis to state the emotional state of each word. The cleaning reduced the number of comments to 2005.

3.3. Preparing the Data

The code used in natural language processing (NLP) was utilized for preparing the data. This process contains five steps (Figure 3).

Figure 3. Preparing the data

In the first step, mail addresses and IP addresses were extracted. Then, punctuation and special characters were cleared from the data. The third step clears the figures and *stopwords* in the data. Finally, all the sentences in the data are converted to lowercase.

3.4. Data Analysis

The data had to be tokenized, i.e., to separate a text into smaller units, and digitized for the sentiment analysis. The words were digitized by assigning frequency numbers, for example, the most frequently used word is assigned 1, the second most often used word digitized as 2, and so on. The frequent words can be translated as, quality, phone, SAR, price/performance, plus, theme, and so on.

3.5. Building the Model

Several classification algorithms were used to select the best algorithm providing the appropriate results and to construct the model accordingly. The classification methods used in this study are the *Random Forest*, the *Support Vector Machines (SVM)*, and the *Recurrent Neural Networks (RNN)*. These three algorithms have different structures. We used RNN firstly in constructing the model. The tokenized words are categorized as test and training data. For the most accurate categorization, the K-fold cross-validation method was used. The method suggested a separation with 80% accuracy. However, a manual trial and error method to increase the efficiency of this categorization is also recommended. So, we used the accuracy levels as 70% training and 30% test; 90% training and 10% test; 80% training and 20%, and 75% training and 25% test data, and a model was constructed for each trial. These rates are intuitive. In the light of the obtained results, it is appropriate to distinguish between 80% training and 20% test data for the data set. The summary of the constructed model is as follows:

Table 1. The summary of the model

Layer (Type)	Output Shape	Param #
Embedding_layer (Embedding)	(None, 80, 50)	500000
gru (GRU)	(None, 80,16)	3216
gru_1 (GRU)	(None, 80, 8)	600
gru_2 (GRU)	(None, 4)	156
dense (Dense)	(None, 1)	5
Total Params: 503,977		
Trainable Params: 503,977		
Non- trainable param: 0		

There are 503.977 variables in the dataset. We established the layers with 16, 8, and 4 neurons after the layer starting with 80-50. The density of the model is 1. A matrix with the size of 1604x80 is obtained as *x* training.

The *Adaptive Moment Estimation* (Adam) was used as the optimization method. Different parameters and combinations have been tried to ensure that the epoch and batch size are selected from the hyperparameters in the most ideal size. For the number of the epoch; 5, 10, 20, 50, and 100; for the batch size, 16, 32, 64, 256, 512 were tried. The ideal combination for the model is balanced in a combination of 50 layers and 256 feeds. The model created has achieved a success rate of 0.802 or approximately 80% suggesting a pretty good level of accuracy.

To test the operability of the model, the comments written were uploaded to and the model was asked to estimate whether they are positive or negative. All comments were tokenized for these sentences. The input to the neural network should be of a specific size. Since the model is determined as a size of 80 for the training and test set, a 10 x 80 matrix was created. The model estimated the new data as presented in Table 2. If results are close to 1, they are considered positive; if results are close to 0, they are estimated negative.

Table 2. Model prediction results

Code	Comments	Real Value	Estimated Value	Accuracy
text1	"I recommend this product to everyone, very good"	1="positive"	0,927988	True
text2	"the cargo is very fast, I got it the same day"	1="positive"	0,927411	True
text3	"I had a big disappointment with this product, it does not fit into the reputation of the brand"	0="negative"	0,324321	True
text4	"excellent"	1="positive"	0,813670	True
text5	"the design is great, but the cargo arrived too late and the product was opened, I wouldn't recommend"	0="negative"	0,895924	False
text6	"it does not look like as in the picture "	0="negative"	0,271155	True
text7	"Although negative comments affected my opinion, I had no bad experience, Thanks!"	1="positive"	0,899600	True
text8	"I have never come across such a bad seller, I am returning the product"	0="negative"	0,175681	True
text9	"just price-performance product"	1="positive"	0,923602	True
text10	"it didn't come out as I expected"	0="negative"	0,859850	False

80% accuracy of the model was tested with new comments. Eight comments out of ten were correctly predicted. The results obtained from the second algorithm, the SVM algorithm are as Table 3:

Table 3. SVM results

	Precision	Recall	F-Measure
0	0.00	0.00	0.00
1	0.78	1.00	0.88
Accuracy			0.78
Macro average	0.39	0.50	0.44
Weighted average	0.61	0.78	0.68

The accuracy rate of the model, 0.78, is very successful. Precision is determined as the ratio of the number of positive samples to the total number of positive samples. The maximum level of recall rate is 1. A maximum sensitivity rate has been achieved for the SVM. The F measure is the harmonic average of the recall and precision rates, and this measurement is obtained by 88%. Table 4 exhibits the result of the confusion matrix.

Table 4. The SVM confusion matrix

	Real Positive	Real Negative
Predicted Positive	[[391	111]
Predicted Negative	[0	0]]

The model predicts the real positive ones correctly. However, it appears to predict some negative comments positively. So, the precision value decreases. The most important reason for this is that the positive comments in the whole data are much more than the negative comments. Therefore, the negative interpretation does not correspond to the test data.

The last classification method used is the Random Forest algorithm. Table 5 displays the results of the Random Forest algorithm.

Table 5. Random forest results

	Precision	Recall	F-Measure
0	0.10	0.03	0.04
1	0.78	0.93	0.85
Accuracy			0.75
Macro average	0.44	0.48	0.45
Weighted average	0.65	0.75	0.69

The accuracy rate of the model created with the random forest algorithm is 0.745. As the rate is above 70% it is considered as successful. The recall rate is very high at 93%. The F-measurement is 88%. The result of the confusion matrix is in table 6.

Table 6. Random forest confusion matrix

	Real Positive	Real Negative
Predicted Positive	[[371	100]
Predicted Negative	[28	3]]

There are 371 positive predictions with the Random Forest algorithm. But, the false negative is 100. Similar to the results obtained from the SVM, the data is affected by the excessive positive comments. Table 7 presents the comparison of the results of the three analyses.

Table 7. Accuracy rates of algorithms

Algorithms	Accuracy Rate
Recurrent Neural Network	0,802
Support Vector Machine	0,779
Random Forest	0,745

The three methods suggest similar ratios but the highest result was obtained from the RNN, which is 80%. These findings are similar to some studies in the literature, for instance, Baldwin et al. (2006), Mitchell et al. (2013), Chang and Taylor (2016), Dabrowski (2018), Ng et al. (2018), and Kiene et al. (2019).

4. MATCHING USERS COMMENTS WITH NPD

In our interview with the manager with the company, they stated that due attention is paid to user feedback in NPD by collecting their data from social media. We examined the positive and the negative comments and their effects on product development and improvement process. We have 1571 positive comments and 434 negative comments. The proliferation of the versions of the smartphone is listed and coded by considering the launch dates.

Figure 4. The versions of the smartphone

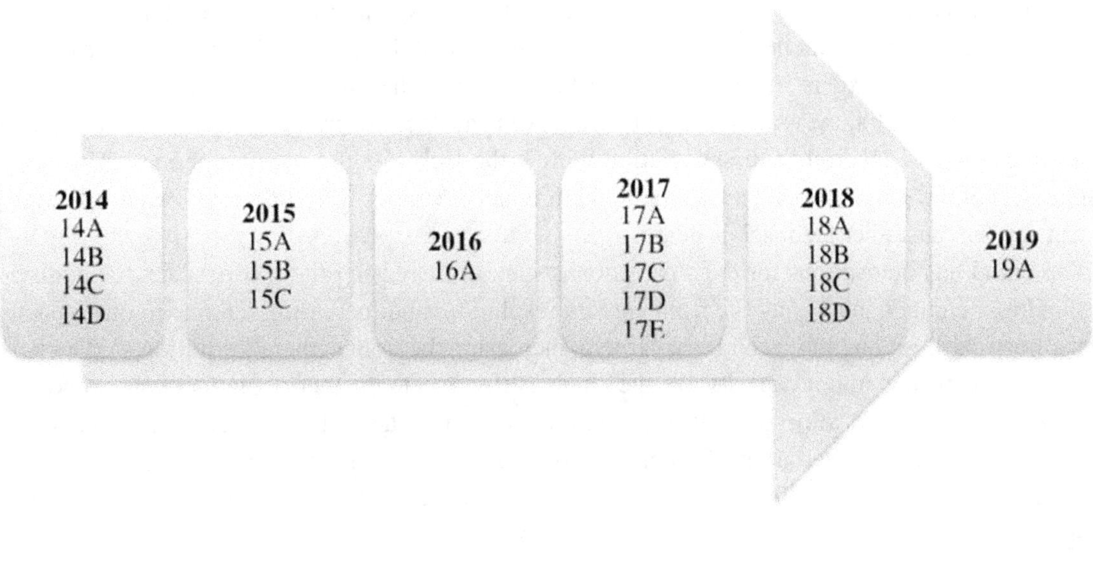

This order of the version of products by years proposes a sort of improvement in the product. The improvements requested by the customers are about the sound quality problem, the heating problem, the high Specific Absorption Rate (SAR) value, the camera problem, the poor night shooting, the lack of radio, and the battery life problem.

One of the main complaints of smartphone users is related to the sound quality of the product. All models except the model z30- USB Type-C have classic 3.5 mm (0.14 in) headphone jacks. Headphones are standard. There are no multimedia improvements such as speakers. In brief, there is no enough improvement for the sound problem.

Rapid discharge of the battery is another complaint of the users. Some consumers complain about the lack of fast charging. The battery capacity of the V4 model is 5000 mAh and of the V7 model is 4050 mAh. The battery capacity of other models is between 2000 and 3400 mAh. A battery capacity above 3000 mAh is recommended for sufficiently long uses. The battery capacity should be over 5000 mAh. The battery capacity is also associated with the screen size and the ram features of the phone. The average length of smartphones is 14.98 cm and the average width is 7.4 cm. The number of cores also decreases the lifetime of the battery. The high number of cores also calls for a processor capable of powering the cores. In sum, we observe that the core improvement is made for a specific brand's product, but instead, increasing the battery capacity is expected to increase customer satisfaction.

The reasons for the warming problem may originate from different issues such as the battery, applications, software, overuse, and other technical problems. The branded phone uses Android as software like many other brands. Therefore, the software problem is not related to the warming issue. Improving the battery and the main processor can relatively solve this problem. Some technology companies produce and use their processors. Qualcomm is one of the most widely used processors for other companies. The brand uses a Cortex processor. A Comparison of Qualcomm with the Cortex processor suggests that the Qualcomm processor is superior. The problem of being incapable of hosting games can also be solved by using a better processor. Although the processor is further versions of the product improved, the company continues to use the same brand as the processor.

Another complaint about the smartphone is related to the SAR value. The SAR value relates to the radiation value emitted by the devices. Therefore, a low SAR value is acceptable and preferable. The SAR value of the smartphone products is at the level below 1.7 which differs in various versions. The upper limit of the SAR value by the European standard is 2 w/kg and by the US Standard is 1.6 w/kg.

The users also pointed out the insufficiency of the camera of the product. The best camera resolution of the phones in the market is 108 mp. The highest resolution of the smartphone is 16 mp. So, the camera resolution of the smartphone is not sufficiently improved and does not provide satisfaction for customers.

Also, customers experience dissatisfaction with is the lack of radio in smartphones. Although the first models of the product do not have a radio feature, later versions were improved by added radio. So, we can say that customer feedback is heard.

The brand has innovations for the smartphones such as increasing the colors scale, the addition of a second rear camera, increasing the front camera resolution, and increasing the maximum capacity of the memory card. Adding the second rear camera, increasing the front camera resolution, and increasing the memory card maximum capacity are the changes that meet consumer complaints to some extent. However, increasing the colors scale is a feature that consumers are not interested in that much.

The smartphone attracted a lot of attention from consumers in Turkey when the product was first introduced to the market in 2014. However, this interest decreased in the following periods. Table 8 compares the consumer demands and product improvements made by the brand.

Table 8. Expected improvements and realization status

Improvement Desired by the Consumer	Realization Status
Camera Problem	Not Improvement
Heating Problem	Not Improvement
Camera Problem	Not Improvement
Sound Quality Problem	Not Improvement
Radio Problem	Improvement
High SAR Value	Improvement

The consumers frequently expressed 6 basic deficiencies. The improvements made by the firm are not very compatible with these demands except the radio on the phone. The SAR value remains compatible with world standards, though.

This study analyzed customers' feedback and the process of NPD relation in the case of the smartphones which belong to a specific brand, by using machine learning algorithms. The mentioned algorithms allowed us to tokenize and digitize the comments as positive or negative. Our findings obtained by the algorithms used suggest that we have an accuracy rate of 80% of the learning model in the analysis. This high success rate ensures us to be clear about the positive and negative feedback of customers on the smartphone. We also examined the product development process of this product by sketching out the features of all versions that have been introduced into the market since 2014. Then, we attempted to investigate any possible user-driven product development given the user comments.

5. DISCUSSION AND CONCLUSION

The literature suggests that the customers' inclusion in the NPD and improvement process is likely to make firms more efficient, increase their market share, provide sustainability and confidence in their customer relations, and strengthen their competitiveness in the market. Diagnosing customer demands and needs are the basic foundation of the marketing concept and critical for the NPD process. Therefore, the strategy of establishing the NPD process in accordance with customer needs would be a rational choice for companies as time and cost savings can be achieved by preventing the loss of customers.

The period between the launch and the decline of a product is accepted to be less than a year in the technology-intensive smartphone industry. So, the R&D activities are very important in technology-intensive enterprises and the NPD and product improvement process must be active for firms to survive. Given this time constraint, it takes a considerable amount of effort to generate new ideas and transform them into products. This process accelerates when users are involved. Besides, it will be possible to obtain cost advantages along with the increases in sales if economies of scale are available.

Our research indicates that the local company in our case actively follows social media. However, the comparison of the user comments with the product development outcomes of the smartphone chronologically shows that the feedback of the consumers did not affect the product improvement. So, we can claim that customers are not heard in the product development process. There can be many reasons why the firm did not or cannot follow a product development strategy in line with customer feedback. We interpret one of the reasons as the imported input-dependency in technology-intensive products. Many developing countries, including Turkey, import many intermediate goods used in the production of technology-intensive goods. This dependency limits the flexibility in creating product specifications. As a matter of fact, the product specifications are imposed by the features of the imported inputs. Firms are left with very limited room to maneuver to determine their marketing as well as a product development strategy. So, the product development process is unresponsive to customer comments and demands.

Considering the results obtained, two strategies are proposed for the firms. A first strategy is a hybrid approach: Given the imported input-dependency, we know that the cost of imported inputs matters for local firms. They prefer the lowest possible imported inputs in production. However, it might also be possible to hear the consumers' wishes and needs without increasing input costs. In this way, both the cost issue and consumer wishes and needs are reconciled. We call this strategy a hybrid approach.

The second strategy is repositioning the brand in the long term. The user profile shows that customers are sensitive to the price. However, we know that there are people who are willing to pay higher prices for technology products, given the prices of the market leader in Turkey. Also, local firms seem to be uncompetitive in the face of the increasing domination of Chinese producers in the world markets with low prices and high-quality products. If the local firms reposition the product in the long-term to a high-quality product without cost concerns, the target market will change and the sales revenue might be enhanced.

As a domestic manufacturer, one of the most important advantages of the companies is the ability to address patriotic feelings. Using the advantage of being a domestic brand, it would be possible to increase the market share for the firm in the long term. Therefore, local firms have a competitive advantage coming primarily from consumer ethnocentrism. However, in our case, the domestic market share of the firm under consideration does not seem to be compatible with this advantage. Global players are upgrading their share in the domestic market whereas those of domestic producers decline. Why foreign products have wide consumer preference is a topic of further research. However, it is possible to say that the reason for local firms to lag behind their competitors in the market is the inadequate strategy of NPD and improvement as well as the inability to meet customer demands and needs.

There are additional suggestions for firms in general, regarding the NPD process. First of all, it is suggested to establish new and updated information systems in accordance with the changing and developing technological infrastructure. It is possible to reach crucial information without wasting any data by using data mining and to build up sustainable customer relationships by this system. Successful models can be constructed with the advanced deep and machine learning algorithms to be used. It is possible to increase competitiveness along with increasing customer loyalty. The establishment of such information for a specific product, like the smartphone, provides accurate and relevant data to be analyzed with algorithms than the data for the whole products of the brand as the firm produces a wide range of products in addition to smartphones. For example, the most commonly encountered sentence in this text mining phase of this study is *price-performance*. Although price-performance may not be relevant for many firms or products, it expresses the opinion of consumers who think that the smartphone product is very good in this respect, so in fact, it is the competitive advantage of this product. If no data is collected and analyzed for this purpose, this vital information could have been overlooked or lost.

These algorithmic systems can be used in many processes. The use of these systems besides microenvironment such as advertising, production, inventory, marketing, management, human resources can also provide advantages for firms in macro areas like at, the nation-wide and worldwide, economy, policy, and markets. Future studies may wish to analyze different products in different sectors. Repeating this analysis is recommended for those products where import dependency is not an issue for product specifications. In addition to a high-tech product, further studies may also conduct such an analysis for medium- and low-technology sectors.

REFERENCES

Agarwal, A., Xie, B., Vovsha, I., Rambow, O., & Passonneau, R. (2011). Sentiment Analysis of Twitter Data. In *Proceedings of the Workshop on Language in Social Media.* Association for Computational Linguistics.

Alli, H. (2018). User Involvement Method in The Early Stage of The New Product Development Process For A Successful Product. *Alam Cipta: International Journal of Sustainable Tropical Design Research and Practice*, *11*(1), 23–28.

Amplayo, R. K., & Song, M. (2017). An Adaptable Fine-Grained Sentiment Analysis For Summarization of Multiple Short Online Reviews. *Data & Knowledge Engineering*, *110*, 54–67. doi:10.1016/j.datak.2017.03.009

Araque, O., Corcuera-Platas, I., Sánchez-Rada, J. F., & Iglesias, C. A. (2017). Enhancing Deep Learning Sentiment Analysis with Ensemble Techniques in Social Applications. *Expert Systems with Applications*, *77*, 236–246. doi:10.1016/j.eswa.2017.02.002

Athanasiou, V., & Maragoudakis, M. (2017). A Novel, Gradient Boosting Framework For Sentiment Analysis in Languages Where NLP Resources Are Not Plentiful: A Case Study For Modern Greek. *Algorithms*, *10*(1), 34. doi:10.3390/a10010034

Baldwin, C., Hienerth, C., & Von Hippel, E. (2006). How User Innovations Become Commercial Products: A Theoretical Investigation and Case Study. *Research Policy*, *35*(9), 1291–1313. doi:10.1016/j.respol.2006.04.012

Bashir, N., Papamichail, K. N., & Malik, K. (2017). Use of Social Media Applications for Supporting New Product Development Processes in Multinational Corporations. *Technological Forecasting and Social Change*, *120*, 176–183. doi:10.1016/j.techfore.2017.02.028

Bhuiyan, N. (2011). A Framework for Successful New Product Development. *Journal of Industrial Engineering and Management*, *4*(4), 746–770. doi:10.3926/jiem.334

Booz, Allen, & Hamilton. (1982). *New Product Management for the 1980's*. Booz, Allen and Hamilton, Inc. doi:10.3926/jiem.334

Carr, J., Decreton, L., Qin, W., Rojas, B., Rossochacki, T., & Wen Yang, Y. (2015). Social Media in Product Development. *Food Quality and Preference*, *40*, 354–364. doi:10.1016/j.foodqual.2014.04.001

Chang, W., & Taylor, S. A. (2016). The Effectiveness of Customer Participation in New Product Development: A meta-analysis. *Journal of Marketing*, *80*(1), 47–64. doi:10.1509/jm.14.0057

Cheng, C. C., & Krumwiede, D. (2018). Enhancing the Performance Of Supplier Involvement in New Product Development: The Enabling Roles of Social Media and Firm Capabilities. Supply Chain Management. *Supply Chain Management*, *3*(23), 1359–8546. doi:10.1108/SCM-07-2017-0230

Cooper, R. G. (1988). The New Product Process: A Decision Guide for Management. *Journal of Marketing Management, 3*(3), 238–255. doi:10.1080/0267257X.1988.9964044

Cooper, R. G. (1990). Stage-Gate Systems: A New Tool for Managing New Products. *Business Horizons, 33*(3), 44–54. doi:10.1016/0007-6813(90)90040-I

Cooper, R. G. (1996). Overhauling the new product process. *Industrial Marketing Management, 25*(6), 465–482. doi:10.1016/S0019-8501(96)00062-4

Cooper, R. G., & Kleinschmidt, E. J. (1987). Success Factors in Product Innovation. *Industrial Marketing Management, 16*(3), 215–223. doi:10.1016/0019-8501(87)90029-0

Cooper, R. G., & Kleinschmidt, E. J. (1995). Benchmarking The Firm's Critical Success Factors in New Product Development. *Journal of Product Innovation Management: An International Publication of the Product Development & Management Association, 12*(5), 374–391. doi:10.1111/1540-5885.1250374

Dabrowski, D. (2018). Sources of Market Information, It's Quality and New Product Financial Performance. *The Engineering Economist, 29*(1), 115–122. doi:10.5755/j01.ee.29.1.13405

De Jong, J. P., & Von Hippel, E. (2009). Transfers of User Process Inovations to Process Equipment Producers: A Study of Dutch High-Tech Firms. *Research Policy, 38*(7), 1181–1191. doi:10.1016/j.respol.2009.04.005

Du, S., Yalcinkaya, G., & Bstieler, L. (2016). Sustainability, Social Media Driven Open Innovation, And New Product Development Performance. *Journal of Product Innovation Management, 33*, 55–71. doi:10.1111/jpim.12334

Epey. (n.d.). https://www.epey.com/

Estévez-Ortiz, F. J., García-Jiménez, A., & Glösekötter, P. (2016). An Application of People's Sentıment From Socıal Media To Smart Cıtıes. *El Profesional de la Información, 25*(6), 851–858. doi:10.3145/epi.2016.nov.02

Fang, E. (2008). Customer Participation and the Trade-Off Between New Product Innovativeness and Speed to Market. *Journal of Marketing, 72*(4), 90–104. doi:10.1509/jmkg.72.4.090

Fang, E., Palmatier, R. W., & Evans, K. R. (2008). Influence of Customer Participation on Creating and Sharing Of New Product Value. *Journal of the Academy of Marketing Science, 36*(3), 322–336. doi:10.100711747-007-0082-9

Fersini, E., Pozzi, F. A., & Messina, E. (2017). Approval Network: A Novel Approach For Sentiment Analysis in Social Networks. *World Wide Web (Bussum), 20*(4), 831–854. doi:10.100711280-016-0419-8

Fundin, A. P., & Bergman, B. L. (2003). Exploring the Customer Feedback Process. *Measuring Business Excellence, 7*(2), 55–65. doi:10.1108/13683040310477995

Gabrovšek, P., Aleksovski, D., Mozetič, I., & Grčar, M. (2017). Twitter Sentiment Around The Earnings Announcement Events. *PLoS One, 12*(2), E0173151. doi:10.1371/journal.pone.0173151 PMID:28235103

Ghiassi, M., Zimbra, D., & Lee, S. (2016). Targeted Twitter Sentiment Analysis for Brands Using Supervised Feature Engineering and The Dynamic Architecture for Artificial Neural Networks. *Journal of Management Information Systems*, *33*(4), 1034–1058. doi:10.1080/07421222.2016.1267526

Giannakis, M., Dubey, R., Yan, S., Spanaki, K., & Papadopoulos, T. (2020). Social Media and Sensemaking Patterns in New Product Development: Demystifying the Customer Sentiment. *Annals of Operations Research*, 1–31.

Goffin, K., & New, C. (2001). Customer Support and New Product Development-An Exploratory Study. *International Journal of Operations & Production Management*. https://trends.google.com/trends/?geo=TR

Grosse, K., González, M. P., Chesñevar, C. I., & Maguitman, A. G. (2015). Integrating Argumentation and Sentiment Analysis For Mining Opinions From Twitter. *AI Communications*, *28*(3), 387–401. doi:10.3233/AIC-140627

Hassan Khan, F., Qamar, U., & Bashir, S. (2015). Building Normalized Sentimı to Enhance Semi-Supervised Sentiment Analysis. *Journal of Intelligent & Fuzzy Systems*, *29*(5), 1805–1816. doi:10.3233/IFS-151658

Hidayanti, I., Herman, L. E., & Farida, N. (2018). Engaging Customers Through Social Media to Improve Industrial Product Development: The Role of Customer Co-Creation Value. *Journal of Relationship Marketing*, *17*(1), 17–28. doi:10.1080/15332667.2018.1440137

Hsu, Y. (2018). Launch Strategy and New Product Innovatıon: An Empırıcal Examınatıon in Taıwan. *International Journal of Organizational Innovation*, *11*(1).

Jain, A., & Jain, M. (2017). Location Based Twitter Opinion Mining Using Common-Sense Information. *Global Journal of Enterprise Information System*, *9*(2), 28–32. doi:10.18311/gjeis/2017/15616

Jay Polonsky, M., & Ottman, J. (1998). Stakeholders' Contribution to the Green New Product Development Process. *Journal of Marketing Management*, *14*(6), 533–557. doi:10.1362/026725798784867707

Jiao, Y., Wu, Y., & Lu, Q. S. (2020). Improving the Performance of Customer Participation in New Product Development: The Moderating Effect of Social Media and Firm Capabilities. *Asian Journal of Technology Innovation*, *28*(2), 284–304. doi:10.1080/19761597.2020.1742749

Joshi, A. W., & Sharma, S. (2004). Customer Knowledge Development: Antecedents and Impact on New Product Performance. *Journal of Marketing*, *68*(4), 47–59. doi:10.1509/jmkg.68.4.47.42722

Ju, M., Jin, J. L., & Zhou, K. Z. (2018). How Can International Ventures Utilize Marketing Capability in Emerging Markets? Its Contingent Effect on New Product Development. *Journal of International Marketing*, *26*(4), 1–17. doi:10.1177/1069031X18809999

Kiene, C., Jiang, J. A., & Hill, B. M. (2019). Technological Frames and User Innovation: Exploring Technological Change in Community Moderation Teams. *Proceedings of the ACM on Human-Computer Interaction*, *3*(CSCW), 44. 10.1145/3359146

Kotler, P., & Armstrong, G. (2010). *Principles of Marketing*. Pearson International Education.

Krishnamurthy, B., Gill, P., & Arlitt, M. (2008). A Few Chirps About Twitter. *Proceedings of the First Workshop On Online Social Networks,* 19-24. 10.1145/1397735.1397741

Lahuerta-Otero, E., & Cordero-Gutiérrez, R. (2016). Looking for The Perfect Tweet. The Use of Data Mining Techniques to Find Influencers on Twitter. *Computers in Human Behavior, 64,* 575–583. doi:10.1016/j.chb.2016.07.035

Li, W., Chen, H., & Nunamaker, J. F. Jr. (2016). Identifying and Profiling Key Sellers in Cyber Carding Community: Azsecure Text Mining System. *Journal of Management Information Systems, 33*(4), 1059–1086. doi:10.1080/07421222.2016.1267528

Li, X., Li, J., & Wu, Y. (2015). A Global Optimization Approach to Multi-Polarity Sentiment Analysis. *PLoS One, 10*(4), E0124672. doi:10.1371/journal.pone.0124672 PMID:25909740

Lilien, G. L., Morrison, P. D., Searls, K., Sonnack, M., & Hippel, E. V. (2002). Performance Assessment of the Lead User Idea-Generation Process for New Product Development. *Management Science, 48*(8), 1042–1059. doi:10.1287/mnsc.48.8.1042.171

Lin, M. J. J., Tu, Y. C., Chen, D. C., & Huang, C. H. (2013). Customer Participation and New Product Development Outcomes: The Moderating Role of Product Innovativeness. *Journal of Management & Organization, 19*(3), 314–337. doi:10.1017/jmo.2013.8

Marshall, G. W., Solomon, M. R., & Stuart, E. W. (2008). *Marketing Real People, Real Choices.* Pearson Education.

Mitchell, L., Frank, M. R., Harris, K. D., Dodds, P. S., & Danforth, C. M. (2013). The Geography of Happiness: Connecting Twitter Sentiment and Expression, Demographics, and Objective Characteristics of Place. *PLoS One, 8*(5), E64417. Advance online publication. doi:10.1371/journal.pone.0064417 PMID:23734200

Morgan, T., Anokhin, S., Kretinin, A., & Frishammar, J. (2015). The Dark Side of the Entrepreneurial Orientation and Market Orientation Interplay: A New Product Development Perspective. *International Small Business Journal, 33*(7), 731–751. doi:10.1177/0266242614521054

Mu, J. (2015). Marketing Capability, Organizational Adaptation and New Product Development Performance. *Industrial Marketing Management, 49,* 151–166. doi:10.1016/j.indmarman.2015.05.003

Ng, J. C. Y., Song, K. K. W., & Tan, Q. (2018). Expanding the Scope Of Application Of User Innovation Theory—A Case Study of The Civil-Military Integration Project in China. *International Journal of Innovation Studies, 2*(1), 33–41. doi:10.1016/j.ijis.2018.03.002

Oh, J., Lee, S., & Yang, J. (2015). A Collaboration Model For New Product Development Through the Integration Of PLM And SCM in The Electronics Industry. *Computers in Industry, 73,* 82–92. doi:10.1016/j.compind.2015.08.003

Otto, K. N. (2003). Product Design: Techniques in Reverse Engineering and New Product Development. Pearson Education Asia Limited, Tsinghua University Press.

Paulose, R., Samy, B. G., & Jegatheesan, K. (2018). Text Mining and Natural Language Processing on Socia Media Data Giving Insights for Pharmacovigilance: A Case Study with Fentanyl. *Indian Journal of Pharmaceutical Sciences*, *80*(4). Advance online publication. doi:10.4172/pharmaceutical-sciences.1000418

Peltola, T., & Mäkinen, S. J. (2014). Influence of The Adoption and Use of Social Media Tools on Absorptive Capacity in New Product Development. *Engineering Management Journal*, *26*(3), 45–51. doi :10.1080/10429247.2014.11432019

Piller, F. T., Vossen, A., & Ihl, C. (2012). From Social Media to Social Product Development: The Impact of Social Media on Co-Creation of Innovation. *Die Unternehmung*, *65*(1).

Pozzi, F. A., Fersini, E., Messina, E., & Liu, B. (2016). *Sentiment Analysis in Social Networks*. Morgan Kaufmann.

Rakshit, S., Mondal, S., Islam, N., Jasimuddin, S., & Zhang, Z. (2021). Social Media and the New Product Development During COVID-19: An Integrated Model for SMEs. *Technological Forecasting and Social Change*, *170*, 120869. doi:10.1016/j.techfore.2021.120869

Rani, S., & Kumar, P. (2017). A Sentiment Analysis System to Improve Teaching and Learning. Computer, *IEEE. Computers & Society*, *50*(5), 36–43. doi:10.1109/MC.2017.133

Rathore, A. K., Das, S., & Ilavarasan, P. V. (2018). Social Media Data Inputs in Product Design: Case of a Smartphone. *Global Journal of Flexible Systems Managment*, *19*(3), 255–272. doi:10.100740171-018-0187-7

Rathore, A. K., & Ilavarasan, P. V. (2020). Pre-and Post-Launch Emotions in New Product Development: Insights From Twitter Analytics of Three Products. *International Journal of Information Management*, *50*, 111–127. doi:10.1016/j.ijinfomgt.2019.05.015

Rautela, S., Sharma, S., & Virani, S. (2020). Influence of Customer Participation in New Product Development: The Moderating Role of Social Media. *International Journal of Productivity and Performance Management*. doi:10.1108/IJPPM-05-2020-0241

Riggs, W., & Von Hippel, E. (1994). Incentives to Innovate and The Sources of Innovation: The Case of Scientific İnstruments. *Research Policy*, *23*(4), 459–469. doi:10.1016/0048-7333(94)90008-6

Roberts, D. L., & Piller, F. T. (2016). Finding the Right Role for Social Media in Innovation. *MIT Sloan Management Review*, *57*(3), 41–47.

Ruhwinaningsih, L., & Djatna, T. (2016). A Sentiment Knowledge Discovery Model in Twitter's TV Content Using Stochastic Gradient Descent Algorithm. *TELKOMNIKA*, *14*(3), 1067–1076. doi:10.12928/telkomnika.v14i3.2671

Salas-Zárate, M. D. P., Medina-Moreira, J., Lagos-Ortiz, K., Luna-Aveiga, H., Rodriguez-Garcia, M. A., & Valencia-García, R. (2017). Sentiment Analysis on Tweets About Diabetes: An Aspect-Level Approach. *Computational and Mathematical Methods in Medicine*, *2017*, 1–9. Advance online publication. doi:10.1155/2017/5140631 PMID:28316638

Schilling, M. A., & Hill, C. W. (1998). Managing the New Product Development Process: Strategic Imperatives. *The Academy of Management Perspectives*, *12*(3), 67–81. doi:10.5465/ame.1998.1109051

Seyyedamiri, N., & Tajrobehkar, L. (2019). Social Content Marketing, Social Media and Product Development Process Effectiveness in High-Tech Companies. *International Journal of Emerging Markets*, *16*(1), 1746–8809. doi:10.1108/IJOEM-06-2018-0323

Sozo, V., & Ogliari, A. (2019). Stimulating Design Team Creativity Based on Emotional Values: A Study on Idea Generation in The Early Stages of New Product Development Processes. *International Journal of Industrial Ergonomics*, *70*, 38–50. doi:10.1016/j.ergon.2019.01.003

Statcounter. (2019). *GlobalStat*. https://gs.statcounter.com/vendor-market-share/mobile/turkey/#monthly-201810-201910-bar

Statista. (n.d.). https://www.statista.com/topics/737/twitter/

Tripathy, C. R., & Katyayn, A. (2021). Product Development Process Concept—Industrial Perspective. In *Advances in Design and Thermal Systems* (pp. 331–339). Springer. doi:10.1007/978-981-33-6428-8_26

Tzokas, N., Hultink, E. J., & Hart, S. (2004). Navigating the New Product Development Process. *Industrial Marketing Management*, *33*(7), 619–626. doi:10.1016/j.indmarman.2003.09.004

Von Hippel, E. (1978). Successful Industrial Products From Customer Ideas. *Journal of Marketing*, *42*(1), 39–49. doi:10.2307/1250327

Von Hippel, E. (1986). Lead users: A source of novel product concepts. *Management Science*, *32*(7), 791–805. doi:10.1287/mnsc.32.7.791

Von Hippel, E. (1988). *The Source of Innovation*. Oxford University Press.

Von Hippel, E. (2001). User Toolkits for Innovation. Journal of Product Innovation Management. *Journal of Product Innovation Management*, *18*(4), 247–257. doi:10.1111/1540-5885.1840247

Von Hippel, E. (2006). *Democratizing Innovation*. The MIT Press.

Von Hippel, E., & Baker, E. (2006). Ideas on the Edge. *CIO Insight*, 12-18.

Von Hippel, E. A. (2009). People Don't Need a Profit Motive to Innovate. *Harvard Business Review*. https://hbr.org/2011/11/people-dont-need-a-profit-motive-to-innovate

Wang, L., Jin, J. L., Zhou, K. Z., Li, C. B., & Yin, E. (2020). Does Customer Participation Hurt New Product Development Performance? Customer Role, Product Newness, and Conflict. *Journal of Business Research*, *109*, 246–259. doi:10.1016/j.jbusres.2019.12.013

We Are Social. (2020). *Digital in 2020*. https://wearesocial.com/digital-2020

Wu, G. J., Tajdini, S., Zhang, J., & Song, L. (2019). Unlocking Value Through an Extended Social Media Analytics Framework: Insights for New Product Adoption. *Qualitative Market Research*, *22*(2), 1352–2752. doi:10.1108/QMR-01-2017-0044

Yorumbudur. (n.d.). https://yorumbudur.com/

Yu, Y., Lin, H., Meng, J., & Zhao, Z. (2016). Visual and Textual Sentiment Analysis of A Microblog Using Deep Convolutional Neural Networks. *Algorithms*, *9*(2), 41. doi:10.3390/a9020041

Zahay, D., Griffin, A., & Fredericks, E. (2004). Sources, Uses, and Forms of Data in The New Product Development Process. *Industrial Marketing Management*, *33*(7), 657–666. doi:10.1016/j.indmarman.2003.10.002

Zahay, D., Hajli, N., & Sihi, D. (2018). Managerial Perspectives on Crowdsourcing in the New Product Development Process. *Industrial Marketing Management*, *71*, 41–53. doi:10.1016/j.indmarman.2017.11.002

Zhan, Y., Tan, K. H., Chung, L., Chen, L., & Xing, X. (2020). Leveraging Social Media in New Product Development: Organisational Learning Processes, Mechanisms and Evidence from China. *International Journal of Operations & Production Management*, *40*(5), 671–694. doi:10.1108/IJOPM-04-2019-0318

Chapter 15
Review and Analysis of Carbon Pricing for Mitigating Climate Change Problems

Manish Kumar
Indian Institute of Management, Indore, India

ABSTRACT

Carbon emission is the most significant driver of pollution and climate change problems. So, this chapter tries to study a recently developed concept to tackle the carbon emission and climate change problem. This chapter has chronologically reviewed the carbon pricing scheme in a supply chain context. This chapter has discussed the carbon pricing scheme in detail and tried to find the applicability and effectiveness of this scheme in the supply chain. With the help of the literature review, this study also tried to explore different ways of effectively operationalizing the scheme in the supply chain. The result of this study shows that carbon pricing is an effective and advanced scheme to tackle the carbon emission and climate change problem. The results further emphasized creating awareness, labelling carbon footprints, setting environmental regulations, implementing a two-part tariff, using data for strategizing, and re-manufacturing used products for better implementation of carbon pricing in the supply chain.

1. INTRODUCTION

In this era of industrialization and globalization, carbon emissions and climate change have become dangerous problems. The use of non-renewable sources of energy everywhere, cutting of trees, and lack of preventive measures are contributing significantly as the cause of these problems. These problems are also causing many health-related issues, which are significantly deteriorating the lifespan of most of the people on this earth. The global and dangerous nature of these problems has drawn significant attention from many government agencies and concerned authorities. Awareness among people has also elevated these problems to the next level. History shows that many preventive and mitigative actions have been taken to tackle these problems. Some of these actions involve shifting towards renewable sources

DOI: 10.4018/978-1-6684-6366-6.ch015

of energy, tree plantation drives, awareness drives, and setting limits for every kind of carbon releaser (Hornibrook et al., 2015). Shifting towards renewable energy, tree plantation drives, and awareness campaigns contributed to the purpose judiciously with the support of concerned authorities in terms of subsidies and grants. New methods and schemes, such as carbon pricing and Industry 4.0 or the fourth industrial revolution (4IR), were also introduced to make these solutions more effective. The use of renewable energy sources for tackling carbon emissions from conventional sources contributed to more complexity in providing reliable energy, and the 4IR solves these issues and helps in providing efficient and reliable energy (Apata et al., 2021). 4IR involves the use of disruptive technologies such as artificial intelligence, machine learning, virtual reality, blockchain, robotics, and 3-D printing in manufacturing processes and supply chains. In comparison, carbon pricing involves putting a cost to the release of carbon for demotivating carbon emitters. This paper chapter particularly focuses on chronically reviewing and discussing carbon pricing in detail and finds its applicability and effectiveness in the supply chain context.

Earlier, to control carbon release, certain limits were set, and all the carbon releasers had to be abiding by these rules/limits (Liang et al., 2014). Although this was a good method to stop concerned carbon-releasing firms or organizations from releasing carbon beyond acceptable limits, it could not effectively minimize the release of carbon. This limit was forcing carbon-releasing firms or organizations to improve the existing technology to get the carbon emission under the limit, but they were mostly making the carbon emission just beside the set limit. Such carbon-releasing firms or organizations that were making a heavy investment to minimize carbon emissions were not getting any kind of incentives. So, this was also demotivating for minimizing carbon emissions.

This paper talks about the new scheme to tackle the above-discussed limitations. This paper discusses the recently developed carbon pricing scheme from a supply chain perspective (He et al., 2012). This paper extensively does a chronological review of the development of the carbon pricing scheme in the supply chain context. This paper also tries to study the carbon pricing instruments such as carbon emission taxes and carbon emission trading schemes for the supply chain (Li et al., 2017; Xu et al., 2016; Zakeri et al., 2015). This paper also tries to find the applicability and effectiveness of the carbon pricing scheme in the supply chain. With the objective of studying this new scheme from a supply chain perspective, the study is divided into the following sections:

- The first section dealt with the introduction of the topic,
- The second section is about the carbon pricing scheme and its development over the years,
- Result and discussion of the study is shown in the third section,
- The fourth section presents the concluding statement,
- Future scope of the study is presented in the last section.

2. CARBON PRICING SCHEME AND ITS DEVELOPMENT

The term carbon pricing was mostly introduced in the early nineties (Barrett, 1990; Burgess, 1990; Cline, 1995; Decanio, 1992; Pena-Torres et al., 1999), but carbon pricing in the supply chain is a very recently developed area. Most of thc research work on this topic is done after 2010 (Du et al., 2013; He et al., 2012; Li et al., 2014; Liang et al., 2014). In recent times, researchers around the world have started to explore the applicability and effectiveness of carbon pricing in the supply chain (Diaz-Trujillo et al., 2019; Ma et al., 2018; Xia et al., 2018). They have also started to explore the different methods and scenarios of applying carbon pricing in the supply chain. They are also exploring the applicability and implementation of the two types of carbon pricing instruments (Diaz-Trujillo et al., 2019).

The carbon pricing scheme was introduced to incentivize carbon emission reduction. The awareness among people and the seriousness of the issue resulted in the devising of this new scheme. Under this scheme, the carbon emission is being charged, and so this incentivizes those who are fewer carbon emitters. The carbon emission can be charged in terms of the carbon tax and the price of permits for carbon emissions (Li et al., 2017; Xu et al., 2016; Zakeri et al., 2015). The permits for carbon emissions can be sold by a less carbon emitter party to a more carbon emitter party, so it is also called a carbon emission trading scheme. The carbon emission taxes and the carbon emission trading schemes are the carbon pricing instruments (Diaz-Trujillo et al., 2019).

As per the literature, carbon pricing started to gain some attention in the early nineties (Barrett, 1990; Decanio, 1992). Barret (1990) first highlights severe problems of the greenhouse effect, mostly due to the emission of carbon dioxide gas. The author also highlights that there are other gases also responsible for the greenhouse effect, but he is focusing on carbon dioxide. The author also notes that forestation is a good method to reduce pollution and the greenhouse effect. The author also highlights that some countries contribute to the most of the emissions, citing the example of the USA (United States of America), which contributes to 21 per cent of the total emissions. The author argues that the solution must involve all parties. Otherwise, some may not reduce emissions for the sake of development, and all will have to suffer. So, the author suggests the carbon tax scheme as a recommendation. The author also tries to give the rationale for the carbon tax solution that the cost of fossil fuels excludes the social cost. The author also stated the use of energy demand elasticity for determining the carbon tax size. The author finally concluded that the proposed solution of an optimal carbon tax scheme should be implemented.

Along with carbon pricing, international cooperation is also needed to avert global warming, and Stephen J. Decanio (1992) discusses this. The paper mainly talks about economic growth, carbon pricing and energy efficiency. The article also tried to develop a simulation model to link population growth, economic growth and carbon dioxide emissions. The developed model of the article also shows that higher carbon dioxide emissions are associated with economic output. The result of the study suggests that economic growth is probably the most effective policy for tackling global warming. The study further suggests that this growth will result in increased wealth which can be used to reduce dependency on fossil fuels and can also be used to improve energy efficiency. The study also suggests that for practical implementation of this policy, carbon pricing should be introduced by charging higher prices for carbon dioxide emissions.

Barrett (1990) and Decanio (1992) were among the first few articles that talked about the need to introduce carbon pricing to combat pollution, global warming and climate change, but carbon pricing in the supply chain context was introduced mostly after 2010 by He et al. (2012), Li et al. (2014), Liang et al. (2014), Xie et al. (2013), and many others. He et al. (2012) used the generation expansion framework of the electric power industry to compare the efficiency and effectiveness of carbon tax and cap-and-trade policies. The authors defined the effectiveness of a policy as its capability to control carbon emissions. The author further defined efficiency from seven criteria which include average emissions price, actual emission, and emission-adjusted economic welfare. The authors used four variations of carbon tax policies and cap-and-trade policy in a game theoretic model to assess their impacts. The authors also conducted a case study on a thirty-bus test system whose results provided the advantages and disadvantages of these policies. Although the results of the study could not explicitly state the best policy, the results showed that the carbon tax policy could be very advantageous in many scenarios.

He et al. (2012) tried to compare the efficiency and effectiveness of two popular policies (carbon tax and cap-and-trade) to combat carbon emissions in the electric power industry context. However, the application of the carbon pricing strategy in the supply chain, which contains one manufacturer and one retailer, is discussed by Liang et al. (2014). In the article, they investigated the amount of emission released in every unit of production. They have also tried to study the effect of customer awareness and environmental regulations on carbon pricing. The result of the study shows that carbon pricing is positively related to environmental regulations, and customer awareness also encourages it. So, the study emphasizes the importance of environmental regulations and customers' awareness of carbon pricing. There must be strict rules and people's support for implementing the carbon pricing scheme so that carbon emissions and climate change can be controlled.

As the awareness among consumers, manufacturers, and retailers regarding climate change has increased tremendously, manufacturers and retailers are taking brave decisions to reduce carbon emissions in the supply chain. Xia et al. (2018) discussed the fact that low carbon awareness of customers and reciprocal preferences of manufacturers and suppliers are reducing the carbon emission level significantly. So, this suggests the fact that awareness regarding carbon emissions, climate change, and the danger due to this encourages manufacturers, retailers, and customers to take appropriate measures like carbon pricing to reduce carbon emissions and mitigate climate change. Li et al. (2017) emphasized the fact that the centralized model is best for optimal decision-making. They studied different types of models and compared them to conclude the results. They also suggested that the return of used products is helping in controlling carbon emissions and climate change. Their study further suggests that returning used products is giving profit to the manufacturers along with tackling the environmental problem, and that further helps in keeping the carbon emission costs low. They also emphasized that low carbon awareness (LCA) is important in implementing the scheme. Nowadays, both online and offline business models are widely spread, and the article by Cheng et al. (2017) tries to study the carbon pricing-related decisions in such supply chains. The paper explores pricing decisions and coordination of manufacturers and retailers in such supply. The paper also talks about two-part tariffs in such a mixed supply chain and concludes that it is an effective pricing tool.

Cheng et al. (2018) discussed a very innovative method of tackling the carbon emission problem. They have talked about carbon labelling, which communicates the carbon footprint of the product. The effectiveness of the carbon labelling scheme is due to the increased awareness among people about global warming, climate change, and the need to mitigate all these. The result of their study shows that the carbon labelling scheme significantly reduces the carbon emissions in the supply chain, and these have a positive impact on manufacturers and retailers in the long run. So, the carbon pricing scheme, along with the carbon labelling scheme, can be a very effective method to tackle carbon emission problems and mitigate climate change. The article by Karstensen et al. (2018) widens the discussion and talks about the role of carbon pricing in dealing with global challenges. In the article, they talk about carbon pricing in the global supply chain. They have investigated the scenario in which carbon pricing is applied while extraction, combustion or production, and consumption of goods and services in which carbon is released. The results of their study also show that carbon prices are spread between sectors and countries due to domestic and global trade. Their study also shows that carbon pricing can be very beneficial for underdeveloped nations and can generate revenue significantly high in comparison to the GDP (gross domestic product) of those nations.

The fact that awareness is necessary to apply any carbon emission reduction and climate change mitigation-related schemes such as carbon pricing are discussed in detail by articles such as Liu et al.

(2018) and Liu (2019). The authors also emphasized that the use of data can be done to know the consumer preferences and performance of a product. They further highlighted that this data could also be used to do targeted advertising. Targeted advertising will help the producers to sell the product to the appropriate consumers. This way, the total cost can be reduced instead of applying carbon pricing. This targeted advertising can also be used to create awareness among all consumers in an effective manner. Studying the impact of carbon pricing on supply chain members' pricing decisions and social welfare is also extremely necessary, and this has been done by Zhou et al. (2018) in their article. They have tried to investigate this with the application of the carbon tax scheme. They have also considered the environmental awareness of consumers and competition among different producers in their study. The study shows that both wholesale price and retail price increase with the implementation of carbon pricing in any kind of market scenario. The study also highlights that the market demand for environmentally friendly retailers will increase in this case. The study also emphasizes the fact that the carbon pricing scheme and regulations significantly improve social welfare.

Another article by Zhang et al. (2019) discussed a socially responsible supply chain and studied centralized and decentralized models of the supply chain. The paper also tried to explore corporate social responsibility and the government's regulations on a socially responsible supply chain. The results of the study suggest that the centralized supply chain possesses higher social responsibility than the decentralized supply chain. Diaz-Trujillo et al. (2019) also mentioned in their article about carbon emission taxes and carbon emission trading schemes as carbon pricing instruments. They have analyzed both the carbon pricing instruments in the Mexican energy sector, where they wanted to analyze the biogas supply chain as a substitute for liquefied petroleum gas. They have also used a multi-objective optimization approach to encourage investment in the biogas supply chain. Results of their study show that these carbon pricing schemes, along with the multi-objective optimization, contribute significantly to the environmental benefits.

As discussed in the article by Li et al. (2017), returning used products helps in tackling environmental problems like carbon emissions and climate change. The article by Gan et al. (2019) further discussed the remanufacturing of used products. They have also discussed the pricing strategy of the remanufactured product. The article also proposed a revenue-sharing model between the manufacturer and retailer in such scenarios. The results of their study also show that the pricing decisions depend upon the willingness to pay and the carbon emission cap. They have also conducted a numerical experiment to show that the carbon cap and trade policies significantly affect pricing decisions.

To obtain the trade-off between emission and cost in the supply chain, Malladi et al. (2020) considered carbon pricing policies and developed a bi-objective optimization model in the bio-mass supply chain. The paper highlights that solving such an optimization model can be very time-consuming and, therefore, develops an algorithm to get the solution of the model in less computational effort. The paper test the developed model and algorithm in a case study involving biomass-fed heater. The results of the study suggest that the carbon tax model is costlier than the carbon offset and carbon cap and trade models. The article finally concludes that the results are applicable to other industries also. Cadavid-Giraldo et al. (2020) highlighted the cement sector as one of the main contributors to carbon emissions, comprising 7% of total carbon emissions. The paper studied the impact of different carbon pricing methods in the reduction of emissions in the cement supply chain. For this, the paper considered a mixed integer linear model and did computational experiments on data from the cement sector to show the benefits of carbon pricing implementation. The results of the study suggest that the carbon tax rate can be between 15 US$ to 150 US$ to effectively reduce the emission, keeping the profit of other stakeholders in mind.

The results further suggest that the suggested tax rate can decrease the emission of carbon by up to 24%. The decision-makers and the cement companies can use the results of this study to reduce carbon emissions and achieve sustainability goals effectively.

Anser et al. (2020) discussed the linkages between carbon pricing, air transportation freight, and combustible renewables and waste by using time series data from 1975 to 2018 in the context of the Arab world. For estimating the relationships like short-term, long-term, and inter-temporal, the study used an innovation accounting matrix and Autoregressive Distributed Lag (ARDL) cointegration technique. The results of the study highlight that combustible renewables and waste shows a negative relationship in the long run and a positive relationship in the short run on the carbon emissions. The results also showed that foreign direct investment (FDI) is directly proportional to carbon emission in the short as well as long run, and fuel imports positively impact carbon emissions. Carbon pricing also positively impacts carbon emissions, and air transport freight also positively influences carbon emissions. The results further showed that carbon pricing, combustible renewables, and waste has a shock of the high intensity on the carbon emission. The study finally concludes that carbon pricing incentivizes firms to look for innovative methods for carbon emission reduction, which can involve ways such as the use of renewable energy sources and increasing the energy efficiency of the system.

A novel model for the network design problem of the supply chain considering economic, environmental, and social dimensions of sustainability is presented by Sherafati et al. (2020). The paper contributes to the literature heavily in terms of applying carbon pricing policies for enhancing regional development levels in addition to environmental improvements. The novel method proposed in the study is in the context of supply chain network design, and it captures the uncertain parameters of the carbon emission problem. The performance and applicability of the proposed model are tested with the help of a case study and numerical examples. The results of the study highlight that all the stakeholders involved can be better off by applying the proposed model. The results of the study will also help the players of the supply chain in taking decisions which will optimize their profit. The paper will also be useful for policymakers and government agencies in taking decisions regarding carbon emission reduction.

Babagolzadeh et al. (2020) discussed the impact of carbon pricing in the cold supply chain, which significantly contributes to carbon emission due to the distribution and storage of products that are temperature sensitive. The paper considers uncertain demand and carbon tax and formulates a stochastic model of two stages for taking decisions regarding the transportation schedule and replenishment policies. The study also aimed to minimize emission cost as well as operational cost and developed a metaheuristic algorithm involving mixed integer programming and iterated local search algorithm. The authors checked the applicability and usefulness of the model and the developed algorithm by testing various instances of various sizes and implementing a case study in Queensland, Australia. The most important results of the study suggested that the high carbon price is not proportional to the efficiency and carbon reduction of cold supply chains. The results further suggested the use of light and medium-duty vehicles for emission reduction and cost saving. The managers and decision-makers can use this study to make the right decisions in cold supply chains for profit maximization and reduction of carbon emissions. Kabadurmus et al. (2020) discussed the cap and trade method of carbon pricing and modelled the problem as a mixed integer linear formulation considering risk factors along with economic and environmental factors. The paper simultaneously considers reliability as well as sustainability factors and analysis the impact of different carbon prices and risk thresholds on the emission, cost and risk of the supply chain. The results of the study suggest that multimodal transportation is an effective way to reduce emissions as well as costs. The results further suggest that the risk averseness increases the carbon emission and cost of the supply chain.

The paper by Mintz-Woo et al. (2021) suggested that the period of COVID-19 was better for introducing carbon pricing than normal time. The authors highlighted that carbon pricing would lead to a new revenue stream for the government which would help in having a safety net at such a difficult time. The authors also highlighted that the oil costs are low, so carbon pricing will not harm consumers much in comparison to the normal time. So, the article highly suggested implementing carbon pricing during the COVID-19 period. Another study by Khan et al. (2022) used carbon damages, renewable energy usage, research and development usage and foreign direct investment to create a climate funding index. The study used cross-panel data from 43 nations to evaluate the impact of carbon pricing, climate funding and financial literacy in exposure to COVID-19 cases. The study uses a robust least square regression model for COVID-19 prediction. The result of the study suggests that climate financing reduces COVID-19 exposure, but carbon pricing and financial literacy do not do so effectively. However, the study further highlighted that carbon pricing and financial literacy improve air quality, and combining this with sustainable healthcare reforms can result in the reduction of COVID-19 infections.

For analyzing a prospering programme among corporations to reduce carbon footprints, Ben-Amer et al. (2022) studied internal carbon pricing (ICP), under which the company assigns a monetary value to emissions, and the money collected can be used in different carbon reduction schemes. The paper tries to explore the motives for such voluntary internal carbon pricing. For this, the article uses data from 1362 firms in the year 2016 to 2018 collected from Carbon Disclosure Project. The result of the study suggests that firm-level climate change exposure contributes to ICP adoption, and opportunity exposure and regulatory shock are the main reasons for internal carbon pricing adoption. The results also suggest board independence as the moderator for the above result. The result of the study can be used by both companies and regulatory bodies for identifying factors affecting ICP adoption and using it for better implementation of ICP and thus contributing to emission reduction. The article by Kannan et al. (2022) emphasizes the need to identify the barriers to implementing carbon regulatory policies and proposes a method to identify and access the same. The study mainly focuses on developing nations and proposes an integrated multi-criteria decision-making approach. The paper first determines the relative significance of barriers by using the Best Worst Method and then uses Evaluation Laboratory and Decision Making Trail to find the interrelationship between barriers. The study uses the proposed framework in manufacturing firms in India having multiple stakeholders, and the results of the case suggest the proposed framework to be very effective.

Rontard et al. (2022) focused on the Emission Trading Scheme (ETS) of the carbon pricing scheme by first examining the establishment of ETS in New Zealand. The authors used policy documents from the New Zealand government and analysis documents by domestic researchers and government officials. The result of the study highlights that the ETS in New Zealand is formed by an exhaustive democratic process involving representatives from all important sectors of the economy. The results further highlighted that the ETS in New Zealand manages the free allocation and supports a low-emission economy. The results finally highlight the difficulty in meeting the theoretical goals in reality for the emission trading schemes. So, the study suggests considering not giving allocation of cap and not keeping low prices for effective implementation of emission trading schemes. Grottera et al. (2022) also showed that the carbon pricing schemes could be very effective in meeting the climate change commitments made in Brazil. The paper studied three greenhouse gas emission pathways involving the emission pricing scenario, sensitive fuel exemption scenario, and command and control policies in the energy sector of

Brazil. The study combines top-down and bottom-up models and thus uses integrated modelling. The results of the study highlight that carbon pricing can be effective, but one needs to consider its scope and applicability as well as coverage in the sector for its effective implementation. For studying the effect of carbon pricing in the European Union, Feindt et al. (2021) conducted a study and examined the incidence of the carbon price on households in the European Union countries. The results suggest that its national-level impact is mainly neutral and sometimes progressive. The results further suggest that the aggregate impact at the European Union level is regressive because of the very strong impact of the carbon price in low-income countries. The results suggest redistribution at the European Union level or national level as an effective scheme for the progressive incidence of carbon pricing. The study also proposed rendering tax burden and providing compensation for the most affected households with the aim of offsetting the regressive effects. The study concludes that the redistribution schemes with the suggested measure will increase the acceptability and usability of carbon pricing in public.

The study by Karthick et al. (2022) analyzed various factors responsible for carbon emission in the production process and maximized the gross profit considering carbon emission reduction as a goal. The study also recognizes the uncertain demand and addresses this with fuzzy numbers. The results of the study propose strategies for effectively managing the inventory of the buyer, and it also helps in taking pricing decisions. Managers and decision-makers can use this study to make decisions regarding pricing and inventory management considering carbon reduction in uncertain environments. For a more holistic understanding of popular carbon pricing policies, Halat et al. (2023) considered carbon regulations in terms of the carbon tax, carbon cap, and carbon cap and trade to study the ordering and pricing competition between two competitive green supply chains. The study also considers the government as a player in the game, which tries to maximize total social welfare and models the problem as a Stackelberg game. The green supply chain considered in the study consists of one manufacturer and multiple retailers. The results of the study provide valuable insights which can be used in taking decisions regarding order size and price. The results also provide insights into how carbon emissions can be reduced, and social welfare can be improved. Abbasi et al. (2023) further the discussion and studied the impact of carbon regulations on green closed-loop supply chain design (GCLSCD) and COVID-19 logistics management. The authors talk about the impact of carbon trading (cap and trade), carbon cap and carbon tax in the GCLSCD in the context of COVID-19 logistics management. The study uses multi-objective mixed integer programming (MOMIP) model for modelling and analyzing the problem. The result of the study can be used by managers to make predictions and take decisions on how to react under carbon regulations to optimize profit.

3. RESULTS AND DISCUSSION

This study tried to review the literature and analyze the applicability of carbon pricing in the supply chain. The result of the study shows that carbon pricing should be used to tackle carbon emissions and climate change problems. The study emphasizes the fact that awareness is very necessary to implement the carbon pricing scheme. The awareness is not only limited to consumers but also applies to manufacturers/producers and retailers. Rational preferences also play an important role in implementing a carbon pricing scheme. For accepting and implementing a carbon pricing scheme, the consumer, manufacturer/producer, and retailer should be rational. The role of the government is also very important in implementing schemes like carbon pricing. The government should impose environmental regulations strictly and ensure the smooth and strict implementation of schemes like carbon pricing. The government should also monitor everything related to the implementation of any scheme, like carbon pricing. The government should also provide support to those manufacturers/producers who want to implement the carbon pricing scheme.

Apart from awareness, environmental regulations, government support, and rational preferences, carbon labelling also helps in implementing the carbon pricing scheme in an effective manner. The study shows that carbon labelling not only shows the carbon footprints of the product but also creates awareness. The awareness created by carbon labelling motivates everyone to look for low-carbon products. The study also shows that implementing carbon labelling and carbon pricing schemes may be non-profitable in the short run, but in the medium and long run, these create a positive impact for manufacturers/producers.

The study also suggests the use of data for implementing these schemes. The study shows that data can be very effectively used for many aspects of smoothly implementing such schemes. The study further suggests that the data can be used to know the awareness among consumers, which is very important for such schemes. The data can also be used to create awareness in an effective manner with the help of targeted advertising. The data can also be used to sell the products to a suitable audience at a suitable price. Targeted advertising or marketing with the help of data helps in keeping the cost low which further helps in keeping the overall cost low after applying carbon pricing.

The study also suggests ways in which pricing can be done. The study suggests some ways of implementing both the instruments of carbon pricing, carbon tax scheme and carbon cap and trade. The study emphasizes the use of two-part tariffs, which involves a tariff from the retailer's side and a tariff from the manufacturer's side. The study also suggests the use of data for deciding these tariffs. The study also talks about the spreading of prices due to domestic and global trade. The study suggests the use of a centralized supply chain for implementing all these pricing schemes. The study shows that the implementation of carbon pricing is the most effective and easy in a centralized supply chain. The study further shows that the collection of data and monitoring is also easy in a centralized supply chain. The study prefers the centralized supply chain over all types of decentralized supply chains for the implementation of such schemes.

The study further emphasizes the fact that these types of schemes significantly impact pricing decisions and social welfare. So, the creation of awareness among consumers for the implementation of these schemes is very important. The awareness can be created in an efficient way with the help of data which is supported by many ways such as targeted advertising and carbon labelling.

The study finally suggests returning used products for better implementation of these schemes to control carbon emission reduction and climate change. The study also suggests the remanufacturing of used products. The study further suggests that remanufacturing used products is an effective method of tackling carbon emission and climate change problems. The study also suggests some ways of implementing this, which involve revenue sharing between both retailer and manufacturer, creating awareness, using data, and incentivizing returning of used products. The study also emphasizes the fact that awareness is extremely necessary for implementing this, along with government and local bodies' support, as this is about making behavioural change. Consumers, in general, have the habit of throwing used products, and collection of these used products is required. These products can be collected after the consumers throw them with the help of local bodies, but this study suggests creating awareness and incentivizing consumers to give the used product in such places so that it can get easily returned to manufacturers. After remanufacturing also, proper pricing needs to be done to sell the product. Targeted marketing with the use of data is another important tool for doing the same.

Overall, this section discussed whether the carbon pricing scheme could be or should be implemented in the supply chain or not and the ways in which it can be effectively implemented. The results suggested that carbon pricing can be and should be implemented in the supply chain. The study also showed many ways of implementing the scheme in an effective manner. The study emphasized the use of carbon labelling, creating awareness and use of data. The study also discussed the importance of government support and environmental regulation in the implementation of such schemes. The study also talked about pricing methods, tariffs, and centralized supply chains. The study also gave emphasis on the remanufacturing of used products. The study also prioritized the social welfare of people for the implementation of such schemes.

4. CONCLUSION

This study was aimed at chronologically reviewing the literature and understanding the applicability and effectiveness of carbon pricing in the supply chain. This study also tried to understand different methods by which it can be applied in the supply chain more effectively. This study also explored the important factors which affect the carbon pricing decisions in the supply chain. From the study, it appears that carbon pricing in the supply chain is an effective method of reducing carbon emissions and mitigating climate change. This study further suggests that carbon pricing in the supply chain should be widely used. This study also emphasizes creating awareness for the smooth and effective implementation of carbon pricing in the supply chain. This study also suggests the use of data for strategizing the implementation of carbon pricing in the supply chain. The strategy involves targeted advertising, taking pricing decisions, ways of creating awareness and many such things. This study also emphasizes carbon labelling for better implementation of such schemes. The study further suggests the use of two-part tariffs and a centralized supply chain for better implementation. This study further emphasized on remanufacturing of used products to lower the price as well as to contribute to carbon emission reduction and climate change mitigation objectives.

5. FUTURE SCOPE

This study dealt with the literature review, and in future research, data from different industries or firms can be taken to establish the results. By using pre and post-data on carbon pricing of a supply chain, the study can be done to get the findings. In future, a case study of any of the results can be done to validate the results. The different methods of pricing decisions can also be studied. Future research can also be done to find different methods of creating awareness and compare them to get the most effective method. Future research can also be intended to explore the applicability and effective operationalization of carbon labelling methods in different contexts and different industries. Future research can also be done by reviewing and analyzing the effect and applicability of 4IR in the supply chain context for tackling carbon emission and climate change problems.

REFERENCES

Abbasi, S., & Erdebilli, B. (2023). Green closed-loop supply chain networks' response to various carbon policies during COVID-19. *Sustainability, 15*(4), 3677. doi:10.3390u15043677

Anser, M. K., Yousaf, Z., Hishan, S. S., Nassani, A. A., Sheikh, A. Z., Vo, X. V., ... Abro, M. M. Q. (2020). Dynamic linkages between transportation, waste management, and carbon pricing: Evidence from the Arab World. *Journal of Cleaner Production, 269*, 122151. doi:10.1016/j.jclepro.2020.122151

Apata, O., Adebayo, A. V., & Ainah, P. K. (2021, August). Renewable energy systems and the fourth industrial revolution. In *2021 IEEE PES/IAS PowerAfrica* (pp. 1-5). IEEE.

Babagolzadeh, M., Shrestha, A., Abbasi, B., Zhang, Y., Woodhead, A., & Zhang, A. (2020). Sustainable cold supply chain management under demand uncertainty and carbon tax regulation. *Transportation Research Part D, Transport and Environment, 80*, 102245. doi:10.1016/j.trd.2020.102245

Barrett, S. (1990). Pricing the environment: The economic and environmental consequences of a carbon tax. *Economic Outlook, 14*(5), 24–33. doi:10.1111/j.1468-0319.1990.tb00670.x

Ben-Amar, W., Gomes, M., Khursheed, H., & Marsat, S. (2022). Climate change exposure and internal carbon pricing adoption. *Business Strategy and the Environment, 31*(7), 2854–2870. doi:10.1002/bse.3051

Burgess, J. C. (1990). The contribution of efficient energy pricing to reducing carbon dioxide emissions. *Energy Policy, 18*(5), 449–455. doi:10.1016/0301-4215(90)90045-6

Cadavid-Giraldo, N., Velez-Gallego, M. C., & Restrepo-Boland, A. (2020). Carbon emissions reduction and financial effects of a cap and tax system on an operating supply chain in the cement sector. *Journal of Cleaner Production, 275*, 122583. doi:10.1016/j.jclepro.2020.122583

Cheng, Y., Sun, H., Jia, F., & Koh, L. (2018). Pricing and Low-Carbon Investment Decisions in an Emission Dependent Supply Chain under a Carbon Labelling Scheme. *Sustainability, 10*(4), 1238. doi:10.3390u10041238

Cheng, Y., & Zhang, P. (2017, July). Pricing strategy and coordination in an O2O mixed channel supply chain with carbon cap-and-trade mechanism. In *2017 36th Chinese Control Conference (CCC)* (pp. 7528-7534). IEEE.

Cline, W. R. (1995). *Pricing Carbon Dioxide Pollution*. Economic Analysis Project Working Paper.

Decanio, S. J. (1992). International cooperation to avert global warming: Economic growth, carbon pricing, and energy efficiency. *Journal of Environment & Development, 1*(1), 41–62. doi:10.1177/107049659200100104

Diaz-Trujillo, L. A., Tovar-Facio, J., Napoles-Rivera, F., & Ponce-Ortega, J. M. (2019). Effective Use of Carbon Pricing on Climate Change Mitigation Projects: Analysis of the Biogas Supply Chain to Substitute Liquefied-Petroleum Gas in Mexico. *Processes (Basel, Switzerland), 7*(10), 668. doi:10.3390/pr7100668

Du, S., Zhu, L., Liang, L., & Ma, F. (2013). Emission-dependent supply chain and environment-policy-making in the 'cap-and-trade'system. *Energy Policy, 57*, 61–67. doi:10.1016/j.enpol.2012.09.042

Feindt, S., Kornek, U., Labeaga, J. M., Sterner, T., & Ward, H. (2021). Understanding regressivity: Challenges and opportunities of European carbon pricing. *Energy Economics*, *103*, 105550. doi:10.1016/j.eneco.2021.105550

Gan, W., Peng, L., Li, D., Han, L., & Zhang, C. (2019). A Coordinated Revenue-Sharing-Based Pricing Decision Model for Remanufactured Products in Carbon Cap and Trade Regulated Closed-Loop Supply Chain. *IEEE Access : Practical Innovations, Open Solutions*, *7*, 142879–142893. doi:10.1109/ACCESS.2019.2943385

Grottera, C., Naspolini, G. F., La Rovere, E. L., Gonçalves, D. N. S., de Farias Nogueira, T., Hebeda, O., ... Lefèvre, J. (2022). Energy policy implications of carbon pricing scenarios for the Brazilian NDC implementation. *Energy Policy*, *160*, 112664. doi:10.1016/j.enpol.2021.112664

Halat, K., Hafezalkotob, A., & Sayadi, M. K. (2023). The green supply chains' ordering and pricing competition under carbon emissions regulations of the government. *International Journal of Systems Science: Operations & Logistics*, *10*(1), 1983884.

He, Y., Wang, L., & Wang, J. (2012). Cap-and-trade vs. carbon taxes: A quantitative comparison from a generation expansion planning perspective. *Computers & Industrial Engineering*, *63*(3), 708–716. doi:10.1016/j.cie.2011.10.005

Hornibrook, S., May, C., & Fearne, A. (2015). Sustainable development and the consumer: Exploring the role of carbon labelling in retail supply chains. *Business Strategy and the Environment*, *24*(4), 266–276. doi:10.1002/bse.1823

Kabadurmus, O., & Erdogan, M. S. (2020). Sustainable, multimodal and reliable supply chain design. *Annals of Operations Research*, *292*(1), 47–70. doi:10.100710479-020-03654-0

Kannan, D., Solanki, R., Kaul, A., & Jha, P. C. (2022). Barrier analysis for carbon regulatory environmental policies implementation in manufacturing supply chains to achieve zero carbon. *Journal of Cleaner Production*, *358*, 131910. doi:10.1016/j.jclepro.2022.131910

Karstensen, J., & Peters, G. (2018). Distributions of carbon pricing on extraction, combustion and consumption of fossil fuels in the global supply-chain. *Environmental Research Letters*, *13*(1), 014005. doi:10.1088/1748-9326/aa94a3

Karthick, B., & Uthayakumar, R. (2022). Impact of carbon emission reduction on supply chain model with manufacturing decisions and dynamic lead time under uncertain demand. *Cleaner Logistics and Supply Chain*, *4*, 100037. doi:10.1016/j.clscn.2022.100037

Khan, H. U. R., Usman, B., Zaman, K., Nassani, A. A., Haffar, M., & Muneer, G. (2022). The impact of carbon pricing, climate financing, and financial literacy on COVID-19 cases: Go-for-green healthcare policies. *Environmental Science and Pollution Research International*, *29*(24), 35884–35896. doi:10.100711356-022-18689-y PMID:35064505

Li, H., Wang, C., Shang, M., & Ou, W. (2017). Pricing, carbon emission reduction, low-carbon promotion and returning decision in a closed-loop supply chain under vertical and horizontal cooperation. *International Journal of Environmental Research and Public Health*, *14*(11), 1332. doi:10.3390/ijerph14111332 PMID:29104268

Li, J., Du, W., Yang, F., & Hua, G. (2014). The carbon subsidy analysis in remanufacturing closed-loop supply chain. *Sustainability*, *6*(6), 3861–3877. doi:10.3390u6063861

Li, W., & Jia, Z. (2017). Carbon tax, emission trading, or the mixed policy: Which is the most effective strategy for climate change mitigation in China? *Mitigation and Adaptation Strategies for Global Change*, *22*(6), 973–992. doi:10.100711027-016-9710-3

Liang, X., Xiong, Y., Hou, Y. Z., & Xia, S. M. (2014). Pricing and emission reduction strategy of manufacturer with carbon emission constraints in low carbon supply chain. *Key Engineering Materials*, *572*, 659–662. doi:10.4028/www.scientific.net/KEM.572.659

Liu, P. (2019). Pricing policies and coordination of low-carbon supply chain considering targeted advertisement and carbon emission reduction costs in the big data environment. *Journal of Cleaner Production*, *210*, 343–357. doi:10.1016/j.jclepro.2018.10.328

Liu, P., & Yi, S. P. (2018). A study on supply chain investment decision-making and coordination in the Big Data environment. *Annals of Operations Research*, *270*(1-2), 235–253. doi:10.100710479-017-2424-4

Ma, X., Ho, W., Ji, P., & Talluri, S. (2018). Coordinated pricing analysis with the carbon tax scheme in a supply chain. *Decision Sciences*, *49*(5), 863–900. doi:10.1111/deci.12297

Malladi, K. T., & Sowlati, T. (2020). Bi-objective optimization of biomass supply chains considering carbon pricing policies. *Applied Energy*, *264*, 114719. doi:10.1016/j.apenergy.2020.114719

Mintz-Woo, K., Dennig, F., Liu, H., & Schinko, T. (2021). Carbon pricing and COVID-19. *Climate Policy*, *21*(10), 1272–1280. doi:10.1080/14693062.2020.1831432

Pena-Torres, A., & Pearson, P. J. (1999). *Pricing carbon emissions in the UK electricity market: A nuclear revival?* Academic Press.

Rontard, B., & Hernandez, H. R. (2022). Political construction of carbon pricing: Experience from New Zealand emissions trading scheme. *Environmental Development*, *43*, 100727. doi:10.1016/j.envdev.2022.100727

Sherafati, M., Bashiri, M., Tavakkoli-Moghaddam, R., & Pishvaee, M. S. (2020). Achieving sustainable development of supply chain by incorporating various carbon regulatory mechanisms. *Transportation Research Part D, Transport and Environment*, *81*, 102253. doi:10.1016/j.trd.2020.102253

Xia, L., Guo, T., Qin, J., Yue, X., & Zhu, N. (2018). Carbon emission reduction and pricing policies of a supply chain considering reciprocal preferences in cap-and-trade system. *Annals of Operations Research*, *268*(1-2), 149–175. doi:10.100710479-017-2657-2

Xie, X. P., & Zhao, D. Z. (2013). Decision mechanism study on product pricing and emission reduction in two level low-carbon supply chain enterprises based on the CDM. *Soft Science, 27*(5), 80-85.

Xu, X., Xu, X., & He, P. (2016). Joint production and pricing decisions for multiple products with cap-and-trade and carbon tax regulations. *Journal of Cleaner Production*, *112*, 4093–4106. doi:10.1016/j.jclepro.2015.08.081

Zakeri, A., Dehghanian, F., Fahimnia, B., & Sarkis, J. (2015). Carbon pricing versus emissions trading: A supply chain planning perspective. *International Journal of Production Economics*, *164*, 197–205. doi:10.1016/j.ijpe.2014.11.012

Zhang, Z. C., Li, K. W., Liu, Z., & Huang, J. (2019). Pricing decisions in a socially responsible supply chain under carbon cap-and-trade regulation. *IFAC-PapersOnLine*, *52*(13), 331–336. doi:10.1016/j.ifacol.2019.11.130

Zhou, Y., Hu, F., & Zhou, Z. (2018). Pricing decisions and social welfare in a supply chain with multiple competing retailers and carbon tax policy. *Journal of Cleaner Production*, *190*, 752–777. doi:10.1016/j.jclepro.2018.04.162

Chapter 16
Existing Implications and Relationship Between Anthropology and Anthropocene Urban Socio-Ecology Planning Resilience

José G. Vargas-Hernandez
Tecnológico Superior de Jalisco Mario Molina, Zapopan, Mexico

Carlos Rodriguez-Maillard
https://orcid.org/0000-0003-2406-196X
Universidad Cristobal Colón, Mexico

Omar C. Vargas-González
https://orcid.org/0000-0002-6089-956X
Tecnológico Nacional de México, Ciudad Guzmán, Mexico

ABSTRACT

This study aims to analyze some of the existing implications between urban anthropology and Anthropocene urban socio-ecology planning resilience. Beginning with the assumption that urban anthropology gives support to create and develop any urban planning based on the Anthropocene urban socio-ecology resilience, the methods employed are the analytical-descriptive based on an ethnographic interpretation and reflection of the theoretical and empirical literature review. The analysis concludes that urban anthropology fundamentals give support to strengthen the Anthropocene orientation of the urban socioecological planning resilience.

DOI: 10.4018/978-1-6684-6366-6.ch016

INTRODUCTION

Kuhn (1962/1996) analyzes the paradigm shifts focusing on the epistemic cycles of methods of inquiry from emergence and decline, which can be referred to the shift from traditional urban anthropology to Anthropocene urban socio-ecology and resilience. Theoretical, strategic and tactical systematization lead to demand paradoxes and contradictions through diagnostics and consultancies providing solutions to problems and explanation of false consciousness in the applications of organizational and urban anthropology to urban socio-ecological resilience planning and green innovation areas, in the current global environmental problems facing the world.

First, the term resilience was introduced by Holling (1973) in the field of socio-ecological research to describe the ability of the socio-ecosystem to return to the original shape after being deformed by the impact of external pressures from the outside environment. Resilience influences urban planning and innovation, construction of urban infrastructure, and development leading to urban and regional economic resilience as one of the contents of urban planning and governance resilience research. Improvement of urban socioecological resilience contributes to disaster prevention, management and reduction leading to strengthen the urban resilience processes (Yang, Lili & Hongchi, 2022). Some approaches to urban anthropology do not consider it as a science and those who prefer theories or the particular over the universal, and those who choose to interpret instead of to explain.

Second, the term urban anthropology is a scientific discipline epistemologically closely linked and interrelated to other scientific theories. Anthropology as any social sciences predict various phenomena (Alvarez, 2016; Brady, 2019; Grimmer et al., 2021; Hindman, 2015; Hofman et al., 2017; Hofman et al., 2021; Taagepera, 2008). The scientific status of anthropology is beyond the conception of a soft, interpretive, and humanistic phenomena to interpret meanings to use quantitative techniques, apply formal methods in experimental designs to postulate scientific theories predicting phenomena and formulating scientific laws linked and interrelated to other scientific theories. Sociology, anthropology, and political science phenomena differs from individual behavior in accordance with Bunge (1999) who sustains that predicting the behavior of an unknow individual is different from the prediction of a system behavior. However, social sciences have moved prediction from small groups to complex social systems through sophisticated computational methods (Bunge, 1999; Kaplan, 1940). A scientific law is a logical-mathematical representation of the relationships between variables (Pfeifer, 2006).

Scientific predictions emerge from the verification and application of a capable theory to explain the specific phenomena. Research in urban anthropology and socio-ecology resilience in urban planning and green innovation areas using interdisciplinary methodologies is changing village societies relevant for research at global level.

Third, since 1945, the post-war period accelerated the changes in human civilizations with negative impacts on the biosphere which has been increasing more since the end of the cold war (Eriksen, 2016; Snow, 1998). This displacement originated in large urban centers of industrialized countries marks the bounding period leading to validate the ethnographic method. Alternatives to ideologies and practices of global capitalism are human lives not dominated by the state and market forces (Scott, 2009; Escobar, 2020). Other natural phenomena are also causing urban ecological and resilience dysfunctionalities such as, severe storms flooding urban communities, immobilizing, and disrupting urban transportation, energy, water, and waste networks. Cities and any urban area are virtually marinated in fossil fuels largely controlled by corporations. The interactions of stakeholders and actors contributes disorderly to identify the inherent complexity of urban socio-ecological meta-problems and contribute to intensify the complexity which requires to deal the cross-linking process.

Anthropomorphic representations in research have not been convincing to favor the functional hypothesis. Science solves testable hypothesis (Hirst, 2019). Socio-economic and organizational sciences research focus in sympathy for going native for subject population moving from the realm of an observer to become an active participant (Bilharz, 1998; Kolb, 1998). Interdisciplinarity and transdisciplinary of urban anthropology applied to urban socio-ecological planning and green innovation areas are limited in space to carry out the development of disciplines in multidimensional crisis including the ethical values and moral issues leading to an ontological crisis of the homo sapiens (Tarantino-Curseri, 2018). Ethnographic writing, interpretivism and perspectives in urban anthropology must be replaced with experimentation, theorization, causality, prediction, mathematical modeling and epistemology. Participant observation is an ethnographic methodology and one of the most popularized and important technique developed by Malinowski (1961) in anthropology research which can be applied to urban anthropology and its impact on urban socio-ecology resilience planning and development (DeWalt, DeWalt & Wayland, 1998; Kawulich, 2005; Jorgenson, 2015).

The conceptual framework of the institutional analysis model is proposed by René Lourau (1979) and Lapassade (1986) which can be applied to analyze the urban anthropology in the Anthropocene of socioecological urban planning and green innovation areas. The ethnomethodological urban anthropology perspective of urban socio-ecology resilience planning and green innovation areas aims to the inherent transformation of socio-historical process affecting the production of non-contemplative knowledge claiming the meaning and challenging to decipher without typologies in an ahistorical way. Unstructured techniques without a specific direction leads to collect deeper and broader answers and the original terms to know the native categories and perspectives of urban anthropology in urban socio-ecological planning (Pizarro, 2014). Interviews through defined questions are recommended as an ethnographic research instrument at the lowest level of structuring to extract greater amount of information linking urban anthropology and urban socio-ecology resilience planning.

This study aims to analyze some of the relationships and implications between the urban anthropology and the Anthropocene urban socioecology planning resilience, in order to create awareness in the impact of the climate change and green innovations. To achieve this goal, it starts in the first term to analyze the organizational and urban anthropology followed by the urban socio-ecology resilience planning and urban green innovation areas. Finally, some conclusions are given.

BACKGROUND

Qualitative-quantitative division of methods and techniques in urban anthropology and urban socio-ecology resilience planning and innovation is objectionable (Aldrich, 2014; Reynoso, 1995b). Recent research addressing the quantitative and qualitative approaches to reveal the consequences of the COVID-19 pandemic crisis on the theoretical framework of urban, society and institutional governance resilience, was conducted by (Olimid, Georgescu & Gherghe, 2022). Quantitative methods and techniques in urban anthropology and urban socio-ecology resilience planning and innovation aim to collect and analyze numerical data (Roni et al., 2020).

Formal methods must meet requirements in operationalization of terms, data analysis, logical consistency, justification of statements and replication of results, although neither formal theories were dominant nor were the dominant theories ever dominant (Reynoso, 1995b). The discipline of urban anthropology is capable of formalizing produced knowledge. Formal computational methods used for the ethnographic analysis in urban anthropology and archeological data based on agent models, cellular automata, diffusion and percolation models, network analysis, etc., proposed for a new research form open to theoretical development and experimentation (Díaz et al. 2007).

Urban anthropological experiments, predictions and theories can be compared with natural and physical sciences to determine the scientific status (Llanos, 2009) but are not sufficient scientific criteria for good science if it is conducted without laws and predictions. The most common experiments in urban anthropology and social sciences are natural experiments (Diamond & Robinson, 2010; Dunning, 2012), field experiments (Baldassarri & Abascal, 2017) and naturalistic experiments (Bernard, 2011).

To carry out this bibliometric review, electronic, literary and scientific sources were used as the basis for recent research on relationship between Anthropology and Anthropocene urban socio-ecology planning resilience. The study was conducted by searching for the keywords "urban anthropology", "urban socio-ecology resilience planning", "Anthropocene urban socio-ecology planning resilience", and "urban green innovations areas" in the main scientific search engines, previous works were reviewed in detail to define those that would be included in this study.

Organizational and Urban Anthropology

Organizational anthropology has close relationships with urban anthropology and that includes the demands for participatory actions in practical anthropology of urban imaginaries and organizational facilitation (Gravano, 1992). This situation integrates the organizational and urban anthropological perspectives and the ethnological methodology tools to the analysis of organizational studies, transversal to research financing demands. Social sciences can make possible scientific predictions on specific events and problems in urban socio-ecological planning and development despite the arbitrariness of human behavior and the assumption that is not useful to predict behavior (Nießen et al., 2020).

The debates in the analysis of organizational and urban anthropology bridging the studies between organizations and urban areas from an anthropological perspective to analyze the interrelationships between urban anthropology and business and work organizations, from a sociohistorical orientation to develop reflections situating the analysis of urban areas and neighborhoods in post-industrial and global societies. Urban anthropology can be analyzed from a socio-historical through its interrelationships with organizations and business.

The work of Warner ratified the analysis of Durkheim from a sociohistorical perspective has an influence on anthropology and sociology related to the social structures of tribal and industrialized complex societies, advancing the analysis of interrelationships between urban anthropology and organizations. Industrial technology and social research become relevant and, consequently, may improve the understanding of technology in organizational anthropology and urban socio- ecological planning and innovation (Roca, 1998).

Organizational and urban anthropology has conceptual, methodological, and theoretical tools aimed to develop the interrelationships between organizations and urban settlements, leading to create the meanings that both workers in organizations and citizens have of their industrial heritage. The studies of Mayo (1945) and collaborators Roethlisberger and Dickson (1939) at the industrial plant of Hawthorne

Wester Electric Co. pioneering the investigations of organizational anthropology applied to work productivity. Job design in urban work settlements and organizations is characterized by the development of multiple skills based on redundancy of a variety of functions instead of redundancy of parts being capable to perform the job.

A sociohistorical perspective analysis of the interrelationships between urban anthropology and work organizations is conducted on the anthropological practices in organizations and urban settings that have challenges affecting research and academic training and development. A multidisciplinary perspective on urban anthropology and green and socio-ecological innovation attempts to understand traditional technology in production process. Technological training of firms reinforces the converging of scientific research on green innovation at universities. Firms can share the green technological innovation training fostered by public green investment. Technological and social capital determine the level of socio-ecological and urban anthropology in green innovation of companies. Accumulation of social, economic, and symbolic capital becomes relevant to undertake activities in urban socio-ecology resilience planning and green innovation (Reguera, 2009). Renewable energy and technological innovation have been growing in the last years, but such technologies developed over time and are accompanied by decay and depreciation effects, need to be tracked to avoid their potential negative consequences (Su et al., 2022).

Organizational performance reflects the problems and dilemmas of the organizational practices linked to the state and urban anthropology in socio-ecological planning control and weighting the anthropological ethnography method used in organizational facilitation in professional practice. Urban anthropology in socio-ecological planning is a negotiation and mobilization process leading to the implementation of incremental action definition (de Melo, 2002). Urban neighborhood, as organizational anthropology in organizational culture exercise controls of symbolic culture which is manipulated as ideology aimed to control the urban anthropology or organizational living conditions and interrelationships of the social subjects.

Urban anthropology leading to socio-ecological planning processes are in evolution as a dialectical process from a centralized, sequenced, and systematic integral rational planning to a complete fragmentation and disjointed incrementalism (Lindblom, 1959). Strengthening the consolidation of the national systems of science, technology, and innovation has effects on competitiveness derived from incremental socio-ecological and green innovations in urban planning. Today it is important to include the covid in our urban anthropology study to understand how cities were being reshaped by the pandemic (Hyatt, 2021), it is imperative that the universities and governments start to study about it.

Urban Socio-Ecology Resilience Planning and Urban Green Innovations Areas

The evolution of scientific knowledge and socio-ecological and technological green innovation in urban planning is driven by the challenges proposed by the design of new goods, services and processes carrying out internal research and development activities in the firms and in cooperation with other agents in efforts to improve efficiency of regions with deficiencies in their scientific, technological and innovation systems. Firms seeking to innovate in urban socio-ecological planning face obstacles in finances. Tax and market incentives leave decisions to firms about which urban development projects to finance (Melo, 2001).

Urban planning of socio-technical and urban socio-ecological innovation in urban regions known as local industrial systems may constitute an example of an economy derived from industrial location self-reinforcing agglomerations of cumulatively socio-technical skills, investments and venture capital, infrastructure, suppliers, knowledge spillovers, services, etc. associated with informal information flows and proximity to research centers and universities (Saxenian, 1994). Urban planning impacts more variables that are related to the environment (Baschera and Schneider,2022).

To stimulate investments in green science, green technology and socio-ecological innovation systems for urban planning requires the necessary conditions to invest in research and development (de Melo, 2001). It is essential that government should stimulate enterprises to strengthen the research of environmental protection technologies (Su et al., 2023). Green technological innovation not tested has high risks and difficulties in evaluating the green high-tech goods and socio-ecological services market and green technology-based firms have difficulties in accessing capital sources. Investment activities in innovation activities have a high level of risk and uncertainty in investments in credit financing and equity financing in research & development of urban green planning innovation in technologies.

Universities provide technical training and development in urban planning and socio-ecological innovation. The university-industry interactions between the research institutes and centers for urban socio-ecological and environmental planning with firms interested in joint research programs sponsored by consortiums or individual government agencies and corporations to exploit results through patents and licenses. Through business incubators, universities and research institutions, technology parks, and science, green technology and socio-ecological innovation agents in university-based local green innovation systems, startups at earlier stages can have the opportunity to access government venture capital.

Government funding for research in green technological innovation for urban planning carried out by firms has a positive effect on business research and development spending (Guellec & van Pottelsbergue, 2000, in Melo, 2001). Government authorities facilitate the formation of local green and urban socio-ecology resilience innovation systems from universities and research centers increasing public investing in foundations of science, green technology, and socio-ecological innovation. Government intervention in university-based local urban green innovation university oriented may regulate the incentives to promote green technology innovation training and strengthen emerging green technology-based firms.

Local urban planning innovation systems for green socio-ecological areas integrate the definitions considered in Lundvall (1992) and the one used by MIT (Lester, 2005). They have some elements that interact with creation, dissemination, transference, and use of knowledge for economic value in specific regions (Lundvall, 1992), and show that urban planning impacts community vitality; ecological diversity and resilience; and most important, living standards of sustainable socio-economic development and environmental conservation. (Baschera and Schneider,2022).

The context of urban green technological innovation transfers from the University-based to the firms is a factor of competitiveness of productive sectors of society. Massive investments are made in infrastructure and equipment by government agencies and institutions at research centers and universities for urban planning and green innovation (Porter, 2001). Creation of technical schools and incentive programs for implementation of urban planning policies and green innovative practices in industry. Several policies and programs have been instituted to manage climate change impacts but institutional capacity of agency systems to ensure the realization of the policies is limited (Asibey et al., 2022).

Urban socio-ecology resilience planning and green innovation networks constitute a similar economy of connections which can be effective in the future of work, feeding the discussions of ideas, thoughts, and sentiments to uncover the native`s point of view in local perspective to obtain, analyze and interpret data. Some actors such as research laboratories, students and researchers' network, firms of the incubator, angel investors and venture capital, etc., participate in urban environmental planning processes, although those who are not involved on the governance of local urban planning and socio-ecological innovation systems are not necessarily less influential. Non-financial institutions in governments are hardly available to complement the green investment of private venture capitalists in green firms.

Declining government participation is a principle applied to mature local urban green innovation systems (OECD, 1997b; Melo, 2001). Anchor organizations and companies focus on urban planning of local innovation systems university-oriented to cooperate in joint efforts of R&D projects directing the focus of cooperation for research to socio-ecological and green areas can get competitive advantage with a competent university. The analysis on morale and evolution of cooperation demonstrated that individual moral models tend to obey the rules and religious beliefs, predict intergroup cooperation and honest behavior (Purzycki et al. 2018).

The set of green firms are involved in an articulating task to support mechanisms to identify, select, and consolidate urban green innovation projects through the articulating instruments (Pimienta-Bueno, 1999). Local urban planning socio-ecological concentration of green firms enables economies of scale, agglomeration, and gains from externalities (Audretsch, 1998). Urban green innovation technologies in emerging technology intensive green firms can be compared to mature firms in terms of achievements and outputs. Green technology-based start-ups have higher costs related to their seed and initial stages that are higher than those of small firms (OECD, 1997b). Today firms and industries should produce, import, and update traditional facilities with green technologies to optimize energy structure (Su et al., 2022)

The urban imaginary phenomenon in the social participation processes can be treated from the anthropological organizational facilitation (Gravano, 2008a), aimed to the treatment of specific problems of these processes. The relationship between metaphorical mediation and interpretation deepens the analysis of urban anthropology contexts to challenge the societal construction of otherness.

The urban planning for green and socio-ecological innovation theory still was marginal during the 1970s until recently innovation acquired the multidisciplinary character encompassing knowledge in fields such as the sociology, economics, psychology, administration, geography, engineering, etc. (Sundbo, 1998; Schlapfer & Marinova, 2001). Throughout the 1980s and 1990s the economy was submitted to technological socio-ecological innovation restructuring and a process of economic internationalization and globalization of markets leading to replace the scientific and technology system to focus on the competitiveness of the industrial sector aimed to meet the international markets (González Alvarez, 1997).

Urban planning for socio-ecological and green innovation areas is the process that begins with the invention of a new element until its development for commercial use. The commercialization of urban socio-ecological planning green innovations generated in public universities and research institutions requires clear regulations for sharing and transferring intellectual property rights to private firms, new forms of association in public-private partnerships for technology development. Green innovation must be practical, possess commercial value and be socially acceptable (Sundbo, 1998). Knowledge sources for planning in urban green areas are embedded in institutional and human forms subject to financial capital and less to geographical movement (Schlapfer & Marinova 2001). Urban planning can adequately address the impacts of climate change while preparing urban areas to be more proactive and adaptive to climate change impacts through effective partnerships (Asibey et al., 2022).

In urban socio-ecological planning, imaginaries are articulated to social participation framed by an anthropological theoretical approach and ethnographic research to disciplinary reflection and practice of organizational facilitation. Conceptual, theoretical, and practical ruptures produce the transformation of urban socio-ecological planning participation processes and their relationships challenged by methodological research and facilitation.

Approaching the professional practice of organizational facilitation and the participatory processes presents theoretical-methodological challenges in metropolitan urban socio-ecological environmental planning (Gravano, 2007b) supported by the work of organizational culture of organizational and economic agents and political actors. Manchester and Chicago schools have been working on urban anthropology (Cucó Giner, 2004; Gravano, 2005; Hannerz, 1986; Signorelli, 1999; Velho, 1978).

Works on entrepreneurship and business anthropology date back to the study of ethnology in urban societies in the 1970s (Durrenberger, 2007). Expansion of infrastructure built by the university through urban green innovation technological planning provides socioeconomic consequences results of entrepreneurial alliance research of the constitution of angel investors as the formal group of employees and students at the university investing in firms at the business incubators and innovation parks.

Urban socio-ecological planning processes occurring with institutional and community participation addressing the circularity of management issues and exchanging institutional power relations based on the anthropological object of cultural otherness (Gravano, 2006). Urban planning and architecture are demanded by specific actors and urban management agents linked to urban imaginaries weighted as an anthropological object for the detection of meanings for the living spaces for social actors. Administration and management practices have questioned effects on equality and the distribution of wealth.

Anthropological contributions to urban socio-ecological and environmental planning are based on community participation in institutional and organizational processes. Individual and corporate profits take precedence over vocations and values of local socio-ecological communities leading to the sacrifice of common good. Community participatory process enforced by law delimit the specific possibilities, difficulties and opportunities based on asymmetric rationalities at stake and the state action in the context or urban socio-ecological environmental planning. A specific situation of demand from the State for an anthropological contribution facilitates the community participatory process. A well-structured state can develop a local green and urban socio-ecology resilience innovation system to support activities of firms, institutions, universities and research centers, private economic sector, etc.

The process of urban planning and green innovation and green technology transfer begins with an idea that the research develops into a technology to be used and applied followed by a disclosure and evaluation of the potential economic value, discuss, and consult with specialists in concerning urban areas of discovery considering the criteria based on scientific and technical relevance, existent patents, and market potential. From this, it follows the registration of the patent and intellectual property. Reports indicate that urban planning efforts have yielded the maximum positive result in building the resilience and adaptive capacities of cities to climate change impacts (Asibey et al., 2022).

The impact of specific risk disasters in urban areas such as earthquakes, hurricanes, floods, public health emergencies, etc., lead to urban socio-ecology planning resilience to prevent the possible impact of these extreme events and provide a valuable diagnose of the weak links in urban resilience and improve the resistance and city resilience of future sustainable development of human society. The case was analyzed by Mu, Fang, Yang, & Guo (2022) based on the population mobility, centrality, and nodality between cities in China was significantly reduced affecting the urban resilience within the network during the COVID-19 epidemic.

The ancient Greeks developed critical thinking on urban socio-ecology resilience based on mutualistic relationships (Egerton, 2001). Socio-ecological change in societies is caused by exogenous factors, such as globalization, colonialism, capitalism, etc. Some colonial societies are compounded of heterogeneous elements with instability due to divergent interests (Radcliffe-Brown, 1940). On the other side, the Spanish applied anthropology concentrated more on the analysis of the legacy left on its colonies and directed to the colonized others by the Spanish empire. Corporations are considered and critiqued as an anthropological colony of the real opposition to knowledge transfer in socio-ecology of work (Mir & Mir, 2009). Individuals and groups with conflicting ecological interests encompasses incommensurable connections between ends and means (Caballero, 1969 in de Melo, 2002a, 2002b).

Anthropocene Urban Socio-Ecology Planning Resilience

Socio ecosystems in urban areas are aimed to maximize the benefits of human needs in an Anthropocene era. The concept of Anthropocene is linked to socio-ecological aspects of climate change and to the analysis of the causes of the collapse of civilization and societies that can be learned from archaeological research when facing crises (Tainter, 1988; Tainter, 2008; McIntosh, et al., 2000; Trawick & Alf, 2015). Western science as a mode of existence is a source of prescriptions to climate change complemented with the interpretation and translational (Latour, 2013; Finucane, 2009). Communities merge global scientific explanations with regional and local knowledge blaming humans for the global warming and climate change (Schnegg et al., 2021).

The global framework for climate services approach has provoked critical reflections on initiatives related to ethical implications, epistemic politics, colonial legacies, etc. (Haines, 2019b; Krauß, 2020; Nost, 2019; Webber, 2019). International policy frameworks underpinned discourses combined with the adverse effects of warming (Baer & Singer, 2014). Consistent with the anthropogenic climate change perspectives, humans may be accountable of weather. People give meaning and respond to change patterns of global warming and climate change (Barnes et al., 2013; Fiske et al., 2014; Peterson & Broad, 2009). The Anthropocene is much more than just a discussion of the impact of climate change and global warming, it is important to study it from the perspectives of sustainable development, regional security, or international collaboration (Chandler et al., 2022).

Urban resilience is more tangible than the invisible greenhouse gases with serious implications. Home insulation is associated with investment and frugality while organic food with risk and pastoral ideals (Press & Arnould, 2011; Wilk & Wilhite, 1985). The ability of human beings to make decisions for nutritious and healthy food choices that increases survival chances (Birch *et al.*, 1998; Cashdan, 1998; Wright, 1991). Pleasure, holistic health and purity are values of purchasing behavior in some European countries (Ditlevsen et al., 2019). Consumption is relational embedded in social relations. Consumers of traditional marketing goods and services are alienated (Lambert, 2019). Marx critics of consumer capitalism leads to naturalistic epistemology (Kramer, 2021; Stoner, 2021; Trujillo, 2021).

The social exchange theory is a dyadic approach in marketing that extends to incorporate philanthropic actors and multiple stakeholders. Marketing theory and practices replace the consumer fetish with the value creation by a collectively of stakeholders (Arnould & Cayla, 2015; Tsing, 2015; Atanasova (2021; Casey et al. 2020), Giesler, 2006, 2008; Figueiredo & Scaraboto, 2016). Without replacement of the consumer construct, consumers can engage in sharing, gifting, bartering and other semiotic consumption processes (Belk, 2014; Wei, et al. 2021).

Consumption is socially learned and individually determined during the primary socialization The urban anthropology of consumption is the study, analysis and interpretation of consumption as social phenomenon that incorporates the acquisition, consumption and above all the appropriation of products and brands that serve as a person's identity reinforcer (Páramo and Ramirez, 2017).

Socio-technical approach has proved to be effective in some studies in coal mines and a project involving operating units at Royal Dutch Shell in England, in a textile factory in India, implying unions, employers and the government involved to improve the quality of working life in Norway and Sweden, and other socio-technical projects in Canada, Netherlands and Australia. Implementation of socio-technical urban projects are carried out in a continuous urban renewal process to achieve completeness as the consequence of revisions and modifications.

Technoscientific urban models of relationality, such as radioecology, enables to visualize the cross-boundary movements in urban areas. The sociotechnical approach is based on semi-autonomous groups with roots in the self-regulated projects and redundancy of functions (Trist, 1976; de Melo, 1997). The socio-technical support or urban systems based on promotions and rewards, promotions, must be congruent with activities and aligned with the objectives of the institution. Communication across the urban metropolitan boundaries of organizational departments links the control and supervision levels. Local urban autonomy and freedom of action is valued that what is prescribed by rules in the urban community. The socio-technical urban approach encourages the autonomy and initiative of employees. Variance must be controlled in sociotechnical urban criterion as the source for each person responsible for controlling learning from work and failures. The urban movements reveal the limits of the state to the post fallout urban socio-ecology resilience.

Locals consider that radioecology offers more than limited human relations and delimits an Eco semiotic of ineffable interrelated of living, decaying, and dying things. Application of Eco semiotics considers humans and other species co-constitute ecological phenomena (Maran & Kull 2014). Radioecology uses specific radioisotopes to investigate tracing and mapping connections radiological contaminants and their transboundary movements between the earth and the atmosphere (Odum, 1959). The resultant production of accumulative nuclear waste bringing urban socio-ecology resilience to collective recovery.

Global transformation has climate change at the heart of its concerns, and the sociological approach is concerned by unpredictability in the work research of Bauman and Beck and social acceleration by Hartmut Rosa (Beck, 2009; Hartmut Resonanz, 2016; W. S. et al., 2007). Evolution of human activities have resulted in temporal urban landscapes perceived as natural but shaped by evolution of generations (Meeus, 1995; Kuna & Dreslerová, 2007). Transformations of urban landscapes by human groups lead to the anthropogenic impacts on local socio-ecological equilibrium. The doughnut economics model proposes knowledges and practices for urban socio-ecology resilience, sustainability, and well-being to be applied everywhere (Strang, 2017).

Cooperative urban socio-ecological networks are characterized by established relationships of interdependence, specialization, communication, reciprocity, and trust (Nahapiet & Ghoshal, 1998). Consistent with the socio-ecological approach, the urban network integration mechanisms are also consisting of collaborating efforts to solve urban socio-ecological and resilience problems. The member´s objectives of the urban socio-ecology resilience network are complementary and not necessarily common integrated by mechanisms consisting of informal and voluntary relationships strengthened by a reticulation process.

DISCUSSION

The findings align with the body of existing research, the analysis of the relationships between the urban anthropology establishment of action is framed by the reflexive dialectic actor and author with specific disciplinary orientation in urban socio-ecology resilience planning and green innovation areas. The analysis of the relationships between the objectivity of inter-subjectivity of urban anthropology turn into methodological problems and dilemmas that require to be addressed in a specific way. Urban anthropological inter-subjectivity is an objective element of constructing reality without relativist ideology but with methodologies and practices based on valid assumptions for urban social-ecological resilience planning and green innovation.

Anthropological postmodernism considered as the anti-scientific narratives has disappeared but continue to prevent from access to scientific methods. Postmodernism argues that it is immoral to seek laws, generalizations, verifications and proofs, which dehumanize by objectifying people in urban planning and ecological innovation. (Barrett, 2009). Relativism is an escapist generalization perspective. Postmodernism in urban anthropology continues influencing academic development.

The urban socio-ecology resilience anthropology approach has shorter interactions in time between social groups aiming to urban socio-ecology resilience planning and green innovations. The development of urban anthropological motivation on creative abilities gives the conditions of the local innovation economy based on urban socio-ecological planning and green innovation areas. International institutions are engaged in studying sideways working with the groups of experts on urban socio-ecological planning and green innovation areas supported by urban anthropology (Hannerz, 2004; Holmes and Marcus, 2006, 2021; Ortner, 2010, 2013).

Change is not a continuous social life dimension considering stability as the norm. (Gluckman, 1968). A structural form for scientific purposes is not static but a dynamic continuity of an organic structure throughout a living body of an organism being constantly renewed and the social life that constantly renews the social structure (Radcliffe-Brown, 1940). Accounting practices on urban socio-ecology resilience and resilience stress accountability on verifiable data and information in annual reports. Changes in reporting require teamwork cooperation of the involved stakeholders. Several studies shows that Urban planning remains an important tool in managing climate change impacts (Asibey et al., 2022).

CONCLUSION

The analysis aimed to find some existing implications and relationships between urban anthropology and Anthropocene urban socio-ecology planning resilience, begins from the assumption that urban anthropology give support to create and develop any urban planning based on the Anthropocene urban socio-ecology resilience. This analysis concludes that urban anthropology fundamentals give support to create and develop urban planning based on the Anthropocene urban socio-ecology resilience.

Urban anthropological research positions between the boundaries of the global reach in connections and assemblages with urban anthropological socio-ecology and resilience and green innovation areas. The anthropoid dominance has led to results in resource shortages of the planet that may be a more acuity problem that climate change. The institutional support of urban anthropology to local urban planning and socio-ecological innovation system is a relevant factor to achieve successful results. Urban planning and green innovation institutions are characterized by their ability to create and develop social capital to facilitate relationships of cooperation and conflict.

There are several levels of urban anthropology applied to urban green planning and socio-ecological innovation. Technological urban green innovation park brings green firms seeking the benefits of agglomeration. Emergence of innovative green and socio-ecological ventures explore opportunities in urban planning innovation from main sources and drivers using the wealth of green knowledge. Generating scientific green knowledge and technological socio-ecological innovation, extension, and consulting activities must be structured through diverse mechanisms for the planning and formation of cooperative urban communities structured by interactions of deliberative processes to stimulate entrepreneurship and innovation.

Urban anthropology at the local dimension of the urban socio-ecological planning and green innovation system university-based must be supported by financial subsidies to guide the efforts of the governance structure. Other mechanism of urban socio-ecological planning and innovation is the indirect support from green infrastructures and constitutions of green assets used by flow firms and a set of projects. The green infrastructures, socio-ecological physical facilities, and green technologies of urban planning and innovation centers are green assets than can be financed.

An effort requiring shared objectives and collaboration supported by a governance structure capable to synchronize mission and objectives of agents and actors from the perspective analysis of urban anthropology to urban planning and socioecological innovation centers and parks. Deliberative actions on urban green innovation areas are triggered by normative and deliberative action carried out by agents and actors. The involved actors focused on the urban green innovation technology park must have autonomy and flexibility and above all must avoid erosions caused by natural occurring regulatory issues.

Research on the mechanisms of urban socio-ecology planning resilience based on cross-field linkages must be strengthened.

Limitations and Future Research

It is necessary to deepen both theoretically and empirically on the association between urban anthropology and its impact on the Anthropocene urban socio-ecology planning resilience in different regions and countries, and their impacts in the new ecological innovations.

REFERENCES

Aldrich, H. (2014). *Stand up and be counted: Why social science should stop using the qualitative/quantitative dichotomy.* https://blogs.lse.ac.uk/impactofsocialsciences/2014/11/28/stand-up-and-becounted-social-science-qualitative-quantitative-dichotom

Alvarez, R. (2016). *Computational social science.* Cambridge University Press. doi:10.1017/CBO9781316257340

Arnould, E. J., & Cayla, J. (2015). Consumer fetish: Commercial ethnography and the sovereign consumer. *Organization Studies, 36*(October), 1361–1386. doi:10.1177/0170840615580012

Asibey, M., Mintah, F., Adutwum, I., Wireko-Gyebi, R., Tagnan, J., Yevugah, L., Agyeman, K., & Abdul-Salam, J. (2022). Beyond rhetoric: Urban planning climate change resilience conundrum in Accra, Ghana. *Cities (London, England), 131*, 1–12. doi:10.1016/j.cities.2022.103950

Atanasova, A. (2021). Re-examining utopia in contemporary consumption: Conceptualization and implications for marketing. *Academy of Marketing Science Review, 11*(1-2), 23–29. doi:10.100713162-021-00193-0

Audretsch, D. B. (1998). *Agglomeration and the Location of Innovative Activity.* Academic Press.

Baer, H., & Singer, M. (2014). *The anthropology of climate change: An integrated critical perspective*. Routledge. doi:10.4324/9781315818702

Baldassarri, D., & Abascal, M. (2017). Field experiments across the social sciences. *Annual Review of Sociology*, *43*(1), 41–73. doi:10.1146/annurev-soc-073014-112445

Barnes, J., Dove, M., Lahsen, M., Mathews, A., McElwee, P., McIntosh, R., Moore, F., O'Reilly, J., Orlove, B., Puri, R., Weiss, H., & Yager, K. (2013). Contribution of anthropology to the study of climate change. *Nature Climate Change*, *3*(6), 541–544. doi:10.1038/nclimate1775

Barrett, S. (2009). *Anthropology*. University of Toronto Press. doi:10.3138/9781442697362

Baschera, W., & Schneider, I. (2022). *Can sustainable urban planning determine people's happiness and well-being?* Brazilian Journal of Management / Revista de Administração da UFSM., doi:10.5902/1983465969433

Beck, U. (2009). *World at Risk*. Polity.

Belk, R. W. (2014). You are what you can access: Sharing and collaborative consumption online. *Journal of Business Research*, *67*(August), 1595–1600. doi:10.1016/j.jbusres.2013.10.001

Bernard, H. (2011). *Research methods in anthropology: Qualitative and quantitative approaches*. AltaMira Press.

Bilharz, J. A. (1998). *The Allegany Senecas and Kinzua Dam: Forced Relocation through Two Generations*. University of Nebraska Press.

Birch, L. L., Gunder, L., Grimm-Thomas, K., & Laing, D. G. (1998). Infants' consumption of a new food enhances acceptance of similar foods. *Appetite*, *30*(3), 283–295. doi:10.1006/appe.1997.0146 PMID:9632459

Brady, H. (2019). The challenge of big data and data science. *Annual Review of Political Science*, *22*(1), 297–323. doi:10.1146/annurev-polisci-090216-023229

Bunge, M. (1999). *Buscar la filosofía en las ciencias sociales*. Siglo XXI.

Casey, K., Lichrou, M., & O'Malley, L. (2020). Prefiguring sustainable living: An ecovillage story. *Journal of Marketing Management*, *36*(17–18), 1658–1679. doi:10.1080/0267257X.2020.1801800

Cashdan, E. (1998). Adaptiveness of food learning and food aversions children. *Social Sciences Information. Information Sur les Sciences Sociales*, *37*(4), 613–632. doi:10.1177/053901898037004003

Chandler, D., Rothe, D., & Muller, F. (n.d.). Relaciones Internacionales en el Antropoceno. *Relaciones Internacionales*, *50*, 107–126. doi:10.15366/relacionesinternacionales2022.50.005

Connolly, C., Keil, R., & Ali, S. H. (2021). Extended urbanisation and the spatialities of infectious disease: Demographic change, infrastructure and governance. *Urban Studies (Edinburgh, Scotland)*, *58*(2), 245–263. doi:10.1177/0042098020910873

de Melo, A. (2001). *The Innovation Systems of Latin America and the Caribbean*. Inter-American Development Bank. Agosto.

de Melo, M. A. C. (1997). *Processo de Planejamento e as Inovações Tecnológicas e Sociais: uma perspectiva sócio-ecológica* (5th ed.). Seminário de Modernização Tecnológica.

de Melo, M. A. C. (2002a). *Enriquecendo a Atuação de Incubadora de Emrpesas. Tecnologia e Inovação: experiências de gestão nas micro e pequenas empresas*. PGT/USP.

de Melo, M.A.C. (2002b). Inovação e Modernização Tecnológica e Organizacional nas MPMEs: o domínio interorganizacional. *Seminário Internacional: Políticas para Sistemas Produtivos Locais de MPME*.

DeWalt, K., DeWalt, B., & Wayland, C. (1998). Participant observation. In H. R. Bernard (Ed.), *Handbook of methods in cultural anthropology* (pp. 259–299). Altamira.

Diamond, J., & Robinson, J. (2010). *Natural experiments of history*. The Belknap Press of Harvard University Press.

Díaz, D., Kristoff, J., Castro, M., Miceli, J., Castro, D., Quinteros, R., & Guerrero, S. (2007). *Exploraciones en antropología y complejidad*. Editorial Sb.

Ditlevsen, K., Sandøe, P., & Lassen, J. (2019). Healthy food is nutritious, but organic food is healthy because it is pure: The negotiation of healthy food choices by Danish consumers of organic food. *Food Quality and Preference*, *71*(January), 46–53. doi:10.1016/j.foodqual.2018.06.001

Dunning, T. (2012). *Natural experiments in the social sciences*. Cambridge University Press. doi:10.1017/CBO9781139084444

Durrenberger, P. (2007) The Anthropology of Organized Labor in the United States. *Ann Review of Anthropology, 36*, 73-88. Disponible en: https://www.annualreviews.org/doi/pdf/10.1146/annurev.anthro.36.081406.094433

Egerton, F. N. (2001). A history of the ecological sciences: Early Greek origins. *Bulletin of the Ecological Society of America*, *82*(1), 93–97.

Eriksen, T. H. (2016). *Overheating: An Anthropology of Accelerated Change*. Pluto. doi:10.2307/j.ctt1cc2mxj

Escobar, A. (2020). *Pluriversal Politics: The Real and the Possible*. Duke University Press.

Figueiredo, B., & Scaraboto, D. (2016). The systemic creation of value through circulation in collaborative consumer networks. *The Journal of Consumer Research*, *43*(4), 509–533. Advance online publication. doi:10.1093/jcr/ucw038

Finucane, M. L. (2009). Why science alone won't solve the climate crisis: Managing climate risks in the Pacific. *Asia Pacific Issues*, *89*, 1–8.

Fiske, S. J., Crate, S. A., Crumley, C., Galvin, K., Lazrus, H., Lucero, L., Oliver-Smith, A., Orlove, B., Strauss, S., & Wilk, R. R. (2014). *Changing the atmosphere. Anthropology and climate change. Final report of the AAA global climate change task force*. American Anthropological Association.

Giesler, M. (2008). Conflict and compromise: Drama in marketplace evolution. *Journal of Consumer Research*, 34(April), Giesler, M. (2006). Consumer gift systems. *The Journal of Consumer Research*, *33*(September), 283–290. doi:10.1086/506309

Giner, C. (2004). J. Antropología urbana. *Ariel.*

Gluckman, M. (1968). The Utility of the Equilibrium Model in the Study of Social Change. *American Anthropologist, 70*(2), 219–237. doi:10.1525/aa.1968.70.2.02a00010

González Álvarez, M. D. (1997). Processos de Planejamento nos Pólos Tecnológicos: um enfoque adaptativo. Tese de Doutorado, Departamento de Engenharia Industrial, PUC-Rio, Rio de Janeiro.

Gravano, A. (1992). Antropología Práctica, muestra y posibilidades de la Antropología Organizacional. *Publicar en Antropología y Ciencias Sociales, Buenos Aires*, (1), 95–126.

Gravano, A. (2005). *El barrio en la teoría social.* Espacio Editorial.

Gravano, A. (2006). Imaginarios regionales y circularidad en la planificación: El caso del TOAR. *Intersecciones*, (7), 305–323.

Gravano, A. (2007b). Claves para la facilitación organizacional del proceso participativo en la planificación urbano-ambiental metropolitana. *Revista Regional de Trabajo Social, 21*(40), 9-22.

Gravano, A. (2008a) *¿Vecinos o ciudadanos? el fenómeno NIMBY (Not In My Back Yard) o SPAN (Sí, Pero No Aquí) del imaginario urbano en un proceso de participación social y su tratamiento desde la facilitación organizacional antropológica.* Trabajo presentado en II Congreso de la Asociación Latinoamericana de Antropología (ALA), Universidad de Costa Rica, Simposio "Imaginarios urbanos y participación social".

Grimmer, J., Roberts, M., & Stewart, B. (2021). Machine learning for social science: An agnostic approach. *Annual Review of Political Science, 24*(1), 395–419. doi:10.1146/annurev-polisci-053119-015921

Haines, S. (2019b). Managing expectations: Articulating expertise in climate services for agriculture in Belize. *Climatic Change, 157*(1), 43–59. doi:10.100710584-018-2357-1

Hanners, U. (1986). *Exploración de la ciudad.* Fondo de Cultura Económica.

Hannerz, U. (2004). *Foreign New: Exploring the World of Foreign Correspondents.* University of Chicago Press.

Hartmut Resonanz, R. (2016). *Eine Soziologie der Weltbeziehung.* Suhrkamp.

Hindman, M. (2015). Building better models: Prediction, replication, and machine learning in the social sciences. *The Annals of the American Academy of Political and Social Science, 659*(1), 48–62. doi:10.1177/0002716215570279

Hirst, K. (2019). *Is anthropology a science?* ThoughtCo. https://www.thoughtco.com/isanthropology-a-science-3971060

Hofman, J., Sharma, A., & Watts, D. (2017). Prediction and explanation in social systems. *Science, 355*(6324), 486–488. doi:10.1126cience.aal3856 PMID:28154051

Hofman, J., Watts, D., Athey, S., Garip, F., Griffiths, T., Kleinberg, J., Margetts, H., Mullainathan, S., Salganik, M., Vazire, S., Vespignani, A., & Yarkoni, T. (2021). Integrating explanation and prediction in computational social science. *Nature*, *595*(7866), 181–188. doi:10.103841586-021-03659-0 PMID:34194044

Holling, C. S. (1973). *Resilience and Stability of Ecological Systems. Annual Review of Ecology and Systematics*, 4. https://www.jstor.org/stable/2096802

Holmes, D. R., & Marcus, G. E. (2006). Fast capitalism: para-ethnography and the rise of the symbolic analyst. In M. Fisher & G. Downey (Eds.), *Frontiers of Capital: Ethnographic Reflections on the New Economy*. Duke University Press.

Holmes, D. R., & Marcus, G. E. (2021). How do we collaborate? An updated manifesto. In D. Boyer & G. Marcus (Eds.), *Collaborative Anthropology Today: A Collection of Exceptions*. Cornell University Press. doi:10.7591/cornell/9781501753343.003.0002

Hyatt, S. (2021). *Teaching Urban Anthropology in a time of COVID*. National Association for the Practice of Anthropology. https://scholarworks.iupui.edu/bitstream/handle/1805/29883/Hyatt2021Teaching-pp.pdf?sequence=1

Jorgensen, D. (2015). Participant observation. In R. Scott & S. Kosslyn (Eds.), *Emerging trends in the social and behavioral sciences* (pp. 1–15). Wiley. doi:10.1002/9781118900772.etrds0247

Kaplan, O. (1940). Prediction in the social sciences. *Philosophy of Science*, *7*(4), 492–498. doi:10.1086/286658

Kawulich, B. (2005). La observación participante como método de recolección de datos. *Forum Qualitative Social Research*, *68*(2), 43.

Kolb, Ch. (1998). Review of Bilharz, Joy A., *The Allegany Senecas and Kinzua Dam: Forced Relocation through Two Generations. H-AmIndian, H-Net Reviews*. https://www.h-net.org/reviews/showrev.php?id=2366

Kosova, R., Qendraj, D.H., Xhafaj, E. (2022). Meta-Analysis ELECTRE III and AHP in Evaluating and Ranking the Urban Resilience. *Journal of Environmental Management and Tourism*, *3*(59), 756 - 768. doi:10.14505/jemt.v13.3(59).15

Kramer, J. M. (2021). From predator to parasite: On private property and our ecological disaster. *Journal of Economic Issues*, *55*(June), 416–422. doi:10.1080/00213624.2021.1908804

Krauß, W. (2020). Narratives of change and the co-development of climate services for action. *Climate Risk Management*, *28*, 100217. doi:10.1016/j.crm.2020.100217

Kuhn, T. S. (1996). The Structure of Scientific Revolutions (3rd ed.). University of Chicago Press. (Original work published 1962)

Kuna, M., & Dreslerová, D. (2007). Landscape archaeology and “community areas” in the archaeology of central Europe. In D. Hicks, L. McAtackney, & G.J. Fairclough (Eds.), Envisioning landscape: Situations and standpoints in archaeology and heritage, Left Coast Press.

Lambert, A. (2019). Psychotic, acritical and precarious? A Lacanian exploration of the neoliberal consumer subject. *Marketing Theory*, *19*(3), 329–346. doi:10.1177/1470593118796704

Lapassade, G. (1986). *Grupos, organizaciones e instituciones: la transformación de la burocracia*. Gedisa.

Latour, B. (2013). *An inquiry into modes of existence: An anthropology of the moderns* (C. Porter, Trans.). Harvard University Press.

Lester, R. K. (2005). *Universities, Innovation, and the Competitiveness of Local Economies A Summary Report from the Local Innovation Systems Project – Phase I*. Industrial Performance Center, Massachusetts Institute of Technology Working Paper 05-010.

Lindblom, C. E. (1959). The Science of Muddling through. *Public Administration Review*, *19*(2), 79–88. doi:10.2307/973677

Llanos, M. (2009). *Epistemología de las ciencias sociales*. UNMSM.

Lourau, R. (1979). *El análisis institucional*. Amorrortu.

Lundvall, B.-Å. (Ed.). (1992). National Systems of Innovation: towards a theory of innovation and interactive learning. Printer Publisher.

Malinowski, B. (1961). *Argonauts of the Western Pacific*. Routledge.

Maran, T., & Kull, K. (2014). Ecosemiotics: Main principles and current developments. *Geografiska Annaler. Series B, Human Geography*, *96*(1), 41–50. doi:10.1111/geob.12035

Mayo, E. (1945). *The social problems of an Industrial Civilization*. Harvard University Press.

McIntosh, R. J., Tainter, J. A., & McIntosh, S. K. (Eds.). (2000). *The Way the Wind Blows: Climate Change, History, and Human Activity*. Columbia University Press.

Meeus, J. H. A. (1995). Pan-European landscapes. *Landscape and Urban Planning*, *31*(1-3), 57–79. doi:10.1016/0169-2046(94)01036-8

Mir, R., & Mir, A. (2009). From the Colony to the Corporation: Studying Knowledge Transfer Across International Boundaries. *Group & Organization Management*, *34*(1), 90–113. doi:10.1177/1059601108329714

Mu, X., Fang, C., Yang, Z., & Guo, X. (2022). Impact of the COVID-19 Epidemic on Population Mobility Networks in the Beijing–Tianjin–Hebei Urban Agglomeration from a Resilience Perspective. *Land (Basel)*, *2022*(11), 675. doi:10.3390/land11050675

Nahapiet, J. E., & Ghoshal, S. (1998). Social Capital, Intellectual Capital, and the Organizational Advantage. *Academy of Management Review*, *33*(2), 242–266. doi:10.2307/259373

Nießen, D., Danner, D., Spengler, M., & Lechner, C. (2020). Big Five personality traits predict successful transitions from school to vocational education and training: A large-scale study. *Frontiers in Psychology*, *11*, 1827. doi:10.3389/fpsyg.2020.01827 PMID:32903700

Nost, E. (2019). Climate services for whom? The political economics of contextualizing climate data in Louisiana's coastal master plan. *Climatic Change*, *157*(1), 27–42. doi:10.100710584-019-02383-z

Odum, E. P. (1957). *Fundamentals of ecology*. W. B. Saunders Company.

OECD. (1997b). Government Venture Capital for Technology-Based Firms. Ducument OCDE/GD (97) 201. Paris: OECD.

Olimid, A. P., Georgescu, C. M., & Gherghe, C. L. (2022). Influences of Covid-19 Crisis on Resilience Theories: An analysis of Community, Societal and Governance Resilience. *Revista de Ştiinţe Politice. Revue des Sciences Politiques*, *73*, 38–51.

Ortner, S. (2010). Access: Reflections on studying up in Hollywood. *Ethnography*, *11*(2), 211–233. doi:10.1177/1466138110362006

Ortner, S. (2013). *Not Hollywood: Independent Film at the Twilight of the American Dream*. Duke University Press.

Páramo, D., & Ramírez, E. (2017). *Etnomarketing, la dimensión cultural del marketing*. Klasse Editorial.

Peterson, N., & Broad, K. (2009). Climate and weather discourse in anthropology: From determinism to uncertain futures. In *Anthropology and climate change: From encounters to actions* (pp. 70–86). Left Coast Press.

Pfeifer, J. (2006). Laws of nature. In S. Sarkar & J. Pfeifer (Eds.), *The philosophy of science: An encyclopedia* (pp. 439–444).

Pimienta-Bueno, J.A. (1999). *Parque de Inovação Tecnológica e Cultural da Gávea: A Visão da PUC-Rio.* Documento Interno, 23 págs.

Pizarro, C. (2014). La entrevista etnográfica como práctica discursiva: Análisis de caso sobre las pistas meta-discursivas y la emergencia de categorías nativas. *Revista de Antropologia*, *57*(1), 461–496. doi:10.11606/2179-0892.ra.2014.87770

Porter, M. (2001). Clusters of Innovation: regional foundations of U.S competitiveness. Council on Competitiveness. Monitor Group.

Press, M., & Arnould, E. J. (2014). Narrative transparency. *Journal of Marketing Management*, *30*(13–14), 1353–1376. doi:10.1080/0267257X.2014.925958

Purzycki, B., Pisor, A., Apicella, C., Atkinson, Q., Cohen, E., Henrich, J., McElreath, R., McNamara, R., Norenzayan, A., Willard, A., & Xygalatas, D. (2018). The cognitive and cultural foundations of moral behavior. *Evolution and Human Behavior*, *39*(5), 490–501. doi:10.1016/j.evolhumbehav.2018.04.004

Radcliffe-Brown, A. R. (1940). On Social Structure. *J. R. Anthropol. Inst. Great Br. Irel.*, *1940*(70), 1–12.

Reguera, A. (2009) Empresarios de ayer, de hoy y de siempre. *Un recorrido latinoamericano por sus formas espacio-temporales América Latina en la historia económica, num 32, jul-dic, México*. Disponible en: https://www.scielo.org.mx/scielo.php?script=sci_arttext&pid=S1405-22532009000200003&lng=es&nrm=iso

Reynoso, C. (1995b). Hacia la perfección del consenso: Los lugares comunes de la antropología. *Intersecciones*, *1*, 51–72.

Roca, A. (2008). *La comprensión de la tecnología en la antropología.* Actas VII ESOCITE UFRJ, Jornadas Latino-Americanas de Estudos Sociais das Ciências e das Tecnologías.

Roethlisberger, F., & Dickson, W. (1939). *Management and the Worker.* Harvard Univ. Press.

Roni, S., Merga, M., & Morris, J. (2020). *Conducting quantitative research in education.* Springer. doi:10.1007/978-981-13-9132-3

Saxenian, A. (1994). *Regional Advantage: culture and competition in Silicon Valley and Route 128.* Harvard Business Press.

Schlapfer, A., & Marinova, D. (2001). Local Innovation Systems: nature, importance and role. *Conference Proceedings: International Summer Academy on Technological Studies.*

Schnegg, M., O'Brian, C. I., & Sievert, I. J. (2021). It's our fault: A global comparison of different ways of explaining climate change. *Human Ecology, 49*(3), 327–339. doi:10.100710745-021-00229-w

Scott, J. (2009). The Art of Not Being Governed: An Anarchist History of Upland Southeast Asia. Yale University Press.

Signorelli, A. (1999). *Antropología Urbana.* UAM.

Snow, C. P. (1998). *The Two Cultures.* Cambridge University Press.

Stoner, A. M. (2021). Things are getting worse on our way to catastrophe: Neoliberal environmentalism, repressive desublimation, and the autonomous ecoconsumer. *Critical Sociology, 47*(May), 491–506. doi:10.1177/0896920520958099

Strang, V. (2017). The Gaia complex: Ethical Challenges to an Anthropocentric 'Common Future. In B. Marc & L. Jerome (Eds.), *The Anthropology of Sustainability* (pp. 255–283). Palgrave. doi:10.1057/978-1-137-56636-2_12

Su, C., Liu, F., Stefea, P., & Umar, M. (2023). Does technology innovation help to achieve carbon neutrality? *Economic Analysis and Policy, 78*, 1–14. doi:10.1016/j.eap.2023.01.010

Su, C., Pang, L., Tao, R., Shao, X. & Umar, M. (2022). Renewable energy and technological innovation: Which one is the winner in promoting net-zero emissions? *Technological Forecasting & Social Change,* 1 – 11. doi:10.1016/j.techfore.2022.121798

Sundbo, J. (1998). *The Theory of Innovation: Entrepreneurs, technology, and strategy.* Edward Elgar.

Taagepera, R. (2008). *Making social sciences more scientific: The need for predictive models.* Oxford University Press. doi:10.1093/acprof:oso/9780199534661.001.0001

Tainter, J. A. (1988). *The Collapse of Complex Societies.* Cambridge University Press.

Tainter, J. A. (2008). Collapse and Sustainability: Rome, the Maya, and the Modern World. *Archeological Papers of the American Anthropological Association, 24*(1), 201–214. doi:10.1111/apaa.12038

Tarantino-Curseri, S. (2018). Antropología, etología, sociología, y psicología, 4 disciplinas útiles en el ámbito empresarial. *Management Review, 3*(3). Advance online publication. doi:10.18583/umr.v3i3.126

Trawick, P., & Alf, H. (2015). Revisiting the Image of Limited Good: On Sustainability, Thermodynamics, and the Illusion of Creating Wealth. *Current Anthropology*, *56*(1), 1–27. doi:10.1086/679593

Trist, E. L. (1976). *A Concept of Organizational Ecology: an invited address to the three Melbourne universities*.

Trujillo, M. L. N. (2021). Notes for a critical and ecological view of patriarchal capitalism in the web of life. *Capital and Class*, *45*(March), 21–32. doi:10.1177/0309816820929115

Tsing, A. L. (2015). *The mushroom at the end of the world*. Princeton University Press.

Velho, G. (1978). Observando o familiar. In A aventura sociológica: objetividade, paixao, improviso e método na pesquisa social. Zahar.

W. S. (2007). The Anthropocene: Are Human Beings Now Overwhelming the Forces of Nature? *Ambio*, *36*, 614–621. doi:10.1579/0044-7447(2007)36[614:TAAHNO]2.0.CO;2 PMID:18240674

Webber, S. (2019). Putting climate services in contexts: Advancing multi-disciplinary understandings: Introduction to the special issue. *Climatic Change*, *157*(1), 1–8. doi:10.100710584-019-02600-9

Wei, X., Lo, C. K. Y., Jung, S., & Choi, T.-M. (2021). From co-consumption to co-production: A systematic review and research synthesis of collaborative consumption practices. *Journal of Business Research*, *129*(May), 282–294. doi:10.1016/j.jbusres.2021.02.027

Wilk, R. R., & Wilhite, H. L. (1985). Why don't people weatherize their homes? An ethnographic solution. *Energy*, *10*(May), 621–629. doi:10.1016/0360-5442(85)90093-3

Wolf, M. (2016). Rethinking urban epidemiology: Natures, networks and materialities. *International Journal of Urban and Regional Research*, *40*(5), 958–982. doi:10.1111/1468-2427.12381 PMID:32336869

Yang, Y., Lili, W., & Hongchi, Z. (2022). Bibliometric research on the evolution of resilience theme from the perspective of Geographical Science. *2022 29th International Conference on Geoinformatics*, 1-9. 10.1109/Geoinformatics57846.2022.9963870

Compilation of References

Abbasi, M. (2017). Towards socially sustainable supply chains–themes and challenges. *European Business Review*, *29*(3), 261–303. doi:10.1108/EBR-03-2016-0045

Abbasi, S., & Erdebilli, B. (2023). Green closed-loop supply chain networks' response to various carbon policies during COVID-19. *Sustainability*, *15*(4), 3677. doi:10.3390u15043677

Abbas, J. R., Kenth, J. J., & Bruce, I. A. (2020). The role of virtual reality in the changing landscape of surgical training. *The Journal of Laryngology and Otology*, *134*(10), 863–866. doi:10.1017/S0022215120002078 PMID:33032666

Abdel Wahed Ahmed, M. M., & Abd El Monem, N. (2020). Sustainable and green transportation for better quality of life case study greater Cairo–Egypt. *HBRC Journal*, *16*(1), 17–37. doi:10.1080/16874048.2020.1719340

Abdelkafi, N., & Makhotin, S., & Posselt. (2013). Business Models Innovation for Electric Mobility- What can be learned from Existing Business Model Patterns. *International Journal of Management*, 1–41.

Acharya, S., Bhatt, A. N., Chakrabarti, A., Delhi, V. S., Diehl, J. C., van Andel, E., & Subra, R. (2021). Problem-Based Learning (PBL) in Undergraduate Education: Design Thinking to Redesign Courses. In *Design for Tomorrow—Volume 2* (pp. 349–360). Springer. doi:10.1007/978-981-16-0119-4_28

Adefila, A., Opie, J., Ball, S., & Bluteau, P. (2020). Students' engagement and learning experiences using virtual patient simulation in a computer supported collaborative learning environment. *Innovations in Education and Teaching International*, *57*(1), 50–61.

Adhikari, A., & Hansen, A. J. (2018). Land use change and habitat fragmentation of wildland ecosystems of the North Central United States. *Landscape and Urban Planning*, *177*, 196–216. doi:10.1016/j.landurbplan.2018.04.014

Advertising, M. (2019). *Mastering the customer journey: applying the predictive power of Artificial Intelligence* [White Paper] Microsoft. https://advertiseonbing-blob.azureedge.net/blob/bingads/media/insight/ebook/2019/10-october/cdj-chapter-2/cdj_mini_ebook_uk.pdf

Aebersold, M., & Dunbar, D. M. (2021). Virtual and Augmented Realities in Nursing Education: State of the Science. *Annual Review of Nursing Research*, *39*(1), 225–242. doi:10.1891/0739-6686.39.225 PMID:33431644

Agarwal, A., Xie, B., Vovsha, I., Rambow, O., & Passonneau, R. (2011). Sentiment Analysis of Twitter Data. In *Proceedings of the Workshop on Language in Social Media.* Association for Computational Linguistics.

Agha, S. (2021). Aligning continuing professional development (CPD) with quality assurance (QA): A perspective of healthcare leadership. *Quality & Quantity*, 1–15.

Agrawal, A., Batra, D., Parikh, D., & Kembhavi, A. (2018). Don't just assume; look and answer: Overcoming priors for visual question answering. In *Proceedings of the IEEE Conference on Computer Vision and Pattern Recognition*, (pp. 4971–4980). IEEE. 10.1109/CVPR.2018.00522

Agrawal, A., Lu, J., Antol, S., Mitchell, M., Zitnick, C. L., Parikh, D., & Batra, D. (2017). VQA: Visual question answering. *International Journal of Computer Vision, 123*(1), 4–31. doi:10.100711263-016-0966-6

Ahi, P., & Searcy, C. (2013). A comparative literature analysis of definitions for green and sustainable supply chain management. *Journal of Cleaner Production, 52*, 329–341. doi:10.1016/j.jclepro.2013.02.018

Ahn, Y. J., & Bessiere, J. (2023). The Relationships between Tourism Destination Competitiveness, Empowerment, and Supportive Actions for Tourism. *Sustainability, 15*(1), 626. doi:10.3390u15010626

Aksenov, V., Semochkin, A., Bendik, A., & Reviakin, A. (2022). Utilizing Digital Twin for Maintaining Safe Working Environment among Railway Track Tamping Brigade. *Transportation Research Procedia, 61*, 600–608. doi:10.1016/j.trpro.2022.01.097

Albert, I., Shakantu, W., & Ibrahim, S. (2021). The effect of poor materials management in the construction industry: A case study of Abuja, Nigeria. *Acta Structilia, 28*(1), 142–167.

Aldrich, H. (2014). *Stand up and be counted: Why social science should stop using the qualitative/quantitative dichotomy.* https://blogs.lse.ac.uk/impactofsocialsciences/2014/11/28/stand-up-and-becounted-social-science-qualitative-quantitative-dichotom

Aleinikov. (2002). *MegaCreativity: 5 Steps to Thinking Like a Genius*. Walking Stick Press.

Aleksandar, J., & Aleksandar, M. (2022, January 20). VoRtex Metaverse Platform for Gamified collaborative learning. *Electronics (Basel), 11*(3), 2–20. doi:10.3390/electronics11030317

Alexander, A., Walker, H., & Naim, M. (2014). Decision theory in sustainable supply chain management: A literature review. *Supply Chain Manag: An International Journal, 19*(5/6), 504–522. doi:10.1108/SCM-01-2014-0007

Allam, Z., Sharifi, A., Bibri, S. E., Jones, D. S., & Krogstie, J. (2022). The Metaverse as a Virtual Form of Smart Cities: Opportunities and Challenges for Environmental, Economic, and Social Sustainability in Urban Futures. *Smart Cities, 5*(3), 771–801. doi:10.3390martcities5030040

Allen, M. (2017). *The sage encyclopedia of communication research methods* (Vol. 1-4). SAGE Publications, Inc., doi:10.4135/9781483381411

Alli, H. (2018). User Involvement Method in The Early Stage of The New Product Development Process For A Successful Product. *Alam Cipta: International Journal of Sustainable Tropical Design Research and Practice, 11*(1), 23–28.

Alnagrat, A. J. A., Ismail, R. C., & Idrus, S. Z. S. (2021, May). Extended Reality (XR) in Virtual Laboratories: A Review of Challenges and Future Training Directions. [). IOP Publishing.]. *Journal of Physics: Conference Series, 1874*(1), 012031. doi:10.1088/1742-6596/1874/1/012031

Alnoor, A., Eneizan, B., Makhamreh, H. Z., & Rahoma, I. A. (2018). The effect of reverse logistics on sustainable manufacturing. *International Journal of Academic Research in Accounting, Finance and Management Sciences, 9*(1), 71–79.

AlNuaimi, B. K., Singh, S. K., Ren, S., Budhwar, P., & Vorobyev, D. (2022). Mastering digital transformation: The nexus between leadership, agility, and digital strategy. *Journal of Business Research, 145*, 636–648. doi:10.1016/j.jbusres.2022.03.038

Alt, A., & Zimmermann, H.-D. (2014). *Electronic Markets and Business Models.*

Alvarez, R. (2016). *Computational social science*. Cambridge University Press. doi:10.1017/CBO9781316257340

Amabile, T. M. (1983). *The social psychology of creativity*. New York: Spring-Verlag. doi:10.1007/978-1-4612-5533-8

Amabile, T. M. (1996). *Componential theory of creativity* (Working Paper 12-096). Harvard Business School. https://www.hbs.edu/ris/Publication%20Files/12-096.pdf

Ameen, N., Tarhini, A., Reppel, A., & Anand, A. (2021). Customer experiences in the age of Artificial intelligence. *Computers in Human Behavior*, *114*, 106548. doi:10.1016/j.chb.2020.106548 PMID:32905175

Amplayo, R. K., & Song, M. (2017). An Adaptable Fine-Grained Sentiment Analysis For Summarization of Multiple Short Online Reviews. *Data & Knowledge Engineering*, *110*, 54–67. doi:10.1016/j.datak.2017.03.009

Anderson, J., & Rainie, L. (2018, December 10). *Artificial intelligence and the future of humans*. Pew Research Centre. https://www.pewresearch.org/internet/2018/12/10/artificial-intelligence-and-the-future-of-humans/

Andrej, Z., & Jilles, H. (2020). Decentralized Network Governance: Blockchain Technology and Future Regulation. *Frontiers in Blockchain*, 1-11.

Annala, J., Lindén, J., Mäkinen, M., & Henriksson, J. (2021). Understanding academic agency in curriculum change in higher education. *Teaching in Higher Education*, 1–18. doi:10.1080/13562517.2021.1881772

Ansell, C., & Gash, A. (2007). Collaborative Governance in Theory and Practice. *Journal of Public Administration: Research and Theory*, *18*(4), 534–571. doi:10.1093/jopart/mum032

Anser, M. K., Yousaf, Z., Hishan, S. S., Nassani, A. A., Sheikh, A. Z., Vo, X. V., ... Abro, M. M. Q. (2020). Dynamic linkages between transportation, waste management, and carbon pricing: Evidence from the Arab World. *Journal of Cleaner Production*, *269*, 122151. doi:10.1016/j.jclepro.2020.122151

Antoniou, P., Arfaras, G., Pandria, N., Ntakakis, G., Bambatsikos, E., & Athanasiou, A. (2020). Real-time affective measurements in medical education, using virtual and mixed reality. In *International Conference on Brain Function Assessment in Learning* (pp. 87-95). Springer, Cham. 10.1007/978-3-030-60735-7_9

Apata, O., Adebayo, A. V., & Ainah, P. K. (2021, August). Renewable energy systems and the fourth industrial revolution. In *2021 IEEE PES/IAS PowerAfrica* (pp. 1-5). IEEE.

Araque, O., Corcuera-Platas, I., Sánchez-Rada, J. F., & Iglesias, C. A. (2017). Enhancing Deep Learning Sentiment Analysis with Ensemble Techniques in Social Applications. *Expert Systems with Applications*, *77*, 236–246. doi:10.1016/j.eswa.2017.02.002

Ardito, L., Cerchione, R., Del Vecchio, P., & Raguseo, E. (2019). Big data in smart tourism: Challenges, issues and opportunities. *Current Issues in Tourism*, *22*(15), 1805–1809. doi:10.1080/13683500.2019.1612860

Argyris, C. (1991). Teaching smart people how to learn. *Harvard Business Review*, *69*(3).

Arnould, E. J., & Cayla, J. (2015). Consumer fetish: Commercial ethnography and the sovereign consumer. *Organization Studies*, *36*(October), 1361–1386. doi:10.1177/0170840615580012

Artto, K., Kujala, J., Dietrich, P., & Martinsuo, M. (2008). What is project strategy? *International Journal of Project Management*, *26*(1), 4–12. doi:10.1016/j.ijproman.2007.07.006

Asibey, M., Mintah, F., Adutwum, I., Wireko-Gyebi, R., Tagnan, J., Yevugah, L., Agyeman, K., & Abdul-Salam, J. (2022). Beyond rhetoric: Urban planning climate change resilience conundrum in Accra, Ghana. *Cities (London, England)*, *131*, 1–12. doi:10.1016/j.cities.2022.103950

Atadana, L., Borbor, S., & Rabbani, M. (2020). Sustainable Development Goals and Clean Energy: A Review. *International Journal of Energy Economics and Policy, 10*(5), 445–452.

Atanasova, A. (2021). Re-examining utopia in contemporary consumption: Conceptualization and implications for marketing. *Academy of Marketing Science Review, 11*(1-2), 23–29. doi:10.100713162-021-00193-0

Athanasiou, V., & Maragoudakis, M. (2017). A Novel, Gradient Boosting Framework For Sentiment Analysis in Languages Where NLP Resources Are Not Plentiful: A Case Study For Modern Greek. *Algorithms, 10*(1), 34. doi:10.3390/a10010034

Atiku, S. O. (2018). Reshaping Human Capital Formation Through Digitalization. In P. Duhan, K. Singh, & R. Verma (Eds.), *Radical Reorganization of Existing Work Structures Through Digitalization* (pp. 52–73). IGI Global. doi:10.4018/978-1-5225-3191-3.ch004

Atiku, S. O., & Anane-simon, R. (2020). Leadership and Innovative Approaches in Higher Education. In N. Baporikar & M. Sony (Eds.), *Quality Management Principles and Policies in Higher Education* (pp. 83–100). IGI Global. doi:10.4018/978-1-7998-1017-9.ch005

Atiku, S. O., & Boateng, F. (2020). Rethinking Education System for the Fourth Industrial Revolution. In S. Atiku (Ed.), *Human Capital Formation for the Fourth Industrial Revolution* (pp. 1–17). IGI Global. doi:10.4018/978-1-5225-9810-7.ch001

Audretsch, D. B. (1998). *Agglomeration and the Location of Innovative Activity*. Academic Press.

Auer, R., & Boehme, R. (2020). The technology of retail central bank digital currency. *BIS Quarterly Review,* 1-16.

Auer, R. A., Cornerlli, G., & Frost, J. (2020). *Rise of the Central Bank Digital Currencies: Drivers*. Approaches and Technology.

Aysan, A., Kayani, F., & Kayani, U. N. (2020). The Chinese Inward RDI and Economic Prospects amid Covid 19 Crisis. *Pakistan Journal of Commerce and Social Science*, 1088-1105.

Aysan, A., Khan, A., Topuz, H., & Tunali, A. S. (2021). Survival of the Fittest: A Natural Experiment from Crypto Exchanges. *The Singapore Economic Review*, 1–20. doi:10.1142/S0217590821470020

Azapagic, A., & Perdan, S. (2000). Indicators of sustainable development for industry: A general framework. *Process Safety and Environmental Protection, 78*(4), 243–261. doi:10.1205/095758200530763

Babagolzadeh, M., Shrestha, A., Abbasi, B., Zhang, Y., Woodhead, A., & Zhang, A. (2020). Sustainable cold supply chain management under demand uncertainty and carbon tax regulation. *Transportation Research Part D, Transport and Environment, 80*, 102245. doi:10.1016/j.trd.2020.102245

Babatunde, O. M., Munda, J. L., & Hamam, Y. (2020). A comprehensive state-of-the-art survey on hybrid renewable energy system operations and planning. *IEEE Access : Practical Innovations, Open Solutions, 8*, 75313–75346. doi:10.1109/ACCESS.2020.2988397

Bacon, J. (2017). The future of creativity in an automated world. *Marketing Week.* https://www.marketingweek.com/future-creativity-automated-world/

Baer, H., & Singer, M. (2014). *The anthropology of climate change: An integrated critical perspective*. Routledge. doi:10.4324/9781315818702

Baines, T., Brown, S., Benedettini, O., & Ball, P. D. (2012). Examining green production and its role within the competitive strategy of manufacturers. *Journal of Industrial Engineering and Management, 5*(1), 53–87. doi:10.3926/jiem.405

Baldassarri, D., & Abascal, M. (2017). Field experiments across the social sciences. *Annual Review of Sociology, 43*(1), 41–73. doi:10.1146/annurev-soc-073014-112445

Baldwin, C., Hienerth, C., & Von Hippel, E. (2006). How User Innovations Become Commercial Products: A Theoretical Investigation and Case Study. *Research Policy, 35*(9), 1291–1313. doi:10.1016/j.respol.2006.04.012

Bao'an Daily. (2021, September 27). *Sea, Land, Air And Rail Three-Dimensional Traffic Reaches All Directions.* Bao'an Daily. http://barb.sznews.com/PC/content/202109/27/content_1099775.html

Barbieri, J. C. (1997). Políticas públicas indutoras de inovações tecnológicas ambientalmente saudáveis nas empresas. *Revista Brasileira de Administração Pública, 31*(2), 135–152.

Barbieri, J. C. (2004). *Gestão ambiental empresarial: conceitos, modelos e instrumentos.* Saraiva.

Barnes, J., Dove, M., Lahsen, M., Mathews, A., McElwee, P., McIntosh, R., Moore, F., O'Reilly, J., Orlove, B., Puri, R., Weiss, H., & Yager, K. (2013). Contribution of anthropology to the study of climate change. *Nature Climate Change, 3*(6), 541–544. doi:10.1038/nclimate1775

Barrett, S. (1990). Pricing the environment: The economic and environmental consequences of a carbon tax. *Economic Outlook, 14*(5), 24–33. doi:10.1111/j.1468-0319.1990.tb00670.x

Barrett, S. (2009). *Anthropology.* University of Toronto Press. doi:10.3138/9781442697362

Bartolini, M., Bottani, E., & Grosse, E. H. (2019). Green warehousing: Systematic literature review and bibliometric analysis. *Journal of Cleaner Production, 226*, 242–258. doi:10.1016/j.jclepro.2019.04.055

Baschera, W., & Schneider, I. (2022). *Can sustainable urban planning determine people's happiness and well-being?* Brazilian Journal of Management / Revista de Administração da UFSM., doi:10.5902/1983465969433

Bashir, N., Papamichail, K. N., & Malik, K. (2017). Use of Social Media Applications for Supporting New Product Development Processes in Multinational Corporations. *Technological Forecasting and Social Change, 120*, 176–183. doi:10.1016/j.techfore.2017.02.028

Batista, L., Bourlakis, M., Smart, P., & Maull, R. (2018). In search of a circular supply chain archetype–a content-analysis-based literature review. *Production Planning and Control, 29*(6), 438–451. doi:10.1080/09537287.2017.1343502

Batra, M. M. (2019). Strengthening Customer Experience through Artificial Intelligence: An Upcoming Trend. *Competition Forum, 17*(2), 223-231. https://login.iris.etsu.edu:3443/login?url=https://www.proquest.com/docview/2343014949?accountid=10771

BeC Glossary. (2020). *Glossary of Thinking, Entrepreneurship, Business, Finance, Sustainable Development, and Environment Terms.* https://1000ventures.com/doc/glossary.html

Beck, U. (2009). *World at Risk.* Polity.

Bekkers, V. J., Edelenbos, J., Nederhand, J., Steijn, A. J., Tummers, L. G., Voorberg, W. H., & Edelenbos, J. (n.d.). *The social innovation perspective in the public sector: Co-creation, self-organization, and meta-governance.*

Belk, R. W. (2014). You are what you can access: Sharing and collaborative consumption online. *Journal of Business Research, 67*(August), 1595–1600. doi:10.1016/j.jbusres.2013.10.001

Ben-Amar, W., Gomes, M., Khursheed, H., & Marsat, S. (2022). Climate change exposure and internal carbon pricing adoption. *Business Strategy and the Environment, 31*(7), 2854–2870. doi:10.1002/bse.3051

Beres, D. (2016, March 28). Microsoft Chat Bot Goes On Racist, Genocidal Twitter Rampage. *Huffpost*. https://www.huffpost.com/entry/microsoft-tay-racist-tweets_n_56f3e678e4b04c4c37615502

Berg, P., Feber, D., Granskog, A., Nordigarden, D., & Ponkshe, S. (2020). *The drive toward sustainability in packaging—beyond the quick wins*. Mckinsey & Company.

Bermejo, C., & Hui, P. (2021). A survey on haptic technologies for mobile augmented reality. *ACM Computing Surveys*, *54*(9), 1–35. doi:10.1145/3465396

Bernard, H. (2011). *Research methods in anthropology: Qualitative and quantitative approaches*. AltaMira Press.

Bernard, M. A., Johnson, A. C., Hopkins-Laboy, T., & Tabak, L. A. (2021). The US National Institutes of Health approach to inclusive excellence. *Nature Medicine*, *27*(11), 1861–1864. doi:10.103841591-021-01532-1 PMID:34764481

Bevins, F., Bryant, J., Krishnan, C., & Law, J. (2020). Coronavirus: How should US higher education plan for an uncertain future. *McKinsey*.

Bhamra, T. A., & Evans, S. (1999). The next step in ecodesign: service-oriented life cycle design. *IEEE International Symposium on Electronics and the Environment-ISEE*, 263–267. . 765887.10.1109/ISEE.1999.765887

Bharadwaj, A., Sawy, O. A., Pavlou, P. A., & Venkatraman, N. (2013). Digital Business Strategy: Toward A Next Generation of Insights. Management Information Systems Quarterly, 37(2), 471–482. doi:10.25300/MISQ/2013/37:2.3

Bhatti, G., Mohan, H., & Singh, R. R. (2021). Towards The Future of Smart Electric Vehicles: Digital Twin Technology. *Renewable & Sustainable Energy Reviews*, *141*, 110801. doi:10.1016/j.rser.2021.110801

Bhuiyan, N. (2011). A Framework for Successful New Product Development. *Journal of Industrial Engineering and Management*, *4*(4), 746–770. doi:10.3926/jiem.334

Bianchi, C. (2015). Enhancing Joined-Up Government and Outcome-Based Performance Management through System Dynamics Modelling to Deal with Wicked Problems: The Case of Societal Ageing. *Systems Research and Behavioral Science*, *32*(4), 502–505. doi:10.1002res.2341

Bianzino, N. M. (2022, June 6). Is AI the start of the truly creative human? *EY Global*. https://www.ey.com/en_gl/ai/is-ai-the-start-of-the-truly-creative-human

Bichou, K. (2014). *Port operations, planning and logistics*. CRC Press. doi:10.4324/9781315850443

Bilharz, J. A. (1998). *The Allegany Senecas and Kinzua Dam: Forced Relocation through Two Generations*. University of Nebraska Press.

Biomet, Z. (2021, December 19). OptiVu™ mixed reality. ZimmerBioMet. https://www.zimmerbiomet.com/en/products-and-solutions/zb-edge/optivu.html

Birch, L. L., Gunder, L., Grimm-Thomas, K., & Laing, D. G. (1998). Infants' consumption of a new food enhances acceptance of similar foods. *Appetite*, *30*(3), 283–295. doi:10.1006/appe.1997.0146 PMID:9632459

Birt, J., Stromberga, Z., Cowling, M., & Moro, C. (2018). Mobile mixed reality for experiential learning and simulation in medical and health sciences education. *Information (Basel)*, *9*(2), 31. doi:10.3390/info9020031

Bishara, A. J., Kruschke, J. K., Stout, J. C., Bechara, A., McCabe, D. P., & Busemeyer, J. R. (2010). Sequential learning models for the Wisconsin card sort task: Assessing processes in substance dependent individuals. *Journal of Mathematical Psychology*, *54*(1), 5–13. doi:10.1016/j.jmp.2008.10.002 PMID:20495607

Blessinger, P., Sengupta, E., & Yamin, T. S. (2018). Human creativity as a renewable resource. *International Journal for Talent Development and Creativity*, *6*(1), 17–26.

Boccia, F. (2018). The importance of green technologies in the fight against climate change. *Energy Policy*, *115*, 634–639.

Bocken, N. M., & Short, S. W. (2021). Unsustainable business models–Recognising and resolving institutionalised social and environmental harm. *Journal of Cleaner Production*, *312*, 127828. doi:10.1016/j.jclepro.2021.127828

Bodamer, D. (2016, November 16). *14 Charts from the EPA's Latest MSW Estimates*. Waste460, Retrieved June 8, 2022, from https://www.waste360.com/waste-reduction/14-charts-epa-s-latest-msw-estimates

Boden, M. (1998). Creativity and artificial intelligence. *Artificial Intelligence*, *103*(1-2), 347–356. doi:10.1016/S0004-3702(98)00055-1

Boes, K., Buhalis, D., & Inversini, A. (2015). Conceptualising smart tourism destination dimensions. In I. Tussyadiah & A. Inversini (Eds.), *Information and Communication Technologies in Tourism 2015* (pp. 391–404). Springer. doi:10.1007/978-3-319-14343-9_29

Bolívar, M. P. R., & Meijer, A. J. (2016). Smart governance: Using a literature review and empirical analysis to build a research model. *Social Science Computer Review*, *34*(6), 673–692. doi:10.1177/0894439315611088

Bonvoisina, J., & Lelaha, A. (2014). An integrated method for environmental assessment and ecodesign of ICT-based optimization services. *Journal of Cleaner Production*, *68*, 144–154. doi:10.1016/j.jclepro.2014.01.003

Borgonovi, E., Bianchi, C., & Rivenbark, W. C. (2019). Pursuing Community Resilience through Outcome-Based Public Policies: Challenges and Opportunities for the Design of Performance Management Systems. *Public Organization Review*, *19*(4), 153–158. doi:10.100711115-017-0395-1

Bosch. (2021). *Home Appliances*. Bosch Invented for Life. https://www.bosch-home.com/us/

Bossink, B. A. G. (2002). A Dutch public-private strategy for innovation in sustainable construction. *Construction Management and Economics*, *20*(7), 633–642. doi:10.1080/01446190210163534

Böttcher, T. P., Weber, M., Weking, J., Hein, A., & Krcmar, H. (2022). *Value Drivers of Artificial Intelligence. 28th Americas Conference on Information Systems (AMCIS)*, Minneapolis, USA

Boulanger, P., Girod, B., Houdret, A., & Lashermes, B. (2020). Aligning Business Strategies with the Sustainable Development Goals: Insights from the Green Technology Sector. *Sustainability*, *12*(14), 5844.

Bovaird, T. (2007). Beyond Engagement and Participation: User and Community Coproduction of Public Services. *Public Administration Review*, *67*(5), 846–860. doi:10.1111/j.1540-6210.2007.00773.x

Bové, A. T., & Swartz, S. (2016). Starting at the source: Sustainability in supply chains. *McKinsey on Sustainability and Resource Productivity*, *4*, 36–43.

Brady, H. (2019). The challenge of big data and data science. *Annual Review of Political Science*, *22*(1), 297–323. doi:10.1146/annurev-polisci-090216-023229

Brandon, E., Freiwirth, R., & Hjersman, J. (2021, May). Special Session—Student Engagement with Reduced Bias in a Virtual Classroom Environment. In *2021 7th International Conference of the Immersive Learning Research Network (iLRN)* (pp. 1-3). IEEE.

Brasil. (1981). *Política Nacional do Meio Ambiente, Lei nº 6.938, de 31 de agosto de 1981*. Retrieved June 25, 2015, from: http://www.planalto.gov.br/ccivil_03/Leis/L6938.htm

Brenner, M. (2018). Why AI is better than A/B Testing. *Concured.* https://www.concured.com/blog/why-ai-is-better-than-a/b-testing

Brent, A. C., Heuberger, R., & Manzini, D. (2005). Evaluating projects that are potentially eligible for Clean Development Mechanism (CDM) funding in the South African context: A case study to establish weighting values for sustainable development criteria. *Environment and Development Economics, 10*(5), 631–649. doi:10.1017/S1355770X05002366

Brent, A. C., & Labuschagne, C. (2007). An appraisal of social aspects in project and technology life cycle management in the process industry. *Management of Environmental Quality, 18*(4), 413–426. doi:10.1108/14777830710753811

Brent, A. C., & Petrick, W. (2007). Environmental Impact Assessment (EIA) during project execution phases: Towards a stage-gate project management model for the raw materials processing industry of the energy sector. *Impact Assessment and Project Appraisal, 25*(2), 111–122. doi:10.3152/146155107X205832

Breuer, H., & Ludeke-Freund, F. (2014). Normative Innovation for Sustainable business Models in Value Networks. In S. C. In K. Huizingh (Ed.), *The proceedings of XXV ISPIM Conference.* Dublin, Ireland: Lappeenranta: University of Technology Press. http://papers.ssrn.com/sol3/papers.cfm?abstract_

Briassoulis, H. (2001). Sustainable Development and its Indicators: Through a (Planner's) Glass Darkly. *Journal of Environmental Planning and Management, 44*(3), 409–427. doi:10.1080/09640560120046142

British Standards Institute. (2007). *Occupational Health and Safety Management Systems — Requirements.* BSI Global.

Brock, J. K. U., & Von Wangenheim, F. (2019). Demystifying AI: What digital transformation leaders can teach you about realistic artificial intelligence. *California Management Review, 61*(4), 110–134. doi:10.1177/1536504219865226

Brones, F., Carvalho, M. M., & Zancul, E. S. (2017). Reviews, action and learning on change management for ecodesign transition. *Journal of Cleaner Production, 142*, 8–22. doi:10.1016/j.jclepro.2016.09.009

Browne, R. (2023, February 23). *Elon Musk, who co-founded firm behind CatGPT, warns AI is one of the biggest risks to civilization.* CNBC. https://www.cnbc.com/2023/02/15/elon-musk-co-founder-of-chatgpt-creator-openai-warns-of-ai-society-risk.html#:~:text=billionaire%20Elon%20Musk.-,%E2%80%9COne%20of%20the%20biggest%20risks%20to%20the%20future%20of%20civilization,great%20capability%2C%E2%80%9D%20Musk%20said

Bruntland, G. H. (1987). *Our Common Future: World Commission on Environment and Development.* Oxford University Press.

Bryant, A. (2017). *Grounded theory and grounded theorizing: Pragmatism in research practice.* Oxford University Press. doi:10.1093/acprof:oso/9780199922604.001.0001

Bryson, J. M., Crosby, B. C., & Bloomberg, L. (2014). Public Value Governance: Moving beyond Traditional Public Administration and the New Public Management. *Public Administration Review, 74*(4), 445–456. doi:10.1111/puar.12238

Bryson, J. M., Crosby, B. C., & Stone, M. M. (2006). The Design and Implementation of Cross-Sector Collaborations: Propositions from the Literature. *Public Administration Review, 66*(s1), 44–55. doi:10.1111/j.1540-6210.2006.00665.x

Buckley, M. F. (2019, October 20). *The value of practical creativity.* Medium. https://medium.com/swlh/the-value-of-objective-creativity-9b4dd4d72d15

Bughin, J., Hazan, E., Ramaswamy, S., Chui, M., Allas, T., & Dahlstrom, P. (2017). *Artificial Intelligence.* McKinsey Global Institute. https://www.mckinsey.com/~/media/McKinsey/Industries/Advanced%20Electronics/Our%20Insights/How%20artificial%20intelligence%20can%20deliver%20real%20value%20to%20companies/MGI-Artificial-Intelligence-Discussion-paper.ashx

Buhalis, D., & Jun, S. H. (2011). E-tourism. *Contemporary Tourism Reviews, 1*, 2-38. www.goodfellowpublishers.com/free_files/Contemporary-Tourism-ReviewEtourism-66769a7ed0935d0765318203b843a64d.pdf

Buhalis, D. (2019). Technology in tourism-from information communication technologies to eTourism and smart tourism towards ambient intelligence tourism: A perspective article. *Tourism Review*.

Buhalis, D., & Amaranggana, A. (2013). Smart tourism destinations. In *Information and communication technologies in tourism 2014* (pp. 553–564). Springer. doi:10.1007/978-3-319-03973-2_40

Buhalis, D., & Leung, R. (2018). Smart hospitality-Interconnectivity and interoperability towards an ecosystem. *International Journal of Hospitality Management, 71*, 41–50. doi:10.1016/j.ijhm.2017.11.011

Buhalis, D., Lin, M. S., & Leung, D. (2022). Metaverse as a driver for customer experience and value co-creation: Implications for hospitality and tourism management and marketing. *International Journal of Contemporary Hospitality Management, 35*(2), 701–716. doi:10.1108/IJCHM-05-2022-0631

Bunge, M. (1999). *Buscar la filosofía en las ciencias sociales*. Siglo XXI.

Burgess, A., van Diggele, C., Roberts, C., & Mellis, C. (2020). Key tips for teaching in the clinical setting. *BMC Medical Education, 20*(2), 1–7. doi:10.118612909-020-02283-2 PMID:33272257

Burgess, J. C. (1990). The contribution of efficient energy pricing to reducing carbon dioxide emissions. *Energy Policy, 18*(5), 449–455. doi:10.1016/0301-4215(90)90045-6

Burke, J. (2021). Reintroducing the open Metaverse OS paper. *Outlier Ventures*. https://outlierventures.io/research/the-open-Metaverse-os/

Burns, E. (2022). What is artificial intelligence (AI)? *TechTarget*. https://www.techtarget.com/searchenterpriseai/definition/AI-Artificial-Intelligence

Byrnes, T. A., & Dunn, R. J. (2020). Boating-and shipping-related environmental impacts and example management measures: A review. *Journal of Marine Science and Engineering, 8*(11), 908. doi:10.3390/jmse8110908

Cadavid-Giraldo, N., Velez-Gallego, M. C., & Restrepo-Boland, A. (2020). Carbon emissions reduction and financial effects of a cap and tax system on an operating supply chain in the cement sector. *Journal of Cleaner Production, 275*, 122583. doi:10.1016/j.jclepro.2020.122583

Campbell, C., Gibbs, A., Maimane, S., & Yugi, N. (2008). Hearing community voices: Grassroots perceptions of an intervention to support health volunteers in South Africa. *Journal of Social Aspects of HIV/AIDS Research Alliance, 5*(4), 162–177. doi:10.1080/17290376.2008.9724916 PMID:19194598

Cardoso, P., Carvalho, J. C., Crespo, L. C., & Arnedo, M. A. (2016). Optimal inventorying and monitoring of taxon, phylogenetic and functional diversity. *Biorxiv*, 060400. doi:10.1101/060400

Car, J., Sheikh, A., Wicks, P., & Williams, M. (2019). Beyond the hype of big data and artificial intelligence: Building foundations for knowledge and wisdom. *BMC Medicine, 17*(143), 143. doi:10.118612916-019-1382-x PMID:31311603

Carr, D. (2021). How Insurtech Is Transforming the Public Sector. *Insurance Journal*. https://www.insurancejournal.com/news/national/2021/01/26/597401.htm

Carr, J., Decreton, L., Qin, W., Rojas, B., Rossochacki, T., & Wen Yang, Y. (2015). Social Media in Product Development. *Food Quality and Preference, 40*, 354–364. doi:10.1016/j.foodqual.2014.04.001

Carson, J. (n.d.). *Why is creativity important and what does it contribute?* National Youth Council of Ireland. https://www.youth.ie/articles/why-is-creativity-important-and-what-does-it-contribute

Carter, B. (n.d.). *The Impact of Packaging on the Environment: Is Plastic the Only Demon?* Eco & Beyond.

Carvalho, M. M., Patah, L. A., & Bido, D. S. (2015). Project management and its effects on project success: Cross-country and cross-industry comparisons. *International Journal of Project Management, 33*(7), 1509–1522. doi:10.1016/j.ijproman.2015.04.004

Carvalho, M. M., & Rabechini, R. (2015). Impact of risk management on project performance: The importance of soft skills. *International Journal of Production Research, 53*(2), 321–340. doi:10.1080/00207543.2014.919423

Case, S. (2016). *The third wave*. Simon & Schuster.

Casey, K., Lichrou, M., & O'Malley, L. (2020). Prefiguring sustainable living: An ecovillage story. *Journal of Marketing Management, 36*(17–18), 1658–1679. doi:10.1080/0267257X.2020.1801800

Cashdan, E. (1998). Adaptiveness of food learning and food aversions children. *Social Sciences Information. Information Sur les Sciences Sociales, 37*(4), 613–632. doi:10.1177/053901898037004003

Central Broadcasting Network. (2022, April 27). *Smart Transportation Observation: Solving Efficiency And Safety Problems Is The Core Of Intelligent Networking The Digital Twin Should Help Urban Traffic Governance*. CNR. http://tech.cnr.cn/techph/20220427/t20220427_525808782.shtml

Chalmers, D., MacKenzie, N. G., & Carter, S. (2020). Artificial Intelligence and Entrepreneurship: Implications for Venture Creation in the Fourth Industrial Revolution. *Entrepreneurship Theory and Practice*. doi:10.1177/1042258720934581

Chamorro-Premuzic, T. (2015, November 24). The dark side of creativity. *Harvard Business Review*. https://hbr.org/2015/11/the-dark-side-of-creativity

Chandler, D., Rothe, D., & Muller, F. (n.d.). Relaciones Internacionales en el Antropoceno. *Relaciones Internacionales, 50*, 107–126. doi:10.15366/relacionesinternacionales2022.50.005

Chang, W., & Taylor, S. A. (2016). The Effectiveness of Customer Participation in New Product Development: A meta-analysis. *Journal of Marketing, 80*(1), 47–64. doi:10.1509/jm.14.0057

ChaniasS.HessT., (2016Understanding Digital Transformation Strategy Formation. Insights from Europe's Automotive Industry. PACIS. .

Chan, S. (2021). *Digitally Enabling'Learning by Doing'in Vocational Education: Enhancing 'Learning as Becoming'Processes*. Springer Nature. doi:10.1007/978-981-16-3405-5

Chen, C., Chaudhary, A., & Mathys, A. (2020). Nutritional and environmental losses embedded in global food waste. *Resources, Conservation and Recycling, 160*, 104912. doi:10.1016/j.resconrec.2020.104912

Cheng, Y., & Zhang, P. (2017, July). Pricing strategy and coordination in an O2O mixed channel supply chain with carbon cap-and-trade mechanism. In *2017 36th Chinese Control Conference (CCC)* (pp. 7528-7534). IEEE.

Chen, G. (2022). Design Scheme and Application of Foundation Treatment for Non-stop Construction of New Connecting Roads in Shenzhen Airport. *Subgrade Engineering, 130-133*. doi:10.13379/j.issn.1003-8825.202101010

Chen, G., Xie, P., Dong, J., & Wang, T. (2019). Understanding Programmatic Creative: The Role of AI. *Journal of Advertising, 48*(4), 347–355. doi:10.1080/00913367.2019.1654421

Cheng, C. C., & Krumwiede, D. (2018). Enhancing the Performance Of Supplier Involvement in New Product Development: The Enabling Roles of Social Media and Firm Capabilities. Supply Chain Management. *Supply Chain Management, 3*(23), 1359–8546. doi:10.1108/SCM-07-2017-0230

Cheng, R., Wu, N., Chen, S., & Han, B. (2022). Will metaverse be nextg internet? vision, hype, and reality. *IEEE Network, 36*(5), 197–204. doi:10.1109/MNET.117.2200055

Cheng, Y., Sun, H., Jia, F., & Koh, L. (2018). Pricing and Low-Carbon Investment Decisions in an Emission Dependent Supply Chain under a Carbon Labelling Scheme. *Sustainability, 10*(4), 1238. doi:10.3390u10041238

Chen, X., Jin, Z., Zhang, Q., Li, P., Zhang, S., Sun, J., Tian, X., Wang, Y., & Zhang, J. (2022). Research on Automatic Driving Simulation Test System Based on Digital Twin. *Journal of Physics: Conference Series, 2170*(1), 012039. doi:10.1088/1742-6596/2170/1/012039

Chesbrough, H., & Rosenbloom, R. (2002). The Role of the Business Model in Capturing Value from Innovation: Evidence from Xerox Corporation's Technology Spin-off Companies. *Industrial and Corporate Change, 11*(11), 529–555. doi:10.1093/icc/11.3.529

Chilvers, B. L., Morgan, K. J., & White, B. J. (2021). Sources and reporting of oil spills and impacts on wildlife 1970–2018. *Environmental Science and Pollution Research International, 28*(1), 754–762. doi:10.100711356-020-10538-0 PMID:32822011

China Automotive News. (2021, October 23). Digital Twin: "Black Technology" Incarnates As An Automotive Industry Booster. *China Automotive News*. http://www.cnautonews.com/

China Economic Network. (2016, February 5). Guizhou Provincial Road Passenger Transport Network Ticketing System was put into Trial Operation. *China Economic Network*. http://district.ce.cn/newarea/roll/201602/05/t20160205_8776229.shtml

China Informatization Weekly. (2019, November 16). Shenzhen Airport Digital Transformation and Exploration. *China Information Weekly.*

Chinese Academy of Information and Communications Technology. (2021, December 20). *Big Data White Paper (2021)*. CAICT.

Chiropa, T. (2020), *Clean Energy for 1 Million People*. http://www.innompics.com/events/ig2020/contests_sbm_tl_zimbabwe_tc_energy.html

Chopra, S. M. (2010). *Supply chain management: Strategy, planning and operations*. Academic Press.

Chou, M. C., & Wang, R. W. (2012). Displayability: An assessment of differentiation design for the findability of bottle packaging. *Displays, 33*(3), 146–156. doi:10.1016/j.displa.2012.06.003

Chukwuma, A. I., Ezenyilimba, E., & Agbara, N. O. (2018). Effect of product packaging on the sales volume of small and medium scale bakery firms in South East Nigeria. *International Journal of Academic Research in Business & Social Sciences, 8*(6), 988–1001.

Civelek, I. (2017). Sustainability in inventory management. In Intelligence, Sustainability, and Strategic Issues in Management (pp. 43-56). Routledge. doi:10.4324/9780203788394-3

Cline, W. R. (1995). *Pricing Carbon Dioxide Pollution*. Economic Analysis Project Working Paper.

Cohen, S. (2021). 5 Key Insurtech Trends to Watch in 2021. *Insurtech Insights*. https://www.insurtechinsights.com/5-key-insurtech-trends-to-watch-in-2021/

Cohen, Y., Faccio, M., Pilati, F., & Yao, X. (2019). Design and management of digital manufacturing and assembly systems in the Industry 4.0 era. *International Journal of Advanced Manufacturing Technology, 105*(9), 3565–3577. doi:10.100700170-019-04595-0

Colom, R., Haier, R. J., Head, K., Álvarez-Linera, J., Quiroga, M. Á., Shih, P. C., & Jung, R. E. (2009). Gray matter correlates of fluid, crystallized, and spatial intelligence: Testing the P-FIT model. *Intelligence, 37*(2), 124–135. doi:10.1016/j.intell.2008.07.007

Colom, R., Jung, R. E., & Haier, R. J. (2007). General intelligence and memory span: Evidence for a common neuroanatomic framework. *Cognitive Neuropsychology, 24*(8), 867–878. doi:10.1080/02643290701781557 PMID:18161499

Communication Message News. (2021, June 4). How Can Digital Twins Empower Smart Transportation? *Communication Message News.* http://www.txxxb.com/yc/yc/2021/0604/247776.shtml

Connolly, C., Keil, R., & Ali, S. H. (2021). Extended urbanisation and the spatialities of infectious disease: Demographic change, infrastructure and governance. *Urban Studies (Edinburgh, Scotland), 58*(2), 245–263. doi:10.1177/0042098020910873

Conservancy, O. (2018). *Fighting for Trash Free Seas: Ending the Flow of Trash at the Source.* Ocean Conservancy. Retrieved March 7, 2019, from https://oceanconservancy.org/trash-free-seas/

Conserve Energy Future. (n.d.). *What is green transportation?* Available from: https://www.conserve-energy-future.com/modes-and-benefits-of-green-transportation.php

Construction Industry Institute (CII). (2006). *Leading indicators during project execution. Resarch Summary 220-1.* The University of Texas in Austin.

Conway, S., & Steward, F. (1998). Networks and interfaces in environmental innovation: A comparative study in the UK and Germany. *The Journal of High Technology Management Research, 9*(2), 239–253. doi:10.1016/S1047-8310(98)90006-2

Cooper, R. (2020, August 25). *Food Waste in America: Facts and Statistics.* RUBICON. https://www.rubicon.com/blog/food-waste-facts/

Cooper, R. G. (1988). The New Product Process: A Decision Guide for Management. *Journal of Marketing Management, 3*(3), 238–255. doi:10.1080/0267257X.1988.9964044

Cooper, R. G. (1990). Stage-Gate Systems: A New Tool for Managing New Products. *Business Horizons, 33*(3), 44–54. doi:10.1016/0007-6813(90)90040-I

Cooper, R. G. (1996). Overhauling the new product process. *Industrial Marketing Management, 25*(6), 465–482. doi:10.1016/S0019-8501(96)00062-4

Cooper, R. G., & Kleinschmidt, E. J. (1987). Success Factors in Product Innovation. *Industrial Marketing Management, 16*(3), 215–223. doi:10.1016/0019-8501(87)90029-0

Cooper, R. G., & Kleinschmidt, E. J. (1995). Benchmarking The Firm's Critical Success Factors in New Product Development. *Journal of Product Innovation Management: An International Publication of the Product Development & Management Association, 12*(5), 374–391. doi:10.1111/1540-5885.1250374

Corder, G. D., McLellan, B. C., & Green, S. (2010). Incorporating sustainable development principles into minerals processing design and operation: SUSOP (R). *Minerals Engineering, 23*(3), 175–181. doi:10.1016/j.mineng.2009.12.003

Cowie, R. H., Bouchet, P., & Fontaine, B. (2022). The Sixth Mass Extinction: Fact, fiction or speculation? *Biological Reviews of the Cambridge Philosophical Society, 97*(2), 640–663. doi:10.1111/brv.12816 PMID:35014169

Craddock, W. T. (2013). How Business Excellence Models Contribute to Project Sustainability and Project Success. In A. J. G. Silvius & J. Tharp (Eds.), *Sustainability Integration for Effective Project Management* (pp. 1–19). IGI Global Publishing. doi:10.4018/978-1-4666-4177-8.ch001

Crawford, L. (2013). Leading Sustainability through Projects. In A. J. G. Silvius & J. Tharp (Eds.), *Sustainability Integration for Effective Project Management* (pp. 235–244). IGI Global Publishing. doi:10.4018/978-1-4666-4177-8.ch014

Crevier, D. (1993). *AI: The Tumultuous Search for Artificial Intelligence*. BasicBooks.

Cronbach, L. J., & Meehl, P. E. (1955). Construct validity in psychological tests. *Psychological Bulletin, 52*(4), 281–302. doi:10.1037/h0040957 PMID:13245896

Cropper, A. (2019). Playgol: Learning programs through play. IJCAI.

Crouch, L., Rolleston, C., & Gustafsson, M. (2021). Eliminating global learning poverty: The importance of equalities and equity. *International Journal of Educational Development, 82*, 102250. doi:10.1016/j.ijedudev.2020.102250

Cruz-Benito, J., Maderuelo, C., García-Peñalvo, F., Therón, R., Pérez, B., Jonás, S., & Martin, A. (2016, July 13). Usalpharma: A Software Architecture to Support Learning in Virtual Worlds. *IEEE Revista Iberoamarica de Technologias del Aprendizaje, 11*(3), 194–204. doi:10.1109/RITA.2016.2589719

Csikszentmihalyi, M., & Sawyer, K. (2014). Shifting the Focus from Individual to Organizational Creativity. In *The systems model of creativity: The collected works of Mihaly Csikszentmihalyi*. Springer. doi:10.1007/978-94-017-9085-7_6

Cuthbertson, A., Smith, A., Massie, G., & Sankaran, V. (2022, May 5). Elon Musk – latest: Tesla boss 'will become Twitter CEO when deal is complete' as billionaires throw cash behind takeover. *Yahoo! Finance*. https://uk.finance.yahoo.com/news/elon-musk-news-latest-tesla-084945674.html

d'Avila, A. (2019). Neural-symbolic computing: An effective methodology for principled integration of machine learning and reasoning. arXiv preprint arXiv:1905.06088. doi:10.1126cience.1102941 PMID:15528409

Dabrowski, D. (2018). Sources of Market Information, It's Quality and New Product Financial Performance. *The Engineering Economist, 29*(1), 115–122. doi:10.5755/j01.ee.29.1.13405

Dameri, R. P., & Cocchia, A. (2013, December). Smart city and digital city: twenty years of terminology evolution. In *X Conference of the Italian Chapter of AIS*. ITAIS.

Dapp, T. H. (2015). *Fintech reloaded- Traditional banks as digital ecosystems: With proven walled garden strategies into the future*. DB Research. https://www.dbresearch.com/PROD/RPS_EN-PROD/PROD0000000000451937/Fintech_reloaded_-_Traditional_banks_as_digital_ec.pdf;REWEBJSESSIONID=6644FE271301B11261BF9B52FFA110E3?undefined&realload=TLdLYj1x74399cki7iNULjWElPlVAayk27uTriyVxZ3mYT0FQ7aed4JHqT7gpdAF

Davenport, T., Guha, A., Grewal, D., & Bressgott, T. (2020). How artificial intelligence will change the future of marketing. *Journal of the Academy of Marketing Science, 28*(1), 24–42. doi:10.100711747-019-00696-0

Daw, N. D., O'Doherty, J. P., Dayan, P., Seymour, B., & Dolan, R. J. (2006). Cortical substrates for exploratory decisions in humans. *Nature, 441*(7095), 876–879. doi:10.1038/nature04766 PMID:16778890

De Jong, J. P., & Von Hippel, E. (2009). Transfers of User Process Inovations to Process Equipment Producers: A Study of Dutch High-Tech Firms. *Research Policy, 38*(7), 1181–1191. doi:10.1016/j.respol.2009.04.005

de Melo, M. A. C. (1997). *Processo de Planejamento e as Inovações Tecnológicas e Sociais: uma perspectiva sócio-ecológica* (5th ed.). Seminário de Modernização Tecnológica.

de Melo, M.A.C. (2002b). Inovação e Modernização Tecnológica e Organizacional nas MPMEs: o domínio interorganizacional. *Seminário Internacional: Políticas para Sistemas Produtivos Locais de MPME.*

de Melo, A. (2001). *The Innovation Systems of Latin America and the Caribbean*. Inter-American Development Bank. Agosto.

de Melo, M. A. C. (2002a). *Enriquecendo a Atuação de Incubadora de Emrpesas. Tecnologia e Inovação: experiências de gestão nas micro e pequenas empresas*. PGT/USP.

De Raedt, L., Kersting, K., Natarajan, S., & Poole, D. (2016). Statistical relational artificial intelligence: Logic, probability, and computation. *Synthesis Lectures on Artificial Intelligence and Machine Learning, 10*(2), 1– 189.

Deary, I. J., Penke, L., & Johnson, W. (2010). The neuroscience of human intelligence differences. *Nature Reviews. Neuroscience, 11*(3), 201–211. doi:10.1038/nrn2793 PMID:20145623

Decanio, S. J. (1992). International cooperation to avert global warming: Economic growth, carbon pricing, and energy efficiency. *Journal of Environment & Development, 1*(1), 41–62. doi:10.1177/107049659200100104

Defy. (2021). Defy world-class kitchen appliances. *Defy*. https://www.defy.co.za/appliances

Dehghani, M., Acikgoz, F., Mashatan, A., & Lee, S. H. (2021). A holistic analysis towards understanding consumer perceptions of virtual reality devices in the post-adoption phase. *Behaviour & Information Technology*, 1–19.

Del Chiappa, G., & Baggio, R. (2015). Knowledge transfer in smart tourism destinations: Analyzing the effects of a network structure. *Journal of Destination Marketing & Management, 4*(3), 145–150. doi:10.1016/j.jdmm.2015.02.001

Deland, D. (2009). Sustainability Through Project Management and Net Impact. In *PMI Global Congress North America*. Philadelphia PA: Project Management Institute.

Dell, R. M., & Rand, D. A. J. (2001). Energy storage—A key technology for global energy sustainability. *Journal of Power Sources, 100*(1), 2–17. doi:10.1016/S0378-7753(01)00894-1

DeMaria, S., & Levine, A. I. (2013). The use of stress to enrich the simulated environment. In *The comprehensive textbook of healthcare simulation* (pp. 65–72). Springer. doi:10.1007/978-1-4614-5993-4_5

Demeke, H. B., Merali, S., Marks, S., Pao, L. Z., Romero, L., Sandhu, P., Clark, H., Clara, A., McDow, K. B., Tindall, E., Campbell, S., Bolton, J., Le, X., Skapik, J. L., Nwaise, I., Rose, M. A., Strona, F. V., Nelson, C., & Siza, C. (2021). Trends in use of telehealth among health centers during the COVID-19 pandemic— United States, June26–November 6, 2020. *Morbidity and Mortality Weekly Report, 70*(7), 240–244. doi:10.15585/mmwr.mm7007a3 PMID:33600385

Deng, T., Zhang, K., & Shen, Z.-J. M. (2021). A Systematic Review of a Digital Twin City: A New Pattern of Urban Governance toward Smart Cities. *Journal of Management Science and Engineering, 6*(2), 125–134. doi:10.1016/j.jmse.2021.03.003

Dennick, R. (2016). Constructivism: Reflections on twenty-five years teaching the constructivist approach in medical education. *International Journal of Medical Education, 7*, 200–205. doi:10.5116/ijme.5763.de11 PMID:27344115

Department of Transportation of Guizhou Province. (2017, February 14). *Promote The Development Of Intelligent "Transportation Cloud"*. Department of Transportation. https://www.guizhou.gov.cn/home/gzyw/202109/t20210913_70356107.html

Department of Transportation of Guizhou Province. (2022, January 27). *Guizhou Province "14th Five-Year" Digital Transportation Development Plan*. Department of Transportation. http://jt.guizhou.gov.cn/xxgkml/ztfl/zcfg/gfxwj/202201/t20220128_72435899.html

DeWalt, K., DeWalt, B., & Wayland, C. (1998). Participant observation. In H. R. Bernard (Ed.), *Handbook of methods in cultural anthropology* (pp. 259–299). Altamira.

Dewett, T. (2007). Linking intrinsic motivation, risk taking, and employee creativity in an R&D environment. *R & D Management, 37*(3), 197–208. doi:10.1111/j.1467-9310.2007.00469.x

Diamantopoulos, A., & Winklhofe, H. M. (2001). Index construction with formative indicators: An alternative to scale development. *JMR, Journal of Marketing Research, 38*(2), 269–277. doi:10.1509/jmkr.38.2.269.18845

Diamond, J., & Robinson, J. (2010). *Natural experiments of history*. The Belknap Press of Harvard University Press.

Díaz, D., Kristoff, J., Castro, M., Miceli, J., Castro, D., Quinteros, R., & Guerrero, S. (2007). *Exploraciones en antropología y complejidad.* Editorial Sb.

Diaz-Trujillo, L. A., Tovar-Facio, J., Napoles-Rivera, F., & Ponce-Ortega, J. M. (2019). Effective Use of Carbon Pricing on Climate Change Mitigation Projects: Analysis of the Biogas Supply Chain to Substitute Liquefied-Petroleum Gas in Mexico. *Processes (Basel, Switzerland), 7*(10), 668. doi:10.3390/pr7100668

Diligenti, M., Gori, M., & Sacca, C. (2017). Semantic-based regularization for learning and inference. *Artificial Intelligence, 244*, 143–165. doi:10.1016/j.artint.2015.08.011

Dincer, I. (2007). Environmental and sustainability aspects of hydrogen and fuel cell systems. *International Journal of Energy Research, 31*(1), 29–55. doi:10.1002/er.1226

Dionisio, J., III, W., & Gilbert, R. (2013, July). 3D virtual worlds and the metaverse: Current status and future possibilities. *ACM Computing Surveys (CSUR), 45*(3), 1-38.

Dionisio, J. D., Burns, W. G., & Gilbert, R. (2013). 3D Virtual worlds and the metaverse. *ACM Computing Surveys, 45*(3), 1–38. doi:10.1145/2480741.2480751

DiPaola, S., Gabora, L., & McCaig, G. (2018). Informing artificial intelligence generative techniques using cognitive theories of human creativity. *Procedia Computer Science, 145*, 158–168. doi:10.1016/j.procs.2018.11.024

Disclosure Insight Action. (2021, February 9). *Environmental supply chain risks to cost companies $120 billion by 2026.* https://www.cdp.net/en/articles/supply-chain/environmental-supply-chain-risks-to-cost-companies-120-billion-by-2026

Ditlevsen, K., Sandøe, P., & Lassen, J. (2019). Healthy food is nutritious, but organic food is healthy because it is pure: The negotiation of healthy food choices by Danish consumers of organic food. *Food Quality and Preference, 71*(January), 46–53. doi:10.1016/j.foodqual.2018.06.001

Domingos, P. (2015). *The Master Algorithm: How the Quest for the Ultimate Learning Machinewill Remake the World.* Basic Books.

Donadello, I., Serafini, L., & d'Avila Garcez, A. (2017). Logic tensor networks for semantic image interpretation. *IJCAI.*

Donaire, D. (1999). *Gestão ambiental na empresa*. Atlas.

Doran, J. W., Sarrantonio, M., & Liebig, M. (1996). *Soil health and sustainability*. Adv. Agron. doi:10.1016/S0065-2113(08)60178-9

Dorcic, J., Komsic, J., & Markovic, S. (2019). Mobile technologies and applications towards smart tourism – state of the art. *Tourism Review, 74*(1), 82–103. doi:10.1108/TR-07-2017-0121

Dornis, T. W. (2020). Artificial creativity: Emergent works and the void in current copyright doctrine. *Yale Journal of Law and Technology, 22*, 1–60.

Douglas, J. (2019). These American workers are the most afraid of A.I. taking their jobs. *CNBC News.* https://www.cnbc.com/2019/11/07/these-american-workers-are-the-most-afraid-of-ai-taking-their-jobs.html

Downe-Wamboldt, B. (1992). Content analysis: Method, applications, and issues. *Health Care for Women International, 13*(3), 313–321. doi:10.1080/07399339209516006 PMID:1399871

DST. (2022). *75 Impactful Startups: DST Incubation Programme*. Vigyan Prasar.

Duffy, M. (2017). Code eats copy for breakfast: Human copywriters are doomed. *Digiday*. https://digiday.com/marketing/humanoid-copywriters-good-as-dead/

Dumancic, S., Gunds, T., Meert, W., Blockeel, H. (2019). Learning relational representations with auto-encoding logic programs. *IJCAI*.

Dunning, T. (2012). *Natural experiments in the social sciences*. Cambridge University Press. doi:10.1017/CBO9781139084444

Durning, S. J., & Artino, A. R. (2011). Situativity theory: a perspective on how participants and the environment can interact: AMEE Guide no. 52. *Medical Teacher, 33*(3), 188–199. doi:10.3109/0142159X.2011.550965 PMID:21345059

Durrenberger, P. (2007) The Anthropology of Organized Labor in the United States. *Ann Review of Anthropology, 36*, 73-88. Disponible en: https://www.annualreviews.org/doi/pdf/10.1146/annurev.anthro.36.081406.094433

Du, S., Yalcinkaya, G., & Bstieler, L. (2016). Sustainability, Social Media Driven Open Innovation, And New Product Development Performance. *Journal of Product Innovation Management, 33*, 55–71. doi:10.1111/jpim.12334

Du, S., Zhu, L., Liang, L., & Ma, F. (2013). Emission-dependent supply chain and environment-policy-making in the 'cap-and-trade'system. *Energy Policy, 57*, 61–67. doi:10.1016/j.enpol.2012.09.042

Dyduch, W., Chudziński, P., Cyfert, S., & Zastempowski, M. (2021). Dynamic capabilities, value creation and value capture: Evidence from SMEs under Covid-19 lockdown in Poland. *PLoS One, 16*(6), e0252423. doi:10.1371/journal.pone.0252423 PMID:34129597

Dygalo, A., Keller, A., & Shcherbin, A. (2020). Principles of Application Of Virtual And Physical Simulation Technology In Production Of Digital Twin Of Active Vehicle Safety Systems. *Transportation Research Procedia, 50*, 121–129. doi:10.1016/j.trpro.2020.10.015

Early, J. (2022, September 30). AI can produce prize-winning art, but it still can't compete with human creativity. *The Conversation*. https://theconversation.com/ai-can-produce-prize-winning-art-but-it-still-cant-compete-with-human-creativity-190279

Ebbesen, J.B., & Hope, A.J. (2013). Re-imagining the Iron Triangle: Embedding Sustainability into Project Constraints. *PM World Journal, 2*(3).

ECO. (1992). Cleaner technologies. *Environmental Conservation, 19*(4), 361-369.

Edelman, D. C., & Singer, M. (2015). Competing on Customer Journeys. *Harvard Business Review*, (November), 88–100.

Edum-Fotwe, F. T., & Price, A. D. F. (2009). A Social Ontology for Appraising Sustainability of Construction Projects and Developments. *International Journal of Project Management, 27*(4), 313–322. doi:10.1016/j.ijproman.2008.04.003

Egerton, F. N. (2001). A history of the ecological sciences: Early Greek origins. *Bulletin of the Ecological Society of America, 82*(1), 93–97.

Eid, M. (2009). *Sustainable Development & Project Management*. Lambert Academic.

Einstein, A. (2011). *Essays in science*. Open Road Media.

Elkington, J. (1997). *Cannibals with Forks: the Triple Bottom Line of 21st Century Business*. Capstone Publishing.

Ellatar, S. M. S. (2009). Towards developing an improved methodology for evaluating performance and achieving success in construction projects. *Scientific Research and Essays, 4*, 549–554.

Eman, G., Kathy, K., & Ibrahim, A. (2013). Metaverse-retail service quality: A future framework for retail service quality in the 3D internet. *Journal of Marketing Management*, *29*(13-14), 1493–1517. doi:10.1080/0267257X.2013.835742

EMMIE. (2022). *Erasmus Mundus Master in Impact Entrepreneurship Web Summit (2022), Impact Startups at Web Summit 2023*. EMMIE.

Encyclopaedia Britannica. (2009, February 9). *Home*. Britannica. https://www.britannica.com/topic/public-enterprise

Epey. (n.d.). https://www.epey.com/

Erev, I., & Roth, A. E. (1998). Predicting how people play games: Reinforcement learning in experimental games with unique, mixed strategy equilibria. *The American Economic Review*, *88*, 848–881.

Eriksen, T. H. (2016). *Overheating: An Anthropology of Accelerated Change*. Pluto. doi:10.2307/j.ctt1cc2mxj

Eriksson, P. E., Olander, S., Szentes, H., & Widén, K. (2013). Managing short-term efficiency and long-term development through industrialized construction. *Construction Management and Economics*, *32*(1–2), 97–108. doi:10.1080/01446193.2013.814920

Escobar, A. (2020). *Pluriversal Politics: The Real and the Possible*. Duke University Press.

Eskerod, P., & Huemann, M. (2013). Sustainable development and project stakeholder management: What standards say. *International Journal of Managing Projects in Business*, *6*(1), 36–50. doi:10.1108/17538371311291017

Esteva, A., Kuprel, B., Novoa, R. A., Ko, J., Swetter, S. M., & Balu, H. M. (2017). Dermatologist-level classification of skin cancer with deep neural networks. *Nature*, *542*(7639), 115–118. doi:10.1038/nature21056 PMID:28117445

Estévez-Ortiz, F. J., García-Jiménez, A., & Glösekötter, P. (2016). An Application of People's Sentıment From Socıal Media To Smart Cıtıes. *El Profesional de la Información*, *25*(6), 851–858. doi:10.3145/epi.2016.nov.02

Evans, M. (2019). Build A 5-star customer experience with artificial intelligence. *Forbes*. https://www.forbes.com/sites/allbusiness/2019/02/17/customer-experience-artificialintelligence/#1a30ebd415bd

Evans, P. C., & Annunziata, M. (2012). *Industrial Internet: Pushing the Boundaries of Minds and Machines*. GE. https://www.ge.com/news/sites/default/files/5901.pdf

Falagas, M.E., Pitsouni, E.I., Malietzis, G.A., & Pappas, G. (2008). Comparison of PubMed, Scopus, web of science, and Google scholar: strengths and weaknesses. *The FASEB Journal, 22*(2), 338-342. doi:10.1096/fj.07-9492LSF

Falchuk, B., Loeb, S., & Neff, R. (2018). The social metaverse: Battle for privacy. *IEEE Technology and Society Magazine*, *37*(2), 52–61. doi:10.1109/MTS.2018.2826060

Falk, R. F., & Miller, N. B. (1992). *A Primer for Soft Modeling*. The University of Akron Press.

Fang, E. (2008). Customer Participation and the Trade-Off Between New Product Innovativeness and Speed to Market. *Journal of Marketing*, *72*(4), 90–104. doi:10.1509/jmkg.72.4.090

Fang, E., Palmatier, R. W., & Evans, K. R. (2008). Influence of Customer Participation on Creating and Sharing Of New Product Value. *Journal of the Academy of Marketing Science*, *36*(3), 322–336. doi:10.100711747-007-0082-9

Faul, F., Erdfelder, E., Lang, A. G., & Buchner, A. (2007). G*Power 3: A flexible statistical power analysis program for the social, behavioral, and biomedical sciences. *Behavior Research Methods*, *39*(2), 175–191. doi:10.3758/BF03193146 PMID:17695343

Feindt, S., Kornek, U., Labeaga, J. M., Sterner, T., & Ward, H. (2021). Understanding regressivity: Challenges and opportunities of European carbon pricing. *Energy Economics*, *103*, 105550. doi:10.1016/j.eneco.2021.105550

Fei-Yue, F. (2004). Parallel system methods for management and control of complex systems. *Control Decis*, *19*(5), 485–489.

Fellows, R., & Liu, A. (2008). Impact of participants' values on construction sustainability. *Proceedings of the Institution of Civil Engineers. Engineering Sustainability*, *161*(4), 219–227. doi:10.1680/ensu.2008.161.4.219

Fernandez, I. (2020). The Impact of Insurtech on the Public Sector. *Fintech Magazine*. https://www.fintechmagazine.com/insurance-and-protection/impact-insurtech-public-sector

Fernandez-Herrero, J., & Lorenzo, G. (2019). An immersive virtual reality educational intervention on people with autism spectrum disorders (ASD) for the development of communication skills and problem-solving. *Education and Information Technologies*, 1689–1722.

Fernandez-Sanchez, G., & Rodriguez-Lopez, F. (2010). A methodology to identify sustainability indicators in construction project management-application to infrastructure projects in Spain. *Ecological Indicators*, *10*(6), 1193–1201. doi:10.1016/j.ecolind.2010.04.009

Fersini, E., Pozzi, F. A., & Messina, E. (2017). Approval Network: A Novel Approach For Sentiment Analysis in Social Networks. *World Wide Web (Bussum)*, *20*(4), 831–854. doi:10.100711280-016-0419-8

Fields, Z., & Atiku, S. O. (2017). Collective Green Creativity and Eco-Innovation as Key Drivers of Sustainable Business Solutions in Organizations. In Z. Fields (Ed.), *Collective Creativity for Responsible and Sustainable Business Practice* (pp. 1–25). IGI Global. doi:10.4018/978-1-5225-1823-5.ch001

Figueiredo, B., & Scaraboto, D. (2016). The systemic creation of value through circulation in collaborative consumer networks. *The Journal of Consumer Research*, *43*(4), 509–533. Advance online publication. doi:10.1093/jcr/ucw038

Finucane, M. L. (2009). Why science alone won't solve the climate crisis: Managing climate risks in the Pacific. *Asia Pacific Issues*, *89*, 1–8.

Fiske, S. J., Crate, S. A., Crumley, C., Galvin, K., Lazrus, H., Lucero, L., Oliver-Smith, A., Orlove, B., Strauss, S., & Wilk, R. R. (2014). *Changing the atmosphere. Anthropology and climate change. Final report of the AAA global climate change task force*. American Anthropological Association.

Fitzgerald, M., Kruschwitz, N., Bonnet, D., & Welch, M. (2013). Embracing Digital Technology: A New Strategic Imperative. *MIT Sloan Management Review*, *55*(2), 1.

Flynn, B. B., Kakibara, S. S., Schroeder, R. G., Bates, K. A., & Flynn, E. J. (1990). Empirical research methods in operations management. *Journal of Operations Management*, *9*(2), 250–284. doi:10.1016/0272-6963(90)90098-X

Focus Guiyang. (2019, July 30). The Coverage Rate Of The Road Passenger Network Ticketing System In Guizhou Province Is More Than Half. *Focus Guiyang*. https://gy.focus.cn/zixun/06d41c6b1f08d13a.html

Følstad, A., Kvale, K., & Halvorsrud, R. (2013). *Customer journey measures – State of the art research and best practice*. Report A24488, Oslo, Norway: SINTEF.

Fornell, C., & Lacker, D. F. (1981). Evaluating structural equation models with unobservable variables and measurement errors. *JMR, Journal of Marketing Research*, *18*(1), 39–50. doi:10.1177/002224378101800104

Forza, C. (2002). Survey research in operations management: A process-based perspective. *International Journal of Operations & Production Management*, *22*(2), 152–194. doi:10.1108/01443570210414310

FosterCapital. (2022). *Use Resources To Build an Impactful Startup*. https://fastercapital.com/content/Use-resources-to-build-an-impactful-startup.html

Frankenfield, J. (2022, July 6). Artificial intelligence: What it is and how it is used. *Investopedia.* https://www.investopedia.com/terms/a/artificial-intelligence-ai.asp

Fthenakis, V. (2009). Sustainability of photovoltaics: The case for thin-film solar cells. *Renewable & Sustainable Energy Reviews*, *13*(9), 2746–2750. doi:10.1016/j.rser.2009.05.001

Fundin, A. P., & Bergman, B. L. (2003). Exploring the Customer Feedback Process. *Measuring Business Excellence*, *7*(2), 55–65. doi:10.1108/13683040310477995

Fung, A., & Wright, E. O. (2001). Deepening Democracy: Innovations in Empowered Participatory Governance. *Politics & Society*, *29*(1), 5–41. doi:10.1177/0032329201029001002

Gabora, L. (2013). Research on creativity. In E. G. Carayannis (Ed.), *Encyclopedia of Creativity, Invention, Innovation, and Entrepreneurship* (pp. 1548–1558). Springer. doi:10.1007/978-1-4614-3858-8_387

Gabrovšek, P., Aleksovski, D., Mozetič, I., & Grčar, M. (2017). Twitter Sentiment Around The Earnings Announcement Events. *PLoS One*, *12*(2), E0173151. doi:10.1371/journal.pone.0173151 PMID:28235103

Gagnon, K., Young, B., Bachman, T., Longbottom, T., Severin, R., & Walker, M. J. (2020). Doctor of physical therapy education in a hybrid learning environment: Reimagining the possibilities and navigating a "new normal". *Physical Therapy*, *100*(8), 1268–1277. doi:10.1093/ptj/pzaa096 PMID:32424417

Galea-Pace, S. (2020, May 17). *Why is inventory management in the supply chain important?* https://supplychaindigital.com/digital-supply-chain/why-inventory-management-supply-chain-important

Galeazzo, A., Ortiz-de-Mandojana, N., & Delgado-Ceballos, J. (2021). Green procurement and financial performance in the tourism industry: The moderating role of tourists' green purchasing behaviour. *Current Issues in Tourism*, *24*(5), 700–716. doi:10.1080/13683500.2020.1734546

Gandolfi, E., Kosko, K. W., & Ferdig, R. E. (2021). Situating presence within extended reality for teacher training: Validation of the extended Reality Presence Scale (XRPS) in preservice teacher use of immersive 360 video. *British Journal of Educational Technology*, *52*(2), 824–841. doi:10.1111/bjet.13058

Gan, W., Peng, L., Li, D., Han, L., & Zhang, C. (2019). A Coordinated Revenue-Sharing-Based Pricing Decision Model for Remanufactured Products in Carbon Cap and Trade Regulated Closed-Loop Supply Chain. *IEEE Access : Practical Innovations, Open Solutions*, *7*, 142879–142893. doi:10.1109/ACCESS.2019.2943385

Gao, C., Hou, H., Zhang, J., Zhang, H., & Gong, W. (2006). Education for regional sustainable development: Experiences from the education framework of HHCEPZ project. *Journal of Cleaner Production*, *14*(9–11), 994–1002. doi:10.1016/j.jclepro.2005.11.043

Gareis Atkinson, R. (1999). Project management: Cost, time and quality, two best guesses and a phenomenon, its time to accept other success criteria. *International Journal of Project Management*, *17*(6), 337–342. doi:10.1016/S0263-7863(98)00069-6

Gareis, R., Huemann, M., & Martinuzzi, R.-A. (2011). What can project management learn from considering sustainability principles? *Project Perspectives*, *33*, 60–65.

Gareis, R., Huemann, M., Martinuzzi, R.-A., Sedlacko, M., & Weninger, C. (2011b). *The SustPM Matrix: Relating sustainability principles to project assignment and project management. EURAM11*. Talinn.

Gartner. (2022). *Gartner Identifies the Top 10 Strategic Technology Trends for 2022*. Gartner. https://www.gartner.com/en/newsroom/press-releases/2021-10-18-gartner-identifies-the-top-strategic-technology-trends-for-2022

Gassmann, O., Frankenberger, K., & Sauer, R. (2016). *Exploring the field of business model innovation: New theoretical perspectives*. Springer. doi:10.1007/978-3-319-41144-6

Gawellek, M. (2021). *Sustainable Scoring Platform for Employing Institutions*. http://www.innompics.com/events/ig2021/sbm-socsys_sustainable-scoring_mg.html

Gentile, C., Spiller, N., & Noci, G. (2007). How to Sustain the Customer Experience: An Overview of Experience Components That Co-create Value with the Customer. *European Management Journal*, *25*(5), 395–410. doi:10.1016/j.emj.2007.08.005

Genus, A., & Theobald, K. (2015). Roles for university researchers in urban sustainability initiatives: The UK Newcastle Low Carbon Neighbourhoods project. *Journal of Cleaner Production*, *106*, 119–126. doi:10.1016/j.jclepro.2014.08.063

Gerup, J., Soerensen, C. B., & Dieckmann, P. (2020). Augmented reality and mixed reality for healthcare education beyond surgery: An integrative review. *International Journal of Medical Education*, *11*, 1–18. doi:10.5116/ijme.5e01.eb1a PMID:31955150

Getoor, L., & Taskar, B. (Eds.). (2007). *An Introduction to Statistical Relational Learning*. MIT Press. doi:10.7551/mitpress/7432.001.0001

Ge, Y., Wang, Y., & Han, Q. (2020). Test Method of Connected and Automated Vehicles Based On Digital Twin. *ZTE Communications.*, *26*(1). doi:10.12142/ZTETJ.202001006

Ghiassi, M., Zimbra, D., & Lee, S. (2016). Targeted Twitter Sentiment Analysis for Brands Using Supervised Feature Engineering and The Dynamic Architecture for Artificial Neural Networks. *Journal of Management Information Systems*, *33*(4), 1034–1058. doi:10.1080/07421222.2016.1267526

Giannakis, E., Serghides, D., Dimitriou, S., & Zittis, G. (2020). Land transport CO2 emissions and climate change: Evidence from Cyprus. *International Journal of Sustainable Energy*, *39*(7), 634–647. doi:10.1080/14786451.2020.1743704

Giannakis, M., Dubey, R., Yan, S., Spanaki, K., & Papadopoulos, T. (2020). Social Media and Sensemaking Patterns in New Product Development: Demystifying the Customer Sentiment. *Annals of Operations Research*, 1–31.

Giesler, M. (2008). Conflict and compromise: Drama in marketplace evolution. *Journal of Consumer Research*, 34(April), Giesler, M. (2006). Consumer gift systems. *The Journal of Consumer Research*, *33*(September), 283–290. doi:10.1086/506309

Gilbert, R., Stevenson, D., Girardet, H., & Stern, R. (1996). *Making Cities Work: The Role of Local Authorities in the Urban Environment*. Earthscan Publications Ltd.

Gimpel, H., Hosseini, S., Huber, R., Probst, L., Röglinger, M., & Faisst, U. (2018). Structuring Digital Transformation: A Framework of Action Fields and its Application at ZEISS. *Journal of Information Technology Theory and Application*, *19*(1), 31–54.

Giner, C. (2004). J. Antropología urbana. *Ariel*.

Glaser, B. G., & Strauss, A. L. (1965). *Awareness of dying*. Aldine.

Glaser, B. G., & Strauss, A. L. (1967). *The discovery of grounded theory*. Aldine.

Glavič, P., & Lukman, R. (2007). Review of sustainability terms and their definitions. *Journal of Cleaner Production*, *15*(18), 1875–1885. doi:10.1016/j.jclepro.2006.12.006

GlossaryW. I. (2012). http://1world1way.com/coach/glossary.html

Gluckman, M. (1968). The Utility of the Equilibrium Model in the Study of Social Change. *American Anthropologist*, *70*(2), 219–237. doi:10.1525/aa.1968.70.2.02a00010

Gobet, F., & Sala, G. (2019). How artificial intelligence can help us understand human creativity. *Frontiers in Psychology*, *10*, 1401. doi:10.3389/fpsyg.2019.01401 PMID:31275212

Goedknegt, D. (2013). Sustainability in Project Management: Perceptions of Responsibility. In A. J. G. Silvius & J. Tharp (Eds.), *Sustainability Integration for Effective Project Management* (pp. 279–287). IGI Global Publishing. doi:10.4018/978-1-4666-4177-8.ch017

Goffin, K., & New, C. (2001). Customer Support and New Product Development-An Exploratory Study. *International Journal of Operations & Production Management*. https://trends.google.com/trends/?geo=TR

Goh, P. S., & Sandars, J. (2020). A vision of the use of technology in medical education after the COVID-19 pandemic. *MedEdPublish*, 9.

Goldemberg, J., Coelho, S. T., & Guardabassi, P. (2008). The sustainability of ethanol production from sugarcane. *Energy Policy*, *36*(6), 2086–2097. doi:10.1016/j.enpol.2008.02.028

Gomez, B. (2021, August 24). *Elon Musk warned of a 'Terminator'-like AI apocalypse — now he's building a Tesla robot*. CNBC Make It. https://www.cnbc.com/2021/08/24/elon-musk-warned-of-ai-apocalypsenow-hes-building-a-tesla-robot.html

González Álvarez, M. D. (1997). Processos de Planejamento nos Pólos Tecnológicos: um enfoque adaptativo. Tese de Doutorado, Departamento de Engenharia Industrial, PUC-Rio, Rio de Janeiro.

Gössling, S., Hansson, C. B., Hörstmeier, O., & Saggel, S. (2002). Ecological footprint analysis as a tool to assess tourism sustainability. *Ecological Economics*, *43*(2), 199–211. doi:10.1016/S0921-8009(02)00211-2

Goyal, Y., Khot, T., Summers-Stay, D., Batra, D., & Parikh, D. (2017). Making the V in VQA matter: Elevating the role of image understanding in visual question answering. CVPR, 6325–6334.

Goyal, M., & Netessine, S. (2011). Volume flexibility, product flexibility, or both: The role of demand correlation and product substitution. *Manufacturing & Service Operations Management*, *13*(2), 180–193. doi:10.1287/msom.1100.0311

Graf, A. C., Jacob, E., Twigg, D., & Nattabi, B. (2020). Contemporary nursing graduates' transition to practice: A critical review of transition models. *Journal of Clinical Nursing*, *29*(15-16), 3097–3107. doi:10.1111/jocn.15234 PMID:32129522

Gravano, A. (2007b). Claves para la facilitación organizacional del proceso participativo en la planificación urbano-ambiental metropolitana. *Revista Regional de Trabajo Social*, *21*(40), 9-22.

Gravano, A. (2008a) *¿Vecinos o ciudadanos? el fenómeno NIMBY (Not In My Back Yard) o SPAN (Sí, Pero No Aquí) del imaginario urbano en un proceso de participación social y su tratamiento desde la facilitación organizacional antropológica*. Trabajo presentado en II Congreso de la Asociación Latinoamericana de Antropología (ALA), Universidad de Costa Rica, Simposio "Imaginarios urbanos y participación social".

Gravano, A. (1992). Antropología Práctica, muestra y posibilidades de la Antropología Organizacional. *Publicar en Antropología y Ciencias Sociales, Buenos Aires*, (1), 95–126.

Gravano, A. (2005). *El barrio en la teoría social*. Espacio Editorial.

Gravano, A. (2006). Imaginarios regionales y circularidad en la planificación: El caso del TOAR. *Intersecciones*, (7), 305–323.

Greenwald, S. W., Corning, W., & Maes, P. (2017). Multi-User Framework for Collaboration and Co-Creation in Virtual Reality. *Proceedings of the 12th International Conference on Computer Supported Collaborative Learning (CSCL)*, (pp. 18-22). ACM.

Gregersen, H.M., Lundgren, A.L., & White, T.A. (1994). *Improving project management for sustainable development.* Midwest Universities Consortium for International Activities, Inc. (MUCIA), Policy Brief No. 7.

Gretzel, U., Sigala, M., Xiang, Z., & Koo, C. (2015). Smart tourism: Foundations and developments. *Electronic Markets, 25*(3), 179–188. doi:10.100712525-015-0196-8

Gretzel, U., Werthner, H., Koo, C., & Lamsfus, C. (2015). Conceptual foundations for understanding smart tourism ecosystems. *Computers in Human Behavior, 50*, 558–563. doi:10.1016/j.chb.2015.03.043

Grewal, D., & Roggeveen, A. (2020). Understanding Retail Experiences and Customer Journey Management. *Journal of Retailing, 96*(1), 3–8. doi:10.1016/j.jretai.2020.02.002

Grimmer, J., Roberts, M., & Stewart, B. (2021). Machine learning for social science: An agnostic approach. *Annual Review of Political Science, 24*(1), 395–419. doi:10.1146/annurev-polisci-053119-015921

Gronau, Q. F., & Wagenmakers, E.-J. (2019). Limitations of Bayesian leave-one-out cross-validation for model selection. *Computational Brain & Behavior, 2*(1), 1–11. doi:10.100742113-018-0011-7 PMID:30906917

Grosse, K., González, M. P., Chesñevar, C. I., & Maguitman, A. G. (2015). Integrating Argumentation and Sentiment Analysis For Mining Opinions From Twitter. *AI Communications, 28*(3), 387–401. doi:10.3233/AIC-140627

Grottera, C., Naspolini, G. F., La Rovere, E. L., Gonçalves, D. N. S., de Farias Nogueira, T., Hebeda, O., ... Lefèvre, J. (2022). Energy policy implications of carbon pricing scenarios for the Brazilian NDC implementation. *Energy Policy, 160*, 112664. doi:10.1016/j.enpol.2021.112664

Gruner, D. T., & Csikszentmihalyi, M. (2019). Engineering creativity in an age of artificial intelligence. In I. Lebuda & V. P. Glăveanu (Eds.), *The Palgrave Handbook of Social Creativity Research.* Palgrave Macmillan., doi:10.1007/978-3-319-95498-1_27

Guangzhou Daily. (2021, November 30). Reveal the technological elements behind The First Insensible Toll Station in China. *Guangzhou Daily*. https://www.gzdaily.cn/amucsite/web/index.html#/detail/1717097 (gzdaily.cn).

Guan, Y., Ren, Y., Sun, Q., Li, E. B., Ma, H., Duan, J., Dai, Y., & Cheng, B. (2022). Integrated Decision and Control: Toward Interpretable and Computationally Efficient Driving Intelligence. *IEEE Transactions on Cybernetics*. doi:1 doi:0.1109/TCYB.2022.3163816 PMID:35439160

Gudergan, S. P., Ringle, C. M., Wende, S., & Will, A. (2008). Confirmatory tetrad analysis in PLS path modeling. *Journal of Business Research, 61*(12), 1238–1249. doi:10.1016/j.jbusres.2008.01.012

Guizhou Daily. (2016) "Intelligent Transportation Cloud" Makes Guizhou More Convenient. *Guangzhou Daily*. http://www.gov.cn/xinwen/2016-03/14/content_5053045.htm

Guizhou Provincial Big Data Bureau. (2021, May 24). *Big Data Deep Mining Integration, Guiyang Pilot Digital Twin Transportation System.* GPBDB. https://www.guizhou.gov.cn/home/gzyw/202109/t20210913_70368737.html

Guo, X. (2021). Suggestions On The Transformation And Development Of Civil Airport Operation And Reflections On The Transformation To A Management Model: Take Shenzhen Airport As An Example. *Aviation Think Tank*. http://att.caacnews.com.cn/zsfw/jcgl/202111/t20211101_59856.html

Haeberle, H. S., Helm, J. M., Navarro, S. M., Karnuta, J. M., Schaffer, J. L., Callaghan, J. J., Mont, M. A., Kamath, A. F., Krebs, V. E., & Ramkumar, P. N. (2019). Artificial intelligence and machine learning in lower extremity arthroplasty: A review. *The Journal of Arthroplasty, 34*(10), 2201–2203. doi:10.1016/j.arth.2019.05.055 PMID:31253449

Haenlein, M., Kaplan, A., Tan, C. W., & Zhang, P. (2019). Artificial intelligence (AI) and management analytics. *Journal of Management Analytics, 6*(4), 341–343. doi:10.1080/23270012.2019.1699876

Haier, R. J., Colom, R., Schroeder, D. H., Condon, C. A., Tang, C., Eaves, E., & Head, K. (2009). Gray matter and intelligence factors: Is there a neuro-g? *Intelligence, 37*(2), 136–144. doi:10.1016/j.intell.2008.10.011

Haines, S. (2019b). Managing expectations: Articulating expertise in climate services for agriculture in Belize. *Climatic Change, 157*(1), 43–59. doi:10.100710584-018-2357-1

Hair, J. F., Anderson, R. E., Tatham, R. L., & Black, W. C. (2005). *Multivariate Data Analysis* (4th ed.). Prentice Hall.

Hair, J. F., Hult, G. T. M., Ringle, C. M., & Sarstedt, M. (2014). *A Primer on Partial Least Squares Structural Equation Modeling (PLS-SEM)*. Sage.

Hakim, N. (2021). *Social Networking Site (SNS) for Students*. http://innompics.com/events/ig2021/sbm-socsys_socnet_nasrul.html

Halat, K., Hafezalkotob, A., & Sayadi, M. K. (2023). The green supply chains' ordering and pricing competition under carbon emissions regulations of the government. *International Journal of Systems Science: Operations & Logistics, 10*(1), 1983884.

Hall, J. (2019). How Artificial Intelligence Is Transforming Digital Marketing. *Forbes*. https://www.forbes.com/sites/forbesagencycouncil/2019/08/21/how-artificial-intelligence-is-transforming-digital-marketing/#379462ee21e1

Hall, J. K., & Vrendenburg, H. (2003). Environmentally sustainable technologies. *Journal of Cleaner Production, 11*(2), 139–149.

Hamilton, K., Mintz, T., Date, P., & Schuman, C. D. (2020). Spike-based graph centrality measures. In *International Conference on Neuromorphic Systems 2020* (pp. 1–8). ACM. 10.1145/3407197.3407199

Hamilton, D., McKechnie, J., Edgerton, E., & Wilson, C. (2021). Immersive virtual reality as a pedagogical tool in education: A systematic literature review of quantitative learning outcomes and experimental design. *Journal of Computers in Education, 8*(1), 1–32. doi:10.100740692-020-00169-2

Hammond, K. R. (2000). *Human judgment and social policy: Irreducible uncertainty, inevitable error, unavoidable injustice*. Oxford University Press on Demand.

Hanners, U. (1986). *Exploración de la ciudad*. Fondo de Cultura Económica.

Hannerz, U. (2004). *Foreign New: Exploring the World of Foreign Correspondents*. University of Chicago Press.

Hannun, A. Y., Rajpurkar, P., Haghpanahi, M., Tison, G. H., Bourn, C., Turakhia, M. P., & Ng, A. Y. (2019). Cardiologist-level arrhythmia detection and classification in ambulatory electrocardiograms using a deep neural network. *Nature Medicine, 25*(1), 65–69. doi:10.103841591-018-0268-3 PMID:30617320

Harmon, S. (2011). *Human perception of gendered artificial entities*. Colby College.

Harrabin, R. (2017). *Ocean plastic a 'planetary crisis' – UN*. BBC. Retrieved March 7, 2019, from https://www.bbc.com/news/science-environment-42225915

Harrison, A., Skipworth, H., van Hoek, R. I., & Aitken, J. (2019). Logistics management and strategy: competing through the supply chain. Academic Press.

Hartman, E., Reynolds, N. P., Ferrarini, C., Messmore, N., Evans, S., Al-Ebrahim, B., & Brown, J. M. (2020). Coloniality-decoloniality and critical global citizenship: Identity, belonging, and education abroad. *Frontiers: The Interdisciplinary Journal of Study Abroad*, *32*(1), 33–59. doi:10.36366/frontiers.v32i1.433

Hartmut Resonanz, R. (2016). *Eine Soziologie der Weltbeziehung*. Suhrkamp.

Hart, S. L. (1997). Beyond greening: Strategies for a sustainable world. *Harvard Business Review*, *75*(1), 66.

Hassabis, D., Kumaran, D., Summerfield, C., & Botvinick, M. (2017). Neuroscience-inspired artificial intelligence. *Neuron*, *95*(2), 245–258. doi:10.1016/j.neuron.2017.06.011 PMID:28728020

Hassan Khan, F., Qamar, U., & Bashir, S. (2015). Building Normalized Sentimı to Enhance Semi-Supervised Sentiment Analysis. *Journal of Intelligent & Fuzzy Systems*, *29*(5), 1805–1816. doi:10.3233/IFS-151658

Haugan, G. (2012). *The New Triple Constraints for Sustainable Projects, Programs, and ortfolios*. CRC Press.

Hawkins, F. H. (2017). *Human factors in flight*. Routledge. doi:10.4324/9781351218580

Hayes, C., & Capper, S. (2020). Illustrating the transcendence of disciplinarity. In *Beyond Disciplinarity* (pp. 40–49). Routledge. doi:10.4324/9781315108377-4

Hayes, C., & Graham, Y. (2020). *Designing a Benchmarking Tool for Testing Posttest Confidence Levels in Emergency Obstetrics Training*. SAGE Publications. doi:10.4135/9781529709285

Hayes, C., Hinshaw, K., & Petrie, K. (2019). Reconceptualizing medical curriculum design in strategic clinical leadership training for the 21st century physician. In *Preparing Physicians to Lead in the 21st Century* (pp. 147–163). IGI Global. doi:10.4018/978-1-5225-7576-4.ch009

Hazaea, S. A., Al-Matari, E. M., Zedan, K., Khatib, S. F., Zhu, J., & Al Amosh, H. (2022). Green Purchasing: Past, Present and Future. *Sustainability*, *14*(9), 5008. doi:10.3390u14095008

He, A. Z., & Zhang, Y. (2022). AI-powered touch points in the customer journey: A systematic literature review and research agenda. *Journal of Research in Interactive Marketing*, 1–20. doi:10.1108/JRIM-03-2022-0082

Head, B., & Alford, J. (2013). Wicked Problems: Implications for Public Policy and Management. *Administration & Society*. doi:10.1177/0095399713481601

Heal, G. (2020). Reflections—What would it take to reduce US greenhouse gas emissions 80 percent by 2050? *Review of Environmental Economics and Policy*.

Hellmut, W. (2018). Public and Personal Social Services in European Countries from Public/Municipal to Private—And Back to Municipal and "Third Sector" Provision. *International Public Management Journal*, 413–431. http://www.tandfonline.com/action/showCitFormats?doi=10.1080/10967494.2018.1428255

Henderson, J. (2005). Google Scholar: A source for clinicians? *Canadian Medical Association Journal*, *172*(12), 1549–1550. doi:10.1503/cmaj.050404 PMID:15939908

Hengboriboon, L., Sayut, T., Srisathan, W. A., & Naruetharadhol, P. (2022). Strengthening a company–customer relationship from sustainable practices: A case study of petrotrade in Laos. *Cogent Social Sciences*, *8*(1), 2038355. doi:10.1080/23311886.2022.2038355

Henriette, E., Feki, M., & Boughzala, I. (2015). The Shape of Digital Transformation: A Systematic Literature Review. *MCIS 2015 Proceedings*. IEEE.

Henseler, J., Ringle, C. M., & Sinkovics, R. R. (2009). The use of partial least squares path modeling in international marketing. *Adv. Int. Mark.*, *20*, 277–319. doi:10.1108/S1474-7979(2009)0000020014

Heong, Y. M., Ping, K. H., Hamdan, N., Ching, K. B., Yunos, J. M., Mohamad, M. M., ... Azid, N. (2020). Integration of Learning Styles and Higher Order Thinking Skills among Technical Students. *Journal of Technical Education and Training*, *12*(3), 171–179.

Heravi, G., Fathi, M., & Faeghi, S. (2015). Evaluation of sustainability indicators of industrial buildings focused on petrochemical projects. *Journal of Cleaner Production*, *109*, 92–107. doi:10.1016/j.jclepro.2015.06.133

Herazo, B., Lizarralde, G., & Paquin, R. (2012). Sustainable development in the building sector: A Canadian case study on the alignment of strategic and tactical management. *Project Management Journal*, *43*(2), 84–100. doi:10.1002/pmj.21258

Hess, T., Matt, C., Benlian, A., & Wiesböck, F. (2016). Options for Formulating a Digital Transformation Strategy. *MIS Quarterly Executive*, *15*(2), 123–139.

He, Y., Wang, L., & Wang, J. (2012). Cap-and-trade vs. carbon taxes: A quantitative comparison from a generation expansion planning perspective. *Computers & Industrial Engineering*, *63*(3), 708–716. doi:10.1016/j.cie.2011.10.005

Hidayanti, I., Herman, L. E., & Farida, N. (2018). Engaging Customers Through Social Media to Improve Industrial Product Development: The Role of Customer Co-Creation Value. *Journal of Relationship Marketing*, *17*(1), 17–28. doi:10.1080/15332667.2018.1440137

Hilburg, R., Patel, N., Ambruso, S., Biewald, M. A., & Farouk, S. S. (2020). Medical education during the coronavirus disease-2019 pandemic: Learning from a distance. *Advances in Chronic Kidney Disease*, *27*(5), 412–417. doi:10.1053/j.ackd.2020.05.017 PMID:33308507

Hill, R. C., & Bowen, P. A. (1997). Sustainable construction: Principles and a framework for attainment. *Construction Management and Economics*, *15*(3), 223–239. doi:10.1080/014461997372971

Hilty, D. M., Parish, M. B., Chan, S., Torous, J., Xiong, G., & Yellowlees, P. M. (2020). A comparison of in-person, synchronous and asynchronous telepsychiatry: Skills/competencies, teamwork, and administrative workflow. *Journal of Technology in Behavioral Science*, *5*(3), 273–288. doi:10.100741347-020-00137-8

Hindman, M. (2015). Building better models: Prediction, replication, and machine learning in the social sciences. *The Annals of the American Academy of Political and Social Science*, *659*(1), 48–62. doi:10.1177/0002716215570279

Hirst, K. (2019). *Is anthropology a science?* ThoughtCo. https://www.thoughtco.com/isanthropology-a-science-3971060

Hoffman, S., Garnaut, J., Izenman, K., Johnson, M., Pascoe, A., Ryan, F., & Thomas, E. (2020). *The Flip side of China's Central Bank Digital Currency.*

Hofman, J., Sharma, A., & Watts, D. (2017). Prediction and explanation in social systems. *Science*, *355*(6324), 486–488. doi:10.1126cience.aal3856 PMID:28154051

Hofman, J., Watts, D., Athey, S., Garip, F., Griffiths, T., Kleinberg, J., Margetts, H., Mullainathan, S., Salganik, M., Vazire, S., Vespignani, A., & Yarkoni, T. (2021). Integrating explanation and prediction in computational social science. *Nature*, *595*(7866), 181–188. doi:10.103841586-021-03659-0 PMID:34194044

Holling, C. S. (1973). *Resilience and Stability of Ecological Systems. Annual Review of Ecology and Systematics*, 4. https://www.jstor.org/stable/2096802

Holmes, D. R., & Marcus, G. E. (2006). Fast capitalism: para-ethnography and the rise of the symbolic analyst. In M. Fisher & G. Downey (Eds.), *Frontiers of Capital: Ethnographic Reflections on the New Economy*. Duke University Press.

Holmes, D. R., & Marcus, G. E. (2021). How do we collaborate? An updated manifesto. In D. Boyer & G. Marcus (Eds.), *Collaborative Anthropology Today: A Collection of Exceptions*. Cornell University Press. doi:10.7591/cornell/9781501753343.003.0002

Holmström, J. (2022). From AI to digital transformation: The AI readiness framework. *Business Horizons*, *65*(3), 329–339. doi:10.1016/j.bushor.2021.03.006

Hood, C. (2005). The Idea of Joined-Up Government. A Historical Perspective. In V. Bogdanor, Joined-Up Government. Oxford University Press: Oxford, British Academy.

Hoogendoorn, R. (2021, September 6). *Genopets combines physical activity with play-to earn gaming*. Play to Earn. https://www.playtoearn.online/2021/09/06/genopets-combines-physical-activity-with-play-to-earn-gaming/

Horne, B. D., Joy, E. A., Hofmann, M. G., Gesteland, P. H., Cannon, J. B., Lefler, J. S., Blagev, D. P., Korgenski, E. K., Torosyan, N., Hansen, G. I., Kartchner, D., & Pope, C. A. III. (2018). Short-term elevation of fine particulate matter air pollution and acute lower respiratory infection. *American Journal of Respiratory and Critical Care Medicine*, *198*(6), 759–766. doi:10.1164/rccm.201709-1883OC PMID:29652174

Hornibrook, S., May, C., & Fearne, A. (2015). Sustainable development and the consumer: Exploring the role of carbon labelling in retail supply chains. *Business Strategy and the Environment*, *24*(4), 266–276. doi:10.1002/bse.1823

Horton, S. (2021). Empathy Cannot Sustain Action in Technology Accessibility. *Frontiers of Computer Science*, *3*, 31.

Howell, H., & Mikeska, J. N. (2021). Approximations of practice as a framework for understanding authenticity in simulations of teaching. *Journal of Research on Technology in Education*, *53*(1), 8–20. doi:10.1080/15391523.2020.1809033

Hsieh, H.-F., & Shannon, S. E. (2005). Three Approaches to Qualitative Content Analysis. *Qualitative Health Research*, *15*(9), 1277–1288. doi:10.1177/1049732305276687 PMID:16204405

Hsu, Y. (2018). Launch Strategy and New Product Innovatıon: An Empırıcal Examınatıon in Taıwan. *International Journal of Organizational Innovation*, *11*(1).

Huang, B., Zhao, J., Geng, Y., Tian, Y., & Jiang, P. (2017). Energy-related GHG emissions of the textile industry in China. *Resources, Conservation and Recycling*, *119*, 69–77. doi:10.1016/j.resconrec.2016.06.013

Huang, C. D., Goo, J., Nam, K., & Yoo, C. W. (2017). Smart tourism technologies in travel planning: The role of exploration and exploitation. *Information & Management*, *54*(6), 757–770. doi:10.1016/j.im.2016.11.010

Hu, B., Huang, S., & Yin, L. (2021). The cytokine storm and COVID-19. *Journal of Medical Virology*, *93*(1), 250–256. doi:10.1002/jmv.26232 PMID:32592501

Hudson, DManning, C. (2019). GQA: A new dataset for real-world visual reasoning and compositional question answering. *Conference on Computer Vision and Pattern Recognition (CVPR)*.

Hugos, M. H. (2018). *Essentials of supply chain management*. John Wiley & Sons. doi:10.1002/9781119464495

Humpherys, S. L., Bakir, N., & Babb, J. (2021). Experiential learning to foster tacit knowledge through a role play, business simulation. *Journal of Education for Business*, 1–7.

Hu, P., Li, H., Fu, H., Cansever, D., & Mohapatra, P. (2015). Dynamic defense strategy against advanced persistent threat with insiders. *IEEE Conference on Computer Communications (INFOCOM)*, (pp. 747-755). IEEE. 10.1109/INFOCOM.2015.7218444

Huyck, C. R., & Ghalib, H. (2008). A Neuropsychological Framework for Advancing Artificial Intelligence. *AAAI Fall Symposium: Biologically Inspired Cognitive Architectures.*

Hwang, B.-G., & Ng, W. J. (2013). Project management knowledge and skills for green construction: Overcoming challenges. *International Journal of Project Management, 31*(2), 272–284. doi:10.1016/j.ijproman.2012.05.004

Hyatt, S. (2021). *Teaching Urban Anthropology in a time of COVID.* National Association for the Practice of Anthropology. https://scholarworks.iupui.edu/bitstream/handle/1805/29883/Hyatt2021Teaching-pp.pdf?sequence=1

Hyken, S. (2009). *The Cult of the Customer.* John Wiley & Sons.

Ibbs, C. W., & Kwak, Y. H. (2000). Assessing Project Management Maturity. *Project Management Journal, 31*(1), 32–43. doi:10.1177/875697280003100106

IBM Cloud Education. (2020). *Artificial intelligence (AI).* IBM. https://www.ibm.com/cloud/learn/what-is-artificial-intelligence

IBM. (2012). *Developing global leadership: how IBM engages the workforce of a globally integrated enterprise.* https://www.ibm.com/downloads/cas/K7EWX39G

Innompics. (2018). *Global Innompic Ecosystem as a Harmonious Mega-Innovation.* http://innompics.com/org/ecosystem.html

Innompics. (2021). *Innompic Planet of Loving Creators as the World's Leading Peace Platform.* http://innompics.com/games/ig-way_peace-nobel.html

Innompiology. (2020). *INNOMPIOLOGY - The social science that examines and explains how to create harmonious mega-innovations and civilizational breakthroughs.* http://innompics.com/org/innompiology.html

Innompirsity. (2019). *Inompic University – growing disruptive innopreneurs and mega-innovators,* http://innompics.com/org/innompirsity.html

Innoteam-Ru. (2017). *INNOMPUS – harmonized all-inclusive multi-functional complex.* Russia Team, 1st World Innompic Games. http://innompics.com/innompics1/contests_bc2_russia-innompus.html

Innovarsity. (2022). *Glossary of Innovation Terms.* Innovation University (Innovarsity). http://www.innovarsity.com/coach/glossary.html

Innovation, S. (2021, November 4). What does the Metaverse hold for health care? *Sagentiainnocation.*https://www.sagentiainnovation.com/insights/what-does-the-metaverse-hold-for-healthcare/https://www.sagentiainnovation.com/insights/what-does-the-metaverse-hold-for-healthcare/

Insights, D. (2020). *Tech Trends 2020* [White Paper]. Deloitte. https://www2.deloitte.com/content/dam/Deloitte/pt/Documents/tech-trends/TechTrends2020.pdf

Instituto Nacional de Propriedade Industrial (INPI). (n.d.). Retrieved September 19, 2015, from www.inpi.gov.br

International Patent Classification (IPC). (2006). *WIPO* (8th ed.). IPC.

International Standard. (2010). ISO 26000:2010. Guidance on Social Responsibility. ISO.

International Standard. (2016). *ISO/CD 45001. Occupational Health and Safety Management Systems — Requirements.* Draft.

International Standards Organization. (2010). *ISO 26000 Guidance on Social Responsibility.* ISO.

IPMA. (2006). International Competency Baseline (3rd ed.). IPMA.

Jaffe, A. B., Newell, R. G., & Stavins, R. N. (2005). A tale of two market failures: Technology and environmental policy. *Ecological Economics*, *54*(2-3), 164–174. doi:10.1016/j.ecolecon.2004.12.027

Jaillon, L., & Chi-Sun, P. (2010). Design issues of using prefabrication in Hong Kong building construction. *Construction Management and Economics*, *28*(10), 1025–1042. doi:10.1080/01446193.2010.498481

Jain, A., & Jain, M. (2017). Location Based Twitter Opinion Mining Using Common-Sense Information. *Global Journal of Enterprise Information System*, *9*(2), 28–32. doi:10.18311/gjeis/2017/15616

Janishewski, J. (2022). *e-Learning on Sustainable Finance and Corporate Governance*. https://www.1000ventures.com/doc/glossary-finance-sustainable_elearning.html

Jarvis, C. B., MacKenzie, S. B., & Podsakoff, P. M. (2003). A critical review of construct indicators and measurement model misspecification in marketing and consumer research. *The Journal of Consumer Research*, *30*(2), 199–218. doi:10.1086/376806

Jay Polonsky, M., & Ottman, J. (1998). Stakeholders' Contribution to the Green New Product Development Process. *Journal of Marketing Management*, *14*(6), 533–557. doi:10.1362/026725798784867707

Jentsch, F., & Curtis, M. (2017). *Simulation in aviation training*. Routledge. doi:10.4324/9781315243092

Jeong, M., & Shin, H. H. (2019). Tourists' Experiences with Smart Tourism Technology at Smart Destinations and Their Behavior Intentions. *Journal of Travel Research*, *004728751988303*. doi:10.1177/0047287519883034

Jiao, Y., Wu, Y., & Lu, Q. S. (2020). Improving the Performance of Customer Participation in New Product Development: The Moderating Effect of Social Media and Firm Capabilities. *Asian Journal of Technology Innovation*, *28*(2), 284–304. doi:10.1080/19761597.2020.1742749

Jobs. (1997). *Here's to the crazy ones*. Steve Jobs speech. http://innompics.com/coach/innopreneur-crazy_sj.html

Johansson, G., & Magnusson, T. (2006). Organising for environmental considerations in complex product development projects: Implications from introducing a "green" sub-project. *Journal of Cleaner Production*, *14*(15–16), 1368–1376. doi:10.1016/j.jclepro.2005.11.014

Johnson, A.-G. (2022). Why are smart destinations not all technology-oriented? Examining the development of smart tourism initiatives based on path dependence. *Current Issues in Tourism*. doi:10.1080/13683500.2022.2053071

Johnson, C. E., Kimble, L. P., Gunby, S. S., & Davis, A. H. (2020). Using deliberate practice and simulation for psychomotor skill competency acquisition and retention: A mixed-methods study. *Nurse Educator*, *45*(3), 150–154. doi:10.1097/NNE.0000000000000713 PMID:31246693

Johnson, C., Lizarralde, G., & Davidson, C. H. (2006). A systems view of temporary housing projects in post-disaster reconstruction. *Construction Management and Economics*, *24*(4), 367–378. doi:10.1080/01446190600567977

Johnson, J. (2015). Image retrieval using scene graphs. In *Proceedings of the IEEE conference on computer vision and pattern recognition*. IEEE.

Jongerius, C., Hessels, R. S., Romijn, J. A., Smets, E. M., & Hillen, M. A. (2020). The measurement of eye contact in human interactions: A scoping review. *Journal of Nonverbal Behavior*, *44*(3), 1–27. doi:10.100710919-020-00333-3

Jorgensen, D. (2015). Participant observation. In R. Scott & S. Kosslyn (Eds.), *Emerging trends in the social and behavioral sciences* (pp. 1–15). Wiley. doi:10.1002/9781118900772.etrds0247

Joshi, A. W., & Sharma, S. (2004). Customer Knowledge Development: Antecedents and Impact on New Product Performance. *Journal of Marketing*, *68*(4), 47–59. doi:10.1509/jmkg.68.4.47.42722

Ju, M., Jin, J. L., & Zhou, K. Z. (2018). How Can International Ventures Utilize Marketing Capability in Emerging Markets? Its Contingent Effect on New Product Development. *Journal of International Marketing*, *26*(4), 1–17. doi:10.1177/1069031X18809999

Juraschek, M., Büth, L., Posselt, G., & Herrmann, C. (2018). Mixed reality in learning factories. *Procedia Manufacturing*, *23*, 153–158. doi:10.1016/j.promfg.2018.04.009

Kabadurmus, O., & Erdogan, M. S. (2020). Sustainable, multimodal and reliable supply chain design. *Annals of Operations Research*, *292*(1), 47–70. doi:10.100710479-020-03654-0

Kagermann, H., Wahlster, W., & Helbig, J. (2013). Acatech–National academy of science and engineering. *Recommendations for implementing the strategic initiative INDUSTRIE, 4.*

Kaisara, G., Atiku, S. O., & Bwalya, K. J. (2022). Structural Determinants of Mobile Learning Acceptance among Undergraduates in Higher Educational Institutions. *Sustainability*, *14*(21), 13934. doi:10.3390u142113934

Kamasak, R. (2015). How Marketing Capabilities Create Competitive Advantage in Turkey. In Marketing and Consumer Behavior: Concepts, Methodologies, Tools, and Applications (pp. 1602-1621). IGI Global. doi:10.4018/978-1-4666-7357-1.ch079

Kanade, V. (2022, March 14). *What is artificial intelligence (AI)? Definition, types, goals, challenges, and trends in 2022*. Spiceworks. https://www.spiceworks.com/tech/artificial-intelligence/articles/what-is-ai

Kandil, N., Battaïa, O., & Hammami, R. (2020). Globalisation vs. Slowbalisation: A literature review of analytical models for sourcing decisions in supply chain management. *Annual Reviews in Control*, *49*, 277–287. doi:10.1016/j.arcontrol.2020.04.004

Kane, G. C., Palmer, D., Phillips, A. N., Kiron, D., & Buckley, N. (2015). Strategy, not technology, drives digital transformation. *MIT Sloan Management Review and Deloitte University Press*, *14*, 1–25.

Kang, J., Diederich, M., Lindgren, R., & Junokas, M. (2021). Gesture patterns and learning in an embodied XR science simulation. *Journal of Educational Technology & Society*, *24*(2), 77–92.

Kannan, D., Solanki, R., Kaul, A., & Jha, P. C. (2022). Barrier analysis for carbon regulatory environmental policies implementation in manufacturing supply chains to achieve zero carbon. *Journal of Cleaner Production*, *358*, 131910. doi:10.1016/j.jclepro.2022.131910

Kaplan, A., & Haenlein, M. (2019). Siri, Siri, in My Hand: Who's the Fairest in the Land? On the Interpretations, Illustrations, and Implications of Artificial Intelligence. *Business Horizons*, *62*(1), 15–25. doi:10.1016/j.bushor.2018.08.004

Kaplan, O. (1940). Prediction in the social sciences. *Philosophy of Science*, *7*(4), 492–498. doi:10.1086/286658

Karstensen, J., & Peters, G. (2018). Distributions of carbon pricing on extraction, combustion and consumption of fossil fuels in the global supply-chain. *Environmental Research Letters*, *13*(1), 014005. doi:10.1088/1748-9326/aa94a3

Karthick, B., & Uthayakumar, R. (2022). Impact of carbon emission reduction on supply chain model with manufacturing decisions and dynamic lead time under uncertain demand. *Cleaner Logistics and Supply Chain*, *4*, 100037. doi:10.1016/j.clscn.2022.100037

Karunathilake, I. M., & Samarasekera, D. D. (2021). Learning In The 21st Century— 'What's All the Fuss about Change?'. In Educate, Train and Transform: Toolkit on Medical and Health Professions Education (pp. 1-14). Routledge.

Kattel, R., Lember, V., & Tõnurist, P. (2020). Collaborative innovation and human-machine networks. *Public Management Review*, *22*(11), 1652–1673. doi:10.1080/14719037.2019.1645873

Kaufman, J. C., & Glăveanu, V. (2022). Positive Creativity in a Negative World. *Education Sciences*, *12*(3), 193. doi:10.3390/educsci12030193

Kawulich, B. (2005). La observación participante como método de recolección de datos. *Forum Qualitative Social Research*, *68*(2), 43.

Keeble, J. J., Topiol, S., & Berkeley, S. (2003). Using Indicators to Measure Sustainability Performance at a Corporate and Project Level. *Journal of Business Ethics*, *44*(2-3), 149–158. doi:10.1023/A:1023343614973

Keeys, L.A. (2012). Emerging Sustainable Development Strategy in Projects: A Theoretical Framework. *PM World Journal, 1*(2).

Kelly, K. (2022, November 17). *Picture limitless creativity at your fingertips*. Wired. https://www.wired.com/story/picture-limitless-creativity-ai-image-generators/

Kennedy, A., & Smith, K. (1995). Soil microbial diversity and the sustainability of agricultural soils. *Plant and Soil*, *170*(1), 75–86. doi:10.1007/BF02183056

Këpuska, V., & Bohouta, G. (2018, January). Next-generation of virtual personal assistants (microsoft cortana, apple siri, amazon alexa and google home). In *2018 IEEE 8th annual computing and communication workshop and conference (CCWC)* (pp. 99-103). IEEE.

Khalfan, M. M. A. (2006). Managing Sustainability within Construction Projects. *Journal of Environmental Assessment Policy and Management*, *8*(1), 41–60. doi:10.1142/S1464333206002359

Khalili-Damghani, K., & Tavana, M. (2014). A comprehensive framework for sustainable project portfolio selection based on structural equation modeling. *Project Management Journal*, *45*(2), 83–97. doi:10.1002/pmj.21404

Khan, H. U. R., Usman, B., Zaman, K., Nassani, A. A., Haffar, M., & Muneer, G. (2022). The impact of carbon pricing, climate financing, and financial literacy on COVID-19 cases: Go-for-green healthcare policies. *Environmental Science and Pollution Research International*, *29*(24), 35884–35896. doi:10.100711356-022-18689-y PMID:35064505

Khorasanizadeh, M., Bazargan, A., & McKay, G. (2018). *An Introduction to Sustainable Materials Management*. Academic Press.

Kiene, C., Jiang, J. A., & Hill, B. M. (2019). Technological Frames and User Innovation: Exploring Technological Change in Community Moderation Teams. *Proceedings of the ACM on Human-Computer Interaction*, *3*(CSCW), 44. 10.1145/3359146

Kietzmann, J., Paschen, J., & Treen, E. (2018). Artificial Intelligence in Advertising: How Marketers Can Leverage Artificial Intelligence Along the Consumer Journey. *Journal of Advertising Research*, *58*(3), 263–267. doi:10.2501/JAR-2018-035

Kietzmann, J., & Pitt, L. (2020). Artificial intelligence and machine learning: What managers need to know. *Business Horizons*, *63*(2), 131–133. doi:10.1016/j.bushor.2019.11.005

Kim, B., Lee, H., Park, H., & Kim, H. (2012). Greenhouse gas emissions from onsite equipment usage in road construction. *Journal of Construction Engineering and Management*, *138*(8), 982–990. doi:10.1061/(ASCE)CO.1943-7862.0000515

King, O., Borthwick, A., Nancarrow, S., & Grace, S. (2018). Sociology of the professions: What it means for podiatry. *Journal of Foot and Ankle Research*, *11*(1), 1–8. doi:10.118613047-018-0275-0 PMID:29942353

Kivimaa, P., & Mickwitz, P. (2006). The challenge of greening technologies: Environmental policy integration in Finnish technology policies. *Research Policy, 35*(5), 729–744. doi:10.1016/j.respol.2006.03.006

Klaus, S. (2016, January 14). *Agenda.* World Economic Forum. http://www.weforum

Klijn, E. H., & Koppenjan, J. (2000). Public Management and Policy Networks: The Theoretical Foundation of the Network Approach to Governance. *Public Management, 2*(2), 135–158. doi:10.1080/14719030000000007

Klotz, L., & Horman, M. (2010). Counterfactual analysis of sustainable project delivery processes. *Journal of Construction Engineering and Management, 136*(5), 595–605. doi:10.1061/(ASCE)CO.1943-7862.0000148

Knight, P., & Jenkins, J. O. (2009). Adopting and applying eco-design techniques: A practitioners perspective. *Journal of Cleaner Production, 17*(5), 549–558. doi:10.1016/j.jclepro.2008.10.002

Knoepfel, H. (2010). *Survival and Sustainability as Challenges for Projects.* International Project Management Association.

Knowledge Base, M. B. A. (2021). *Warehousing Function of Logistics.* https://www.mbaknol.com/logistics-management/warehousing-function-of-logistics/

Kolar, J. L. (2000). Alternative energy technologies. *Environmental Quality Management Journal, 10*(3), 45–53.

Kolb, Ch. (1998). Review of Bilharz, Joy A., *The Allegany Senecas and Kinzua Dam: Forced Relocation through Two Generations. H-AmIndian, H-Net Reviews.* https://www.h-net.org/reviews/showrev.php?id=2366

Kolotzek, C., Helbig, C., Thorenz, A., Reller, A., & Tuma, A. (2018). A company-oriented model for the assessment of raw material supply risks, environmental impact and social implications. *Journal of Cleaner Production, 176*, 566–580. doi:10.1016/j.jclepro.2017.12.162

Kometa, S., Olomolaiye, P., & Harris, F. (1995). An evaluation of clients' needs and responsibilities in the construction process. *Engineering, Construction, and Architectural Management, 2*(1), 57–76. doi:10.1108/eb021003

Komninos, N., Pallot, M., & Schaffers, H. (2013). Special issue on smart cities and the future internet in Europe. *Journal of the Knowledge Economy, 4*(2), 119–134. doi:10.100713132-012-0083-x

Konecki, K. (2011). Visual Grounded Theory: A Methodological Outline and Examples from Empirical Work. *Revija za Sociologiju, 41*, 131–160. doi:10.5613/rzs.41.2.1

Koo, C., Mendes-Filho, L., & Buhalis, D. (2019). Smart tourism and competitive advantage for stakeholders: Guest editorial. *Tourism Review, 74*(1), 1–4. doi:10.1108/TR-02-2019-208

Kopp, B., Steinke, A., Bertram, M., Skripuletz, T., & Lange, F. (2019). Multiple levels of control processes for Wisconsin Card Sorts: An observational study. *Brain Sciences, 9*(6), 141. doi:10.3390/brainsci9060141 PMID:31213007

Korkmaz, S., Riley, D., & Horman, M. (2010). Piloting evaluation metrics for sustainable high-performance building project delivery. *Journal of Construction Engineering and Management, 136*(8), 877–885. doi:10.1061/(ASCE)CO.1943-7862.0000195

Kosova, R., Qendraj, D.H., Xhafaj, E. (2022). Meta-Analysis ELECTRE III and AHP in Evaluating and Ranking the Urban Resilience. *Journal of Environmental Management and Tourism, 3*(59), 756 - 768. doi:10.14505/jemt.v13.3(59).15

Kotelnikov, D. (2020). *My Clean City Team.* http://www.innompics.com/events/ig2020/contests_sbm_tl_ru-in_denko.html

Kotelnikov, V. (2008). *Virtuous Entrepreneur: 6+6 Engines.* http://kotelnikov.biz/coach/entrepreneur_12drivers.html

Kotelnikov, V. (2010). *KoRe 10 Innovative Thinking Tools (10 KITT).* http://www.kotelnikov.biz/coach/creativity_10magictools.html

Kotelnikov, V. (2011). *Three Levels of Individual Creativity: Conscious, Subconscious, and Divine*. http://www.kotelnikov.biz/coach/creativity_3levels.html

Kotelnikov, V. (2012). *Create Breakthrough by Playing INNOBALL Simulation Game*. http://www.innoball.com / http://kotelnikov.biz/coach/innogames.html

Kotelnikov, V. (2013). *3Bs of Strategic Creativity: Brainstilling, Brainstorming, Brainstilling*. http://kotelnikov.biz/coach/creativity_strategic_3b.html

Kotelnikov, V. (2014). *White Marketing – Enjoy Both Nobel Joy and High Revenues*. http://kotelnikov.biz/coach/marketing_white.html

Kotelnikov, V. (2016). *Business Design of Innompic Games: Harmonising the Five Basic Elements*. IG Way. http://innompics.com/games/ig_about_5be.html

Kotelnikov, V. (2017). *Innompic Games as a Cilizational Breakthrough*. http://innompics.com/games/innompics_civbreak.html

Kotelnikov, V. (2018). *How To Develop Serendipity: KoRe 10 Tips*. http://www.kotelnikov.biz/coach/discovery_serendipity.html

Kotelnikov, V. (2019). *Business Design of Innompic Games*. http://www.innompics.com/org/ig_biz-design.html

Kotelnikov, V. (2020). *Harmonious Mega-Innovation*. http://innompics.com/coach/innovation-mega-harmonious.html

Kotelnikov, V. (2021). *Holistic Innovation (HI): What, Why, and How*. http://innompics.com/coach/innovation-holistic.html

Kotelnikov. (2009). *Learning SWOT Questions*. 1000ventures.com

Kotelnikov. (2022). *Gamification 10+*. Innompirsity.

Kotelnikova, K. (2020). *HealthBiotics Startup Success Story*. http://innompics.com/events/ig2020/contests_sbm_msiw_ru-in_ksu.html

Kotler, P., & Armstrong, G. (2010). *Principles of Marketing*. Pearson International Education.

Kotrikla, A. M., Zavantias, A., & Kaloupi, M. (2021). Waste generation and management onboard a cruise ship: A case study. *Ocean and Coastal Management*, *212*, 105850. doi:10.1016/j.ocecoaman.2021.105850

Koufidis, C., Manninen, K., Nieminen, J., Wohlin, M., & Silén, C. (2021). Unravelling the polyphony in clinical reasoning research in medical education. *Journal of Evaluation in Clinical Practice*, *27*(2), 438–450. doi:10.1111/jep.13432 PMID:32573080

Krajcovic, M., Gabajová, G., Furmannová, B., Vavrík, V., Gašo, M., & Matys, M. (2021). A Case Study of Educational Games in Virtual Reality as a Teaching Method of Lean Management. *Electronics (Basel)*, *10*(7), 838. doi:10.3390/electronics10070838

Kramer, J. M. (2021). From predator to parasite: On private property and our ecological disaster. *Journal of Economic Issues*, *55*(June), 416–422. doi:10.1080/00213624.2021.1908804

Krauß, W. (2020). Narratives of change and the co-development of climate services for action. *Climate Risk Management*, *28*, 100217. doi:10.1016/j.crm.2020.100217

Krishnamurthy, B., Gill, P., & Arlitt, M. (2008). A Few Chirps About Twitter. *Proceedings of the First Workshop On Online Social Networks,* 19-24. 10.1145/1397735.1397741

Krishna, R., Zhu, Y., Groth, O., Johnson, J., Hata, K., Kravitz, J., Chen, S., Kalantidis, Y., Li, L.-J., Shamma, D. A., Bernstein, M. S., & Fei-Fei, L. (2017). Visual genome: Connecting language and vision using crowdsourced dense image annotations. *International Journal of Computer Vision, 123*(1), 32–73. doi:10.100711263-016-0981-7

Križaj, D., Bratec, M., Kopić, P., & Rogelja, T. (2021). A technology-based innovation adoption and implementation analysis of European smart tourism projects: Towards a smart actionable classification model (SACM). *Sustainability, 13*(18), 10279. doi:10.3390u131810279

Kruglikov, A. A., Lazorenko, G. I., Morozov, A. V., & Yavna, V. A. (2013). Designing Intelligent Systems And Monitoring Of Transport Infrastructure By Elastic Waves. *9th EAGE International Scientific and Practical Conference and Exhibition on Engineering and Mining Geophysics*, Gelendzhik, Russia.

Kuehr, R. (2007). Environmental technologies: From a misleading interpretations to an operational categorization and definition. *Journal of Cleaner Production, 15*(13–14), 1316–1320. doi:10.1016/j.jclepro.2006.07.015

Kuei, C., Madu, C. N., Chow, W. S., & Chen, Y. (2015). Determinants and associated performance improvement of green supply chain management in China. *Journal of Cleaner Production, 95*, 163–173. doi:10.1016/j.jclepro.2015.02.030

Kuhn, T. S. (1996). The Structure of Scientific Revolutions (3rd ed.). University of Chicago Press. (Original work published 1962)

Kumar, S. (2019, November 25). Advantages and disadvantages of artificial intelligence. *Medium.* https://towardsdatascience.com/advantages-and-disadvantages-of-artificial-intelligence-182a5ef6588c

Kumar, S. (2022, October 3). 10 AI-powered tools for designers and creative entrepreneurs in 2022. *Medium.* https://uxplanet.org/10-extraordinary-ai-powered-tools-for-designers-and-creative-entrepreneurs-in-2022-1edee00c7cb5

Kumar, A., Jigyasu, D. K., Subrahmanyam, G., Mondal, R., Shabnam, A. A., Cabral-Pinto, M. M. S., ... Bhatia, A. (2021). Nickel in terrestrial biota: Comprehensive review on contamination, toxicity, tolerance and its remediation approaches. *Chemosphere, 275*, 129996. doi:10.1016/j.chemosphere.2021.129996 PMID:33647680

Kumaraswamy, M. M., & Thorpe, A. (1996). Systematizing construction project evaluations. *Journal of Management Engineering, 12*(1), 34–39. doi:10.1061/(ASCE)0742-597X(1996)12:1(34)

Kumar, N., Agrahari, R. P., & Roy, D. (2015). Review of green supply chain processes. *IFAC-PapersOnLine, 48*(3), 374–381. doi:10.1016/j.ifacol.2015.06.110

Kuna, M., & Dreslerová, D. (2007). Landscape archaeology and "community areas" in the archaeology of central Europe. In D. Hicks, L. McAtackney, & G.J. Fairclough (Eds.), Envisioning landscape: Situations and standpoints in archaeology and heritage, Left Coast Press.

Kuper, M., Dionnet, M., Hammani, A., Bekka, Y., Garin, P., & Bluemling, B. (2009). Supporting the shift from state water to community water: Lessons from a social learning approach to designing joint irrigation projects in Morocco. *Ecology and Society, 14*(1), art19. doi:10.5751/ES-02755-140119

Kusku, H. (2020). Acoustic sound–induced stress response of Nile tilapia (Oreochromis niloticus) to long-term underwater sound transmissions of urban and shipping noises. *Environmental Science and Pollution Research International, 27*(29), 36857–36864. doi:10.100711356-020-09699-9 PMID:32577967

Labuschagne, C., & Brent, A. C. (2008). An industry perspective of the completeness and relevance of a social assessment framework for project and technology management in the manufacturing sector. *Journal of Cleaner Production, 16*(3), 253–262. doi:10.1016/j.jclepro.2006.07.028

Laegreid, P., & Rykkja, L. (2014). Governance for Complexity – How to Organize for the Handling of wicked Issues. *Stein Rokkan Centre for Social Studies*. https://bora.uib.no/handle/1956/9384

Lahuerta-Otero, E., & Cordero-Gutiérrez, R. (2016). Looking for The Perfect Tweet. The Use of Data Mining Techniques to Find Influencers on Twitter. *Computers in Human Behavior*, *64*, 575–583. doi:10.1016/j.chb.2016.07.035

Lambert, A. (2019). Psychotic, acritical and precarious? A Lacanian exploration of the neoliberal consumer subject. *Marketing Theory*, *19*(3), 329–346. doi:10.1177/1470593118796704

Lam, E. W. M., Chan, A. P. C., & Chan, D. W. M. (2007). Benchmarking the performance of design-build projects: Development of project success index. *BIJ*, *14*(5), 624–638. doi:10.1108/14635770710819290

Lange, F., & Dewitte, S. (2019). Cognitive flexibility and pro-environmental behaviour: A multimethod approach. *European Journal of Personality*, *56*(4), 46–54. doi:10.1002/per.2204

Lapassade, G. (1986). *Grupos, organizaciones e instituciones: la transformación de la burocracia*. Gedisa.

Larina, E., Bottjer, D. J., Corsetti, F. A., Thibodeau, A. M., Berelson, W. M., West, A. J., & Yager, J. A. (2021). Ecosystem change and carbon cycle perturbation preceded the end-Triassic mass extinction. *Earth and Planetary Science Letters*, *576*, 117180. doi:10.1016/j.epsl.2021.117180

Latour, B. (2013). *An inquiry into modes of existence: An anthropology of the moderns* (C. Porter, Trans.). Harvard University Press.

Laws, D., & Loeber, A. (2011). Sustainable development and professional practice. *Proceedings of the Institution of Civil Engineers. Engineering Sustainability*, *164*(1), 25–33. doi:10.1680/ensu.2011.164.1.25

Le Breton, S., Lamberti, M. J., Dion, A., & Getz, K. A. (2020). COVID-19 and Its impact on the future of clinical trial execution. *Applied Clinical Trials*. https://www.appliedclinicaltrialsonline.com/view/covid-19-and-its-impact-on-the-future-of-clinical-trialexecution

Leal-Rodríguez, A. L., Roldán, J. L., Ariza-Montes, J. A., & Leal-Millán, A. (2014). From potential absorptive capacity to innovation outcomes in project teams through integrated contracts: Experiences with inclusiveness in Dutch infrastructure projects. *International Journal of Project Management*, *31*(4), 615–627.

Leal-Rodríguez, A. L., Roldán, J. L., Ariza-Montes, J. A., & Leal-Millán, A. (2014, August). the conditional mediating role of the realized absorptive capacity in a relational learning context. *International Journal of Project Management*, *32*(6), 894–907. doi:10.1016/j.ijproman.2014.01.005

Lee, J., Bazilian, M., Sovacool, B., & Greene, S. (2020). Responsible or reckless? A critical review of the environmental and climate assessments of mineral supply chains. *Environmental Research Letters*, *15*(10), 103009. doi:10.1088/1748-9326/ab9f8c

Leenes, R. (2008). Privacy in the Metaverse: Regulagting a Complex Social: Construct in a Virtual World. In The Future of Identity in the Information Society (pp. 95-112).

Lemon, K., & Verhoef, P. (2016). Understanding Customer Experience Throughout the Customer Journey. *Journal of Marketing*, *80*(6), 69–96. doi:10.1509/jm.15.0420

Leos, D. (2022, December 12). Is AI a risk to creativity? The answer is not so simple. *Entrepreneur*. https://www.entrepreneur.com/science-technology/is-ai-a-risk-to-creativity-the-answer-is-not-so-simple/439525

Lester, R. K. (2005). *Universities, Innovation, and the Competitiveness of Local Economies A Summary Report from the Local Innovation Systems Project – Phase I.* Industrial Performance Center, Massachusetts Institute of Technology Working Paper 05-010.

Leurs, M. T. W., Mur-Veeman, I. M., Sar, R., Schaalma, H. P., & Vries, N. K. (2008). Diagnosis of sustainable collaboration in health promotion: A case study. *BMC Public Health, 8*(1), 382. doi:10.1186/1471-2458-8-382 PMID:18992132

Levitt, H. M. (2021). Qualitative generalization, not to the population but to the phenomenon: Reconceptualizing variation in qualitative research. *Qualitative Psychology, 8*(1), 95–110. doi:10.1037/qup0000184

Levy, O., Beechler, S., Taylor, S., & Boyacigiller, N. A. (2014). *Global Mindset. In Wiley Encyclopedia of Management.* John Wiley & Sons, Ltd.

Lewis, H., Gertsakis, J., Grant, T., Morelli, N., & Sweatman, A. (2001). *Design + Environment, A Global Guide to Designing Greener Goods.* Routledge.

Liang, C. (2020). Application of Bentley Digital Twin Technology in Highway Quality Engineering Construction. *China ITS Journal, 6.*

Liang, X., Xiong, Y., Hou, Y. Z., & Xia, S. M. (2014). Pricing and emission reduction strategy of manufacturer with carbon emission constraints in low carbon supply chain. *Key Engineering Materials, 572*, 659–662. doi:10.4028/www.scientific.net/KEM.572.659

Liao, S., Wu, J., Bashir, A. K., Yang, W., Li, J., & Tariq, U. (2021). Digital Twin Consensus for Blockchain-Enabled Intelligent Transportation Systems in Smart Cities. *IEEE Transactions on Intelligent Transportation Systems.* doi:10.1109/TITS.2021.3122566

Li, H. (2019). Special Section Introduction: Artificial Intelligence and Advertising. *Journal of Advertising, 48*(4), 333–337. doi:10.1080/00913367.2019.1654947

Li, H., Wang, C., Shang, M., & Ou, W. (2017). Pricing, carbon emission reduction, low-carbon promotion and returning decision in a closed-loop supply chain under vertical and horizontal cooperation. *International Journal of Environmental Research and Public Health, 14*(11), 1332. doi:10.3390/ijerph14111332 PMID:29104268

Li, J., Du, W., Yang, F., & Hua, G. (2014). The carbon subsidy analysis in remanufacturing closed-loop supply chain. *Sustainability, 6*(6), 3861–3877. doi:10.3390u6063861

Lilien, G. L., Morrison, P. D., Searls, K., Sonnack, M., & Hippel, E. V. (2002). Performance Assessment of the Lead User Idea-Generation Process for New Product Development. *Management Science, 48*(8), 1042–1059. doi:10.1287/mnsc.48.8.1042.171

Li, M., & Tuunanen, T. (2022). Information Technology-Supported value Co-Creation and Co-Destruction via social interaction and resource integration in service systems. *The Journal of Strategic Information Systems, 31*(2), 101719. doi:10.1016/j.jsis.2022.101719

Lim, C., & Mohamed, M. Z. (1999). Criteria of project success: An exploratory reexamination. *International Journal of Project Management, 17*(4), 243–248. doi:10.1016/S0263-7863(98)00040-4

Lindblom, C. E. (1959). The Science of Muddling through. *Public Administration Review, 19*(2), 79–88. doi:10.2307/973677

Lin, M. J. J., Tu, Y. C., Chen, D. C., & Huang, C. H. (2013). Customer Participation and New Product Development Outcomes: The Moderating Role of Product Innovativeness. *Journal of Management & Organization, 19*(3), 314–337. doi:10.1017/jmo.2013.8

Liu, C. H., Zhang, K., & Zhang, J. M. (2010). Sustainable utilization of regional water resources: Experiences from the Hai Hua ecological industry pilot zone (HHEIPZ) project in China. *Journal of Cleaner Production*, *18*(5), 447–453. doi:10.1016/j.jclepro.2009.11.011

Liu, C., Ren, Z., Zhuang, K., He, L., Yan, T., Zeng, R., & Qiu, J. (2021). Semantic association ability mediates the relationship between brain structure and human creativity. *Neuropsychologia*, *151*, 107722. doi:10.1016/j.neuropsychologia.2020.107722 PMID:33309677

Liu, D.-Y., Chen, S.-W., & Chou, T.-C. (2011). Resource fit in digital transformation: Lessons learned from the CBC Bank global e-banking project. *Management Decision*, *49*(10), 1728–1742. doi:10.1108/00251741111183852

Liu, P. (2019). Pricing policies and coordination of low-carbon supply chain considering targeted advertisement and carbon emission reduction costs in the big data environment. *Journal of Cleaner Production*, *210*, 343–357. doi:10.1016/j.jclepro.2018.10.328

Liu, P., & Yi, S. P. (2018). A study on supply chain investment decision-making and coordination in the Big Data environment. *Annals of Operations Research*, *270*(1-2), 235–253. doi:10.100710479-017-2424-4

Li, W., Chen, H., & Nunamaker, J. F. Jr. (2016). Identifying and Profiling Key Sellers in Cyber Carding Community: Azsecure Text Mining System. *Journal of Management Information Systems*, *33*(4), 1059–1086. doi:10.1080/07421222.2016.1267528

Li, W., & Jia, Z. (2017). Carbon tax, emission trading, or the mixed policy: Which is the most effective strategy for climate change mitigation in China? *Mitigation and Adaptation Strategies for Global Change*, *22*(6), 973–992. doi:10.100711027-016-9710-3

Li, X., Li, J., & Wu, Y. (2015). A Global Optimization Approach to Multi-Polarity Sentiment Analysis. *PLoS One*, *10*(4), E0124672. doi:10.1371/journal.pone.0124672 PMID:25909740

Li, Y., Han, W., Shen, H., & Chen, H. L. (2021). Research on the Application of AR Visualization in Urban Traffic Management. *Journal of Transportation Engineering*, *21*(2), 57–61, 67.

Li, Y., Hu, C., Huang, C., & Duan, L. (2017). The concept of smart tourism in the context of tourism information services. *Tourism Management*, *58*, 293–300. doi:10.1016/j.tourman.2016.03.014

Llanos, M. (2009). *Epistemología de las ciencias sociales*. UNMSM.

Loeng, S. (2018). Various ways of understanding the concept of andragogy. *Cogent Education*, *5*(1), 1496643. doi:10.1080/2331186X.2018.1496643

Logeswaran, A., Munsch, C., Chong, Y. J., Ralph, N., & McCrossnan, J. (2021). The role of extended reality technology in healthcare education: Towards a learner-centred approach. *Future Healthcare Journal*, *8*(1), e79–e84. doi:10.7861/fhj.2020-0112 PMID:33791482

Lorek, S., & Fuchs, D. (2013). Strong sustainable consumption governance–precondition for a degrowth path? *Journal of Cleaner Production*, *38*, 36–43. doi:10.1016/j.jclepro.2011.08.008

Lourau, R. (1979). *El análisis institucional*. Amorrortu.

LucasH. C.AgarwalR.ClemonE. K.SawyO. A.WeberB. (2013)

Lucas, H. C. Jr, Agarwal, R., Clemons, E. K., El Sawy, O. A., & Weber, B.Impact Research on Transformational Information Technology. (2013, February 2). An Opportunity to Inform New Audiences. *Management Information Systems Quarterly*, *37*(2), 371–382. doi:10.25300/MISQ/2013/37.2.03

Luce, R. D. (1959). *IndividualChoiceBehaviour*. JohnWiley&SonsInc.

Luctkar-Flude, M., & Tyerman, J. (2021). The Rise of Virtual Simulation: Pandemic Response or Enduring Pedagogy? *Clinical Simulation in Nursing*, *57*, 1–2. doi:10.1016/j.ecns.2021.06.008

Lui, H., Bowman, M., Adams, R., Hurliman, J., & Lake, D. (2010). Scaling virtual worlds: Simulation requirements and challenges. *In Proceedings of the Winter Simulation Conference (WSC). The WSC Foundation*, (pp. 778-798). ACM.

Lund, B., Omame, I., Tijani, S., & Agbaji, D. (2020). Perceptions toward Artificial Intelligence among Academic Library Employees and Alignment with the Diffusion of Innovations' Adopter Categories. *College & Research Libraries*, *865*, 865. doi:10.5860/crl.81.5.865

Lundvall, B.-Å. (Ed.). (1992). National Systems of Innovation: towards a theory of innovation and interactive learning. Printer Publisher.

Luo, C., Lan, Y., Luo, X. R., & Li, H. (2021). The effect of commitment on knowledge sharing: An empirical study of virtual communities. *Technological Forecasting and Social Change*, *163*, 120438. doi:10.1016/j.techfore.2020.120438

Lv, Z. H. & Xie, S.X. (2021). Artificial Intelligence in the Digital Twins: State Of The Art, Challenges, and Future Research Topics. *Digital Twin*, 1-12.

Lv, Z., Zhang, S., & Xiu, W. (2020). Solving the Security Problem of Intelligent Transportation System with Deep Learning. *IEEE Transactions on Intelligent Transportation Systems*, *22*(7), 4281–4290. doi:10.1109/TITS.2020.2980864

Lwin, M. O., Wirtz, J., & Stanaland, A. J. S. (2016). The privacy dyad. *Internet Research*, *26*(4), 919–941. doi:10.1108/IntR-05-2014-0134

Lyytinen, K., Yoo, Y., & Boland, R. J. Jr. (2016). Digital Product Innovation within Four Classes of Innovation Networks. *Information Systems Journal*, *26*(1), 47–75. doi:10.1111/isj.12093

Madden, P. B., & Morawski, J. D. (2011). The future of the Canadian oil stands: Engineering and project management advances. *Energy & Environment*, *22*(5), 579–596. doi:10.1260/0958-305X.22.5.579

Magretta, J. (2002). Why Business Models Matter. *Harvard Business Review*, (80(5)), 86–92. PMID:12024761

Maibach, M. (2008). *Handbook on estimation of external costs in the transport sector - Produced within the study Internalisation Measures and Policies for All external Cost of Transport*. IMPACT.

Malinowski, B. (1961). *Argonauts of the Western Pacific*. Routledge.

Malladi, K. T., & Sowlati, T. (2020). Bi-objective optimization of biomass supply chains considering carbon pricing policies. *Applied Energy*, *264*, 114719. doi:10.1016/j.apenergy.2020.114719

Maltzman, R., & Shirley, D. (2013). Project Manager as a Pivot Point for Implementing Sustainability in an Enterprise. In A. J. G. Silvius & J. Tharp (Eds.), *Sustainability Integration for Effective Project Management* (pp. 262–278). IGI Global Publishing. doi:10.4018/978-1-4666-4177-8.ch016

Marai, O. E., Taleb, T., & Song, J. S. (2020). Roads Infrastructure Digital Twins: A Step Toward Smarter Cities Realization. *IEEE Network*, (99), 1–8.

Maran, T., & Kull, K. (2014). Ecosemiotics: Main principles and current developments. *Geografiska Annaler. Series B, Human Geography*, *96*(1), 41–50. doi:10.1111/geob.12035

Marcelino-Sádaba, S., González-Jaen, L. F., & Pérez-Ezcurdia, A. (2015). Using project management as a way to sustainability. From a comprehensive review to a framework definition. *Journal of Cleaner Production, 99*, 1–16. doi:10.1016/j.jclepro.2015.03.020

Mariani, M., Bresciani, S., & Dagnino, G. B. (2021). The competitive productivity (CP) of tourism destinations: An integrative conceptual framework and a reflection on big data and analytics. *International Journal of Contemporary Hospitality Management, 33*(9), 2970–3002. doi:10.1108/IJCHM-09-2020-1102

Marklund, J., & Berling, P. (2017). Green inventory management. In *Sustainable supply chains* (pp. 189–218). Springer. doi:10.1007/978-3-319-29791-0_8

Marr, B. (2020, February 28). Can machines and artificial intelligence be creative? *Forbes*. https://www.forbes.com/sites/bernardmarr/2020/02/28/can-machines-and-artificial-intelligence-be-creative

Marrone, R., Taddeo, V., & Hill, G. (2022). Creativity and artificial intelligence: A student perspective. *Journal of Intelligence, 10*(65), 65. doi:10.3390/jintelligence10030065 PMID:36135606

Marshall, G. W., Solomon, M. R., & Stuart, E. W. (2008). *Marketing Real People, Real Choices*. Pearson Education.

Martens, M. M., & Carvalho, M. M. (2017). Key factors of sustainability in project management context: A survey exploring the project managers' perspective. *International Journal of Project Management, 35*(6), 1084–1102. doi:10.1016/j.ijproman.2016.04.004

Martens, P. (2006). Sustainability: science or fiction? Sustainability: Science, Practice, &. *Policy, 2*(1), 36–41.

Martinsons, M. G., So, S. K. K., Tin, C., & Wong, D. (1997). Hong Kong and China: Emerging markets for environmental products and technologies. *Long Range Planning, 30*(2), 277–290. doi:10.1016/S0024-6301(96)00119-7

Mathew, P. S., & Pillai, A. S. (2020). Role of Immersive (XR) Technologies in Improving Healthcare Competencies: A Review. *Virtual and Augmented Reality in Education, Art, and Museums*, 23-46.

Mathur, V. N., Price, A. D. F., & Austin, S. (2008). Conceptualizing stakeholder engagement in the context of sustainability and its assessment. *Construction Management and Economics, 26*(6), 601–609. doi:10.1080/01446190802061233

Matinise, S. N. (2019). *Understanding waste management practices in the commercial food service sector* [Doctoral dissertation]. North-West University.

Matt, C., Hess, T., & Benlian, A. (2015). Digital transformation strategies. *Business & Information Systems Engineering, 57*(5), 339–343. doi:10.100712599-015-0401-5

Ma, U. (2011). *No Waste: Managing Sustainability in Construction*. Gower Publishing.

Ma, X., Ho, W., Ji, P., & Talluri, S. (2018). Coordinated pricing analysis with the carbon tax scheme in a supply chain. *Decision Sciences, 49*(5), 863–900. doi:10.1111/deci.12297

Mayo, E. (1945). *The social problems of an Industrial Civilization*. Harvard University Press.

Mazon, R. (1992). Em direção a um novo paradigma de gestão ambiental: Tecnologias limpas ou prevenção da poluição. *Revista de Administração de Empresas, 32*(2), 78–98. doi:10.1590/S0034-75901992000200009

Mbugua, S. (2020, Jan 23). *2 Years Ago, Kenya Set The World's Strictest Plastic Bag Ban. Did It Work?* https://www.huffpost.com/entry/plastic-bag-ban-works-kenya_n_5e272713c5b63211761a4698

McDermott & Jadd. (2001). *NLP Coach. A Comprehensive Guide to Personal Well-being and Professional Success*. Judy Platkus (Publishers) Limited.

McDonough, W., & Braungart, M. (2002). *Cradle To Cradle: Remaking The Way We Make Things.* North Point Press.

McEntee, J. C., & Ogneva-Himmelberger, Y. (2008). Diesel particulate matter, lung cancer, and asthma incidences along major traffic corridors in MA, USA: A GIS analysis. *Health & Place, 14*(4), 817–828. doi:10.1016/j.healthplace.2008.01.002 PMID:18280198

McGrath, J. L., Taekman, J. M., Dev, P., Danforth, D. R., Mohan, D., Kman, N., & Won, K. (2018). Using virtual reality simulation environments to assess competence for emergency medicine learners. *Academic Emergency Medicine, 25*(2), 186–195. doi:10.1111/acem.13308 PMID:28888070

McIntosh, R. J., Tainter, J. A., & McIntosh, S. K. (Eds.). (2000). *The Way the Wind Blows: Climate Change, History, and Human Activity.* Columbia University Press.

McKinlay, M. (2008). *Where is Project Management running to …?* Keynote address delivered at the 22nd World Congress of the International Project Management Association, Rome, Italy.

McKinsey & Company. (2018 July 26). *Artificial intelligence: Why a digital base is critical.* McKinsey Quarterly. https://www.mckinsey.com/business-functions/mckinsey-analytics/our-insights/artificial-intelligence-why-a-digital-base-is-critical#

McNaughton, M. L., McLeod, M. T., McNaughton, M., & Walcott, J. (2016). Open data as a catalyst for problem solving: empirical evidence from a small island developing states (SIDS) context. In 2016 *Open Data Research Symposium, Madrid, Spain.* https://drive.google.com/ open?id=0B4TpC6ecmrM7OEN6OVlIUXh1d1U

Meech, J. A., McPhie, M., Clausen, K., Simpson, Y., Lang, B., Campbell, E., Johnstone, S., & Condon, P. (2006). Transformation of a derelict mine site into a sustainable community: The Britannia project. *Journal of Cleaner Production, 14*(3–4), 349–365. doi:10.1016/j.jclepro.2004.08.009

Meeus, J. H. A. (1995). Pan-European landscapes. *Landscape and Urban Planning, 31*(1-3), 57–79. doi:10.1016/0169-2046(94)01036-8

Meherishi, L., Narayana, S. A., & Ranjani, K. S. (2019). Sustainable packaging for supply chain management in the circular economy: A review. *Journal of Cleaner Production, 237*, 117582. doi:10.1016/j.jclepro.2019.07.057

Mehraliyev, F., Choi, Y., & Köseoglu, M. A. (2019). Progress on smart tourism research. *Journal of Hospitality and Tourism Technology, 10*(4), 522–538. doi:10.1108/JHTT-08-2018-0076

Meier, J. D. (2022). *Innovation Explained – The Big Ideas of Innovation All In One Place.* http://innompics.com/people/meier-jd.html

Melnyk, B. M., Tan, A., Hsieh, A. P., Gawlik, K., Arslanian-Engoren, C., Braun, L. T., Dunbar, S., Dunbar-Jacob, J., Lewis, L. M., Millan, A., Orsolini, L., Robbins, L. B., Russell, C. L., Tucker, S., & Wilbur, J. (2021). Critical care nurses' physical and mental health, worksite wellness support, and medical errors. *American Journal of Critical Care, 30*(3), 176–184. doi:10.4037/ajcc2021301 PMID:34161980

Menouar, H., Güvenc, I., Akkaya, K., Uluagac, A. S., Kadri, A., & Tuncer, A. (2017). UAV-Enabled Intelligent Transportation Systems for the Smart City: Applications and Challenges. *IEEE Communications Magazine, 55*(3), 22–28. doi:10.1109/MCOM.2017.1600238CM

Merriam, S. (1998). *Qualitative research and case study applications in education.* Jossey-Bass.

Michlowicz, E. (2013). Logistics in production processes. *Journal of Machine Engineering, 13*(4), 5–17.

Mikalef, P., & Gupta, M. (2021). Artificial intelligence capability: Conceptualization, measurement calibration, and empirical study on its impact on organizational creativity and firm performance. *Information & Management, 58*(3), 103434. doi:10.1016/j.im.2021.103434

Minervini, P., Demeester, T., Tocktaschel, & Riedel, S. (2017). Adversarial sets for regularising neural link predictors. UAI.

Ministry of Transport. (2017, September 26). *Smart Transportation Makes Mobility Easier Action Plan (2017-2020).* Ministry of Transport. https://xxgk.mot.gov.cn/2020/jigou/kjs/202006/t20200623_3317082.html

Ministry of Transport. (2019, July 25). *Outline of Digital Transport Development Plan.* Ministry of Transport. https://xxgk.mot.gov.cn/2020/jigou/zhghs/202006/t20200630_3321233.html

Mintzberg, H. (1987). The strategy concept I: Five Ps for strategy. *California Management Review, 30*(1), 11–24. doi:10.2307/41165263

Mintz-Woo, K., Dennig, F., Liu, H., & Schinko, T. (2021). Carbon pricing and COVID-19. *Climate Policy, 21*(10), 1272–1280. doi:10.1080/14693062.2020.1831432

Mir, R., & Mir, A. (2009). From the Colony to the Corporation: Studying Knowledge Transfer Across International Boundaries. *Group & Organization Management, 34*(1), 90–113. doi:10.1177/1059601108329714

Mishra, P., Dangayach, G. S., & Mittal, M. L. (2011). An Ethical approach towards sustainable project Success. *Procedia - Social and Behavioral Sciences, 25*, 338-344. 10.1016/j.sbspro.2011.10.552

Mishra, S., & Maiti, A. (2019). Applicability of enzymes produced from different biotic species for biodegradation of textile dyes. *Clean Technologies and Environmental Policy, 21*(4), 763–781. doi:10.100710098-019-01681-5

Mitchell, L., Frank, M. R., Harris, K. D., Dodds, P. S., & Danforth, C. M. (2013). The Geography of Happiness: Connecting Twitter Sentiment and Expression, Demographics, and Objective Characteristics of Place. *PLoS One, 8*(5), E64417. Advance online publication. doi:10.1371/journal.pone.0064417 PMID:23734200

Mitchell, R., & Boyle, B. (2021). Understanding the role of profession in multidisciplinary team innovation: Professional identity, minority dissent and team innovation. *British Journal of Management, 32*(2), 512–528. doi:10.1111/1467-8551.12419

MithasS.TaftiA.MitchellW. (2013)

Mithas, S., Tafti, A., & Mitchell, W. (2013, February 2). How a Firm's Competitive Environment and Digital Strategic Posture Influence Digital Business Strategy. *Management Information Systems Quarterly, 37*(2), 511–536. doi:10.25300/MISQ/2013/37.2.09

Mobilize. (2014). *No Brasil, transporte é o vilão das emissões de CO2.* Retrieved from http://www.mobilize.org.br/noticias/7345/transporte-e-o-vilao-das-emissoes-de-co2.html

Mochal, T., & Krasnoff, A. (2013). GreenPM®: The Basic Principles for Applying an Environmental Dimension to Project Management. In A. J. G. Silvius & J. Tharp (Eds.), *Sustainability Integration for Effective Project Management* (pp. 39–57). IGI Global Publishing. doi:10.4018/978-1-4666-4177-8.ch003

Mogaji, E., Olaleye, S., & Ukpabi, D. (2020). *Using AI to Personalise Emotionally Appealing Advertisement.* Digital and Social Media Marketing. doi:10.1007/978-3-030-24374-6_10

Moghaddam, F. M., & Covalucci, L. (2016). Macro, Meso, and Micro Creativity: The Role of Cultural Carriers. In The Palgrave Handbook of Creativity and Culture Research (pp. 721-741). Palgrave.

Mohannad, A. M. A. D., Smoudy, A. K. A. (2019). The Role of Artificial Intelligence on Enhancing Customer Experience. *International Review of Management and Marketing, 9*(4), 22-31. https://www.proquest.com/docview/2288760601?accountid=10771

Mokhlesian, S., & Holmén, M. (2012). Business model changes and green construction processes. *Construction Management and Economics, 30*(9), 761–775. doi:10.1080/01446193.2012.694457

Moore, S., Bulmer, S., & Elms, J. (2022). The social significance of AI in retail on customer experience and shopping practices. *Journal of Retailing and Consumer Services, 64*, 102755. doi:10.1016/j.jretconser.2021.102755

Morakanyane, R., Grace, A., & O'Reilly, P. (2017). Conceptualizing Digital Transformation in Business Organizations: A Systematic Review of Literature. In *Digital Transformation – From Connecting Things to Transforming Our Lives* (pp. 427–443). University of Maribor Press., doi:10.18690/978-961-286-043-1.30

Moreira, D. (2020). Virtual networks and asynchronous communities: methodological reflections on the digital. In *Ethnography in Higher Education* (pp. 177–196). Springer VS. doi:10.1007/978-3-658-30381-5_11

Morfaw, J. N. (2012). *Fundamentals of Project Sustainability: Strategies, Processes and Plans*. CreateSpace Independent Publishing Platform.

Morgan, B. (2018). 3 Use Cases of Artificial Intelligence for Customer Experience. *Forbes*. https://www.forbes.com/sites/ blakemorgan/2018/08/01/3-use-cases-of-artificial-intelligence-for-customer-experience/#1f084b6e5e34

Morgan, T., Anokhin, S., Kretinin, A., & Frishammar, J. (2015). The Dark Side of the Entrepreneurial Orientation and Market Orientation Interplay: A New Product Development Perspective. *International Small Business Journal, 33*(7), 731–751. doi:10.1177/0266242614521054

MoriokaS. N.CarvalhoM. M. 2015. Sustainability and management of projects: a bibliometric study. *Production*. doi:10.1590/0103-6513.058912

Morris, P. W. (2013). *Reconstructing Project Management*. John Wiley & Sons. doi:10.1002/9781118536698

Mortimore, G., Reynolds, J., Forman, D., Brannigan, C., & Mitchell, K. (2021). From expert to advanced clinical practitioner and beyond. *British Journal of Nursing (Mark Allen Publishing), 30*(11), 656–659. doi:10.12968/bjon.2021.30.11.656 PMID:34109817

Moruzzi, C. (2020). Artificial creativity and General intelligence. *Journal of Science and Technology of the Arts, 12*(3), 84–99.

Muggleton, S., De Raedt, L., Poole, D., Bratko, I., Flach, P., Inoue, K., & Srinivasan, A. (2012). Ilp turns 20. *Machine Learning, 86*(1), 3–23. doi:10.100710994-011-5259-2

Mu, J. (2015). Marketing Capability, Organizational Adaptation and New Product Development Performance. *Industrial Marketing Management, 49*, 151–166. doi:10.1016/j.indmarman.2015.05.003

Mulder, J., & Brent, A. C. (2006). Selection of Sustainable Rural Agriculture Projects in South Africa: Case Studies in the LandCare Programme. *Journal of Sustainable Agriculture, 28*(2), 55–84. doi:10.1300/J064v28n02_06

Müller-Pelzer, F. (2009). *Sustainability Management in CDM Project Activities: How to demonstrate and assess the contribution to sustainable development of Clean Development Mechanism (CDM) project activities*. SVH-Verlag.

Müller, R., Pemsel, S., & Shao, J. (2014). Organizational enablers for governance and governmentality of projects: A literature review. *International Journal of Project Management, 32*(8), 1309–1320. doi:10.1016/j.ijproman.2014.03.007

Musk. (2022). *Lessons from Elon Musk: Be a Genius Innovator.* Innovarsity. http://www.innovarsity.com/coach/quotes_a_musk.html

Mu, X., Fang, C., Yang, Z., & Guo, X. (2022). Impact of the COVID-19 Epidemic on Population Mobility Networks in the Beijing–Tianjin–Hebei Urban Agglomeration from a Resilience Perspective. *Land (Basel)*, *2022*(11), 675. doi:10.3390/land11050675

Myers, I. B., McCaulley, M. H., & Most, R. (1985). *Manual, a guide to the development and use of the Myers-Briggs type indicator.* consulting psychologists press.

Mystakidis, S. (2020). *Distance Education Gamification in Social Virtual Reality: A Case Study on Student Engagement, 11th International Conference on Information, Intelligence, Systems and Applications.* Piraeus, Greece. 10.1109/IISA50023.2020.9284417

Nahapiet, J. E., & Ghoshal, S. (1998). Social Capital, Intellectual Capital, and the Organizational Advantage. *Academy of Management Review*, *33*(2), 242–266. doi:10.2307/259373

Nakamoto, S. (2008). Bitcoin: A Peer-to-Peer Electronic Cash System. *Bitcoin.* https://bitcoin.org/bitcoin

Nalbant, K. G., & Aydin, S. (2023). Development and Transformation in Digital Marketing and Branding with Artificial Intelligence and Digital Technologies Dynamics in the Metaverse Universe. *Journal of Metaverse*, *3*(1), 9–18. doi:10.57019/jmv.1148015

Nass, C., & Moon, Y. (2000, January). Machines and mindlessness: Social responses to computers. *The Journal of Social Issues*, *56*(1), 81–103. doi:10.1111/0022-4537.00153

Natural Resources Canada. (2018). *Greenhouse Gas Emissions from the Plastics Processing Industry.* Government of Canada. Retrieved March 7, 2019, from https://www.nrcan.gc.ca/energy/efficiency/industry/technical-info/benchmarking/plastics/5211

Negev, M., Dahdal, Y., Khreis, H., Hochman, A., Shaheen, M., Jaghbir, M. T., Alpert, P., Levine, H., & Davidovitch, N. (2021). Regional lessons from the COVID-19 outbreak in the Middle East: From infectious diseases to climate change adaptation. *The Science of the Total Environment*, *768*, 144434. doi:10.1016/j.scitotenv.2020.144434 PMID:33444865

Netemeyer, R. G., Bearden, W. O., & Sharma, S. (2003). *Scaling Procedures: Issues and Applications.* Sage Publications. doi:10.4135/9781412985772

Newman, D., & McClimans, F. (2019). *EXPERIENCE 2030: The Future of Customer Experience is … NOW!* [White Paper]. Futurum. file:///C:/Users/tmtyp/OneDrive/2nd%20Year%20Fall/Thesis/Articles%20Journals/futurum-experience-2030-110966.pdf

Ng, J. C. Y., Song, K. K. W., & Tan, Q. (2018). Expanding the Scope Of Application Of User Innovation Theory—A Case Study of The Civil-Military Integration Project in China. *International Journal of Innovation Studies*, *2*(1), 33–41. doi:10.1016/j.ijis.2018.03.002

Nguyen Duc, A., & Chirumamilla, A. (2019, September). Identifying security risks of digital transformation-an engineering perspective. In *Conference on e-Business, e-Services and e-Society* (pp. 677-688). Springer, Cham. 10.1007/978-3-030-29374-1_55

Nguyen, T. M., Quach, S., & Thaichon, P. (2022). The effect of AI quality on customer experience and brand relationship. *Journal of Consumer Behaviour*, *21*(3), 481–493. doi:10.1002/cb.1974

Nicole, S. (2021, August). Biden administration sanctions virtual currency exchange following spike in ransonware attacks. *CBS News.*

Nicolescu, L., & Tudorache, M. T. (2022). Human-Computer Interaction in Customer Service: The Experience with AI Chatbots—A Systematic Literature Review. *Electronics (Basel), 11*(10), 1579. doi:10.3390/electronics11101579

Nießen, D., Danner, D., Spengler, M., & Lechner, C. (2020). Big Five personality traits predict successful transitions from school to vocational education and training: A large-scale study. *Frontiers in Psychology, 11*, 1827. doi:10.3389/fpsyg.2020.01827 PMID:32903700

Ning, C., Zhang, S., & Li, L. (2009). Sustainable Project Management: A Balance Analysis Model of Effect. *International Conference on Management and Service Science*, Wuhan, China. 10.1109/ICMSS.2009.5302357

Nitta, I. (2003, Winter). Green Patent System: An invention for inventions toward sustainable development. *GIP Progress*.

Nitta, I. (2005c). Patents and essential medicines: An application of the green intellectual property project, on the Submission site of the Commission of Intellectual Property Rights, Innovation and Public Health. WHO.

Nitta, I. (2005a). Proposal for a green patent system: Implications for sustainable development and climate change. *Sustainable Development Law and Policy, 5*, 61–65.

Nitta, I. (2005b). Green Intellectual Property: A tool for greening a society. *Ecological Economics*.

Nost, E. (2019). Climate services for whom? The political economics of contextualizing climate data in Louisiana's coastal master plan. *Climatic Change, 157*(1), 27–42. doi:10.100710584-019-02383-z

Novak, M. (2020). The Top 10 Fintech Trends in 2020. *Forbes*. https://www.forbes.com

Nunes, F., Herpich, F., Amaral, É., Voss, G., Zunguze, M., Medina, R., & Tarouco, L. (2017). A dynamic approach for teaching algorithms: Integrating immersive environments and virtual learning environments. *Computer Applications in Engineering Computing*, 1-20.

Nwaiwu, F. (2018). Review and Comparison of Conceptual Frameworks on Digital Business Transformation. *Journal of Competitiveness, 10*(3), 86–100. doi:10.7441/joc.2018.03.06

Nye, M. (2020). Learning compositional rules via neural program synthesis. *Advances in Neural Information Processing Systems, 33*, 10832–10842.

Ny, H., Hallstedt, S., Robèrt, K.-H., & Broman, G. (2008). Introducing templates for sustainable product development: A case study of televisions at the Matsushita Electric Group. *Journal of Industrial Ecology, 12*(4), 600–623. doi:10.1111/j.1530-9290.2008.00061.x

Obinger, H. Schmitt, C., & Traub, S. (2016). The Emergence of Public Enterprises in Historical Perspective. In The Political Economy of Privatization in Rich Democracies (pp. 6-25). Oxford University Press. from doi:10.1093/acprof:oso/9780199669684.003.0002

Obrad, C. (2020). Constraints and consequences of online teaching. *Sustainability, 12*(17), 6982. doi:10.3390u12176982

Ochoa, J. J. (2014). Reducing plan variations in delivering sustainable building projects. *Journal of Cleaner Production, 85*, 276–288. doi:10.1016/j.jclepro.2014.01.024

Ocloo, J., Garfield, S., Franklin, B. D., & Dawson, S. (2021). Exploring the theory, barriers and enablers for patient and public involvement across health, social care and patient safety: A systematic review of reviews. *Health Research Policy and Systems, 19*(1), 1–21. doi:10.118612961-020-00644-3 PMID:33472647

O'Dair, M., & O'Dair, M. (2019). Blockchain: the internet of value. *Distributed Creativity: How Blockchain Technology will Transform the Creative Economy*, 15-30.

Odum, E. P. (1957). *Fundamentals of ecology*. W. B. Saunders Company.

OECD. (1997b). Government Venture Capital for Technology-Based Firms. Ducument OCDE/GD (97) 201. Paris: OECD.

Office of Government Commerce. (2010). *Management of Risk: Guidance for Practitioners*. HMSO.

Oh, J., Lee, S., & Yang, J. (2015). A Collaboration Model For New Product Development Through the Integration Of PLM And SCM in The Electronics Industry. *Computers in Industry*, *73*, 82–92. doi:10.1016/j.compind.2015.08.003

Okoye, K., Rodriguez-Tort, J. A., Escamilla, J., & Hosseini, S. (2021). Technology-mediated teaching and learning process: A conceptual study of educators' response amidst the Covid-19 pandemic. *Education and Information Technologies*, *26*(6), 1–33. doi:10.100710639-021-10527-x PMID:34025205

Olander, S., & Landin, A. (2005). Evaluation of stakeholder influence in the implementation of construction projects. *International Journal of Project Management*, *23*(4), 321–328. doi:10.1016/j.ijproman.2005.02.002

Oleinik, A. (2019). What are neural networks not good at? On artificial creativity. *Big Data & Society*, *6*(1). Advance online publication. doi:10.1177/2053951719839433

Olimid, A. P., Georgescu, C. M., & Gherghe, C. L. (2022). Influences of Covid-19 Crisis on Resilience Theories: An analysis of Community, Societal and Governance Resilience. *Revista de Ştiinţe Politice. Revue des Sciences Politiques*, *73*, 38–51.

Oliveira Neto, G. C. D., Ruiz, M. S., Correia, A. J. C., & Mendes, H. M. R. (2018). Environmental advantages of the reverse logistics: A case study in the batteries collection in Brazil. *Production*, *28*(0), 28. doi:10.1590/0103-6513.20170098

Olson, R. L. (1991). The greening of high tech. *The Futurist*, *25*(3), 28–34.

Organização das Nações Unidas (ONU). (1992). *Agenda 21*. Rio de Janeiro: Author.

Orr, N., Matthews, B., See, Z. S., Burrell, A., Day, J., & Seengal, D. (2021). Transdisciplinarity in extended reality (XR) research design: Technological transformation and social good (co-creation session at XR+ Creativity Symposium, University of Newcastle, 2020). *Virtual Creativity, 11*(1), 163-179.

Orr, R. J., & Scott, W. R. (2008). Institutional exceptions on global projects: A process model. *Journal of International Business Studies*, *39*(4), 562–588. doi:10.1057/palgrave.jibs.8400370

Ortiz, O., Castells, F., & Sonnemann, G. (2009). Sustainability in the construction industry: A review of recent developments based on LCA. *Construction & Building Materials*, *23*(1), 28–39. doi:10.1016/j.conbuildmat.2007.11.012

Ortner, S. (2010). Access: Reflections on studying up in Hollywood. *Ethnography*, *11*(2), 211–233. doi:10.1177/1466138110362006

Ortner, S. (2013). *Not Hollywood: Independent Film at the Twilight of the American Dream*. Duke University Press.

Osborne, S. P. (2021). *Public Service Logic. Creating Value for Public Service Users, Citizens, and Society through Public Service Delivery.*

Osborne, S. P. (2010). The (New) Public Governance: A Suitable Case for Treatment? In S. P. Osborne (Ed.), *In the New Public Governance? Emerging Perspectives on the Theory and Practice of Public Governance* (pp. 1–16). Routledge. doi:10.4324/9780203861684

Ostrom, E. (2009). A general framework for analyzing sustainability of socialecological systems. *Science*, *325*(5939), 419–422. doi:10.1126cience.1172133 PMID:19628857

Otto, K. N. (2003). Product Design: Techniques in Reverse Engineering and New Product Development. Pearson Education Asia Limited, Tsinghua University Press.

Overgoor, G., Chica, M., Rand, W., & Weishampel, A. (2019). Letting the Computers Take Over: Using AI to Solve Marketing Problems. *California Management Review*, *61*(4), 156–185. doi:10.1177/0008125619859318

Owens, K. P. (2021, July). Competency-Based Experiential-Expertise and Future Adaptive Learning Systems. In *International Conference on Human-Computer Interaction* (pp. 93-109). Springer, Cham. 10.1007/978-3-030-77873-6_7

Oyedeji, T. (2022). Harnessing artificial intelligence for educational creativity. *Ife PsychologIA*, *30*(1), 103–114.

Pade, C. I., Mallinson, B., & Sewry, D. (2006). An exploration of the categories associated with ICT project sustainability in rural areas of developing countries: a case study of the Dwesa project. *Proceedings of the 2006 annual research conference of the South African institute of computer scientists and information technologists on IT research in developing countries (SAICSIT),* 100-106. 10.1145/1216262.1216273

Pade, C., Mallinson, B., & Sewry, D. (2008). An Elaboration of Critical Success Factors for Rural ICT Project Sustainability in Developing Countries: Exploring the Dwesa Case. *The Journal of Information Technology Case and Application*, *10*(4), 32–55. doi:10.1080/15228053.2008.10856146

Pade-Khene, C.I., Mallinson, B., & Sewry, D. (2011). Sustainable rural ICT project management practice for developing countries: investigating the Dwesa and RUMEP projects. *Information Technology for Development, 17*(3), 187-212. doi:10.1080/02681102.2011.568222

Palminteri, S., Lebreton, M., Worbe, Y., Grabli, D., Hartmann, A., & Pessiglione, M. (2009). Pharmacological modulation of subliminal learning in Parkinson's and Tourette's syndromes. *Proceedings of the National Academy of Sciences of the United States of America*, *106*(45), 19179–19184. doi:10.1073/pnas.0904035106 PMID:19850878

Panda, G., Upadhyay, A. K., & Khandelwal, K. (2019). Artificial Intelligence: A Strategic Disruption in Public Relations. *Journal of Creative Communications*, *14*(3), 196–213. doi:10.1177/0973258619866585

Papies, D., & Clement, M. (2008). Adoption of New Movie Distribution Services on the Internet. *Journal of Media Economics, 21*(3), 131–157. doi:10.1080/08997760802300530

Paradiso, C. (2016). Artificial Intelligence in Digital Marketing. *Insurance Advocate*, *127*(14), 12–14.

Páramo, D., & Ramírez, E. (2017). *Etnomarketing, la dimensión cultural del marketing*. Klasse Editorial.

Park, C., & Kim, D. G. (2020). Exploring the roles of social presence and gender difference in online learning. *Decision Sciences Journal of Innovative Education*, *18*(2), 291–312. doi:10.1111/dsji.12207

Park, Y., & Chen, J. (2007). Acceptance and adoption of the innovative use of smartphone. *Industrial Management & Data Systems*, *107*(9), 1349–1365. doi:10.1108/02635570710834009

Parn, E., & Edwards, D. (2019). Cyber Threats confronting the digital built environment common data environment vunerabilities and blockchain deterence. *Engineering, Construction, and Architectural Management*, *26*(2), 245–266. doi:10.1108/ECAM-03-2018-0101

Passos, C., Silva, M. H., Abreu Mol, A. C., & Carvalho, P. V. (2017). Design of a collaborative virtual environment for training security agents in big events. *Cognition Technology and Work*, *19*(2-3), 315–328. doi:10.100710111-017-0407-5

Patanakul, P., & Shenhar, A.J. (2012). *What Project Strategy Really Is: the Fundamental Building Block in Strategic Project Management*. doi:10.1002/pmj

Paulose, R., Samy, B. G., & Jegatheesan, K. (2018). Text Mining and Natural Language Processing on Socia Media Data Giving Insights for Pharmacovigilance: A Case Study with Fentanyl. *Indian Journal of Pharmaceutical Sciences, 80*(4). Advance online publication. doi:10.4172/pharmaceutical-sciences.1000418

Pauly, D., Christensen, V., Guénette, S., Pitcher, T. J., Sumaila, U. R., Walters, C. J., Watson, R., & Zeller, D. (2002). Towards sustainability in world fisheries. *Nature, 418*(6898), 689–695. doi:10.1038/nature01017 PMID:12167876

Pearce, A. R. (2008). Sustainable capital projects: Leapfrogging the first cost barrier. *Civil Engineering and Environmental Systems, 25*(4), 291–300. doi:10.1080/10286600802002973

Pecherskiy, V. (2017). Will AI Replace Creative Jobs? *Forbes*. https://www.forbes.com/sites/forbescommunicationscouncil/2017/10/11/will-ai-replace-creative-jobs/#6802eca296a2

Pedersen, T. A., Glomsrud, J. A., Ruud, E. L., Simonsen, A., & Erikson, B. O. H. (2020). Towards Simulation-Based Verification of Autonomous Navigation Systems. *Safety Science*, 129.

PEGA. (2020). *The future of work: New perspectives on disruption & transformation*. PEGA. https://www.pega.com › pega-future-of-work-report

Peltola, T., & Mäkinen, S. J. (2014). Influence of The Adoption and Use of Social Media Tools on Absorptive Capacity in New Product Development. *Engineering Management Journal, 26*(3), 45–51. doi:10.1080/10429247.2014.11432019

Pena-Torres, A., & Pearson, P. J. (1999). *Pricing carbon emissions in the UK electricity market: A nuclear revival?* Academic Press.

Peng, D. X., & Lai, F. (2012). Using partial least squares in operations management research: A practical guideline and summary of past research. *Journal of Operations Management, 30*(6), 467–480. doi:10.1016/j.jom.2012.06.002

Pennypacker, J. S., & Grant, K. P. (2003). Project management maturity: An industry benchmark. [March]. *Project Management Journal, 34*(1), 4–9. doi:10.1177/875697280303400102

People's Daily. (2022, May 2). The Transportation Industry Has Made Great Strides Towards A Transportation Power, And The Transportation Industry Has Achieved Leapfrog Development. *People's Daily*. http://paper.people.com.cn/rmrb/html/2022-05/02/nbs.D110000renmrb_06.htm

Pereira, C. R., Christopher, M., & Da Silva, A. L. (2014). Achieving supply chain resilience: The role of procurement. *Supply Chain Management*.

Perrini, F., & Tencati, A. (2006). Sustainability and Stakeholder Management: The Need for New Corporate Performance Evaluation and Reporting Systems. *Business Strategy and the Environment, 15*(5), 286–308. doi:10.1002/bse.538

Perry, L. D., Seltzer, K., & Stoker, G. (2002). *Towards Holistic Governance*. Palgrave Macmillan. doi:10.5040/9781350391246

Peterson, N., & Broad, K. (2009). Climate and weather discourse in anthropology: From determinism to uncertain futures. In *Anthropology and climate change: From encounters to actions* (pp. 70–86). Left Coast Press.

Petrowski, M. J. (2000). Creativity research: Implications for teaching, learning and thinking. *Emerald, 28*(4), 304–312. doi:10.1108/00907320010359623

Pfeifer, J. (2006). Laws of nature. In S. Sarkar & J. Pfeifer (Eds.), *The philosophy of science: An encyclopedia* (pp. 439–444).

Piccinini, E., Gregory, R. W., & Kolbe, L. M. (2015). *Changes in the producer-consumer relationship-towards digital transformation*.

Pietro, D. R., & Cresci, S. (2021, December 12-15). Metaverse: Security and Privacy Issues. *IEEE TPS*, 1-8.

Pilbeam, C. (2013). Coordinating temporary organizations in international development through social and temporal embeddedness. *International Journal of Project Management*, *31*(2), 190–199. doi:10.1016/j.ijproman.2012.06.004

Piller, F. T., Vossen, A., & Ihl, C. (2012). From Social Media to Social Product Development: The Impact of Social Media on Co-Creation of Innovation. *Die Unternehmung*, *65*(1).

Pimienta-Bueno, J.A. (1999). *Parque de Inovação Tecnológica e Cultural da Gávea: A Visão da PUC-Rio*. Documento Interno, 23 págs.

Pizarro, C. (2014). La entrevista etnográfica como práctica discursiva: Análisis de caso sobre las pistas meta-discursivas y la emergencia de categorías nativas. *Revista de Antropologia*, *57*(1), 461–496. doi:10.11606/2179-0892.ra.2014.87770

Plucker, J. A., Beghetto, R. A., & Dow, G. T. (2004). Why isn't creativity more important to educational psychologists? Potentials, pitfalls, and future directions in creativity research. *Educational Psychologist*, *39*(2), 83–96. doi:10.120715326985ep3902_1

PMI. (2013). *A Guide to the Project Management Body of Knowledge (PMBOK® Guide)*. Project Management Institute, Incorporated.

Pocock, J. B., Hyun, C. T., Liu, L. Y., & Kim, M. K. (1996). Relationship between project interaction and performance indicators. *Journal of Construction Engineering and Management*, *122*(2), 165–176. doi:10.1061/(ASCE)0733-9364(1996)122:2(165)

Poppenheimer, L. (2014, July 7). *Aluminum Beverage Cans – Environmental Impact*. Green Groundswell. Retrieved March 7, 2019, from https://greengroundswell.com/aluminum-beverage-cans-environmental-impact/2014/07/17/

Porter, M. (2001). Clusters of Innovation: regional foundations of U.S competitiveness. Council on Competitiveness. Monitor Group.

Porter, M.E. (1996). What is strategy? *Harvard Business Review*, *74*(6), 61–78.

Porter, M., & Krammer, M. (2011). Creating Shared Value. *Harvard Business Review*, (89(1/2)), 62–67.

Pozzi, F. A., Fersini, E., Messina, E., & Liu, B. (2016). *Sentiment Analysis in Social Networks*. Morgan Kaufmann.

Prasad, S., Tata, J., Herlache, L., & McCarthy, E. (2013). Developmental project management in emerging countries. *Oper. Manag. Res.*, *6*(1-2), 53–73. doi:10.100712063-013-0078-1

Press, M., & Arnould, E. J. (2014). Narrative transparency. *Journal of Marketing Management*, *30*(13–14), 1353–1376. doi:10.1080/0267257X.2014.925958

Prieto, B. (2011). *Sustainability on Large, Complex Engineering & Construction Programs Utilizing a Program Management Approach. PMWorldToday, 13(7).*

Project Management Institute. (2013). *A Guide to Project Management Body of Knowledge (PMBOK® Guide)* (5th ed.). Project Management Institute Publishing.

Promsri, C. (2019). The developing model of digital leadership for a successful digital transformation. *GPH-International Journal of Business Management (IJBM), 2*(08), 01-08.

Pulaski, M. H., & Horman, M. J. (2005). Continuous value enhancement process. *Journal of Construction Engineering and Management*, *131*(12), 1274–1282. doi:10.1061/(ASCE)0733-9364(2005)131:12(1274)

Puryear, J. S., & Lamb, K. N. (2020). Defining creativity: How far have we come since Plucker, Beghetto, and Dow? *Creativity Research Journal*, *32*(3), 206–214. doi:10.1080/10400419.2020.1821552

Purzycki, B., Pisor, A., Apicella, C., Atkinson, Q., Cohen, E., Henrich, J., McElreath, R., McNamara, R., Norenzayan, A., Willard, A., & Xygalatas, D. (2018). The cognitive and cultural foundations of moral behavior. *Evolution and Human Behavior, 39*(5), 490–501. doi:10.1016/j.evolhumbehav.2018.04.004

Quiniou, M. (2019). Blockchain: The Advent of Disintermediation. In *Hoboken John*. Wiley. doi:10.1002/9781119629573

Radcliffe-Brown, A. R. (1940). On Social Structure. *J. R. Anthropol. Inst. Great Br. Irel., 1940*(70), 1–12.

Raedt, D. (2008). *Luc. Logical and relational learning*. Springer Science & Business Media. doi:10.1007/978-3-540-68856-3

Raff, S., Wentzel, D., & Obwegeser, N. (2020). Smart Products: Conceptual Review, Synthesis, and Research Directions. *Journal of Product Innovation Management, 37*(5), 379–404. doi:10.1111/jpim.12544

Rai, A., Constantinides, P., & Sarker, S. (2019). Next generation digital platforms: Toward human-ai hybrids. *Management Information Systems Quarterly, 43*(1), iii–ix.

Rajagopal, P., Kaliani Sundram, V. P., & Maniam Naidu, B. (2015). Future directions of reverse logistics in gaining competitive advantages: A review of literature. *International Journal of Supply Chain Management, 4*(1), 39-48.

Rakshit, S., Mondal, S., Islam, N., Jasimuddin, S., & Zhang, Z. (2021). Social Media and the New Product Development During COVID-19: An Integrated Model for SMEs. *Technological Forecasting and Social Change, 170*, 120869. doi:10.1016/j.techfore.2021.120869

Rani, S., & Kumar, P. (2017). A Sentiment Analysis System to Improve Teaching and Learning. Computer, *IEEE. Computers & Society, 50*(5), 36–43. doi:10.1109/MC.2017.133

Rathore, A. K., Das, S., & Ilavarasan, P. V. (2018). Social Media Data Inputs in Product Design: Case of a Smartphone. *Global Journal of Flexible Systems Managment, 19*(3), 255–272. doi:10.100740171-018-0187-7

Rathore, A. K., & Ilavarasan, P. V. (2020). Pre-and Post-Launch Emotions in New Product Development: Insights From Twitter Analytics of Three Products. *International Journal of Information Management, 50*, 111–127. doi:10.1016/j.ijinfomgt.2019.05.015

Rautela, S., Sharma, S., & Virani, S. (2020). Influence of Customer Participation in New Product Development: The Moderating Role of Social Media. *International Journal of Productivity and Performance Management*. doi:10.1108/IJPPM-05-2020-0241

Raven, R., Jolivet, E., Mourik, R. M., & Feenstra, C. F. J. (2009). ESTEEM: Managing societal acceptance in new energy projects a toolbox method for project managers. *Technological Forecasting and Social Change, 76*(7), 963–977. doi:10.1016/j.techfore.2009.02.005

Raz, T., Shenhar, A. J., & Dvir, D. (2002). Risk Management, Project Success, and Technological Uncertainty. *R & D Management, 32*(2), 101–109. doi:10.1111/1467-9310.00243

Reguera, A. (2009) Empresarios de ayer, de hoy y de siempre. *Un recorrido latinoamericano por sus formas espaciotemporales América Latina en la historia económica, num 32, jul-dic, México*. Disponible en: https://www.scielo.org.mx/scielo.php?script=sci_arttext&pid=S1405-22532009000200003&lng=es&nrm=iso

Remane, G., Hanelt, A., Nickerson, R. C., & Kolbe, L. M. (2017). Discovering digital business models in traditional industries. *The Journal of Business Strategy, 38*(2), 41 51. doi:10.1108/JBS-10-2016-0127

Reynoso, C. (1995b). Hacia la perfección del consenso: Los lugares comunes de la antropología. *Intersecciones, 1*, 51–72.

Rhodes, R. A. (2017). *Network Governance and the Differentiated Polity*.

Rhodes, R. A. (1990). Policy Networks: A British Perspective. *Journal of Theoretical Politics*, *2*(3), 293–317. doi:10.1177/0951692890002003003

Riedl, M. O. (2016). *Computational narrative intelligence: A human-centered goal for artificial intelligence*. Research Gate. https://www.researchgate.net/publication/301844558_Computational_Narrative_Intelligence_A_Human-Centered_Goal_for_Artificial_Intelligence

Ries, J. M., Grosse, E. H., & Fichtinger, J. (2017). Environmental impact of warehousing: A scenario analysis for the United States. *International Journal of Production Research*, *55*(21), 6485–6499. doi:10.1080/00207543.2016.1211342

Riggs, W., & Von Hippel, E. (1994). Incentives to Innovate and The Sources of Innovation: The Case of Scientific İnstruments. *Research Policy*, *23*(4), 459–469. doi:10.1016/0048-7333(94)90008-6

Ringle, C. M., Wende, S., & Becker, J.-M. (2015). *SmartPLS 3*. SmartPLS GmbH. http://www.smartpls.com

Ripple, W. J., Wolf, C., Newsome, T. M., Barnard, P., & Moomaw, W. R. (2019). World scientists' warning of a climate emergency. *Bioscience*, biz088. Advance online publication. doi:10.1093/biosci/biz088

Ritchie, J. R., & Crouch, G. I. (2010). A model of destination competitiveness/sustainability: Brazilian perspectives. *Revista de Administração Pública*, *44*(5), 1049–1066. doi:10.1590/S0034-76122010000500003

Ritika, M., & Montu, B. (2018). Business Sustainability: Exploring the Meaning and Significance. *IMI Konnect*, 8-9.

Robert, I. V. (2021). Formation and development of digital transformation of domestic education on the basis of systemic convergence of pedagogical science and technology. In *SHS Web of Conferences* (*Vol. 101*, p. 03017). EDP Sciences. 10.1051hsconf/202110103017

Roberts, D. L., & Piller, F. T. (2016). Finding the Right Role for Social Media in Innovation. *MIT Sloan Management Review*, *57*(3), 41–47.

Robichaud, L. R., & Anantatmula, V. S. (2011). Greening Project Management Practices for Sustainable Construction. *Journal of Management Engineering*, *27*(148), 48–57. doi:10.1061/(ASCE)ME.1943-5479.0000030

Robinson, J. (2004). Squaring the circle? Some thoughts on the idea of sustainable development. *Ecological Economics*, *48*(4), 369–384. doi:10.1016/j.ecolecon.2003.10.017

Roca, A. (2008). *La comprensión de la tecnología en la antropología*. Actas VII ESOCITE UFRJ, Jornadas Latino-Americanas de Estudos Sociais das Ciências e das Tecnologías.

Rocktaschel T., & Riedel, S. (2017). End-to-end differentiable proving. *NIPS*.

Rocktaschel, T., & Reidel, S. (2016). Lifted rule injection for relation embeddings. EMNLP.

Rodríguez-Abitia, G., & Bribiesca-Correa, G. (2021). Assessing digital transformation in universities. *Future Internet*, *13*(2), 52. doi:10.3390/fi13020052

Roethlisberger, F., & Dickson, W. (1939). *Management and the Worker*. Harvard Univ. Press.

Rogers, E. (2003). *Diffusion of Innovations* (5th ed.). Free Press.

Roni, S., Merga, M., & Morris, J. (2020). *Conducting quantitative research in education*. Springer. doi:10.1007/978-981-13-9132-3

Rontard, B., & Hernandez, H. R. (2022). Political construction of carbon pricing: Experience from New Zealand emissions trading scheme. *Environmental Development*, *43*, 100727. doi:10.1016/j.envdev.2022.100727

Rosen, M. A., Dincer, I., & Kanoglu, M. (2008). Role of exergy in increasing efficiency and sustainability and reducing environmental impact. *Energy Policy*, *36*(1), 128–137. doi:10.1016/j.enpol.2007.09.006

Ross, D. F. (2004). Warehousing. In *Distribution Planning and Control* (pp. 535–608). Springer. doi:10.1007/978-1-4419-8939-0_11

Ross, N., Bowen, P. A., & Lincoln, D. (2010). Sustainable housing for low-income communities: Lessons for South Africa in local and other developing world cases. *Construction Management and Economics*, *28*(5), 433–449. doi:10.1080/01446190903450079

Rostamzadeh, R., Govindan, K., Esmaeili, A., & Sabaghi, M. (2015). Application of fuzzy VIKOR for evaluation of green supply chain management practices. *Ecological Indicators*, *49*, 188–203. doi:10.1016/j.ecolind.2014.09.045

Rouse, M. (2022, April 4). What does narrow artificial intelligence (narrow AI) mean? *Technopedia*. https://www.techopedia.com/definition/32874/narrow-artificial-intelligence-narrow-ai

Roussin, C. J., & Weinstock, P. (2017). SimZones: An organizational innovation for simulation programs and centers. *Academic Medicine*, *92*(8), 1114–1120. doi:10.1097/ACM.0000000000001746 PMID:28562455

Roy, K. (2019, November). Towards Spike-Based Machine Intelligence with Neuromorphic Computing. *Nature*, *27*. www.nature.com PMID:31776490

Roy, R., Chattopadhyay, S., Mitra, S. K., & Ghosh, S. (2020). Green technologies for sustainable development: A review. *Journal of Cleaner Production*, *255*, 120247.

Rudskoy, A., Ilin, I., & Prokhorov, A. (2020). Digital Twins in the Intelligent Transport Systems. *Transportation Research Procedia 54* (2021), 927-935.

Ruhl, C. (2023, February 14). *Theory of mind in psychology: People thinking*. Simple Psychology. https://simplypsychology.org/theory-of-mind.html

Ruhwinaningsih, L., & Djatna, T. (2016). A Sentiment Knowledge Discovery Model in Twitter's TV Content Using Stochastic Gradient Descent Algorithm. *TELKOMNIKA*, *14*(3), 1067–1076. doi:10.12928/telkomnika.v14i3.2671

Rundh, B. (2016). The role of packaging within marketing and value creation. *British Food Journal*, *118*(10), 2491–2511. doi:10.1108/BFJ-10-2015-0390

Russell, J. (2008). Corporate social responsibility: what it means for the project manager. In *PMI Global Congress EMEA*. Project Management Institute.

Russell, S. (2023, April 2). AI has must to offer humanity. It could also wreak terrible harm. It must be controlled. *The Guardian*. https://www.theguardian.com/commentisfree/2023/apr/02/ai-much-to-offer-humanity-could-wreak-terrible-harm-must-be-controlled?CMP=share_btn_link

Russell, S. (2015). Unifying logic and probability. *Communications of the ACM*, *58*(7), 88–97. doi:10.1145/2699411

Ruziyev, B., & Bakhriddinova, Y. (2022). Logistics: Types of transport. *Science Progress*, *3*(2), 456–462.

Salas-Zárate, M. D. P., Medina-Moreira, J., Lagos-Ortiz, K., Luna-Aveiga, H., Rodriguez-Garcia, M. A., & Valencia-García, R. (2017). Sentiment Analysis on Tweets About Diabetes: An Aspect-Level Approach. *Computational and Mathematical Methods in Medicine*, *2017*, 1–9. Advance online publication. doi:10.1155/2017/5140631 PMID:28316638

Salkind, N. (2010). Internal Consitency Reliability. Encyclopedia of Research Design. doi:10.4135/9781412961288.n191

Sanchez, M. A. (2015). Integrating sustainability issues into project management. *Journal of Cleaner Production, 96*, 319–330. doi:10.1016/j.jclepro.2013.12.087

Sandoval, M. C., Veiga, M. M., Hinton, J., & Sandner, S. (2006). Application of sustainable development concepts to an alluvial mineral extraction project in lower Caroni River, Venezuela. *Journal of Cleaner Production, 14*(3–4), 415–426. doi:10.1016/j.jclepro.2004.10.007

Saporovskaya, T. Y., Prohorov, S. V., & Timakov, E. A. (2020, July). Composite materials based on non-recyclable polyethylene. *IOP Conference Series. Materials Science and Engineering, 896*(1), 012078. doi:10.1088/1757-899X/896/1/012078

Sarathy, V., & Scheutz, M. (2018). MacGyver problems: AI challenges for testing resourcefulness and creativity. *Advances in Cognitive Systems, 6*, 31–44.

Saunders, R., & Gero, J. S. (2002). How to study artificial creativity. In *C&C '02: Proceedings of the 4th conference on Creativity & cognition* (pp. 80–87). Association for Computing Machinery. 10.1145/581710.581724

Saunders, M., Lewis, P., & Thornhill, A. (2012). *Research Methods for Business Students* (6th ed.). Pearson Education.

Saxenian, A. (1994). *Regional Advantage: culture and competition in Silicon Valley and Route 128*. Harvard Business Press.

Scanlon, J., & Davis, A. (2011). The role of sustainability advisers in developing sustainability outcomes for an infrastructure project: Lessons from the Australian urban rail sector. *Impact Assessment and Project Appraisal, 29*(2), 121–133. doi:10.3152/146155111X12913679730836

Schaltegger, S., & Burritt, R. (2005). Corporate Sustainability. In H. Folmer & T. Tietenberg (Eds.), *International Yearbook of Environment and Resource Economics* (pp. 185–222). Edward Elgar.

Schaltegger, S., Freund, F. L., & Hansen, E. G. (2012). Business cases for sustainability: The role of business model innovation for corporate sustainability. *International Journal of Innovation and Sustainable Development, 6*(2), 95–119. doi:10.1504/IJISD.2012.046944

Schaltegger, S., Hansen, E. G., & Ludeke-Freund, F. (2016). Business Models for Sustainability: Origins, Present Research and Future Avenues. *Organization & Environment, 29*(1), 3–10. doi:10.1177/1086026615599806

Schieg, M. (2009). The model of corporate social responsibility in project management. *Business: Theory and Practice, 10*(4), 315–321. doi:10.3846/1648-0627.2009.10.315-321

Schilling, M. A., & Hill, C. W. (1998). Managing the New Product Development Process: Strategic Imperatives. *The Academy of Management Perspectives, 12*(3), 67–81. doi:10.5465/ame.1998.1109051

Schipper, J. (2017). The importance of a holistic approach to the development of green technologies. *Technological Forecasting and Social Change, 124*, 215–222.

Schlapfer, A., & Marinova, D. (2001). Local Innovation Systems: nature, importance and role. *Conference Proceedings: International Summer Academy on Technological Studies.*

Schnegg, M., O'Brian, C. I., & Sievert, I. J. (2021). It's our fault: A global comparison of different ways of explaining climate change. *Human Ecology, 49*(3), 327–339. doi:10.100710745-021-00229-w

Schreckling, E., & Steiger, C. (2017). *Digitalize or drown. Shaping the digital enterprise: Trends and use cases in digital innovation and transformation*, 3-27.

Schultz, W. (2017). Reward prediction error. *Current Biology, 27*(10), 369–371. doi:10.1016/j.cub.2017.02.064 PMID:28535383

Schultz, W., Dayan, P., & Montague, P. R. (1997). A neural substrate of prediction and reward. *Science*, *275*(5306), 1593–1599. doi:10.1126cience.275.5306.1593 PMID:9054347

Schuman, C. D., Mitchell, J. P., Patton, R. M., Potok, T. E., & Plank, J. S. (2020) Evolutionary optimization for neuromorphic systems. In *Proceedings of the Neuro-inspired Computational Elements Workshop*, (pp. 1–9). IEEE.

Schwab, K. (n.d.). *3 reasons why AI will never match human creativity*. Fast Company. https://www.fastcompany.com/90339590/3-reasons-why-ai-will-never-match-human-creativity

Scott, J. (2009). The Art of Not Being Governed: An Anarchist History of Upland Southeast Asia. Yale University Press.

Scott, D. (2013). *The New Rules of Lead Generation : Proven Strategies to Maximize Marketing ROI*. AMACOM.

Seglen, P. O. (1994). Causal Relationship between Article Citedness and Journal Impact. *Journal of the American Society for Information Science*, *45*(1), 1–11. doi:10.1002/(SICI)1097-4571(199401)45:1<1::AID-ASI1>3.0.CO;2-Y

Seyyedamiri, N., & Tajrobehkar, L. (2019). Social Content Marketing, Social Media and Product Development Process Effectiveness in High-Tech Companies. *International Journal of Emerging Markets*, *16*(1), 1746–8809. doi:10.1108/IJOEM-06-2018-0323

Shabani, Z. D., & Shahnazi, R. (2019). Energy consumption, carbon dioxide emissions, information and communications technology, and gross domestic product in Iranian economic sectors: A panel causality analysis. *Energy*, *169*, 1064–1078. doi:10.1016/j.energy.2018.11.062

Shafiee, S., Ghatari, A. R., Hasanzadeh, A., & Jahanyan, S. (2019). Developing a model for sustainable smart tourism destinations: A systematic review. *Tourism Management Perspectives*, *31*, 287–300. doi:10.1016/j.tmp.2019.06.002

Shahraeeni, M., Ahmed, S., Malek, K., Van Drimmelen, B., & Kjeang, E. (2015). Life cycle emissions and cost of transportation systems: Case study on diesel and natural gas for light duty trucks in municipal fleet operations. *Journal of Natural Gas Science and Engineering*, *24*, 26–34. doi:10.1016/j.jngse.2015.03.009

Shang, J., Chen, S., Wu, J., & Yin, S. (2022, February). ARSpy: Breaking location-based multi-player augmented reality application for user location tracking. *IEEE Transactions on Mobile Computing*, *21*(2), 433–447. doi:10.1109/TMC.2020.3007740

Shank, D. B., Graves, C., Gott, A., Gamez, P., & Rodriguez, S. (2019). Feeling our way to machine minds: People's emotions when perceiving mind in artificial intelligence. *Computers in Human Behavior*, *98*, 256–266. doi:10.1016/j.chb.2019.04.001

Shan, W., & Wang, J. (2018). Mapping the landscape and evolutions of green supply chain management. *Sustainability*, *10*(3), 597. doi:10.3390u10030597

Sharma, B. (2023, April 7). Elon Musk, Steve Wozniak Lead Call For 6-Month Halt On Development Of AI Systems. *Indiatimes*. https://www.indiatimes.com/technology/science-and-future/elon-musk-steve-wozniak-halt-on-development-of-ai-systems-597580.html#:~:text=%22We%20call%20on%20all%20AI,moratorium%2C%22%20the%20letter%20said

Sharma, S., Devreaux, P., Scribner, D., Grynovicki, J., & Grazaitis, P. (2017). Megacity: A Collaborative Virtual Reality Environment for Emergency Response, Training, and Decision Making. *Electronic Imaging*, 70-77.

Sharma, S., & Henriques, I. (2005). Stakeholder influences on sustainability practices in the Canadian forest products industry. *Strategic Management Journal*, *26*(2), 159–180. doi:10.1002mj.439

Shearer, C. B., & Karanian, J. M. (2017). The neuroscience of intelligence: Empirical support for the theory of multiple intelligences? *Trends in Neuroscience and Education*, *6*, 211–223. doi:10.1016/j.tine.2017.02.002

Shearston, J. A., Johnson, A. M., Domingo-Relloso, A., Kioumourtzoglou, M. A., Hernández, D., Ross, J., Chillrud, S. N., & Hilpert, M. (2020). Opening a large delivery service warehouse in the South Bronx: Impacts on traffic, air pollution, and noise. *International Journal of Environmental Research and Public Health, 17*(9), 3208. doi:10.3390/ijerph17093208 PMID:32380726

Sheehan, L., Vargas-Sánchez, A., Presenza, A., & Abbate, T. (2016). The use of intelligence in tourism destination management: An emerging role for DMOs. *International Journal of Tourism Research, 18*(6), 549–557. doi:10.1002/jtr.2072

Shenhar, A. J., & Dvir, D. (2007). *Reinventing Project Management: The Diamond Approach to Successful Growth and Innovation*. Harvard Business School Press.

Shen, L. Y., Tam, V. W. Y., Tam, L., & Ji, Y. B. (2010). Project feasibility study: The key to successful implementation of sustainable and socially responsible construction management practice. *Journal of Cleaner Production, 18*(3), 254–259. doi:10.1016/j.jclepro.2009.10.014

Shenzhen Bao'an International Airport Airport News. (2019, December 19). *Shenzhen Airport Builds The Most Experienced Digital Airport*. Shenzhen Bao'an International Airport. https://www.szairport.com/szairport/kgxw/201912/9360ae6092c94540bcc9518a9b685994.shtml

Sherafati, M., Bashiri, M., Tavakkoli-Moghaddam, R., & Pishvaee, M. S. (2020). Achieving sustainable development of supply chain by incorporating various carbon regulatory mechanisms. *Transportation Research Part D, Transport and Environment, 81*, 102253. doi:10.1016/j.trd.2020.102253

Shi, D., & Ye, D. (2017). Shenzhen Airport Will Build An International Aviation Hub Facing The Asia-Pacific Region And Radiating The World. *Air Transport and Business*, 36-37.

Shiferaw, A. T., & Klakegg, O. J. (2012). Linking policies to projects: The key to identifying the right public investment projects. *Project Management Journal, 43*(4), 14–26. doi:10.1002/pmj.21279

Shiferaw, A. T., Klakegg, O. J., & Haavaldsen, T. (2012). Governance of public investment projects in Ethiopia. *Project Management Journal, 43*(4), 52–69. doi:10.1002/pmj.21280

Shi, Q., Zuo, J., & Zillante, G. (2012). Exploring the management of sustainable construction at the programme level: A Chinese case study. *Construction Management and Economics, 30*(6), 425–440. doi:10.1080/01446193.2012.683200

Shreay, S., Chouinard, H. H., & McCluskey, J. J. (2016). Product differentiation by package size. *Agribusiness, 32*(1), 3–15. doi:10.1002/agr.21425

Shrivastava, P. (1995). The role of corporations in achieving ecological sustainability. *Academy of Management Review, 20*(4), 936–960. doi:10.2307/258961

Siebert, A., Gopaldas, A., Lindridge, A., & Simões, C. (2020). Customer Experience Journeys: Loyalty Loops Versus Involvement Spirals. *Journal of Marketing, 84*(4), 45–66. doi:10.1177/0022242920920262

Sigala, M., Christou, E., & Gretzel, U. (Eds.). (2012). *Social media in travel, tourism and hospitality: Theory, practice and cases*. Ashgate Publishing, Ltd.

Siggelkow, N. (2007). Persuasion with case studies. *Academy of Management Journal, 50*(1), 20–24. doi:10.5465/amj.2007.24160882

Signorelli, A. (1999). *Antropología Urbana*. UAM.

Silén, C., Wirell, S., Kvist, J., Nylander, E., & Smedby, Ö. (2008). Advanced 3D visualization in student-centred medical education. *Medical Teacher, 30*(5), e115–e124. doi:10.1080/01421590801932228 PMID:18576181

Sillanpää, M. (1999). A new deal for sustainable development in business. In M. Bennet & P. James (Eds.), *Sustainable Measures*. Greenleaf Publishing.

Silver, D., Schrittwieser, J., Simonyan, K., Antonoglou, I., Huang, A., Guez, A., ... & Hassabis, D. (2017). Mastering the game of go without human knowledge. *nature, 550*(7676), 354-359.

Silvius, A. J. G., & Schipper, R. (2014). Sustainability in project management competencies: Analyzing the competence gap of project managers. *Journal of Human Resource and Sustainability Studies*, *2*(02), 40–58. doi:10.4236/jhrss.2014.22005

Simchi-Levi, D., Kaminsky, P., Simchi-Levi, E., & Shankar, R. (2008). *Designing and managing the supply chain: concepts, strategies and case studies*. Tata McGraw-Hill Education.

Simon, H. A. (1992). What is an "explanation" of behavior? *Psychological Science*, *3*(3), 150–161. doi:10.1111/j.1467-9280.1992.tb00017.x

Singapore Airshow. (2022), *A Year in Review: Developments in Sustainable Aviation*. Singapore Airshow 2022 Report.

Singh, D., & Verma, A. (2018). Inventory management in supply chain. *Materials Today: Proceedings*, *5*(2), 3867–3872. doi:10.1016/j.matpr.2017.11.641

Singh, R. K., Murty, H. R., Gupta, S. K., & Dikshit, A. K. (2012). An overview of sustainability assessment methodologies. *Ecological Indicators*, *15*(1), 281–299. doi:10.1016/j.ecolind.2011.01.007

Sjödin, D., Parida, V., Palmié, M., & Wincent, J. (2021). How AI capabilities enable business model innovation: Scaling AI through co-evolutionary processes and feedback loops. *Journal of Business Research*, *134*, 574–587. doi:10.1016/j.jbusres.2021.05.009

Slater, M. e. (2020). The ethics of realism in virtual and augmented reality. *Frontiers in Virtual Reality*. doi:10.3389/frvir.2020.00001

Slater, R. (2003). *Jack Welch and the GE Way*. Tata McGraw-Hill Edition.

Small, S. L., Cottrell, G. W., & Tanenhaus, M. K. (Eds.). (2013). *Lexical Ambiguity Resolution: Perspective from Psycholinguistics, Neuropsychology and Artificial Intelligence*. Elsevier.

Smith, V. (2001). Ecotechnology. *Environmental Engineering Science*, *18*(5), 295–305.

Snow, C. P. (1998). *The Two Cultures*. Cambridge University Press.

Sozo, V., & Ogliari, A. (2019). Stimulating Design Team Creativity Based on Emotional Values: A Study on Idea Generation in The Early Stages of New Product Development Processes. *International Journal of Industrial Ergonomics*, *70*, 38–50. doi:10.1016/j.ergon.2019.01.003

Spalding, M. J. (2016). The new blue economy: The future of sustainability. *Journal of Ocean and Coastal Economics*, *2*(2), 8. doi:10.15351/2373-8456.1052

Spencer, A., Buhalis, D., & Moital, D. (2012). A hierarchical model of technology adoption for small owner-managed travel firms: An organizational decision-making and leadership perspective. *Tourism Management*, *33*(5), 1195–1208. doi:10.1016/j.tourman.2011.11.011

Springett, P. (2020, July 29). *Artificial intelligence isn't the future. Augmented intelligence is*. DLA Ignite. https://digital-leadership-associates.passle.net/post/102gcgk/artificial-intelligence-isnt-the-future-augmented-intelligence-is

Srinivas, H. (2022). *Sustainable Development: Concepts*. GDRC Reseaarch Output E-008. Global Development Research Center. Retrieved from https://www.gdrc.org/sustdev/concepts.html

Statcounter. (2019). *GlobalStat.* https://gs.statcounter.com/vendor-market-share/mobile/turkey/#monthly-201810-201910-bar

State-Owned Assets Supervision and Administration Commission of Shenzhen Municipal People's Government. (2021, November 29). *Shenzhen Airport Has Created A "Heart" For 30 Years And Promoted Digital Transformation.* GZW. http://gzw.sz.gov.cn/gkmlpt/content/9/9408/post_9408956.html#1904

Statista. (n.d.). https://www.statista.com/topics/737/twitter/

Steffen, W., Richardson, K., Rockström, J., Cornell, S. E., Fetzer, I., Bennett, E. M., Biggs, R., Carpenter, S. R., de Vries, W., de Wit, C. A., Folke, C., Gerten, D., Heinke, J., Mace, G. M., Persson, L. M., Ramanathan, V., Reyers, B., & Sörlin, S. (2015). Planetary boundaries: Guiding human development on a changing planet. *Science*, *347*(6223), 1259855. doi:10.1126cience.1259855 PMID:25592418

Steingroever, H., Wetzels, R., & Wagenmakers, E.-J. (2013). Validatingthe PVL-Delta model for the Iowa gambling task. *Frontiers in Psychology*, *4*, 898. doi:10.3389/fpsyg.2013.00898 PMID:24409160

Steinke, A., Lange, F., & Kopp, B. (2020). Parallel Model-Based and Model-Free Reinforcement Learning for Card Sorting Performance. *Scientific Reports*, *10*(1), 15464. doi:10.103841598-020-72407-7 PMID:32963297

Stendal, K., & Balandin, S. (2015). Virtual worlds for people with autism spectrum disorder: A case study in Second Life. *Disability and Rehabilation,* 1-8.

Stephenson, N. (2003). *Snow Crash: A Novel.* Random House Publishing Group.

Stichter, J., Laffey, J., Galyen, K., & Herzog, M. (2013). iSocial: Delivering the Social Competence Intervention for Adolescents (SCI-A) in a 3D Virtual Learning Environment for Youth with High Functioning Autism. *Journal of Autism and Developmental Disorders*, 417–430. PMID:23812663

Stoner, A. M. (2021). Things are getting worse on our way to catastrophe: Neoliberal environmentalism, repressive desublimation, and the autonomous ecoconsumer. *Critical Sociology*, *47*(May), 491–506. doi:10.1177/0896920520958099

Strang, V. (2017). The Gaia complex: Ethical Challenges to an Anthropocentric 'Common Future. In B. Marc & L. Jerome (Eds.), *The Anthropology of Sustainability* (pp. 255–283). Palgrave. doi:10.1057/978-1-137-56636-2_12

Straub, E. T. (2009). Understanding Technology Adoption: Theory and Future Directions for Informal Learning. *Review of Educational Research*, *79*(2), 625–649. www.proquest.com/docview/214121884?accountid=10771. doi:10.3102/0034654308325896

Strauss, A. L. (1987). *Qualitative Analysis for Social Scientists.* Cambridge University Press. doi:10.1017/CBO9780511557842

Su, C., Pang, L., Tao, R., Shao, X. & Umar, M. (2022). Renewable energy and technological innovation: Which one is the winner in promoting net-zero emissions? *Technological Forecasting & Social Change,* 1 – 11. doi:10.1016/j.techfore.2022.121798

Su, C., Liu, F., Stefea, P., & Umar, M. (2023). Does technology innovation help to achieve carbon neutrality? *Economic Analysis and Policy*, *78*, 1–14. doi:10.1016/j.eap.2023.01.010

Sundbo, J. (1998). *The Theory of Innovation: Entrepreneurs, technology, and strategy*. Edward Elgar.

Sun, Y., Kuai, R., Li, X., & Tang, W. (2020). Latency Performance Analysis For Safety-Related Information Broadcasting In VEMAC. *Transactions on Emerging Telecommunications Technologies*, *31*(5). doi:10.1002/ett.3751

Surry, D. W., & Farquhar, J. D. (1997). Diffusion theory and instructional technology. *Journal of Instructional Science and Technology*, *2*(1), 24–36.

Suryanti, S., Sutaji, D., Arifani, Y., Muyasaroh, M., & Zamzamy, M. (2020). Improved learning accessibility and professionalism of teachers in remote areas through mentoring development of teaching materials based on Augmented Reality. [Research Dissemination for Community Development]. *Kontribusia*, *3*(1), 224–232. doi:10.30587/kontribusia.v3i1.1032

Sutton, R. S., & Barto, A. G. (1998). *Reinforcement Learning: An introduction*. MIT Press.

Svanberg, M., Ellis, J., Lundgren, J., & Landälv, I. (2018). Renewable methanol as a fuel for the shipping industry. *Renewable & Sustainable Energy Reviews*, *94*, 1217–1228. doi:10.1016/j.rser.2018.06.058

Swaminathan, A., Wu, M., & Liu, K. R. (2008, March). Digital image forensics via intrinsic fingerprints. *IEEE Transactions on Information Forensics and Security*, *3*(1), 101–117. doi:10.1109/TIFS.2007.916010

Taagepera, R. (2008). *Making social sciences more scientific: The need for predictive models*. Oxford University Press. doi:10.1093/acprof:oso/9780199534661.001.0001

Tabassi, A. A., Roufechaei, K. M., Ramli, M., Abu Bakar, A., Ismail, R., & Pakir, A. H. K. (2016). Leadership competences of sustainable construction project managers. *Journal of Cleaner Production*, *124*, 339–349. doi:10.1016/j.jclepro.2016.02.076

Tabatabai, S. (2020). COVID-19 impact and virtual medical education. *Journal of Advances in Medical Education & Professionalism*, *8*(3), 140–143. PMID:32802908

Taddy, M. (2018). The technological elements of artificial intelligence. In *The economics of artificial intelligence: An agenda* (pp. 61–87). University of Chicago Press.

Tainter, J. A. (1988). *The Collapse of Complex Societies*. Cambridge University Press.

Tainter, J. A. (2008). Collapse and Sustainability: Rome, the Maya, and the Modern World. *Archeological Papers of the American Anthropological Association*, *24*(1), 201–214. doi:10.1111/apaa.12038

Talbot, R., & Venkataraman, R. (2013). Managing Sustainability on Projects Using Indicators. In A. J. G. Silvius & J. Tharp (Eds.), *Sustainability Integration for Effective Project Management* (pp. 194–211). IGI Global Publishing. doi:10.4018/978-1-4666-4177-8.ch012

Tam, G. (2010). The program management process with sustainability considerations. Journal of Project. *Program & Portfolio Management*, *1*(1), 17–27. doi:10.5130/pppm.v1i1.1574

Tang, K. S., Cheng, D. L., Mi, E., & Greenberg, P. B. (2020). Augmented reality in medical education: A systematic review. *Canadian Medical Education Journal*, *11*(1), e81. PMID:32215146

Tao, F., Liu, W., Liu, J. H., Liu, X., Liu, Q., Qu, T., Hu, T., Zhang, Z., Xiang, F., Xu, W., Wang, J., Zhang, Y., Liu, Z., Li, H., Cheng, J., Qi, Q., Zhang, M., Zhang, H., Sui, F., & Cheng, H. (2018). Digital Twin and Its Potential Application Exploration. *Jisuanji Jicheng Zhizao Xitong*, *24*(1), 1–18.

Tao, F., & Zhang, M. (2017). Digital Twin Shop-Floor: A New Shop-Floor Paradigm Towards Smart Manufacturing. *IEEE Access : Practical Innovations, Open Solutions*, *5*, 20418–20427. doi:10.1109/ACCESS.2017.2756069

Tarantino-Curseri, S. (2018). Antropología, etología, sociología, y psicología, 4 disciplinas útiles en el ámbito empresarial. *Management Review*, *3*(3). Advance online publication. doi:10.18583/umr.v3i3.126

Tarek, R. d'Avila Garcez, A., Bader, S., Bowman, H., Domingos, P., Hitzler, P., Kühnberger, K., Lamb, L., Lowd, D., & Lima, P. (2017). Neural-symbolic learn- ing and reasoning: A survey and interpretation. arXiv preprint arXiv:1711.03902.

Taylor, T. (2010). *Sustainability Interventions - for Managers of Projects and Programmes. Centre for Education in the Built Environment*. The Higher Education Academy.

Ten Cate, O., & Billett, S. (2014). Competency-based medical education: Origins, perspectives and potentialities. *Medical Education*, *48*(3), 325–332. doi:10.1111/medu.12355 PMID:24528467

Tenenhaus, M., Esposito Vinzi, V., Chatelin, Y.-M., & Lauro, C. (2005). PLS path modeling. *Computational Statistics & Data Analysis*, *48*(1), 159–205. doi:10.1016/j.csda.2004.03.005

Terman, J., & Smith, C. (2018). Putting your money where your mouth is: green procurement as a form of sustainability. *Journal of Public Procurement*.

Tharp, J. (2013). Sustainability in Project Management: Practical Applications. In A. J. G. Silvius & J. Tharp (Eds.), *Sustainability Integration for Effective Project Management* (pp. 182–193). IGI Global Publishing. doi:10.4018/978-1-4666-4177-8.ch011

The Central Committee of the Communist Party of China and the State Council. (2019, September 19). *Outline For the Construction of a Transportation Powerhouse*. CN. http://www.gov.cn/zhengce/2019-09/19/content_5431432.htm

The General Office of the Ministry of Transport. (2017, January 22). *Action Plan for Promoting the Development of Smart Transportation (2017-2020)*. XXKG. https://xxgk.mot.gov.cn/2020/jigou/zhghs/202006/t20200630_3319779.html

The Ministry of Industry and Information Technology. (2017, November 14). *A Three-Year Action Plan to Promote the Development of a New Generation of Artificial Intelligence Industry (2018-2020)*. MIIT. https://www.miit.gov.cn/jgsj/kjs/jscx/gjsfz/art/2020/art_291b5e6bc13f415494e84a0e9eac78f1.html

The Ministry of Science and Technology of the People's Republic of China. (July 12, 2019). Comprehensive Transportation and Intelligent Transportation. *Key Special Project Large-scale Networked Vehicle Collaborative Service Platform, Beijing*.

The Paper. (2020, August 18). China's Space Industry Trend Report: High-Precision Maps Reconstruct the World and Form A Digital Twin Closed Loop. *The Paper*. https://www.thepaper.cn/newsDetail_forward_8773349

The Paper. (2021, March 1). What Exactly Does The National Integrated Three-Dimensional Transportation Network Look Like? 4 Poles, 6 Axes, 7 Corridors, 8 Channels. *The Paper*. https://www.thepaper.cn/newsDetail_forward_11505409

Thomas, G., & Fernandez, W. (2007). The Elusive Target of IT Project Success. In *Proceedings of the 2nd International Research Workshop on IT Project Management (IRWITPM)* (pp. 93-102). Association of Information Systems, Special Interest Group for Information Technology Project Management.

Thomason, J. (2021). MetaHealth - How will the Metaverse Change Health care? *Journal of Metaverse*, *1*(1), 13–16.

Thomson, C. S., El-Haram, M. A., & Emmanuel, R. (2011). Mapping sustainability assessment with the project life cycle. *Proceedings of the Institution of Civil Engineers. Engineering Sustainability*, *164*(2), 143–157. doi:10.1680/ensu.2011.164.2.143

Thoumy, M., & Vachon, S. (2012). Environmental projects and financial performance: Exploring the impact of project characteristics. *International Journal of Production Economics*, *140*(1), 28–34. doi:10.1016/j.ijpe.2012.01.014

Thrun, S. B. (1992). The role of exploration in learning control. In D. White & D. Sofge (Eds.), *Handbook for Intelligent Control: Neural, Fuzzy and Adaptive Approaches* (pp. 527–559). Van Nostrand Reinhold.

Tilman, D., Cassman, K. G., Matson, P. A., Naylor, R., & Polasky, S. (2002). Agricultural sustainability and intensive production practices. *Nature*, *418*(6898), 671–677. doi:10.1038/nature01014 PMID:12167873

Tingström, J., & Karlsson, R. (2006). The relationship between environmental analyses and the dialogue process in product development. *Journal of Cleaner Production, 14*(15–16), 1409–1419. doi:10.1016/j.jclepro.2005.11.012

Tiron-Tudor, A., & Dragu, I.-M. (2013). Project Success by Integrating Sustainability in Project Management. In A. J. G. Silvius & J. Tharp (Eds.), *Sustainability Integration for Effective Project Management* (pp. 106–127). IGI Global Publishing.

Tlili, A., Huang, R., Shehata, B., Liu, D., Zhao, J., Metwally, A. H. S., Wang, H., Denden, M., Bozkurt, A., Lee, L.-H., Beyoglu, D., Altinay, F., Sharma, R. C., Altinay, Z., Li, Z., Liu, J., Ahmad, F., Hu, Y., Salha, S., & Burgos, D. (2022). Is Metaverse in education a blessing or a curse: A combined content and bibliometric analysis. *Smart Learning Environments, 9*(1), 1–31. doi:10.118640561-022-00205-x

Toor, S., & Ogunlana, S. O. (2010). Beyond the "iron triangle": Stakeholder perception of key performance indicators (KPIs) for large-scale public sector development projects. *International Journal of Project Management, 28*(3), 228–236. doi:10.1016/j.ijproman.2009.05.005

Topolšek, D., Čižiūnienė, K., & Ojsteršek, T. C. (2018). Defining transport logistics: A literature review and practitioner opinion based approach. *Transport, 33*(5), 1196–1203. doi:10.3846/transport.2018.6965

Torfing, J., Peters, B., Pierre, J., & Sørensen, E. (2012). *Interactive Governance: Advancing the Paradigm.*

Torfing, J., & Ansell, C. (2017). Strengthening Political Leadership and Policy Innovation through the Expansion of Collaborative Forms of Governance. *Public Management Review, 19*(1), 37–54. doi:10.1080/14719037.2016.1200662

Townsend, A. M. (2013). *Smart cities: Big data, civic hackers, and the quest for a new utopia.* WW Norton & Company.

Tranfield, D., Denyer, D., & Smart, P. (2003). Towards a Methodology for Developing Evidence-Informed Management Knowledge by Means of Systematic Review. *British Journal of Management, 14*(3), 207–222. doi:10.1111/1467-8551.00375

Trawick, P., & Alf, H. (2015). Revisiting the Image of Limited Good: On Sustainability, Thermodynamics, and the Illusion of Creating Wealth. *Current Anthropology, 56*(1), 1–27. doi:10.1086/679593

Trawnih, A., Al-Masaeed, S., Alsoud, M., & Alkufahy, A. (2022). Understanding artificial intelligence experience: A customer perspective. *International Journal of Data and Network Science, 6*(4), 1471–1484. doi:10.5267/j.ijdns.2022.5.004

Tripathy, C. R., & Katyayn, A. (2021). Product Development Process Concept—Industrial Perspective. In *Advances in Design and Thermal Systems* (pp. 331–339). Springer. doi:10.1007/978-981-33-6428-8_26

Trist, E. L. (1976). *A Concept of Organizational Ecology: an invited address to the three Melbourne universities.*

Trujillo, M. L. N. (2021). Notes for a critical and ecological view of patriarchal capitalism in the web of life. *Capital and Class, 45*(March), 21–32. doi:10.1177/0309816820929115

Tsaih, R.-H., & Hsu, C. C. (2018). *Artificial Intelligence in Smart Tourism: A Conceptual Framework.* ICEB 2018 Proceedings, Guilin, China. 89. https://aisel.aisnet.org/iceb2018/89

Tsing, A. L. (2015). *The mushroom at the end of the world.* Princeton University Press.

Tsinghua University Suzhou Automotive Research Institute. (2020, October 23). *In-vehicle Binocular System.* TSARI. https://www.tsari.tsinghua.edu.cn/scientific/znwlCar/2020-10-23/348.html

Turner, J. R. (2010). Responsibilities for Sustainable Development in Project and Program Management. In H. Knoepfel (Ed.), *Survival and Sustainability as Challenges for Projects Zurich.* International Project Management Association.

Tu, Z., Qiao, L., Nowak, R., Lv, H., & Lv, Z. (2022). Digital Twins-Based Automated Pilot for Energy-Efficiency Assessment of Intelligent Transportation Infrastructure. *IEEE Transactions on Intelligent Transportation Systems*, *23*(11), 22320–22330. Advance online publication. doi:10.1109/TITS.2022.3166585

Tzokas, N., Hultink, E. J., & Hart, S. (2004). Navigating the New Product Development Process. *Industrial Marketing Management*, *33*(7), 619–626. doi:10.1016/j.indmarman.2003.09.004

Ulewicz, R., Siwiec, D., Pacana, A., Tutak, M., & Brodny, J. (2021). Multi-criteria method for the selection of renewable energy sources in the polish industrial sector. *Energies*, *14*(9), 2386. doi:10.3390/en14092386

UN SDG. (2015). *United Nations Sustainable Development Goals*. United Nations Department of Economic and Social Affairs.

UNEP. (2002). *Environmentally interesting technologies*. United Nations Environment Programme. Retrieved from https://www.unep.org/

UNEP. (2019). *The emissions gap report 2019*. United Nations Environment Programme.

United Nations Industrial Development Organization. (2013). *UNIDO ANNUAL REPORT 2012*. UN.

Upton, H. F., & Folger, P. F. (2018). *Ocean Acidification*. Congressional Research Service.

US Environmental Protection Agency. (2014). *Reducing Wasted Food & Packaging: A Guide for Food Services and Restaurants*. EPA. Retrieved March 7, 2019, from https://www.epa.gov/sites/production/files/201508/documents/reducing_wasted_food_pkg_tool.pdf

van den Brink, J. (2009). Duurzaam projectmanagement: verder kijken dan je project lang is [Sustainable project management: looking beyond the project]. *Projectie, 4*.

van der Niet, A. G., & Bleakley, A. (2021). Where medical education meets artificial intelligence: 'Does technology care?'. *Medical Education*, *55*(1), 30–36. doi:10.1111/medu.14131 PMID:32078175

Van der Voet, E., Van Oers, L. F., & De Haan, P. (2019). The importance of green technologies for a sustainable future. *Sustainability*, *11*(19), 5351.

van Marrewijk, M. (2003). Concepts and definitions of CSR and corporate sustainability: Between agency and communion. *Journal of Business Ethics*, *44*(2-3), 95–105. doi:10.1023/A:1023331212247

van Pelt, M. J. F. (1993). *Ecological Sustainability and Project Appraisal: Case Studies in Developing Countries*. Avebury.

Varghese, B. (2016). A Strategic Evaluation on Competency of Karnataka Destinations through Destination Management Organizations. *American Journal of Industrial and Business Management*, *06*(02), 102–108. doi:10.4236/ajibm.2016.62010

Vehtari, A., Gelman, A., & Gabry, J. (2017). Practical Bayesian model evaluation using leave-one-out cross-validation and WAIC. *Statistics and Computing*, *27*(5), 1413–1432. doi:10.100711222-016-9696-4

Velho, G. (1978). Observando o familiar. In A aventura sociológica: objetividade, paixao, improviso e método na pesquisa social. Zahar.

Velik, R. (2010). Towards human-like machine perception 2. 0. *International Review on Computers and Software*, *5*(4), 476–488.

Venkatraman, N., & Ramanujam, V. (1986). Measurement of business performance in strategy esearch: A comparison of approaches. *Academy of Management Review*, *11*(4), 801–814. doi:10.2307/258398

Verma, M., & Manoj, M., & Verma. (2016). Analysis of the influences of attitudinal factors on car ownership decisions among urban young adults in developing Country like India,Transportation Research Part F:Traffic psychology and Behaviour. *Vulume*, *42*(1), 90–103.

Verrier, B., Rose, B., & Caillaud, E. (2016). Lean and Green strategy: The Lean and Green House and maturity deployment model. *Journal of Cleaner Production*, *116*, 150–156. doi:10.1016/j.jclepro.2015.12.022

Vezzoli, C., & Sciama, D. (2006). Life cycle design: from general methods to product type specific guidelines and checklists: a method adopted to develop a set of guidelines/checklist handbook for the eco-efficient design of NECTA vending machines. *Journal of Cleaner Production*, *14*(15–16), 1319–1325. doi:10.1016/j.jclepro.2005.11.011

Viglialoro, R. M., Condino, S., Turini, G., Carbone, M., Ferrari, V., & Gesi, M. (2021). Augmented Reality, Mixed Reality, and Hybrid Approach in Healthcare Simulation: A Systematic Review. *Applied Sciences (Basel, Switzerland)*, *11*(5), 2338. doi:10.3390/app11052338

Voinov, A. (2007). Understanding and communicating sustainability: Global versus regional perspectives. *Environment, Development and Sustainability*, *10*(4), 487–501. doi:10.100710668-006-9076-x

Von Hippel, E. A. (2009). People Don't Need a Profit Motive to Innovate. *Harvard Business Review*. https://hbr.org/2011/11/people-dont-need-a-profit-motive-to-innovate

Von Hippel, E., & Baker, E. (2006). Ideas on the Edge. *CIO Insight*, 12-18.

Von Hippel, E. (1978). Successful Industrial Products From Customer Ideas. *Journal of Marketing*, *42*(1), 39–49. doi:10.2307/1250327

Von Hippel, E. (1986). Lead users: A source of novel product concepts. *Management Science*, *32*(7), 791–805. doi:10.1287/mnsc.32.7.791

Von Hippel, E. (1988). *The Source of Innovation*. Oxford University Press.

Von Hippel, E. (2001). User Toolkits for Innovation. Journal of Product Innovation Management. *Journal of Product Innovation Management*, *18*(4), 247–257. doi:10.1111/1540-5885.1840247

Von Hippel, E. (2006). *Democratizing Innovation*. The MIT Press.

Voštinár, P., Horváthová, D., Mitter, M., & Bako, M. (2021). The look at the various uses of VR. *Open Computer Science*, *11*(1), 241–250. doi:10.1515/comp-2020-0123

Vygotsky, L. S. (1978). Zone of proximal development: A new approach. *Mind in society: The development of higher psychological processes*, 84-91.

W. S. (2007). The Anthropocene: Are Human Beings Now Overwhelming the Forces of Nature? *Ambio*, *36*, 614–621. doi:10.1579/0044-7447(2007)36[614:TAAHNO]2.0.CO;2 PMID:18240674

Wang, F. (2004). Computational theory and method on complex system. *China Basic Sci*, 5-12.

Wang, G.T. (2015). Urban Transportation and Informatization. *Urban transport of China*, *13*(3), 1-4.

Wang, Y., Zhou, S., Zhang, N., Xing, R., Lui, D., & Luan, T. H. (2022, August). *A Survey on Metaverse: Fundamentals, Security and Privacy*. doi:10.48550/arXiv.2203.02662

Wang, D., Li, X. R., & Li, Y. (2013). China's "smart tourism destination" initiative: A taste of the service-dominant logic. *Journal of Destination Marketing & Management*, 2(2), 59–61. doi:10.1016/j.jdmm.2013.05.004

Wang, F. (2010, July/August). The emergence of intelligent enterprises: From CPS to CPSS. *IEEE Intelligent Systems*, *25*(4), 85–88. doi:10.1109/MIS.2010.104

Wang, F. Y. (2004). Computational experiments for behavior analysis and decision evaluation of complex systems. *Journal of Systems Simulation*, *16*(5), 893–987.

Wang, F. Y., Qin, R., Wang, X., & Hu, B. (2022). Metasocieties in metaverse: Metaeconomics and metamanagement for metaenterprises and metacities. *IEEE Transactions on Computational Social Systems*, *9*(1), 2–7. doi:10.1109/TCSS.2022.3145165

Wang, F., & Lansing, J. (2004). From artificial life to artificial societies: New methods for studies of complex social systems. *Complex Sys. Complex. Sci*, *1*(1), 33–41.

Wang, L., Jin, J. L., Zhou, K. Z., Li, C. B., & Yin, E. (2020). Does Customer Participation Hurt New Product Development Performance? Customer Role, Product Newness, and Conflict. *Journal of Business Research*, *109*, 246–259. doi:10.1016/j.jbusres.2019.12.013

Wang, W., Li, X., Xie, L., Lv, H., & Lv, Z. (2022). Unmanned Aircraft System Airspace Structure and Safety Measures Based on Spatial Digital Twins. *IEEE Transactions on Intelligent Transportation Systems*, *23*(3), 2809–2818. doi:10.1109/TITS.2021.3108995

Wapner, P. (2003). World Summit on Sustainable Development: Toward a post-Jo'burg environmentalism. *Global Environmental Politics*, *3*(1), 1–10. doi:10.1162/152638003763336356

Warburton, S. (2009). Second Life in higher education: Assessing the potential for and the barriers to deploying virtual worlds in learning and teaching. *British Journal of Educational Technology*, *40*(3), 414–426. doi:10.1111/j.1467-8535.2009.00952.x

WCED. (1987). Our Common Future. Oxford University Press.

We Are Social. (2020). *Digital in 2020*. https://wearesocial.com/digital-2020

Webber, S. (2019). Putting climate services in contexts: Advancing multi-disciplinary understandings: Introduction to the special issue. *Climatic Change*, *157*(1), 1–8. doi:10.100710584-019-02600-9

Weber, M., Engert, M., Schaffer, N., Weking, J., & Krcmar, H. (2022). Organizational Capabilities for AI Implementation—Coping with Inscrutability and Data Dependency in AI. *Information Systems Frontiers*, 1–21. doi:10.100710796-022-10297-y

Wei, X., Lo, C. K. Y., Jung, S., & Choi, T.-M. (2021). From co-consumption to co-production: A systematic review and research synthesis of collaborative consumption practices. *Journal of Business Research*, *129*(May), 282–294. doi:10.1016/j.jbusres.2021.02.027

Welie, J. V. (2004). Is dentistry a profession? Part 3. Future challenges. *Journal - Canadian Dental Association*, *70*(10), 675–678. PMID:15530264

Wellman, H. M., & Liu, D. (2004). Scaling of theory-of-mind tasks. *Child Development*, *75*(2), 523–541. doi:10.1111/j.1467-8624.2004.00691.x PMID:15056204

WEneurs. (2022). *WEneurs Forum, Annual Newsletter 2022*. Author.

Weninger, C., Huemann, M., Oliveira, J. C., Barros Filho, L. F. M., & Weitlaner, E. (2013). Experimenting with project stakeholder analysis: a case study. In A. J. G. Silvius & J. Thap (Eds.), *Sustainability Integration for Effective Project Management*. IGI Global. doi:10.4018/978-1-4666-4177-8.ch023

Werthner, H., & Ricci, F. (2004). E-commerce and tourism. *Communications of the ACM*, *47*(12), 101–105. doi:10.1145/1035134.1035141

White, D., & Fortune, J. (2002). Current practice in project management— An empirical study. *International Journal of Project Management*, *20*(1), 1–11. doi:10.1016/S0263-7863(00)00029-6

Whiteman, G., Walker, B., & Perego, P. (2013). Planetary boundaries: Ecological foundations for corporate sustainability. *Journal of Management Studies*, *50*(2), 307–336. doi:10.1111/j.1467-6486.2012.01073.x

Wikström, F., Verghese, K., Auras, R., Olsson, A., Williams, H., Wever, R., Grönman, K., Kvalvåg Pettersen, M., Møller, H., & Soukka, R. (2019). Packaging strategies that save food: A research agenda for 2030. *Journal of Industrial Ecology*, *23*(3), 532–540. doi:10.1111/jiec.12769

Wilk, R. R., & Wilhite, H. L. (1985). Why don't people weatherize their homes? An ethnographic solution. *Energy*, *10*(May), 621–629. doi:10.1016/0360-5442(85)90093-3

William, W. (2018). Cohen, Fan Yang, and Kathryn Rivard Mazaitis. Tensorlog: Deep learning meets probabilistic databases. *Journal of Artificial Intelligence Research*, *1*, 1–15.

Wilson, I., & Shankar, P. R. (2021). The COVID-19 pandemic and undergraduate medical student teaching/learning and assessment. *MedEdPublish*, *10*(1), 10. doi:10.15694/mep.2021.000044.1

Wimberley, R. C. (1993). Policy perspectives on social, agricultural, and rural sustainability. *Rural. Sociol.*, *58*(1), 1–29. doi:10.1111/j.1549-0831.1993.tb00480.x

Wingström, R., Hautala, J., & Lundmana, R. (2021). Redefining creativity in the era of AI? Perspectives of computer scientists and new media artists. *Creativity Research Journal*. https://www.tandfonline.com/doi/full/10.1080/1040041 9.2022.2107850

Winnall, J.-L. (2013). Social Sustainability to Social Benefit: Creating Positive Outcomes through a Social Risk-based approach. In A. J. G. Silvius & J. Tharp (Eds.), *Sustainability Integration for Effective Project Management* (pp. 95–105). IGI Global Publishing. doi:10.4018/978-1-4666-4177-8.ch006

Wirtz, B., Pistoia, A., Ullrich, S., & Gottel, V. (2015). Business Models: Origin, Development and Future Research Perspectives. *Long Range Planning*. doi:10.1016/j

Wissink, G. (2002). *Creativity and cognition: A study of creativity within the framework of cognitive science, artificial intelligence and the dynamical system theory*. https://www.academia.edu/29365431/Creativity_and_Cognition_A_study_of_creativity_within_the_framework_of_cognitive_science_artificial_intelligence_and_the_dynamical_system_theory

Wit, A. (1988). Measurement of Project Success. *International Journal of Project Management*, *6*(3), 164–170. doi:10.1016/0263-7863(88)90043-9

Wodecki, A., Wodecki, H., & Harrison. (2019). *Artificial intelligence in value creation*. Springer International Publishing.

Wolf, M. (2016). Rethinking urban epidemiology: Natures, networks and materialities. *International Journal of Urban and Regional Research*, *40*(5), 958–982. doi:10.1111/1468-2427.12381 PMID:32336869

Wood, M., Mathieux, F., Brissaud, D., & Evrard, D. (2010). Results of the first adapted design for sustainability project in a South Pacific small island developing state: Fiji. *Journal of Cleaner Production*, *18*(18), 1775–1786. doi:10.1016/j.jclepro.2010.07.027

World Business Council for Sustainable Development. (2012). Public Policy Options to Scale and Accelerate Business and Actions Towards Vision 2050. Geneva, Switzerland.

World Economic Forum. (2009). *Supply Chain Decarbonization.* World Economic Forum.

World Economic Forum. (2015, August). Collaborative Innovation: Transforming Business, Driving Growth. *Regional Agenda*, 1-44.

World Economic Forum. (2016). Building the healthcare system of the future. WEF. https://reports.weforum.org/digital-transformation/building-the-healthcare-system-of-the-future/

World Health Organisation. (2021) International Data Online Updates. WHO. https://www.who.int/data)

Wu, Z., Ji, D., Yu, K., Zeng, X., Wu, D., & Shidujaman, M. (2021). AI creativity and the human-AI co-creation model. In M. Kurosu (Ed.), *Human-Computer Interaction. Theory, Methods and Tools, Thematic Area, HCI 2021, Held as Part of the 23rd HCI International Conference, , Proceedings, Part I* (pp. 171–190). IEEE. 10.1007/978-3-030-78462-1_13

Wu, G. J., Tajdini, S., Zhang, J., & Song, L. (2019). Unlocking Value Through an Extended Social Media Analytics Framework: Insights for New Product Adoption. *Qualitative Market Research*, *22*(2), 1352–2752. doi:10.1108/QMR-01-2017-0044

Wu, Y., Zhang, K., & Zhang, Y. (2021). Digital Twin Networks: A Survey. *IEEE Internet of Things Journal*, *8*(18), 13789–13804. doi:10.1109/JIOT.2021.3079510

Wu, Z., Wu, X., & Wang, L. (2019). Prospect of Development Trend of Smart Transportation under the Background of Building China into a Country with Strong Transportation Network. *Transportation Research*, *5*(4), 26–36. doi:10.16503/j.cnki.2095-9931.2019.04.003

Xia, L., Guo, T., Qin, J., Yue, X., & Zhu, N. (2018). Carbon emission reduction and pricing policies of a supply chain considering reciprocal preferences in cap-and-trade system. *Annals of Operations Research*, *268*(1-2), 149–175. doi:10.100710479-017-2657-2

Xiang, Z., & Fesenmaier, D. R. (2017). Big data analytics, tourism design and smart tourism. *Analytics in smart tourism design: concepts and methods*, 299-307. doi:10.1177/0047287514522883

Xie, X. P., & Zhao, D. Z. (2013). Decision mechanism study on product pricing and emission reduction in two level low-carbon supply chain enterprises based on the CDM. *Soft Science, 27*(5), 80-85.

Xing, B., & Marwala, T. (2018). *Creativity and artificial intelligence: A digital art perspective.* https://arxiv.org/ftp/arxiv/papers/1807/1807.08195.pdf

Xinhua Net. (2019, September 24). *China's Construction Of A Transportation Power Has Opened A New Chapter.* Xinhua Net.

Xinhua News Agency. (2019, September 19). *Outline for the Construction of a Transportation Powerhouse.* State Council of People's Republic of China. http://www.gov.cn/zhengce/2019-09/19/content_5431432.htm

Xinhua News Agency. (2021, February 24). *Outline of the National Comprehensive Three-Dimensional Transportation Network Plan.* State Council of People's Republic of China. http://www.gov.cn/zhengce/2021-02/24/content_5588654.htm

Xinmin Evening News. (2020, October 17). Intelligent Transportation Facilitates People's Travel. *Xinmin Evening News.*

Xu, J., Zhang, Z. Friedman, T., Liang, Y., & Broeck, G. (2018). A semantic loss function for deep learning with symbolic knowledge. ICML.

Xu, X., Xu, X., & He, P. (2016). Joint production and pricing decisions for multiple products with cap-and-trade and carbon tax regulations. *Journal of Cleaner Production, 112*, 4093–4106. doi:10.1016/j.jclepro.2015.08.081

Xu, Y., Shieh, C., van Esch, P., & Ling, I. (2020). AI customer service: Task complexity, problem-solving ability, and usage intention. *Australasian Marketing Journal*, *28*(4), 189–199. doi:10.1016/j.ausmj.2020.03.005

Yadav, P., Singh, J., Srivastava, D. K., & Mishra, V. (2021). Environmental pollution and sustainability. In *Environmental Sustainability and Economy* (pp. 111–120). Elsevier. doi:10.1016/B978-0-12-822188-4.00015-4

Yang, G. (2006). AI Winter and its lessons. *History of Computing* [PDF file]. https://courses.cs.washington.edu/courses/csep590/06au/projects/history-ai.pdf

Yang, Y., Lili, W., & Hongchi, Z. (2022). Bibliometric research on the evolution of resilience theme from the perspective of Geographical Science. *2022 29th International Conference on Geoinformatics*, 1-9. 10.1109/Geoinformatics57846.2022.9963870

Yang, D., Guo, J., Gu, Y., Zhao, Y., Zhang, X., Ding, Q., Qian, L., Shao, D., Chen, X., Zhou, T., Bai, F., Cui, Y., Zhang, Y., & Wang, G. (2021). Digital Transformation of Urban Comprehensive Transportation Planning. *Urban Transport of China*, *19*(6), 107–113. doi:10.13813/j.cn11-5141/u.2021.0605

Yang, S., Su, Y., Wang, W., & Hua, K. (2019). Research on developers' green procurement behavior based on the theory of planned behavior. *Sustainability*, *11*(10), 2949. doi:10.3390u11102949

Yarbrough, Q. (2021, November 16). *Production Planning in Manufacturing: Best Practices for Production Plans.* Planning, Project Management. https://www.projectmanager.com/blog/production-planning

Ye, B. H., Ye, H., & Law, R. (2020). Systematic review of smart tourism research. *Sustainability*, *12*(8), 3401. doi:10.3390u12083401

Ye, D. (2020). *Shenzhen Airport Group And Huawei Signed A Deepening Strategic Cooperation Agreement.* Civil Aviation Resource Network.

Yigitcanlar, T., Butler, L., Windle, E., Desouza, K. C., Mehmood, R., & Corchado, J. M. (2020). Can building "artificially intelligent cities" safeguard humanity from natural disasters, pandemics, and other catastrophes? An urban scholar's perspective. *Sensors (Basel)*, *20*(10), 2988. doi:10.339020102988 PMID:32466175

Yorumbudur. (n.d.). https://yorumbudur.com/

Young, A., & Verhulst, S. (2017). Jamaica's interactive community mapping: open data and crowdsourcing for tourism. In S. G. Verhulst & A. Young (Eds.), *Open Data in Developing Economies: Toward Building an Evidence Base on What Works and How* (pp. 206–223). African Minds. https://muse.jhu.edu/chapter/2062325

Yu, Z. (2021). Reflections and Practices of Digital Transformation at Shenzhen Airport. *Civil Aviation Resource Network.* http://news.carnoc.com/list/566/566097.html

Yunus, R., & Yang, J. (2014). Improving ecological performance of industrialized building systems in Malaysia. *Construction Management and Economics*, *32*(1/2), 183–195. doi:10.1080/01446193.2013.825373

Yu, Y., Lin, H., Meng, J., & Zhao, Z. (2016). Visual and Textual Sentiment Analysis of A Microblog Using Deep Convolutional Neural Networks. *Algorithms*, *9*(2), 41. doi:10.3390/a9020041

Zahay, D., Griffin, A., & Fredericks, E. (2004). Sources, Uses, and Forms of Data in The New Product Development Process. *Industrial Marketing Management*, *33*(7), 657–666. doi:10.1016/j.indmarman.2003.10.002

Zahay, D., Hajli, N., & Sihi, D. (2018). Managerial Perspectives on Crowdsourcing in the New Product Development Process. *Industrial Marketing Management*, *71*, 41–53. doi:10.1016/j.indmarman.2017.11.002

Zakeri, A., Dehghanian, F., Fahimnia, B., & Sarkis, J. (2015). Carbon pricing versus emissions trading: A supply chain planning perspective. *International Journal of Production Economics, 164*, 197–205. doi:10.1016/j.ijpe.2014.11.012

Zelin, M. (2020). *Education 2030 Forecast: Online & Fun*. Innompics USA. http://innompics.com/events/ig2020/contests_sbm_futh_zelin.html

Zenezini, G., & Ghajargar, M., & Fiore, E., & De Marco, A. (2016). *The Smart Home Services Diffusion Process: A System Dynamics Model.*

Zeng, S. X., Ma, H. Y., Lin, H., Zeng, R. C., & Tam, V. W. Y. (2015). Social responsibility of major infrastructure projects in China. *International Journal of Project Management, 33*(3), 537–548. doi:10.1016/j.ijproman.2014.07.007

Zetzsche, D. A., Arner, D. W., & Buckley, R. P. (2020, September 30). Decentralized Finance. *Journal of Financial Regulation*, 172-203. doi:10.1093/jfr/fjaa010

Zhang, X., Wu, Y., Skitmore, M., & Jiang, S. (2015). Sustainable infrastructure projects in balancing urban–rural development: towards the goal of efficiency and equity. *J. Clean. Prod., 107*, 445–454. .2014.09.068 doi:10.1016/j.jclepro

Zhang, Z., Zhang, M., Chang, Y., Aziz, E., Esche, S., & Chassapis, C. (2018, April 27). Collaborative virtual laboratory environments with hardware in the loop. *Cyber-Physical Laboratories in Engineering and Science Education*, 363-402.

Zhang, B., Matchinski, E. J., Chen, B., Ye, X., Jing, L., & Lee, K. (2019). Marine oil spills—oil pollution, sources and effects. In *World seas: An environmental evaluation* (pp. 391–406). Academic Press. doi:10.1016/B978-0-12-805052-1.00024-3

Zhang, Z. C., Li, K. W., Liu, Z., & Huang, J. (2019). Pricing decisions in a socially responsible supply chain under carbon cap-and-trade regulation. *IFAC-PapersOnLine, 52*(13), 331–336. doi:10.1016/j.ifacol.2019.11.130

Zhan, Y., Tan, K. H., Chung, L., Chen, L., & Xing, X. (2020). Leveraging Social Media in New Product Development: Organisational Learning Processes, Mechanisms and Evidence from China. *International Journal of Operations & Production Management, 40*(5), 671–694. doi:10.1108/IJOPM-04-2019-0318

Zhao, L., Liu, H., Zhang, X., & Wang, F. (2021). Research on Vehicle-Infrastructure Collaboration Virtual Simulation Platform Based on Digital Twin. *Mobile Communications, 45*(6), 7–12.

Zhao, P., & Zhang, Y. (2018). Travel behaviour and life course:Examining changes in car use after residential relocation in Beijing. *Journal of Transport Geography, 7*, 41–53. doi:10.1016/j.jtrangeo.2018.10.003

Zhong, A. (2021, September 24). *4 Steps In Successful Supply Chain Inventory Management*. https://www.gep.com/blog/technology/4-steps-in-successful-supply-chain-inventory-management

Zhong, Y., & Wu, P. (2015). Economic sustainability, environmental sustainability and constructability indicators related to concrete- and steel-projects. *Journal of Cleaner Production, 108*, 748–756. doi:10.1016/j.jclepro.2015.05.095

Zhou, Y., Hu, F., & Zhou, Z. (2018). Pricing decisions and social welfare in a supply chain with multiple competing retailers and carbon tax policy. *Journal of Cleaner Production, 190*, 752–777. doi:10.1016/j.jclepro.2018.04.162

Zhu, W., Zhang, L., & Li, N. (2014). Challenges, function changing of government and enterprises in Chinese smart tourism. *Information and communication technologies in tourism, 10*, 553-564.

Zizza, C., Starr, A., Hudson, D., Nuguri, S., Calyam, P., & He, Z. (2018). Towards a Social Virtual Reality Learning Environment in High Fidelity. *Proceedings of the 15th IEEE Annual Consumer Communications & Networking Conference (CCNC)*. IEEE. 10.1109/CCNC.2018.8319187

Zulkifli, A. F. (2019). Student-centered approach and alternative assessments to improve students' learning domains during health education sessions. *Biomedical Human Kinetics*, *11*(1), 80–86. doi:10.2478/bhk-2019-0010

Zuo, K., Potangaroa, R., Wilkinson, S., & Rotimi, J. O. (2009). A project management prospective in achieving a sustainable supply chain for timber procurement in Banda Aceh, Indonesia. *International Journal of Managing Projects in Business*, *2*(3), 386–400. doi:10.1108/17538370910971045

Zwikael, O., & Ahn, M. (2011). The Effectiveness of Risk Management: An Analysis of Project Risk Planning across Industries and Countries. *Risk Analysis*, *31*(1), 25–37. doi:10.1111/j.1539-6924.2010.01470.x PMID:20723146

About the Contributors

Ziska Fields is an Associate Professor in the College of Business and Economics, Department of Business Management at the University of Johannesburg (UJ). Prof Fields is an Alumna at the University of KwaZulu-Natal (UKZN) in Durban and an external researcher and habilitator at the Chair of Innovation Research and Technology Management, Chemnitz University of Technology. She has recently qualified as a Neethling Brain Instruments (NBI) Practitioner focusing on identifying and developing peoples preferred thinking styles and to help them think more creatively at school, university and in the workplace. She has been appointed as the Ambassador for the World Creativity and Innovation Week (WCIW) and World Creativity and Innovation Day (WCID) in South Africa, as well as some southern African countries like Namibia and Botswana for 2021. The purpose of this role is to connect cities, regions and countries to the WCIW creative community and to foster creativity by promoting, empowering and supporting the WCIW's creativity mission. Professor Fields has 13 years' experience in the Financial Industry and 12 years' experience in Higher Education experience. Her qualifications include a BA degree (Communication Sciences), Diploma in Management Studies, MBA (Cum Laude) and PhD (Business Administration). She teaches and supervises various Management and Entrepreneurship focus area at undergraduate and postgraduate levels. Her research and supervisory focus areas include Entrepreneurship, Higher Education, Responsible Management and Human Resources with a focus on the Fourth Industrial Revolution. Her main research focus area and passion is theoretical and applied creativity across various disciplines and contexts. She developed two theoretical models to measure creativity in South Africa, focusing on youth and tertiary education. She edited five books published by IGI Global titled "Incorporating Business Models and Strategies into Social Entrepreneurship", "Collective creativity for responsible and sustainable business practice", "Handbook of Research on Information and Cyber Security in the Fourth Industrial Revolution", "Responsible, Sustainable, and Globally Aware Management in the Fourth Industrial Revolution" and "Imagination, Creativity and Responsible Management in the Fourth Industrial Revolution". She is currently busy with a book titled "Using Global Collective Intelligence and Creativity to Solve Wicked Problems: Emerging Research and Opportunities" which should be published by the end of 2020. She also has various other research areas, which include Entrepreneurship, Management, Innovation, Higher Education, Responsible Management Education, Human Resources, Future Studies and Information Technology. She edited four books titled "Incorporating Business Models and Strategies into Social Entrepreneurship", "Collective creativity for responsible and sustainable business practice", "Handbook of Research on Information and Cyber Security in the Fourth Industrial Revolution", "Responsible, Sustainable, and Globally Aware Management in the Fourth Industrial Revolution" and "Imagination, Creativity, and Responsible Management in the Fourth Industrial Revolution". She has published various book chapters and papers and presented various conference papers. Professor Fields developed two theoretical models to measure creativity in South Africa, focusing on youth and tertiary education. Professor Fields is a driven, dedicated and passionate person who loves animals, learning new things and is constantly inspired by the positive and life-changing impact that education can have on people. Her objective is to make a difference by empowering people and helping them to reach their full potential. She strongly believes that creativity and entrepreneurial action can help people to create a better and sustainable future for themselves and society.

* * *

Richmond Anane-Simon is a corporate executive, researcher, activist in civil society, and critical thinker with experience working for the government, for-profit companies, and non-profit organizations. He possesses outstanding research, entrepreneurial/intrapreneurial, and leadership skills that are crucial for corporate success. His areas of interest in research include strategic management, sustainability, entrepreneurship, and technological innovation. His current positions include managing partner at Eco Electric Africa Limited and advisory board member of Naturopaths Without Borders USA.

Sulaiman Olusegun Atiku is currently a Professor (Human Resources) at Harold Pupkewitz Graduate School of Business (HP-GSB), Namibia University of Science and Technology, Namibia. He is a pragmatic researcher specializing in Strategic Human Resource Management. His current research area of interest includes; Human Capital Formation for the Fourth Industrial Revolution, and Human Resource Management Practices for Promoting Sustainability. He has over 15 years' experience in Higher Education. A native of Lagos, Sulaiman graduated from the University of KwaZulu-Natal with a PhD degree in Human Resource Management. His Master of Science degree was awarded in Human Resources and Industrial Relations at Lagos State University, Nigeria. His Bachelor of Science (Honours) degree was also awarded at Lagos State University in the field of Industrial Relations and Personnel Management. He has lectured several courses in his field home/abroad and published many scholarly articles in international journals. He is a member of International Labour and Employment Relations Association (ILERA), Nigerian Institute of Management (NIM), and Institute of People Management (IPM) South Africa.

Rohit Bansal is working as Associate Professor in Department of Management Studies in Vaish College of Engineering, Rohtak. He obtained Ph.D. in Management from Maharshi Dayanand University, Rohtak. With a rich experience of 13 years, he has achieved growth through robust and proactive academic initiatives. He has authored & edited 14 books with renowned publishers. In addition, he has published 105 research papers and chapters in Journals of repute as well as edited books. He has also presented papers in 35 conferences and seminars. His area of interest includes marketing management, organizational behaviour, digital marketing, services marketing, blockchain technology and human resource management. He is on Editorial Advisory Board as a member in 110 national and international peer reviewed journals. He is Managing Editor of International Journal of 360° Management Review & International Journal of Techno – Management Research. He Received "Excellence in Teaching Award" in Edge India Times Award organized by Edge India Publications Private Limited held on 15th November 2020 through virtual mode. He has acted as session chair in many conferences.

Aykan Candemir graduated from Dokuz Eylül University, Department of Tourism and Hotel Management in 1992, and from the Faculty of Business Administration (English Business Administration) of the same university in 1996. Mr. Candemir, who received the title of Doctor of Business Administration in 2005, gained the title of Associate Professor in the field of Marketing in 2012 and was promoted to full Professorship in 2021.

Aslı Diyadin Lenger graduated from the Department of Business Administration of Celal Bayar University. She completed her master degree in the same universty. She got her PhD from Ege universty and worked there as part time lecturer. Dr. Lenger's academic interest covers marketing research, digitalization and its effect on marketing. She is currently an Assistant Professor of Business at the Istanbul Gelisim Universty.

Dwijendra Nath Dwivedi is a professional with 20+ years of subject matter expertise creating right value propositions for analytics and AI. A post-Graduate in Economics from Indira Gandhi Institute of Development and Research and perusing PHD from crackow university of economics Poland. He has presented in more than 20 international conference and published a number of Scopus indexed paper on AI adoption in many areas. Academic interest includes econometrics, climate risk, machine learning and risk management . Author has contributed to more than 6 books by springer and other publishers. Here is the ORCID link: .

Sandhya H. is currently pursuing PhD in Tourism Management from Christ University, Bangalore. She successfully completed her Mphil and Masters in Tourism Administration (MTA) from Christ University Bangalore by securing the first rank. She has industry experience of one year having worked with one of the leading multinational tour operations company, Kuoni SOTC, Bangalore and teaching experience of over 6 years. She has worked as Research Associate of the Major Research Project on the lines of Destination Management Organisations. Her research interests include tourism marketing, destination management and destination branding, Smart Tourism. She is currently working as Academic Auditor and Deputy Academic Coordinator at Kairos Institute, Cochin, Kerala.

Emanuela Hanes is an independent researcher. Her cooperations include the Vienna University FH BFI Campus Wien, University of Graz, University of Salzburg, and University of Applied Arts Bremen. Her research interests include China-EU business strategies, RegTech, FinTech, Cryptocurrencies, Geopolitics, Chinese Strategic Planning and Development Policies.

Catherine Hayes is Professor of Health Professions Pedagogy and Scholarship at the University of Sunderland. She qualified in Podiatric Medicine in 1992 and has worked within Higher Education for the past twenty-five years. Catherine is a UK National Teaching Fellow and Principal Fellow of the Higher Education Academy. In 2012 she became a Founding Fellow of the Faculty of Podiatric Medicine at the Royal College of Physicians and Surgeons (Glasgow), following the award of Fellowship of the Royal College of Podiatry in 2010.

Manish Kumar is a Doctoral Scholar in the area of Operations Management and Quantitative Techniques at the Indian Institute of Management Indore. He holds a B.Tech in Electrical Engineering from IITRAM, Ahmedabad, India. His research interests include retail operations, consumer behaviour, sustainable supply chain and contract design. He has presented papers at National and International Conferences.

Ghanashyama Mahanty leads MENA Data science, UAE and Global Data Science Center of Excellence, India for Visa Inc. He is a data enthusiast with expertise in mobilizing the organization to deliver value out of data and specializing in business management, data transformation strategy, and machine learning and artificial intelligence solutions for Retail Banking and Payment industry. He has worked extensively across North America, Asia, Europe and Middle East regions. Currently pursuing PhD. in Economics from Department of Analytical and Applied Economics, Utkal University, Bhubaneswar, India. His research area of interests are machine learning, artificial intelligence, macro econometric and structural data modelling, panel modelling, money, banking and payments

Tariro Munyengeterwa has an MA in Brand and Media Strategy from East Tennessee State University, and was the recipient of the Media and Communications award in Graduate Research. She currently serves as a Marketing Director in SC. She enjoys doing marketing technology research, and is a member of the Digital Marketing Institute (DMI). She is a marketing communications professional who has demonstrated a proven record of driving brand awareness and customer engagement through digital channels such as social media, content marketing, and paid search. Proficient in utilizing customer relationship management (CRM) tools to create effective marketing campaigns. Also, adept at using artificial intelligence to analyze data for email marketing, social media campaigns, and website traffic. Her future research interests center around artificial intelligence and cognitive dissonance, marketing communications, and brand strategy. Tariro likes to play basketball and spend time with her family when she isn't working.

Enock Gideon Musau is a Senior Lecturer in the Department of management science at, Kisii University-Kenya specializing in Transport and Supply Chain Management. His current position is a programme leader in Purchasing and Supplies Management. Currently, he is affiliated with the Department of Transport and supply chain Management, University of Johannesburg South Africa as a post-doctoral Research fellow undertaking research in mobility and well-being. He holds a Doctor of Philosophy (Ph.D.) degree in Supply Chain Management and a Master of Science in Procurement and Logistics both awarded at Jomo Kenyatta University of Agriculture and Technology, Nairobi Kenya. He pursued his Bachelor of Purchasing and Supplies Management (Honors) degree award at the Moi University Eldoret, Kenya in the field of business Management and a Professional Diploma in Procurement awarded by Kenya Institute of Supplies Examination Board, KISEB, Kenya. He has over eight years of experience in higher education lecturing courses in his field of specialization at both undergraduate and postgraduate levels. He has supervised students in his area of research interest including procurement, inventory management, transportation, and supply chain management. He has presented papers at national and international conferences and published many scholarly articles in peer-reviewed international journals. He has outstanding administrative and management experience that plays a vital role in organizational success. He is a member of the Kenya Institute of supplies management (KISM-Kenya) and the Chartered Institute of Purchasing and Supplies, UK. He is a Certified Procurement and Supply Professional of Kenya (CPSP-K). He has a strong spirit of creating a positive impact by empowering and elevating societal groups with the aim of establishing an enabling environment to exploit the potential of their expectations. Dr. Musau believes that innovativeness, self-driven, teamwork, integrity, and utilization of his current work experience, skills, and education coupled with further training progressively is a path to organizational competence and efficiency. Additional details regarding Him are presented in .

Nishita Pruthi is a full-time research scholar at Institute of Management Studies and Research, Maharshi Dayanand University. She has published 2 research articles, 5 chapters and presented papers in many national and international conferences.

Melanie B. Richards is the Associate Chair of the East Tennessee State University Department of Media & Communication and leads the department's Advertising and Public Relations program. She has been working in the research, analytics, and account planning world for over 20 years and prior to ETSU, spent the majority of her career working for both Fortune 500 companies and major nonprofit organizations in various leadership roles. Her academic research focuses on brand experience (especially in the nonprofit and cause space), public health communication, intergenerational communication and dynamics, and on the scholarship of teaching and learning. She is the co-author of the department's published experiential approach to teaching and student learning, the Applied Marketing & Media Education Norm.

Michael Sampat is an Independent Researcher working in Asian International Business and Biblical Studies.

Luan Carlos Santos Silva is an Associate Professor (since 2016) at the Faculty of Business Administration, Accounting Sciences and Economics of the Federal University of Grande Dourados (UFGD), Brazil. Doctor of Philosophy (Ph.D.), Production Engineering the Federal University of Rio Grande do Sul (UFRGS), Porto Alegre, Brazil. Master's degree, Production Engineering at the Federal University of Paraná (UTFPR), Ponta Grossa, Brazil. Bachelor's degree (Undergraduate degree), Business Administration at the Faculty of Discovery (FACESCO), Santa Cruz Cabrália, Brazil.

Jiarui Song is an independent researcher. Her research interests include Chinese Economy, FinTech and Marketing.

Carla Schwengber Ten Caten is an Associate Professor at the Production Engineering Departmentof the Federal University of Rio Grande do Sul (UFRGS), Porto Alegre, Brazil. Doctor of Philosophy (Ph.D.), Materials Engineering the Federal University of Rio Grande do Sul (UFRGS), Porto Alegre, Brazil. Postdoctoral fellow at the University of Southern California (USC) in the USA.

Satvik Tripathi is a student at Loyola International School and is currently studying in 12th grade. He recently attended Stanford Pre-Collegiate Summer Institute to study Artificial Intelligence course. To gain a deeper understanding of AI and ML, he has taken over 20 graduate and undergraduate level OpenCourseWare from Stanford, MIT, Cornell, and Harvard. He has completed various certification exams and courses from Google, Harvard, and the University of Helsinki. He has developed two AI-based apps on Google Assistant with more than a million views and has published an international book chapter on Artificial Intelligence. He has also served as Peer Reviewer at IGI Global Publication. He is well experienced in developing and implementing Machine learning techniques for biomedical applications. His main research interests include Graph Kernels, Pattern Recognition, and Deep Learning.

Zeynep Merve Ünal received her PhD degree in organizational behavior from the Marmara University in 2017. Her research interests involve meaningful work, humor at workplace, workplace spirituality, employee engagement, person-job fit/person-organization fit. She has published various book chapters that are related to corporate social responsibility, knowledge management, entrepreneurship, meaningful job design and contemporary leadership in the new era of COVID-19.

Omar Vargas-González is Professor and Head of Systems and Computing Department at Tecnologico Nacional de Mexico Campus Ciudad Guzman, professor at Telematic Engineering at Centro Universitario del Sur Universidad de Guadalajara with a master degree in Computer Systems. Has been trained in Innovation and Multidisciplinary Entrepreneurship at Arizona State University (2018) and a Generation of Ecosystems of Innovation, Entrepreneurship and Sustainability for Jalisco course by Harvard University T.H. Chan School of Health. At present conduct research on diverse fields such as Entrepreneurship, Economy, Statistics, Mathematics and Information and Computer Sciences. Has collaborated in the publication of over 20 scientific articles and conducted diverse Innovation and Technological Development projects.

Bindi Varghese is a Doctorate in Commerce, specializing in Tourism. As an academician and tourism professional, she has over 18 years of Academic and one year of Industrial experience. Currently, she is affiliated with Christ University, as an Associate Professor and is the Research Coordinator at School of Business Studies and Social Sciences. She has served many educational institutions in South India and has served as a national and international expert, for a decade among the educational institutions of India. Currently, she is actively associated with Indian Tourism Congress (ITC) and Kerala Development Society (KDS), New Delhi. The active researches undertaken include Impact Assessment Studies, Medical Tourism, Destination Management Organization and Ecological Studies. Dr. Bindi completed a major research project on the title "Strategic Intervention of Destination Management Organizations to Enhance Competitiveness of Tourism Destinations– A Model for Karnataka" funded by Christ University. Along with her academic expertise, she is also an Section editor for 'ATNA- Journal of Tourism Studies', published by Christ University, Bengaluru. She has authored one book; on Medical Tourism in India: by an international publisher in Germany, and has also contributed chapters to edited books and has published several articles in areas of Destination Management, Governance, Medical Tourism, E-Tourism etc. To her credit, she has edited a book on "Evolving Paradigms in Tourism and Hospitality in Developing Countries: A Case Study of India". The book is published by CRC Press, Taylor and Francis group - international publisher in US and released in 2018.

Ramakrishnan Vivek has an MBA (Rajarata University) and BBA (Jaffna University). He is Assistant Lecturer at Faculty of Business Studies, Sri Lanka Technological Campus.

Poshan (Sam) Yu is a consultant, adjunct professor, keynote speaker, research award winner, editor, author, ad-hoc reviewer and venture capitalist. He also serves as a guest financial news commentator on more than thirty media platforms in China. Sam lectures at the International Cooperative Education Program of Soochow University (China). He is also an external professor of financial technology and finance at SKEMA Business School (China), a visiting professor at Krirk University (Thailand) and a visiting researcher at the Australian Studies Centre of Shanghai University (China). Besides, Sam is a regional partner (China) at FasterCapital (Dubai, United Arab Emirates), a vice president at Tinbowah Investment (Beijing), an honorary vice president at Belt and Road Blockchain Association, a business mentor of AIC RAISE (Coimbatore, India) and University College London. His research interests include public-private partnerships, financial technology, regulatory technology, mergers and acquisitions, private equity, venture capital, start-ups, intellectual property, art finance, wine and China's "One Belt One Road" policy.

Index

G

H

I

L

M

N

P

R

S

T

U

V

www.ingramcontent.com/pod-product-compliance
Lightning Source LLC
LaVergne TN
LVHW060403300926
849162LV00006B/1333